Production of Recombinant Proteins

Edited by
Gerd Gellissen

Further Titles of Interest

G. Walsh

Proteins

Biochemistry and Biotechnology

2001
ISBN 0-471-89906-2

G. Walsh

Biopharmaceuticals

Biochemistry and Biotechnology

2003
ISBN 0-470-84326-8

Jörg Knäblein and Rainer H. Müller (Eds.)

Modern Biopharmaceuticals

Design, Development and Optimization

2005
ISBN 3-527-31184-X

M. Schleef (Ed.)

DNA-Pharmaceuticals

**Formulation and Delivery in Gene Therapy
and DNA Vaccination**

2005
ISBN 3-527-31187-4

G. Gellissen (Ed.)

Hansenula polymorpha

Biology and Applications

2002
ISBN 3-527-30341-3

H.J. Rehm, G. Reed, A. Pühler,
P. Stadler (Eds.)

Biotechnology

Second, Completely Revised Edition
Volume 2, Genetic Fundamentals and Genetic
Engineering
1992
ISBN 3-527-28312-9

H.-J. Rehm, G. Reed, A. Pühler, P. Stadler,
A. Mountain, U.M. Ney, D. Schomburg
(Eds.)

Biotechnology

Second, Completely Revised Edition
Volume 5a, Recombinant Proteins, Monoclonal
Antibodies, and Therapeutic Genes
1998
ISBN 3-527-28315-3

R.D. Schmid, R. Hammelehle

**Pocket Guide to Biotechnology
and Genetic Engineering**

2003
ISBN 3-527-30895-4

Production of Recombinant Proteins

Novel Microbial and Eukaryotic Expression Systems

Edited by
Gerd Gellissen

WILEY-VCH

WILEY-VCH Verlag GmbH & Co. KGaA

Edited by

Prof. Dr.Gerd Gellissen
Ringstrasse 30
42489 Wülfrath
Germany

Library of Congress Card No.: applied for

**British Library Cataloguing-in-Publication
Data:** A catalogue record for this book is
available from the British Library.

**Bibliographic information published by
Die Deutsche Bibliothek**
Die Deutsche Bibliothek lists this publication
in the Deutsche Nationalbibliografie;
detailed bibliographic data is available in the
Internet at <http://dnb.ddb.de>

Printed in the Federal Republic of Germany
Printed on acid-free paper

Composition   ProSatz Unger, Weinheim
Printing   Strauss GmbH, Mörlenbach
Bookbinding   J. Schäffer GmbH i. G.,
Grünstadt

ISBN   3-527-31036-3

*This book is dedicated to my wife Gabi
and my sons Benedikt, Georg, and Ulrich.*

Preface

Gene technology has invaded the production of proteins, and especially production processes for pharmaceuticals. At the beginning of this new technology only a limited number of microorganisms was employed for such processes, namely the bacterium *Escherichia coli*, followed by the baker's yeast *Saccharomyces cerevisiae* as a microbial eukaryote. For both organisms a wealth of information was available which stemmed from a long tradition of safe use in science and, in case of the yeast, also from food manufacturing. However, certain limitations and restrictions urged the search for alternatives that were able to meet the requirements and demands for the expression of an ever-growing number of target genes. As a consequence, a plethora of microbial and cellular expression platforms were developed. Nonetheless, the range of launched products still leans for the most part on production in a restricted set of organisms, with most of the newly identified microbes being applied to research in academia.

Despite superior characteristics of some industrially employed platforms, limitations and restrictions are still encountered in particular process developments. In a publicly funded program, Rhein Biotech has set out with academic partners in the recent past to identify additional microbes with attractive capabilities that could supplement its key system, *Hansenula polymorpha*. As such, the Gram-positive *Staphylococcus carnosus*, the thermo- and osmotolerant dimorphic yeast species *Arxula adeninivorans*, the filamentous fungi *Aspergillus sojae*, and the nonsporulating species *Sordaria macrospora*, were developed. This development was supplemented by tools such as the definition of fermentation conditions and a "universal vector" that can be employed to target a range of fungi for the identification of the most suited platform in particular process developments. The application of these platforms and tools is included in the business concept of a new German biotech start-up company, MedArtis Pharmaceuticals GmbH, Aachen.

The present book is aimed at providing a comprehensive view of these newly identified and defined systems, and comparing them with a range of established and new alternatives. The book includes the description of two Gram-negative organisms (*E. coli* and *Pseudomonas fluorescens*), the Gram-positive *Staphylococcus carnosus*, four yeast species (*Arxula adeninivorans*, *Hansenula polymorpha*, *Pichia pastoris* and *Yarrowia lipolytica*), and the two filamentous fungi *Aspergillus sojae* and *Sordaria macrospora*. The description of these microbial platforms is further supplemented by an overview on expression in mammalian and plant cells.

I would like to thank all academic partners who co-operated in the development of these new platforms. I gratefully acknowledge funding by the Ministry of Economy NRW, Germany (TPW-9910v08). I would also like to thank D. Ellens, M. Piontek, and F. Ubags, who inspired me to edit this book.

I also express my gratitude to all authors for their fine efforts and contributions, and thank Dr. Paul Hardy, Düsseldorf, for carefully reading some of the manuscripts. I also acknowledge the continuous support of Dr. A. Pillmann and her staff at Wiley-VCH.

Aachen, October 2004   *Gerd Gellissen*

Foreword

The availability of ever-increasing numbers of eukaryotic, prokaryotic, and viral genomes facilitates the rapid identification, amplification, and cloning of coding sequences for technical enzymes and pharmaceuticals, including vaccines. To take advantage of the treasures of information contained in these sequences, elegant multiplatform expression systems are needed that fulfill the specific requirements demanded by each potential application; for example, economy in the case of technical enzyme production, or safety and authenticity in the case of pharmaceutical production. Therefore, while *Escherichia coli* and other bacteria may be perfectly suited for technical enzyme production or the production of selected pharmaceuticals requiring no special modification, eukaryotic organisms may be advisable for applications where safety (e.g., no endotoxin), contamination, or authenticity (e.g., proper protein modification by glycosylation) are of concern. While the choices of microbial and eukaryotic expression systems for the production of recombinant proteins are many in number, most researchers in academic and industrial settings do not have ready access to pertinent biological and technical information as it is usually scattered in the scientific literature. This book aims to close this gap by providing, in each chapter, information on the general biology of the host organism, a description of the expression platform, a methodological section (with strains, genetic elements, vectors and special methods, where applicable), and finally some examples of proteins expressed with the respective platform. The described systems are well balanced by including three prokaryotes (two Gram-negative and one Gram-positive), four yeasts, two filamentous fungi, and two higher eukaryotic cell systems (mammalian and plant cells). The book is rounded off by providing valuable practical and theoretical information about criteria and schemes for selection of the appropriate expression platform, about the possibility and practicality of a universal expression vector, and about comparative industrial-scale fermentation. The production of a recombinant Hepatitis B vaccine is chosen to illustrate an industrial example. As a whole, this book is a valuable and overdue resource for a varied audience. It is a practical guide for academic and industrial researchers who are confronted with the design of the most suitable expression platform for their favorite protein for technical or pharmaceutical purposes. In addition, the book is also a valuable study resource for professors and students in the fields of applied biology and biotechnology.

Fort Collins, Colorado, U.S.A., June 2004 *Herbert P. Schweizer, Ph.D.*

Contents

Production of Recombinant Proteins. Novel Microbial and Eucaryotic Expression Systems. Edited by Gerd Gellissen
Copyright © 2005 WILEY-VCH Verlag GmbH & Co. KGaA, Weinheim
ISBN: 3-527-31036-3

List of Contributors

Josef Altenbuchner
Institute for Industrial Genetics
University Stuttgart
Allmandring 31
70569 Stuttgart
Germany

Oliver Bartelsen
Rhein Biotech GmbH
Eichsfelder Str. 11
40595 Düsseldorf
Germany

Erik Böer
IPK Gatersleben
Correnstr. 3
06466 Gatersleben
Germany

René Brecht
ProBiogen AG
Goethestr. 54
13086 Berlin
Germany

Pascale Brocke
Rhein Biotech GmbH
Eichsfelder Str. 11
40595 Düsseldorf
Germany

Jochen Büchs
Biochemical Engineering
RWTH Aachen University
Worringer Weg 1
52056 Aachen
Germany

Lawrence C Chew
The Dow Chemical Company
Biotechnology, Research and
Development
5501 Oberlin Dr.
San Diego, CA 92121
USA

James M Cregg
Professor and Director of Research
Keck Graduate Institute of Applied
Sciences
535 Watson Drive
Claremont, CA 91711
USA

Ulrike Dahlems
Rhein Biotech GmbH
Eichsfelder Str. 11
40595 Düsseldorf
Germany

Production of Recombinant Proteins. Novel Microbial and Eucaryotic Expression Systems. Edited by Gerd Gellissen
Copyright © 2005 WILEY-VCH Verlag GmbH & Co. KGaA, Weinheim
ISBN: 3-527-31036-3

Jürgen Drossard
Institute of Molecular Biotechnology
RWTH Aachen University
Worringer Weg 1
52056 Aachen
Germany

Rainer Fischer
Institute of Molecular Biotechnology
RWTH Aachen University
Worringer Weg 1
52056 Aachen
Germany

Roland Freudl
Institut für Biotechnologie 1
Forschungszentrum Jülich GmbH
52425 Jülich
Germany

Claude Gaillardin
Microbiologie et Génétique Moleculaire
UMR1238 INAPG-INRA-CNRS
Institut National Agronomique
Paris-Grignon
78850 Thiverval Grignon
France

Gerd Gellissen
MedArtis Pharmaceuticals GmbH
Pauwelsstr. 19
52047 Aachen
Germany

Margreet Heerikhuisen
TNO Nutrition and Food Research
Department of Microbiology
P.O. Box 360
3700 AJ Zeist
The Netherlands

Stephan Hellwig
Fraunhofer Institute for
Molecular Biology and Ecology (IME)
Worringerweg 1
52074 Aachen
Germany
and
MedArtis Pharmaceuticals GmbH
Pauwelsstr. 19
52047 Aachen
Germany

Cornelis P Hollenberg
Institut für Mikrobiologie
Heinrich-Heine-Universität
Universitätsstr. 1
40225 Düsseldorf
Germany

Christine Ilgen
Keck Graduate Institute of Applied
Sciences
535 Watson Drive
Claremont, CA 91711
USA

Zbigniew A Janowicz
Rhein Biotech GmbH
Eichsfelder Str. 11
40595 Düsseldorf
Germany

Volker Jenzelewski
Rhein Biotech GmbH
Eichsfelder Str. 11
40595 Düsseldorf
Germany

Hyun Ah Kang
Korea Research Institute
of Bioscience and Biotechnology (KRIBB)
52 Eoen-dong
Yusong-gu, Daejeon 305–333
Korea

Jens Klabunde
Institut für Mikrobiologie
Heinrich-Heine-Universität
Universitätsstr. 1
40225 Düsseldorf
Germany

Ulrich Kück
Lehrstuhl für
Allgemeine und Molekulare Botanik
Ruhr Universität
44780 Bochum
Germany

Gotthard Kunze
IPK Gatersleben
Corrensstr. 3
06466 Gatersleben
Germany

Joan Lin-Cereghino
Department of Biological Sciences
University of the Pacific
Stockton, CA 95211
USA

Catherine Madzak
Microbiologie et Génétique Moleculaire
UMR1238 INAPG-INRA-CNRS
Institut National Agronomique
Paris-Grignon
78850 Thiverval Grignon
France

Ralf Mattes
Institute for Industrial Genetics
University Stuttgart
Allmandring 31
70569 Stuttgart
Germany

Karl Melber
Rhein Biotech GmbH
Eichsfelder Str. 11
40595 Düsseldorf
Germany

Georg Melmer
MedArtis Pharmaceuticals
Pauwelsstr. 19
52047 Aachen
Germany

Frank Müller
Rhein Biotech GmbH
Eichsfelder Str. 11
40595 Düsseldorf
Germany

Jean-Marc Nicaud
Microbiologie et Génétique Moleculaire
UMR1238 INAPG-INRA-CNRS
Institut National Agronomique
Paris-Grignon
78850 Thiverval Grignon
France

Kyung-Nam Park
227–3, Kuga-li, Giheung-Eup
Yongin-Shi, Kyunggi-do
Korea

Stefanie Poeggeler
Lehrstuhl für Allgemeine und
Molekulare Botanik
Ruhr Universität
44780 Bochum
Germany

Peter J Punt
TNO Nutrition and Food Research
Department of Microbiology
P.O. Box 360
3700 AJ Zeist
The Netherlands

Tom M. Ramseier
The Dow Chemical Company
Biotechnology, Research and
Development
5501 Oberlin Dr.
San Diego, CA 92121
USA

Diane M. Retallack
The Dow Chemical Company
Biotechnology, Research and
Development
5501 Oberlin Dr.
San Diego, CA 92121
USA

Thomas Rose
ProBiogen AG
Goethestr. 54
13086 Berlin
Germany

Volker Sandig
ProBiogen AG
Goethestr. 54
13086 Berlin
Germany

Stephan Schaefer
University Rostock
Department of Medical Microbiology
and Virology
Schillingallee 70
18055 Rostock
Germany

Stefan Schillberg
Fraunhofer Institute for
Molecular Biology and Applied Ecology
(IME)
Worringerweg 1
52074 Aachen
Germany

Jane C. Schneider
The Dow Chemical Company
Biotechnology, Research and
Development
5501 Oberlin Dr.
San Diego, CA 92121
USA

Charles H. Squires
The Dow Chemical Company
Biotechnology, Research and
Development
5501 Oberlin Dr.
San Diego, CA 92121
USA

Christoph Stöckmann
Biochemical Engineering
RWTH Aachen University
Worringer Weg 1
52056 Aachen
Germany

Alexander WM Strasser
Ringelsweide 16
40223 Düsseldorf
Germany

Manfred Suckow
Rhein Biotech GmbH
Eichsfelder Str. 11
40595 Düsseldorf
Germany

Henry W Talbot
The Dow Chemical Company
Biotechnology, Research and
Development
5501 Oberlin Dr.
San Diego, CA 92121
USA

Richard M Twyman
Department of Biological Sciences
University of York
Heslington
York
YO10 5DD, UK
England

Cees AMJJ van den Hondel
TNO Nutrition and Food Research
Department of Microbiology
3700 AJ Zeist
The Netherlands

Karsten Winkler
ProBiogen AG
Goethestr. 54
13086 Berlin
Germany

1
Key and Criteria to the Selection of an Expression Platform

GERD GELLISSEN, ALEXANDER W.M. STRASSER, and MANFRED SUCKOW

The production of recombinant proteins has to follow an economic and qualitative rationale, which is dictated by the characteristics and the anticipated application of the compound produced. For the production of technical enzymes or food additives, gene technology must provide an approach which has to compete with the mass production of such compounds from traditional sources. As a consequence, production procedures have to be developed that employ highly efficient platforms and that lean on the use of inexpensive media components in fermentation processes. For the production of pharmaceuticals and other compounds that are considered for administration to humans, the rationale is dominated by safety aspects and a focus on the generation of authentic products. The demand for suitable expression systems is increasing as the emerging systematic genomics result in an increasing number of gene targets for the various industrial branches (for pharmaceuticals, see Chapter 16). So far, the production of approved pharmaceuticals is restricted to *Escherichia coli*, several yeasts, and mammalian cells. In the present book, a variety of expression platforms is described ranging from Gram-negative and Gram-positive prokaryotes, over several yeasts and filamentous fungi to mammalian and plants cells, thus including greatly divergent cell types and organisms. Some of the systems presented are distinguished by an impressive track record as producers of valuable proteins that have already reached the market, while others are newly defined systems that have yet to establish themselves but demonstrate a great potential for industrial applications. All of them have special favorable characteristics, but also limitations and drawbacks – as is the case with all known systems applied to the production of recombinant proteins. As there is clearly no single system that is optimal for all possible proteins, predictions for a successful development can only be made to a certain extent, and as a consequence misjudgments leading to costly time- and resource-consuming failures cannot be excluded. It is therefore advisable to assess several selected organisms or cells in parallel for their capability to produce a particular protein in desired amounts and quality (see also Chapter 13).

The competitive environment of the considered platforms is depicted in Table 1.1. A cursory correlation exists between the complexity of a particular protein and the complexity and capabilities of an expression platform. Single-subunit proteins can easily be produced in bacterial hosts, whereas proteins that require an authentic complex mammalian glycosylation or the presence of several disulfide bonds neces-

Table 1.1   Some key parameters for the choice of a particular expression system. The column "Expression system" provides the list of the systems described in the various chapters of this book. The column "Classification" provides a rough classification of these organisms. The coloring of the fields indicates the complexity of the respective organism, increasing in the order light gray, medium gray, dark gray. In the following columns, positive and negative aspects are distinguished by the coloring of the fields. Light gray indicates negative, and dark gray positive features. Fields in medium gray indicate an intermediate grading. The column "Development of system" distinguishes between "early stages" and "completely developed". The latter indicates that the full spectrum of methods and elements for genetic manipulations, target gene expression, and handling is available. "Early stages" shall indicate a yet incomplete development. In "Disulfide bonds" and "Glycosylation", two examples of post-translational modification are addressed which may be especially important for heterologous protein production. Prokaryotes have, in general, a strongly limited capability of forming disulfide bonds. If one or more disulfide bonds is necessary for the target protein's activity, a eukaryotic system would be the better choice. If the target protein requires *N*- or *O*-glycosylation for proper function, prokaryotic systems are also disqualified. The production of a glycoprotein for the administration to humans requires special care. So far, only mammalian cells are capable of producing human-compatible glycoproteins. Glycoproteins produced by two methylotrophic yeasts, *Hansenula polymorpha* and *Pichia pastoris*, have been shown not to contain terminal α1,3-linked mannose, which are suspected to be allergenic. For the other yeasts and fungi listed, the particular composition of the glycosylation has yet not been determined, which here is valued as a negative feature. "Secretion" of target protein can be achieved with all systems shown in the list. However, in case of the two Gram-negative bacteria, *Escherichia coli* and *Pseudomonas fluorescens*, "Secretion" means that the product typically accumulates in the periplasm; the complete release requires the degradation of the outer membrane. The following three columns, "Costs of fermentation", "Use of antibiotics", and "Safety costs" refer to a subset of practical aspects for production of a target protein. In general, the "Costs of fermentation" in mammalian cells are much higher than in plant cells, fungi, yeasts, or prokaryotes, due mainly to the costs of the media. However, the use of isopropyl-thiogalactopyranoside (IPTG)-inducible promoters can increase the costs of target protein production in *E. coli* and *P. fluorescens*, as indicated by the medium gray fields. The use of antibiotics in fermentation processes is becoming increasingly undesired. If a therapeutic protein is to be produced in *E. coli* or *Staphylococcus carnosus*, a plasmid/host system should be chosen that allows plasmid maintenance without the use of antibiotics. "Safety costs" refers to the capability of the production system of carrying human pathogenic agents. In this regard, the mammalian-derived cell systems display the highest risks, for example as carriers of retroviruses. "Processes developed" indicates whether processes based on a particular system have already entered the pilot or even the industrial scale, associated with the respective knowledge. "Products on market" indicates which systems have already passed this final barrier. ▶

Expression system	Classification	Development of system	Disulfide bonds	Glycosylation	Secretion	Costs of fermentation	Use of antibiotics	Safety costs	Processes developed	Products on market
Mammalian cells	higher eukaryote	completely developed	yes	yes; typically human-like	possible	high	not required	high costs	industrial scale	yes
Plant cells	higher eukaryote	completely developed	yes	yes; terminal fucose	possible; size-restrictions	moderate	not required	low costs	pilot scale	no
Sordaria macrospora	filamentous fungus	earlys stages	yes	yes; exact features yet unknown	possible	low	not required	low costs expected	lab scale	no
Aspergillus sojae	filamentous fungus	completely developed	yes	yes; exact features yet unknown	possible	low	not required	low costs	pilot scale	no
Arxula adeninivorans	dimorphic yeast	early stages	yes	yes; exact features yet unknown	possible	low	not required	low costs expected	lab scale	no
Yarrowia lipolytica	dimorphic yeast	early stages	yes	yes; exact features yet unknown	possible	low	not required	low costs expected	lab scale	no
Pichia pastoris	methylotrophic yeast	completely developed	yes	yes; no terminal α1, 3 mannose	possible	low	not required	low costs	industrial scale	yes
Hansenula polymorpha	methylotrophic yeast	completely developed	yes	yes; no terminal α1, 3 mannose	possible	low	not required	low costs	industrial scale	yes
Staphylococcus carnosus	gram-positive bacterium	completely developed	limited	no	possible	low	typically required	low costs	pilot scale	no
Pseudomonas fluorescens	gram-negative bacterium	completely developed	(yes); in the periplasm	no	periplasmic secretion	promoter-dependent low to moderate	not required	low costs	pilot scale	no
Escherichia coli	gram-negative bacterium	completely developed	(yes); in the periplasm	no	periplasmic secretion	promoter-dependent low to moderate	typically required	low costs	industrial scale	yes

sitate a higher eukaryote as host. However, ongoing research and ongoing platform development and improvements might render alternative microbes of lower systematic position suitable to produce such sophisticated compounds. For instance, *E. coli*-based production systems have successfully been applied to a tissue plasminogen activator (t-PA) production process (see Chapter 2); system components are now available for the methylotrophic yeast species *Pichia pastoris* and *Hansenula polymorpha* to synthesize core-glycosylated proteins or those with a "humanized" N-glycosylation pattern (see Chapter 6 on *H. polymorpha*, and Chapter 7 on *P. pastoris*).

Microbial system provide in general easy access to process monitoring and validation as compared to the systems based on higher eukaryotes.

The Gram-negative bacterium *E. coli* was the first organism to be employed for recombinant protein production because of its long tradition as a scientific organism, the ease of genetic manipulations, and the availability of well-established fermentation procedures. However, the limitations in secretion and the lack of glycosylation impose restrictions on general use. Furthermore, recombinant products are often retained as inclusion bodies. Although inclusion bodies sometimes represent a good starting material for purification and downstream procedures, they often contain the recombinant proteins as insoluble, biologically inactive aggregates. This requires in these instances a very costly and sophisticated renaturation of the inactive product. Nevertheless, it still provides the option to produce even complex proteins (as described in Chapter 2), and a range of *E. coli*-derived pharmaceuticals have successfully entered the market.

Pseudomonas fluorescens represents a newly defined system based on an alternative Gram-negative bacterium. Some of the advantageous characteristics of this organism are summarized in Chapter 3, including refraining from antibiotics, improved secretion capabilities, and an improved production of soluble, active target proteins.

Staphylococcus carnosus is a representative of Gram-positive bacteria that are capable of secretion into the culture medium. The platform avoids system-specific limitations frequently encountered with Gram-positive organisms. This includes pronounced proteolytic degradation of products by secreted host-derived proteases, as is the case with commonly applied *Bacillus subtilis* strains. In the case of *S. carnosus*, proteases reside within the cell wall. Potential degradation during cell wall passage can be prevented by using a protective *S. hyicus*-derived lipase leader for export targeting. Additionally, it is possible to secrete lipophilic heterologous proteins that were found to be retained in the insoluble intracellular fraction when using yeasts such as *H. polymorpha*. Another possible application of great potential is the option to tether exported proteins to the surface of the host via C-terminal sorting signal sequences. Recombinant microbes exhibiting such a surface display could be applied to the generation of live vaccines and of biocatalysts (see Chapter 4).

Fungi combine the advantages of a microbial system such as a simple fermentability with the capability of secreting proteins that are modified according to a general eukaryotic scheme. Filamentous fungi such as *Aspergillus* sp. efficiently secrete genuine proteins, but the secretion of recombinant proteins turned out be a difficult task in particular cases. Foreign proteins have to be produced as fusion proteins from which the desired product must be released by subsequent proteolytic processing.

Furthermore, *Aspergillus* usually generate spores that are undesirable in the production of pharmaceuticals. Nevertheless, *Aspergillus* sp. have successfully been used for the production of phytase or for lactoferrin (see Chapter 9). The newly defined *Sordaria macrospora* platform is free of these undesired spores, thereby offering a great potential for the production of recombinant pharmaceuticals (see Chapter 10).

This book also covers a selection of divergent yeast systems. The traditional baker's yeast, *Saccharomyces cerevisiae*, has been used for the production of FDA-approved HBsAg and insulin. Again, severe drawbacks are encountered in the application of this system, and it was therefore excluded from this book: *S. cerevisiae* tends to hyperglycosylate recombinant proteins; N-linked carbohydrate chains are terminated by mannose attached to the chain via a $\alpha1,3$ bond, which is considered to be allergenic. In contrast, the two methylotrophs harbor N-linked carbohydrate chains with a terminal $\alpha1,2$-linked mannosyl residue which is not allergenic. Furthermore, the extent of hyperglycosylation is lower as compared to the situation in baker's yeast. Both methylotrophs are established producers of foreign proteins; in particular, *H. polymorpha* is distinguished by a growing track record as production host for industrial and pharmaceutical proteins. Tools have been established in these two species to produce glycoproteins that exhibit a "humanized" glycosylation pattern or that secrete core-glycosylated proteins (see Chapters 6 and 7). More recently, the two dimorphic species *Arxula adeninivorans* and *Yarrowia lipolytica* have been defined as expression platforms. The newly defined systems have yet to demonstrate their potential for industrial processes. Both organisms exhibit a temperature-dependent dimorphism, with hyphae being formed at elevated temperatures. For *A. adeninivorans*, it has been shown that O-glycosylation is restricted to the budding yeast status of the host (see Chapters 5 and 8).

All yeasts – and probably all filamentous fungi – could be addressed in parallel by a wide-range vector for assessment of suitability in a given product development (see Chapter 13).

Mammalian cells [e.g., Chinese hamster ovary cells (CHO) and baby hamster kidney cells (BHK)] are capable of faithfully modifying heterologous compounds according to a mammalian pattern. However, the fermentation procedure is expensive and yields are much lower than those reported for various microbial systems. In addition, mammalian cells are potential targets of infectious viral agents. This forces a vigorous control of all fermentation and purification steps. This situation can be eased to some extent when using hollow-fiber bioreactors, as presented in Chapter 11. To date, the production of industrial compounds is thus restricted to high-price drugs. Nevertheless, very successful pharmaceutical products such as antibodies and their derivatives, or pharmaceuticals such as factor VIII, with its demand for authentic glycosylation, are based on production in mammalian cell cultures.

Plant suspensions cell cultures carry most of the advantages of terrestrial plants, and can be used at present for the production of low or medium amounts of proteins. Benefits include the ability to produce proteins under GMP conditions, the ability to isolate proteins continuously from the culture medium, and the use of sterile conditions. However, further improvements in yield and optimization in downstream processing are required before this platform becomes commercially feasible (see Chapter 12).

2
Escherichia coli
Josef Altenbuchner and Ralf Mattes

List of Genes

Gene	Encoded gene product or function
adhE	alcohol dehydrogenase
araA,B,D	L-arabinose-specific metabolism, kinase, isomerase, epimerase
araC,I	L-arabinose-dependent regulators
araE	L-arabinose-specific transport
argU (dnaY)	arginine $tRNA_5^{[AGA/AGG]}$
atpE	membrane-bound ATP synthase, subunit c
cer	recognition sequence for the site-specific recombinase XerCD
dnaK,J	HSP-70-type molecular chaperone, with DnaJ chaperone
glyT	glycine $tRNA_2$, UGA suppression
grpE	GrpE heat shock protein; stimulates DnaK ATPase; nucleotide exchange function
hok	post-replicational killing by the gene product of the *parB* system of plasmid R1
ileX,Y	Isoleucine $tRNA_2$ and variant
int	integrase
lacZ,Y,I	lactose-specific β-galactosidase, permease and regulator (repressor)
leuW	leucine $tRNA_3$
lysT	lysine tRNA (multiple loci, *lysQTVWYZ*)
ompA	outer membrane protein 3a
ori (*oriC*)	origin of DNA replication (*E. coli* chromosome origin of replication)
parB	stability locus of plasmid R1 consisting of *hok* and *sok* genes
pelB	pectate lyase of *Erwinia carotovora*
proL	proline $tRNA_2$
recA	enzyme for general recombination and DNA repair; pairing and strand exchange
rhaA,B,D	L-rhamnose-specific metabolism, isomerase, kinase, aldolase
rhaR,S	L-rhamnose-dependent regulators
rhaT	L-rhamnose-specific transport
rop (*rom*)	repressor of primer (RNA organizing protein) of ColE1-type plasmids

Production of Recombinant Proteins. Novel Microbial and Eucaryotic Expression Systems. Edited by Gerd Gellissen
Copyright © 2005 WILEY-VCH Verlag GmbH & Co. KGaA, Weinheim
ISBN: 3-527-31036-3

Gene	Encoded gene product or function
rrnB,D,E	operons encoding ribosomal RNA and tRNAs
sacB	levan sucrase of *Bacillus subtilis*
sok	suppressor of post-replicational killing by *hok* gene product of plasmid R1
ssrA	tmRNA or 10Sa RNA
supE,F	(amber suppression); glutamine tRNA$_2$ (*glnV*); tandemly duplicated tyrosine tRNA$_1$ (*tyrTV*)
ter	terminus of DNA replication
trpR	regulator of *trp* operon and *aroH*
trxA,B	thioredoxin and thioredoxin reductase

2.1
Introduction

The Gram-negative bacterium *Escherichia coli* was not only the first microorganism to be subjected to detailed genetic and molecular biological analysis, but also the first to be employed for genetic engineering and recombinant protein production. Our knowledge of its genetics, molecular biology, growth, evolution and genome structure has grown enormously since the first compilation of a linkage map in 1964. Its current status is reviewed in the standard reference "*Escherichia coli* and *Salmonella*" (Neidhardt et al. 1996), now available as *EcoSal* online (www.asmpress.org; http://www.ecosal.org/).

From a model organism for laboratory-based basic research, *E. coli* has evolved into an industrial microorganism, and is now the most frequently used prokaryotic expression system. It has become the standard organism for the production of enzymes for diagnostic use and for analytical purposes, and is even used for the synthesis of proteins of pharmaceutical interest, provided that the desired product does not consist of different multiple subunits or require substantial post-translational modification.

A huge body of knowledge and experience in fermentation and high-level production of proteins has grown up during the past 40 years. Many strains are available which are adapted for the production of proteins in the cytoplasm or periplasm, and hundreds of expression vectors with differently regulated promoters and tags for efficient protein purification have been constructed. Nevertheless, high-level gene expression and fermentation of recombinant strains cannot be regarded as a routine task. Due to the unique structural features of individual genes and their products, optimization of *E. coli*-based processes is quite often tedious, time-consuming and costly. Major drawbacks include the instability of vectors (especially during large-scale fermentation), inefficient translation initiation and elongation, instability of mRNA, toxicity of gene products, and instability, heterogeneity, inappropriate folding and consequent inactivity of protein products.

This chapter describes some of the main features of *E. coli* as a host cell, and focuses on some of the problems mentioned above and recent advances in our attempts to overcome them.

2.2
Strains, Genome, and Cultivation

Following the first description of *E. coli* by T. Escherich in 1885, a line of *Escherichia coli* K12 was isolated in 1922 and deposited as "K-12" at Stanford University in 1925. In the early 1940s, E.L. Tatum began his work on bacteria, and isolated the first auxotrophic mutants. The nomenclature used to designate loci, mutation sites, plasmids and episomes, sex factors, phenotypic traits and bacterial strains developed over time, and was codified by Demerec et al. (1966). A useful reference for the terminology, with a compilation of (older) alternate symbols, is given by Berlyn (1998). This last compilation of the traditional linkage map was complemented by the appearance of the corresponding physical map (Rudd 1998).

A pedigree and description of various standard strains, including the *E. coli* B strain, commonly used in the laboratories all over the world, was first published in 1972 (Bachmann 1972). Salient features of the most relevant and popular strains used today are summarized in the Appendix of this chapter.

The first complete genome sequence of an *E. coli* strain was established by Blattner et al. (1997) for the K-12 strain MG1655 (4,639,221 bp). Sequencing of the closely related strain W3110 has not yet been completed, but about 2 Mb are available for comparison. Based on this comparison, the rate of nucleotide changes was estimated to be less than 10^{-7} per site per year (Itoh et al. 1999). Thus, the degree of difference between the two strains is remarkably low in light of their differing histories. According to laboratory records, these sub-strains originated from W1485 approximately 40 years ago (Bachmann 1972). To elucidate the differences between the reference genome MG1655 and currently employed strains, an alternative method for whole-genome sequencing has been applied. Using whole-genome arrays as a tool to identify deletions, this study revealed the exact nature of three previously unresolved deletions in a particular selected strain, and a fourth deletion which was completely unknown (Peters et al. 2003).

Ongoing sequencing efforts have also permitted genomic sequence comparisons with various other species and relatives of *E. coli*. This work has resulted in an initial view of the evolutionary forces that shape bacterial genomes. The comparisons revealed a startling pattern in which hundreds of strain-specific "islands" are found inserted in a common "backbone" that is highly conserved (98% sequence identity) between strains. Gene loss and horizontal gene transfer have been the major genetic processes that shaped the ancestral *E. coli* genome, resulting in a spectrum of divergent present-day strains, which possess very different arrays of genes (Riley and Serres 2000).

Clearly, *E. coli* today possesses many dispensable functions and genes that are not required for laboratory or technical purposes. Transposable elements and cryptic prophages may even compromise genome stability in industrial strains. Consequently,

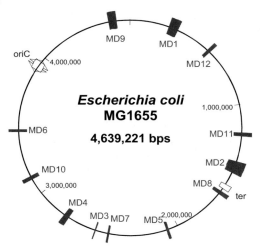

Fig. 2.1 Localization of deletions MD1 through MD12 on the circular map of the MG1655 genome. The replication origin (*oriC*) and terminus (ter) are indicated. Redrawn after Kolisnychenko et al. 2002).

our knowledge of the genome has led to attempts to create new strains using precise genome engineering tools (Zhang et al. 1998). The first result of such approaches was the generation of the *E. coli* strain MDS12, which was the outcome of 12 rounds of deletion formation. This resulted in a genome of 4.263 Mb, equivalent to an 8.1% reduction in size, a 9.3% reduction in gene count, and the elimination of 24 of the 44 transposable elements (Figure 2.1) (Kolisnychenko et al. 2002). Partial characterization of MDS12 revealed only a few phenotypic changes compared to the parental strain. Growth characteristics and transformability were essentially identical to those of MG1655. This strategy opens up new opportunities for the design of strains with favorable characteristics.

Cultivation of *E. coli* was first established on a laboratory scale in academic laboratories. Differences in behavior observed in various strains and mutants were exploited to develop special genetic approaches to create strains adapted to certain conditions. The more recent evolution of *E. coli* into an industrial microorganism has entailed much work in basic research and development to establish today's standards of high-cell-density (HCD) cultivation (for reviews, see Yee and Blanch 1992; Lee 1996; Riesenberg and Guthke 1999). As various currently popular expression plasmids (see below) require special mutant strains as hosts, fermentation protocols have been developed for strains of *E. coli*, which differ considerably from each other. Detailed descriptions are available in particular for *E. coli* K-12 W3110, HB101 and *E. coli* B BL21 (and its mutants) (Hewitt et al. 1999; Rothen et al. 1998; Åkesson et al. 2001). One of the major obstacles during the development of HCD cultivation has been the propensity of *E. coli* for acetate formation and the resulting inhibition of growth (Luli and Strohl 1990). The basis for this tendency has been elucidated in greater detail (Kleman and Strohl 1994; van de Walle and Shiloach 1998), and several approaches are now available to combat it. For example, the use of methyl-α-glucoside as an alternative carbon source (Chou et al. 1994) allows one to bypass this problem. Finally, metabolic engineering using the acetolactate synthetase from *Bacillus subtilis* (Aristidou et al. 1999) or feed-back control of glucose feeding have recently become standard tools (Åkesson et al. 2001).

2.3
Expression Vectors

An expression vector usually contains an origin of replication (*ori*), an antibiotic resistance marker, and an expression cassette for regulated transcription and translation of a target gene. Additional features might include plasmid stability functions, and genes or DNA structures for mobilization and transfer to other strains. The most frequently used vector system is derived from the ColE1-like plasmid pMB1 (Betlach et al. 1976).

2.3.1
Replication of pMB1-derived Vectors

The plasmid pMB1 requires RNA polymerase and DNA polymerase I for replication. Replication is controlled by an anti-sense RNA (RNA I) that binds to the precursor of the primer RNA (RNA II), thus inhibiting RNase H-mediated primer maturation. A small protein called Rop encoded by the plasmid binds to the complex of RNA I and RNA II and stabilizes it (Tomizawa 1990). Deletion of the *rop* gene, as in the case of the pUC series of plasmids (Yanisch-Perron et al. 1985), increases the plasmid copy number in cells from about 50 copies to 150 copies. Increases in copy number have also been observed during overproduction of recombinant proteins. In such cases, excessive consumption of amino acids obviously leads to the accumulation of uncharged tRNAs. Due to sequence homologies between these tRNAs and RNA I (Yavachev and Ivanov 1988), the interaction between RNA I and RNA II is impaired, and this results in the amplification of plasmid DNA. The resulting increase in copy number (up to 250 plasmid copies per cell and more) further raises the dosage of the recombinant gene. This may eventually exhaust the cell's metabolic capacity and in turn cause a breakdown of protein synthesis. Strong expression systems, such as the T7 system, are known to be susceptible to this phenomenon. Recently, a new ColE1-type vector has been constructed in which the region of RNA I that most probably interacts with uncharged tRNAs have been altered (Figure 2.2). Indeed, with this new plasmid, the copy number remained constant during high-level protein synthesis (Grabherr et al. 2002).

2.3.2
Plasmid Partitioning

Many *E. coli* strains that have been selected for high transformation rates are characterized by low biomass formation during HCD fermentations (Lee 1996). In more robust strains like *E. coli* W3110, very high cell densities can be obtained; however, ColE1-derived expression vectors tend to be unstable (Wilms et al. 2001a). Most of

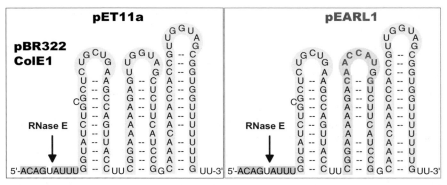

Fig. 2.2 Stem–loop structure of RNA I, the anti-sense repressor of the primer RNA II. RNA I half-life is determined by the indicated cleavage site for RNase E. Uncharged tRNAs are believed to interact with RNA I or II, leading to deterioration of the controlling complex and resulting in hyperproliferation of plasmid DNA. Changes in loop 2 (pEARL1) result in stabilizing the copy number of ColE1-derived vectors. (From Grabherr et al. 2002)

this instability can be attributed to the *recA*[+] status of the host strain. Plasmids present in high copy numbers are generally subject to homologous recombination in *rec*-proficient strains. This converts them into head-to-tail dimeric plasmids. The dimers are either resolved into monomers again, or become substrates for further multimerization. Since the probability of replication of ColE1-type plasmids is assumed to increase in proportion to the number of *ori* sequences per molecule, dimers will replicate twice as often as monomers and finally dominate the plasmid population. The plasmid multimers disrupt control circuits, eventually resulting in copy number depression (the dimer catastrophe) (Summers 1998). In systems which do not have their own partition mechanism and are randomly distributed to the daughter cells, like the ColE1 plasmids, a lower copy number results in higher plasmid instability. This means that multimers accumulate clonally and create a sub-population of cells that show higher rates of plasmid loss. To avoid plasmid loss, ColE1 carries the *cer* recognition sequence for the chromosomally encoded, site-specific recombinase XerCD which, together with some accessory proteins, efficiently resolves chromosomal or plasmid dimers into monomers (Colloms et al. 1996). In addition, a promoter located inside the 240-bp *cer* region directs the synthesis of a 95-nt transcript, Rcd, which is assumed to delay the division of multimer-containing cells (Sharpe et al. 1999). Most of the ColE1-type cloning and expression vectors lack a functional *cer* sequence. This does not affect *recA*-deficient strains, in which multimerization is not possible. However, in Rec[+] strains like W3110, this can have dramatic effects – as demonstrated in the following example. In case of a recombinant W3110 strain with an ʟ-rhamnose-inducible expression vector, more than 50 % of the cells were found to have lost the plasmid at the end of a fed-batch fermentation without antibiotic selection pressure when a construct without the *cer* sequence was used. In contrast, more than 90 % of the cells retained the corresponding *cer*-bearing plasmid. All the plasmids without *cer* isolated from this biomass were multimeric, whereas more than 90 % of the plasmids with *cer* were monomers (Wilms et al. 2001 a).

Another means of stabilizing plasmids lies in the use of post-segregational killing systems or addiction modules, which are frequently found on low-copy number plasmids, and even on chromosomes. This system of plasmid maintenance, exemplified by the *parB* stability locus of plasmid R1, consists of two genes, *hok* (host killing) and *sok* (suppressor of host killing). The *hok* gene encodes a highly toxic protein, while *sok* specifies an anti-sense RNA which is complementary to the *hok* mRNA and prevents its translation. The *hok* mRNA is stable, the *sok* mRNA unstable. In case of plasmid loss, the *sok* product is degraded more rapidly than the *hok* mRNA, leading to production of the toxin which eventually kills the plasmid-free cell (Nagel et al. 1999). Overall, this system ensures that no plasmid-free cells arise, but it does not stabilize the plasmid. The use of such addiction modules in unstable expression systems might therefore lead to slow growth and even impose an additional metabolic burden on the cells.

If a promoter is not tightly regulated and/or gene products are detrimental to the cell, plasmids that are maintained in lower copy numbers may help to minimize the metabolic stress, especially when strong promoters are used. Frequently used ColE1-type plasmids of lower copy number are derived from p15A – for example, the plasmids pACYC177 and pACYC184 (Chang and Cohen 1978). For gene expression in different species, broad-host-range vectors based on RSF1010 with moderate copy numbers and a different type of replication control can be used (Scholz et al. 1989). Other vectors with even lower copy numbers include derivatives of pSC101 (Tait and Boyer 1978), RK2 (Scott et al. 2003) or the F-plasmid (Jones and Keasling 1998). These plasmids are also mutually compatible, and may also be useful for the expression of several genes in a single cell.

If two or more genes have to be expressed in a single cell, the genes can be inserted into the plasmid in a tandem arrangement downstream of the promoter. If the translation efficiency of the selected genes or the activities of the encoded enzymes are unbalanced, the easiest way to ensure optimally balanced production is to use vectors with different copy numbers carrying similar expression modules. For selection, the plasmids must harbor different antibiotic resistance genes. For example, for enantioselective production of amino acids from racemic hydantoins, a hydantoinase, a carbamoylase and a racemase which differed by up to tenfold in specific activities had to be produced in the same cell (Wilms et al. 2001 b). The various genes in question were introduced into an L-rhamnose-inducible expression cassette present in derivatives of pACYC184, pSC101, and pBR322. Various combinations of these plasmids were introduced into recipient strains, and whole-cell reactors with an optimal reaction cascade were eventually developed.

2.3.3
Genome Engineering

In many cases, chromosomal integration of target genes may be preferable to the use of plasmids, especially when very strong promoters are used to compensate for low gene dosage. For *E. coli*, several integration systems are now available (Martin et al. 2002). For example, the vector pKO3 may be used for integration via homologous recombination in *rec*-proficient strains. The vector is temperature-sensitive in its replication, and contains an antibiotic-resistance gene and a *sacB* gene encoding levan sucrase. An expression cassette and the gene of interest, flanked by chromosomal targeting DNA, are integrated into the pKO3 plasmid and introduced into the *E. coli* strain. Selection for the antibiotic resistance during growth at a nonpermissive temperature leads to cells with integrated plasmids. In a second step, the cells are selected for loss of the vector sequences by growth on sucrose, which is lethal in the presence of the levan sucrase. Half of the plasmid-free colonies should have the gene of interest stably integrated together with the expression cassette, but with no further vector sequences (Link et al. 1997). Another *rec*-independent integration system is based on λ site-specific recombination. The expression cassette together with the target gene is inserted into a plasmid containing the λ attachment site *attP*. The replication region of this plasmid is removed by restriction digestion, and after re-ligation, the fragment is introduced into an *E. coli* strain carrying the λ integrase gene on a temperature-sensitive plasmid. The DNA circle is integrated into the chromosome *attB* site, and then the helper plasmid is removed by growth at a nonpermissive temperature (Atlung et al. 1991). Finally, a novel way to engineer DNA in *E. coli* and to integrate DNA into the chromosome or into plasmids independently of restriction sites and *recA* employs the *recET* system. *E. coli* strains that express *recET*, due to a mutation in *sbcA* or because *recET* is placed under the control of another promoter, are able to take up linear PCR fragments and integrate them into the chromosome if the fragments are flanked by short sequences (40–60 bp) homologous to a chromosomal target region. This means that genes of interest can be stably integrated into any region of the chromosome, or into low- or high-copy number vectors, and brought under the control of the regulatory system present at the integration site without having to use any restriction sites (Zhang et al. 1998). The antibiotic resistance genes which have to be used for selection of integration can subsequently be removed, for example by site-specific integrases. Furthermore, this method may also be very useful for engineering host strains for increased production of recombinant proteins; for example, by targeted inactivation of protease or RNase genes (compare the construction of MDS12 in Figure 2.1; Kolisnychenko et al. 2002).

2.3.4
E. coli Promoters

Promoters are DNA sequences which direct RNA polymerase binding and transcription initiation. They usually consist of the two –10 and –35 hexameric sequences, separated by a spacer of 16–19 bp. The sigma subunit confers promoter specificity on RNA polymerase (deHaseth et al. 1998). *E. coli* has seven different sigma factors and, accordingly, seven different types of promoter. The most widely used promoter type is recognized by the sigma 70 factor. The initiation of transcription can be divided into four major steps (Kammerer et al. 1986): (i) recognition of the promoter sequences by the RNA polymerase holoenzyme; (ii) isomerization of the initial complex into a conformation capable of initiation; (iii) initiation of RNA synthesis; and (iv) transition to an elongation complex and promoter clearance. The initial contact between RNA polymerase and promoter results in an open complex in which the DNA strands are separated in the region flanking the start site of RNA synthesis, referred as the +1 position. In the open complex the RNA polymerase covers the region from −50 and +20 (reviewed by Mooney et al. 1998). Strong promoters of the sigma-70 type have motifs in the −35 and −10 regions that are most similar to the consensus sequences TTGACA and TAATAT, respectively. Another important feature is the spacing between the −10 and −35 regions; 17 bp is the optimal length. The nucleotide sequence of the spacer itself is of minor importance. Other regions that influence promoter strength are an AT-rich region upstream of the −35 sequence around position −43, and the region +1 to +20, which seem to participate in promoter recognition and promoter clearance, respectively.

Each of the four steps in transcription initiation can be rate-limiting, which means that promoters of similar strength can have quite different sequences depending upon which steps are optimized. Actually, most promoters found in *E. coli* differ considerably from the hexameric consensus sequences. One obvious reason for these differences is that gene products are needed in quite different amounts. Even more importantly, promoters often overlap with regulatory sequences. Two or more promoters, which may be recognized by either the same or different sigma factors, may be arranged in tandem to allow the cell to respond to specific signals, as well as to the physiological condition of the whole cell.

Many efforts have been made in the past to adapt natural promoters for use in expression vectors, with the aim of generating optimal elements that combine high efficiency and tight regulation, thereby promoting maximal protein production and avoiding plasmid instability.

A completely different type of promoter architecture has been found in some lytic phages, such as phages T3, T5 or T7, and SP6. These phages encode their own RNA polymerases, which are much simpler in structure than the host enzyme. They are highly processive and recognize conserved sequences covering a region between positions −17 and +6 bp relative to the mRNA start site. These are the strongest promoters described for microorganisms so far. They have become very popular for use in expression vectors and in-vitro transcription when coupled with regulatory sequences from natural *E. coli* promoters (Dubendorff and Studier 1991; Sagawa et al. 1996).

2.4
Regulation of Gene Expression

Constitutive heterologous gene expression that results in product yields equivalent to about 30 % of total cell protein will obviously lead to high genetic instability. Therefore, promoters employed in expression vectors must be very tightly regulated during bacterial growth, and be switched on only when the cells have reached a high cell density. Many different transcriptional regulatory mechanisms are found in nature. Binding of a regulatory protein to a promoter is probably the most common principle, but there are other control mechanisms, such as transcriptional attenuation, anti-sense RNA, anti-termination, changes in sigma factors and anti-sigma-factors. The conditions which may lead to changes in promoter activity are countless. Arbitrary examples are changes in the availability of carbon, nitrogen, phosphate and other mineral sources, growth temperature, pH, oxygen supply, osmolarity and mutagenic conditions (Sawers and Jarsch 1996). Many of these options have been assessed for use in expression systems. In practice, however, regulation of promoter activity by regulatory proteins in response to carbon sources or to growth temperature is most often used. In principle, DNA-binding proteins can regulate promoter activity in two different ways. In negatively controlled systems a repressor protein binds in or just downstream to the promoter region and directly inhibits transcription. Positively regulated promoters either exhibit sub-optimal spacing of the −10 and −35 hexameric sequences, or the −35 sequence is quite different from the ideal consensus sequence of strong promoters. In these cases, activator proteins are necessary to bind the RNA polymerase. Both negatively and positively regulated promoters can be controlled either by induction or repression. In negatively controlled inducible systems, an effector molecule binds to the repressor and inhibits its binding to the operator sequence, the binding site of the repressor. In positively controlled inducible systems, activators only bind to their target in the presence of effector molecules. Thus, in the case of mercury resistance genes, the activator MerR is already bound to the operator in the absence of mercury ions, but is only rendered active when Hg^{2+} binds to it (O'Halloran and Walsh 1987). In systems with underlying repression, the situation is exactly the opposite: the inactive repressor becomes active in the presence of the effector and binds to its operator, and the activator is inactivated by the effector.

2.4.1
Negative Control

Many promoters – especially those of operons involved in carbohydrate catabolism – are both negatively and positively controlled. The best known example is the promoter of the *E. coli lac*-operon for consumption of lactose. The lactose repressor LacI binds downstream of the *lac* promoter (position +1 to +21) in the absence of lactose, and inhibits transcription of the genes *lacZYA*. In the presence of allolactose, which is synthesized from lactose by the β-galactosidase LacZ, or following addition of the synthetic inducer isopropyl-thiogalactopyranoside (IPTG), the repressor loses its affinity for the operator. In addition, transcription is positively controlled by the catabolite activator protein. Efficient transcription is only possible in the presence of the activator complex CAP-cAMP, which binds upstream of the −35 region (around position −65). Only in the absence of glucose is the cAMP level in the cell high enough to allow *lac* transcription, leading to a preference for glucose and a diauxic growth pattern when both carbohydrates are added simultaneously to the cells (an additional effect is exclusion of the inducer lactose). Derivatives of the *lac* promoter are still among the most frequently used promoters in *E. coli* expression vectors. A marked improvement in the *lac* promoter was achieved by fusing the −35 region of the promoter of the tryptophan operon (*trp*) with the –10 region of the *lac* promoter (actually the already improved *lac*UV5 promoter was used). In the new *tac* promoter, the spacer between the –10 and –35 regions was 16 bp long, and in the *trc* promoter 17 bp (Brosius et al. 1985). These two promoters are about tenfold more efficient in transcription initiation compared to the wild-type promoter, especially on multicopy plasmids. They also enable production of recombinant proteins in large quantities, independently of catabolite activation.

These promoters serve as good examples for the problems that one may encounter when engineering negatively regulated promoters. The chromosomal *lacI* gene gives rise to only a very few *lac* repressor molecules (on average about 10). If the *lac* promoter–operator sequences are inserted into ColE1-type multicopy plasmids with 40 or more copies per chromosome, most of the *lac* operators will not be occupied by repressor molecules. This results in constitutive transcription from these promoters. Furthermore, there are two additional *lac* operators (also called pseudo-operators), one at the 3'-end of *lacI* upstream of the *lac* promoter and another downstream, inside the *lacZ* coding sequence. Only the presence of all three operators with the correct spacing provides for full repression of the promoter (Oehler et al. 1994). Many attempts have been made to increase the repression of *lac* promoter derivatives. The first step was the isolation of the *lacI*^{q} mutation, a promoter-up mutation of the *lacI* gene which increased production of LacI by tenfold (Calos 1978). This gene is either provided on an F' plasmid or by a derivative of phage Φ80. This approach restricts the use of *lac* promoter-based expression vectors to particular *E. coli* strains. Other vectors carry the *lacI* gene or even the *lacI*^{q} gene, and are less dependent on host genes (Amann et al. 1988). These strains with high repressor content provided by the plasmid are no longer fully inducible with the cheap but weak inducer lactose, but are still fully inducible with the nonhydrolyzable IPTG. On the other hand, this compound is not recommended for production of therapeutic proteins due to its toxicity and cost. An alternative strategy involves the use of a thermo-sensitive *lac* repressor. Here, the system is inducible by a shift in the growth temperature from 30 °C to 42 °C. Using the *lacI*^{ts} gene on the vector, there is still a high basal level of expression, whereas with a *lacI*^{qts} gene on the vector protein production is fivefold less (Hasan and Szybalski 1995; Andrews et al. 1996). Furthermore, an increase in growth temperature favors inclusion body formation and induces heat-shock proteases. Finally, it has been demonstrated that the level of basal expression can be altered by changing the position of the operator within the promoter region (Lanzer and Bujard 1988). Repression was strongly increased when the *lac* operator was positioned between the –10 and –35 hexameric sequences instead of its original position downstream of the –10 sequence or upstream of the –35 sequence.

The *lac* regulatory system is also used in another very efficient expression system. The pET vectors contain the very strong T7 late promoter, which is transcribed by the highly processive T7 RNA polymerase. The RNA polymerase is supplied in *trans*, either by infecting the host with a T7 phage – a procedure which is not practicable for large-scale fermentation – or by using the prophage λDE3 in which the RNA polymerase gene is under the control of the *lac*UV5 promoter. Again, leakiness is a major problem in this system. This can be counteracted to some extent by adding a *lacI* gene to the expression vector, or a *lac* operator downstream of the T7 promoter or by using a plasmid encoding a T7 lysozyme, which degrades the T7 RNA polymerase (Studier and Moffatt 1986).

Another frequently used negatively regulated system is based on the very strong leftwardly oriented p_L promoter of phage λ. This promoter is very tightly regulated by the λ cI repressor. One limitation of this promoter is that it can only be induced using a thermo-labile repressor (λ cI$_{857}$), with all the resulting disadvantages when one wishes to increase the growth temperature (Remaut et al. 1981). Another mode of down-regulation has been described by Hasan and Szybalski (1987). This employs an invertible *tac* promoter. The promoter is oriented away from the target gene during cell growth. For gene expression the promoter is inverted by the λ integrase acting on the *attB* and *attP* sites, which flank the invertible promoter. The *int* gene is placed on a temperature-inducible defective λ phage, and expression is induced by a brief heat shock. The inversion is rapid and over 95% efficient.

Another frequently used and negatively regulated (this time by repression) strong promoter is the *trp* promoter derived from the *E. coli* tryptophan operon. The tryptophan repressor binds to the *trp* operator in the presence of excess tryptophan. Induction is achieved by depletion of tryptophan. Drawbacks of this system are, again, leakiness of the promoter, a limited choice of growth media, and the fact that tryptophan limitation is needed at a time when protein synthesis should be maximal. Such conditions are difficult to define in large-scale fermentations. Alternatively, the promoter can be induced by the addition of the inducer β-indoleacrylic acid, but this compound is expensive (Bass and Yansura 2000).

In general, negatively controlled promoters are difficult to handle in down-regulation, and require a balanced repressor to operator ratio. Induction, especially with strong promoters on multicopy vectors and nondegradable inducers, leads to high mRNA levels which might be toxic, as well to rapid synthesis of proteins which often results in the formation of inclusion bodies.

2.4.2
Positive Control

2.4.2.1 L-Arabinose Operon
Positively regulated systems are characterized by a slower, but more reliable, response, with a very low basal activity. Here, the most popular system is the L-arabinose system of *E. coli*. L-arabinose can be used by *E. coli* as its sole carbon source. It is taken up by two different transport systems (*araE* and *araFGH*) and metabolized to xylulose-5-phosphate by the enzymes encoded by *araB*, *araA*, and *araD*. The genes *araBAD* are organized as a single operon, as are *araFGH* and *araE*. These genes form a regulon that is regulated by AraC. The *araC* gene is located upstream of, and in opposite orientation to, the *araBAD* operon (Schleif 1996). AraC belongs to the AraC/XylS family, one of the most common types of positive regulators (Gallegos et al. 1997). The noncoding region between *araBAD* and *araC* is highly complex. There are three operators, *araI*, *araO1*, and *araO2*, and two binding sites (promoters), pc and p*BAD*, for RNA polymerase and the CAP-cAMP complex, since L-arabinose utili-

zation is dependent on catabolite activation. The AraC binding sites are unusual in that they consist of two direct repeats of 17 bp, instead of shorter, inverted repeated DNA sequences. According to the current model, the homodimeric AraC binds in the absence of ʟ-arabinose to one half-site of *araI* and of *araO2*, leading to the formation of a double-stranded DNA loop of 210 bp. In this loop conformation, transcription of the p*c* promoter is inhibited and the p*BAD* promoter is not active as the second half-site of *araI* is not occupied. The binding of arabinose to AraC together with CAP-cAMP opens the loop. AraC loses its ability to bind to distant sites. Instead, it binds to the adjacent half-sites of *araI* and activates transcription of p*BAD*. In addition, transcription of p*c* increases about tenfold until sufficient AraC is synthesized (within about 10 min) and occupies *araO1*, thereby inhibiting further transcription.

In expression vectors, the *ara* p*BAD* promoter from *E. coli* or *Salmonella typhimurium* is normally used. Most vectors also carry the *araC* gene. Obviously, AraC becomes limiting when produced from a single chromosomal copy, taking into account the many operator elements present on multicopy plasmids. The ratio of induction to repression is between 250- and 1300-fold in such vectors depending on the medium used, and the basal level is very low, especially in the presence of glucose (Guzman et al. 1995). The promoter is about 2.5- to 4.5-fold weaker than the *tac* promoter. Nevertheless, depending on translation efficiency, proteins can be produced in amounts constituting up to 30% of total cell protein. The induction can be modulated by supplementation with sub-optimal concentrations of arabinose. Only at very low arabinose concentrations is a mixed population of induced and noninduced cells found (Siegele and Hu 1997).

2.4.2.2 ʟ-Rhamnose Operon

The catabolism of ʟ-rhamnose in *E. coli* is another example for positive regulation. ʟ-rhamnose is taken up via the product of the *rhaT* gene, isomerized to ʟ-rhamnulose (*rhaB*, phosphorylated to rhamnulose-5-phosphate (*rhaA*), and hydrolyzed by an aldolase (*rhaD*). The genes *rhaBAD* form an operon, whereas *rhaT* is located close by and has its own promoter. Upstream of *rhaBAD* are the two genes *rhaR* and *rhaS*, which are arranged in opposite orientations. The *rhaR* and *rhaS* gene products also belong to the AraC/XylS family of activators (Figure 2.3). Both RhaR and RhaS bind in the noncoding region between *rhaR* and *rhaS* in the presence of rhamnose. RhaR activates transcription of *rhaR* and *rhaS*. RhaS activates transcription of the structural genes *rhaBAD* by binding to the *rhaBAD* promoter. In addition, ʟ-rhamnose consumption requires the catabolite activation complex CAP-cAMP, which binds upstream of the *rhaS* promoter (Badia et al. 1989; Via et al. 1996). The *rhaBAD* promoter is even more tightly regulated than the *araBAD* promoter. The basal level of transcription (determined using a transcriptional fusion with *lacZ*) is about tenfold lower, whereas the induced levels of *rhaBAD* and *araBAD* expression are similar (Haldimann et al. 1998).

Expression modules have been constructed containing just the *rhaBAD* promoter together with the CAP-cAMP binding site, a ribosomal binding site, several restriction sites for the insertion of target genes, and a transcription terminator (Figure 2.4). This module was inserted into various plasmids and used for expression of

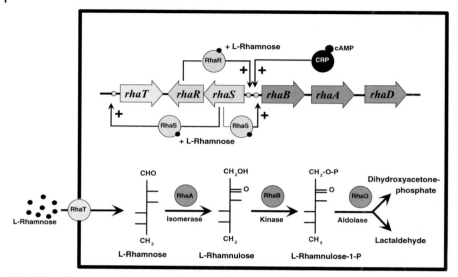

Fig. 2.3 L-Rhamnose pathway and involved genes and pathways. Transcription of *rhaBAD* genes is controlled by RhaS activator (see text for details).

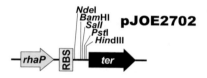

Fig. 2.4 Expression modules for vectors contain just the *rhaBAD* promoter, together with the CAP-cAMP binding site (*rhaP*), a ribosomal binding site (RBS), several restriction sites for insertion of target genes, and a transcription terminator (*ter*). RhaS and RhaR are provided *in trans* by the host strain.

several different genes. Generally, proteins could be produced in levels up to 30 % of total cell protein with this regulatory system. In contrast to the case with the *ara* system, the regulatory genes *rhaRS* are dispensable; the chromosomal copy seems to produce enough regulatory proteins even to saturate the binding sites on multicopy plasmids. The induction times for maximal protein yields varied between 4 h and 10 h. These very slow response rates were found to be beneficial for proteins that tend to form inclusion bodies, as well as for enzymes which have to be transferred into the periplasm. In comparison to the *tac* promoter, up to fourfold higher activities were obtained in the case of the *E. coli* pencillin amidase, and up to 20-fold higher activities in the case of a sucrose isomerase from *Protaminobacter rubrum* (unpublished observations). In addition, due to its low basal expression level, this regulatory system can be used successfully in HCD fermentation. Proteins were produced in amounts comparable to those obtainable in shake-flask experiments, and without having to employ selective conditions with antibiotics (Wilms et al. 2001a).

2.5
Transcription and Translation

2.5.1
Translation Initiation

For efficient expression of foreign genes in *E. coli*, not only transcription but also the translation efficiency of the cloned gene must be considered, with translation initiation being the most important step. Initiation is mainly governed by the ribosomal binding site or Shine–Dalgarno (SD) sequence, a region of four to nine bases which is complementary to the 3'-end of the 16S RNA and is located 5–11 nt from the start codon. Most *E. coli* genes begin with a AUG start codon, but some use a GUG or a UUG codon (Table 2.1) (Blattner et al. 1997). The latter two are frequently found in genes that are weakly expressed, and translation initiation at these triplets is indeed two- to threefold less efficient. For heterologous gene expression it is therefore advisable to replace a GUG or UUG codon by a AUG codon. In some mRNAs the adenine of the start codon is the first 5' nucleotide. It is therefore not surprising that the region immediately downstream – called the downstream box – also influences translation initiation (Etchegaray and Inouye 1999). The secondary structure formed during translation initiation is of crucial importance. Pairing of the SD sequence and/or start codon with other parts of the mRNA makes the mRNA inaccessible to the ribosome. It has been shown that single nucleotide exchanges which affect such stem–loop structures that "hide" the SD sequence and AUG can result in up to a 500-fold difference in translation efficiency (de Smit and van Duin 1990). The most straightforward way to tackle such problems is the use of the translation initiation region of a gene which is highly expressed in *E. coli*, such as the T7 gene 10, T4 gene 32 or the *atpE* gene from *E. coli*, and to insert the target gene about five to ten codons downstream of the start codon of such a gene. If the N-terminal region of the target gene must be left unchanged, it is inserted immediately downstream of the start codon. Many expression vectors already contain a well-functioning SD sequence

Tab. 2.1 Frequency of codons (DNA sequence) used for start and stop of translation in *Escherichia coli* (Blattner et al. 1997).

Codon	aa	Count	%
ATG	met	3542	82.6
GTG	val	612	14.3
TTG	leu	130	3.0
ATT	ile	1	
CTG	leu	1	
Total		4286	
TAA	ochre	2705	63.1
TGA	opal	1257	29.3
TAG	amber	326	7.6
Total		4288	

and allow insertion of the gene exactly at the start codon. There are several restriction enzymes, the recognition sites of which overlap ATG codons. These are *Sph*I (GCATG▾C), *Nco*I (C▾CATGG), and *Nde*I (CA▾TATG). Only the *Nde*I site allows the insertion of any possible codon after the start codon and is therefore preferred for use in expression vectors. The *Nde*I site can be introduced into the target sequence by the oligonucleotide used for PCR amplification of the gene. To avoid the formation of stem–loop structures and optimize the downstream sequence without changing the amino acid sequence, an oligonucleotide may be used which has a variable sequence at the third position of the first five or more codons, in accordance with the wobble theory. After insertion of the gene into the expression vector, the most efficient sequence is selected by screening the clones.

A different strategy for enhancing translation efficiency is based on the translational coupling of genes. In this case, two successive genes share overlapping start/stop codons (TGATG or ATGA), and the SD sequence of the downstream gene is located within the upstream gene. In some expression vectors employing this strategy, the translational initiation region of an efficiently translated gene is inserted downstream of a promoter. After 20–30 codons, the open reading frame (ORF) is interrupted in frame by a translational coupling sequence – that is, by a SD sequence and the TGATG sequence. The target sequence is finally integrated immediately downstream of the start codon. Following this construction scheme it is expected that the translating ribosome will disrupt any secondary structure at the second start site, thus allowing uniform translational re-initiation (Schoner et al. 1990).

There are no functional SD sequences in eukaryotic mRNAs. However, cryptic translational start sites may be found within eukaryotic genes, as has been observed for example in the sequence encoding the human parathyroid hormone. After expression of the gene in *E. coli* two gene products were obtained – a full-length species, and a truncated protein which was seven amino acids shorter. Clearly, in this case there were two competing translational start sites, one at the authentic AUG start codon and one at the methionine codon at position 8. This internal ribosomal binding site could be removed by site-directed mutagenesis without changing the amino acid sequence (Sung et al. 1991), thus resulting in a homogeneous product.

2.5.2
Codon Usage

Once translation has started, efficient elongation is required. It has been shown that so-called "rare codons" can slow down the translation process. It emerged that expression of heterologous genes in *E. coli* can result in the immediate cessation of cell division and in the accumulation of modified or truncated forms of the desired protein products (Brinkmann et al. 1989; Tu et al. 1995; Zahn 1996). This effect was found to be mediated by a remarkable RNA molecule – alternatively designated as SsrA RNA, tmRNA, or 10Sa RNA – that acts both as a tRNA and as an mRNA to direct the modification of proteins of which biosynthesis has stalled or has been interrupted (Tu et al. 1995; Keiler et al. 1996; Muto et al. 1998). The *ssrA*-encoded peptide-tagging system marks nascent polypeptide products originating from damaged

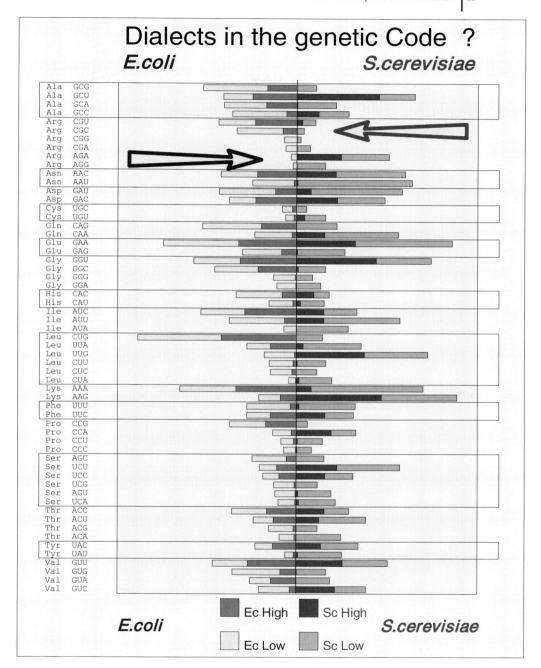

Fig. 2.5 Comparison of codon usage frequencies of *Escherichia coli* and *Saccharomyces cerevisiae*. The arrow points to the codon of lowest frequency for arginine.

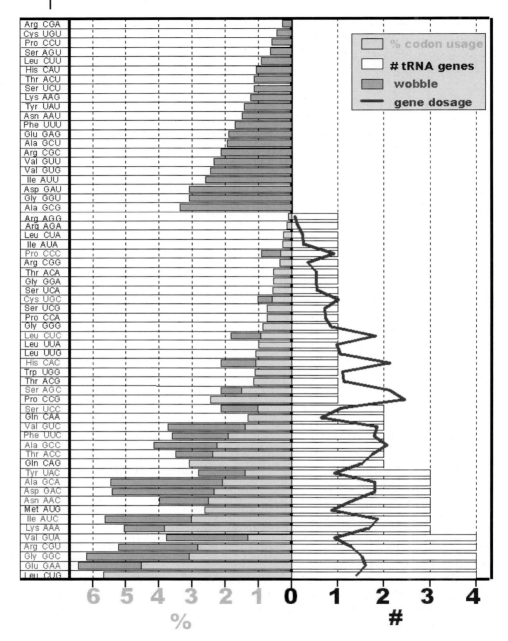

Fig. 2.6 Comparison of codon usage and number of genes known for the respective cognate tRNAs of *Escherichia coli*. Seven tRNA genes are duplicates or triplicates, whereas five are quadruplicates.

mRNAs. A specific (AANDENYALAA) peptide tag is added by co-translational template switching of the ribosome from the damaged mRNA to the *ssrA* tmRNA. The tagged polypeptide is then rapidly degraded. This mechanism constitutes a novel post-translational proof-reading system and has now been described for various bacteria.

Coding sequences in pro- and eukaryotes display species-specific usage of synonymous codons; each species uses what may be termed a "codon dialect". Thus, the codons AGG/AGA (Arg), AUA (Ile), CUA (Leu), and CCC (Pro) are seldom used in *E. coli*, in contrast to the situation in *Saccharomyces cerevisiae* (Figure 2.5), for example. Often, the frequency of synonymous codons reflects the relative abundance of their cognate tRNAs. Thus, for *E. coli* a rather stringent correlation between the codon usage and the numbers of genes encoding the cognate tRNAs can be observed (Figure 2.6). The examination of fractionated tRNAs (Dong et al. 1996) has established for *E. coli* that the frequency of rare codons often correlates with low levels of the cognate tRNAs.

For the codon AGG, in particular, it was observed early on that translation efficiency could be decreased by newly introduced codons in test genes (Robinson et al. 1984; Bonekamp et al. 1985). It has been shown that the provision of additional copies of the tRNA$_{arg}$ gene *argU* (formerly called *dnaY*) results in relief of growth inhibition, and concomitantly provides enhanced product formation, when genes with multiple AGG/AGA codons are expressed in *E. coli* (Brinkmann et al. 1989; Spanjaard et al. 1990; Chen and Inouye 1990).

It has also been observed that these rare AGG/AGA codons are prone to preferential mis-incorporation of lysine instead of arginine by misreading. Supplemental tRNA$_{arg}$ ensured the production of the correct and homogeneous protein product here also (Calderone et al. 1996; Forman et al. 1998; You et al. 1999).

Numerous reports have described successful product development using additional *argU* genes, while others have indicated that the same strategy can also be used successfully for other codons that are rare in *E. coli*. Del Tito and colleagues (1995) first successfully used an multi-augmentation strategy combining *argU* with *ileX*. For the expression of genes from AT-rich organisms such as *Plasmodium* and other protozoans, or helminths (Baca and Hol 2000) and clostridia (Zdanovsky and Zdanovskaia 2000), or GC-rich organisms such as archeal bacteria (Sorensen et al. 2003), the so-called Codon$^+$ strains have been used successfully. These combine *argU* on a helper plasmid with various other tRNA genes, *ileY* for AUA, *leuW* for CUA, *proL* for CCC and sometimes *glyT* for GGA, or *lysT* for AAG. Reports on the use of this multi-augmentation approach are still rare and, as some of these rare tRNAs require additional steps for their maturation (e.g., specific base modifications), it is at present not clear if the addition of tRNA genes alone will always be sufficient, as in the case of *argU*. HCD fermentation may be a problem with such strains, because it is known that the use of rich media may be indispensable for the success of this strategy (Del Tito Jr., et al. 1995). Thus – especially for large-scale use – synthetic re-design of the coding sequence remains a potentially useful, but tedious, alternative (Pan et al. 1999). The use of fusion genes (if possible) opens an attractive means of bypassing the problem, as described recently (Wu et al. 2004).

2.5.3
Translation Termination

Product heterogeneity can also be caused by incorrect translation termination. Frequently, the stop codon UGA is used for translation termination (see Table 2.1); this can lead to translational read-through even in strains without suppressor tRNAs. Two natural UGA-decoding tRNAs insert selenocysteine into proteins. These require a special RNA stem–loop structure downstream of the UGA codon (Böck et al. 1991). During expression of chorismate mutase genes containing UGA stop codons from *E. coli* and *Bacillus subtilis,* translational read-through and tryptophan insertion at the UGA codon was observed in 5 % of the translation products (MacBeath and Kast 1998). Translational +1 frameshifting at UGA codons has also been described by Poole et al. (1995). These authors tested the strength of stop signals and found that the next base following the stop signal has a very strong impact on termination efficiency. The strongest termination was found with the tetranucleotide UAAU, and the weakest with UGAC. Therefore, use of the UAAU stop sequence is strongly recommended.

2.5.4
Transcription Termination and mRNA Stability

As well as a strong stop codon, the use of a strong transcription terminator is recommended. The transcription terminator inhibits read-through by the RNA polymerase into other genes on the vector, or into the replication region. For example, deleting the TΦ terminator in the pET vectors leads to read-through by the T7 RNA polymerase into the β-lactamase gene located downstream in the same orientation, thereby resulting in marked instability of the expression system (unpublished observations). Usually, Rho-independent transcription terminators which are able to form stem–loop structures are used, such as the two from the *rrnB* operon, from phage fd or T7 (TΦ). For some transcripts, higher mRNA stability was observed when these terminators were used. Presumably they protect the RNA from the major $3' \rightarrow 5'$ exonucleases RNase II and polynucleotide phosphorylase.

Degradation of most mRNAs is probably initiated by an endonucleolytic cleavage catalyzed by RNase E (Kushner 2002). The products of this initial cleavage are further degraded by additional RNase E cleavages and by $3' \rightarrow 5'$-exonucleases. RNase E is essential for the maturation of 5S RNA. It is a very large enzyme of 1061 residues, and is active as a homodimer. The catalytic activity is located in the N-terminal half of each subunit. The C-terminal half can be deleted without affecting the maturation of 5S RNA, whereas the half-life of individual mRNAs and the bulk mRNA in the cell is considerably increased by this mutation. This is due to the fact that the C-terminal part of RNase E contains binding sites for polynucleotide phosphorylase, RNase II, enolase and the RNA helicase RhlB. These proteins together form a multiprotein complex called the "degradasome", with RNase E as a scaffold (Cohen and McDowall 1997). The degradasome binds preferentially to single-stranded mono- and triphosphorylated 5'-ends of mRNAs, and it is suggested that after looping of the mRNA–RNase E complex a sequence downstream is recognized and cleaved. The cleavage occurs upstream of the dinucleotide AU. However, other structures are required for efficient cleavage in addition to the AU dinucleotide, the nature of which is not fully understood. When the 5′-end of the mRNA is buried within a stem–loop structure, the end cannot be recognized by the degradasome (Baker and Mackie 2003). This explains the high stability of some transcripts, like *ompA*, which have a stem–loop structure at their 5′-end. There is no sequence specificity of the stem–loop structure, and the addition of such secondary RNA structures to other mRNAs such as *lacZ* also stabilizes these. The half-life of stabilized mRNA can be increased by up to sixfold (Arnold et al. 1998). Nevertheless, this mRNA is also degraded in an RNase E-dependent fashion, and it is postulated that RNase E has a second, internal entry site which is less efficient. Factors other than the 5′-end also influence the stabilization of mRNA. The most important factor seems to be the translational efficiency of the mRNA. For example, changing the start codon UUG to the more efficient AUG codon or improving the ribosomal binding site of the *rpsT* gene has a positive effect on mRNA stability: higher translation rates resulted in higher mRNA stability. It is assumed that the translating ribosomes mask the RNase E cleavage site and therefore stabilize the mRNA.

In addition to RNase E there is an RNase G, which shows sequence similarity to RNase E at its N-terminus. This enzyme is essential for the processing of 16S RNA. In only a single case – the *adhE* gene for alcohol dehydrogenase – has it been shown that RNase G is responsible for a short transcript life time: an RNase G mutant produced about fivefold higher amounts of the *adhE* gene product than the corresponding wild-type strain (Kaga et al. 2002).

2.6
Protein Production

2.6.1
Inclusion Body Formation

One of the main problems encountered in heterologous gene expression using strong promoters is the misfolding and aggregation of proteins in inclusion bodies. For many pharmaceutical proteins, inclusion body formation and subsequent refolding into an active conformation can be employed as an important first purification step, incidentally providing protection of the protein against proteolytic degradation by host proteases (Lilie et al. 1998). For proteins of lower commercial value, such as technical enzymes, purification and re-folding is too expensive and is not possible when the enzymes are used without purification in whole-cell biocatalysts. The key to solving this problem is an understanding of protein folding during protein synthesis and of the effects of misfolded proteins on the cell. In contrast to the initial assumption that inclusion bodies resulted from unspecific coagulation of incompletely folded proteins, later investigations suggested specific interactions between defined domains of partially folded proteins, for example in the case of P22 tail-spike and P22 coat protein (Speed et al. 1996). Thus, two proteins which both show a tendency to form inclusion bodies mainly aggregate in separate inclusion bodies when co-produced in a single cell. High-level expression of genes leads to a substantial change in the overall expression profile of cells. On average, about 6% of genes show a more than threefold change in expression, regardless of the solubility of the recombinant proteins. The misfolding of proteins leads to an additional cellular stress reaction similar to a heat-shock response. Comparison of expression patterns identified 52 genes with a more than threefold change in response to inclusion body formation (Lesley et al. 2002). Some of the most strongly induced genes encode chaperones such as DnaK, DnaJ, GrpE, GroEL, GroES, IbpA, IbpB, trigger factor and ClpB, ClpP and La, which facilitate de-novo protein folding, inhibit aggregation of denatured proteins, bind to inclusion bodies, dissolve aggregated proteins, or act as proteases.

Probably the most efficient way to avoid inclusion body formation is to decrease the growth temperature during the production phase to 20 °C and even lower, depending on the protein. The lower temperature, with the resulting reduction in the rate of protein synthesis, may simply provide more time for proper folding of the nascent chain, or may induce cold shock proteins which assist in folding.

2.6.1.1 Chaperones as Facilitators of Folding

The strategy just described has two obvious disadvantages: low growth rate; and additional energy costs incurred in cooling the fermentor. Therefore, many alternative approaches have been tested. One of the more successful solutions involves the reduction of gene dosage by employing expression vectors with lower copy numbers, or promoters of lower strength such as the positively regulated *araBAD* and *rhaBAD* promoters, or a combination of both. A fundamentally different approach is the co-production of chaperones. Plasmids that are compatible with expression vectors have been constructed, which for example contain *dnaK*, *dnaJ*, *grpE*, or *mopA* and *mopB* (GroEL/ES) genes in tandem, under the control of a different promoter. However, these plasmids are not always effective, and sometimes a combination of different chaperones is necessary to enhance the production of recombinant proteins in a soluble form. For example, co-expression of trigger factor (a cytoplasmic peptidyl-prolyl *cis/trans* isomerase which associates with the ribosome) enhanced the solubility of mouse endostatin (which was otherwise deposited in inclusion bodies), whereas no effect was observed upon co-expression of GroEL-GroES. Only when GroEL-GroES were produced together with DnaK-DnaJ-GrpE was a significant amount of endostatin rendered soluble. Solubilization was only achieved in case of human lysozyme upon co-production of trigger factor and GroEL-GroES. Activity testing of the lysozyme showed that only a small fraction of the soluble protein was enzymatically active. This emphasizes that solubility *per se* does not necessarily imply correct folding (Nishihara et al. 2000). Another disadvantage of chaperone co-production is that these proteins have to be made in large quantities, thereby reducing the fraction of the cell's metabolic capacity that can be devoted to production of the target protein.

2.6.1.2 **Fusion Protein Technology**

Another way of reducing inclusion body formation is to fuse the target gene to the 3'-end of a gene which is highly expressed in *E. coli* and encodes a soluble gene product. Examples of established fusion partners are the *E. coli* maltose-binding protein MalE (without the signal peptide for export), *Schistosoma* glutathione transferase, thioredoxin, or *Staphylococcus* A protein. For example, interleukins IL-2, IL-3, IL-4 and IL-6 could only be produced in inclusion bodies even at 15 °C. However, after fusion of the ORF to the thioredoxin gene *trxA*, each of the fusion proteins could be recovered in soluble form, even when it accounted for up to 20% of the total cell protein (LaVallie and McCoy 1995). Besides their higher solubility, such fusions can be advantageous for productivity and yield in providing improved efficiency of translation initiation and in allowing one to use the fusion partner as the basis for purification by affinity chromatography. However, the success of these fusions is not predictable. The degree of misfolding of the fusion protein may be even higher than that of the native protein, or the fusion protein may be soluble but inactive. Finally, in the case of therapeutic products, such as factor Xa or enteropeptidase, the fusion partner must be cleaved off, and very often this is a quite inefficient process.

2.6.2
Methionine Processing

Recombinant proteins produced in the cytoplasm of *E. coli* can retain the initiation methionine at their N-terminal end, and this may affect the immunogenicity of the product. *E. coli* possesses an enzyme activity in the cytoplasm which has the capacity to remove N-terminal methionine. However, in-vivo and in-vitro experiments have demonstrated that the enzyme has some limitations, and that efficient processing depends on the amino acid adjacent to the methionine (Table 2.2). N-terminal product heterogeneity due to partial processing has also been observed

Tab. 2.2 Cleavage of N-terminal methionine from intracellular prokaryotic proteins. The effect of the penultimate amino-acid on methionine removal is shown (Ben-Bassat 1991).

Met cleaved	Variable	Retained
Ala	Ile	Lys
Gly	Val	Arg
Pro	Cys	Leu
Ser	Thr	Asp
		Asn
		Phe
		Met
		Trp
		Tyr
		Glu
		Gln
		His

(Hirel et al. 1989; Ben-Bassat 1991). The gene for the methionine aminopeptidase (*map*) has been cloned and analyzed (Ben-Bassat et al. 1987), and enhancement of processing by concurrent expression of *map* has also been demonstrated (Sandman et al. 1995).

2.6.3
Secretion into the Periplasm

Proteins destined for the periplasmic space and outer membrane must pass through the cytoplasmic membrane. Most of these proteins are exported by the general secretory pathway (Sec) consisting of a membrane-embedded complex of the proteins SecY, SecE, SecG, and SecA. The proteins to be exported usually need an extended, loosely folded conformation and an N-terminal extension – an 18- to 30-amino acid signal peptide which is cleaved by the signal peptidases during translocation of the polypeptide. Folding of the polypeptide in the cytoplasm is inhibited by binding of SecB and other cytoplasmic chaperones.

The periplasmic space also has its own chaperones to assist in protein folding and to generate disulfide bonds.

There are several practical reasons for targeting recombinant proteins to the periplasm. Some proteins need disulfide bonds, which are only formed in the periplasm and not under the reducing conditions found in the cytoplasm. Other proteins, such as the penicillin amidase which must be processed auto-catalytically into a heterodimer, do not fold into the active conformation in the cytoplasm even after removal of the signal peptide.

Secretion can also serve to protect the desired product from cytoplasmic proteases; the periplasm itself has a different set of proteases which are present in lower abundance. Therefore, isolation of intact proteins from the periplasm by osmotic shock may be easier than recovery from crude extracts of whole cells. Finally, some proteins produced in the cytoplasm still contain the methionine encoded by the start codon. Processing by the signal peptidase may be used to generate the mature N terminus. On the other hand, translocation of a protein into the periplasm is a bottleneck for high-level production of recombinant proteins. Usually, large cytoplasmic proteins cannot be exported when they are fused to a signal peptide.

Export of recombinant proteins seems to be most promising route if the protein is exported in its original host. For efficient secretion, signal sequences are used from *E. coli* proteins like the outer membrane proteins OmpT or OmpA, from β-lactamases or from the pectate lyase PelB of *Erwinia carotovora*. In many cases, however, pre-proteins are not readily exported, and form inclusion bodies in the cytoplasm, or are degraded. In other cases the proteins seem to be transported to the Sec complex but are trapped during translocation, leading to lethality and cell lysis. It is not yet clear why some proteins are exported and others not, or which signal peptides might be optimal. Presumably several factors are crucial for export: the recognition of the nascent polypeptide by SecB or the signal recognition particle and other chaperones, maintenance of its unfolded conformation, the time needed for transport to the translocation complex, and the overall hydrophobicity of the protein.

Even when a protein is exported into the periplasm there are still potential pitfalls to be overcome before an active protein is obtained. These obstacles include the formation of incorrect disulfide bridges, misfolding and inclusion body formation, and proteolytic degradation. As in the cytoplasm, where misfolded proteins induce the σ^H-dependent heat-shock genes, the periplasm also possesses a heat-shock system regulated by σ^E and the Cpx two-component system, which respond to extracytoplasmic stress by synthesizing proteases and chaperones (Raivio and Silhavy 2001).

2.6.4
Disulfide Bond Formation and Folding

Some of the problems encountered with periplasmic proteins can be solved by co-production of periplasmic chaperones. Thus, it has been shown that overproduction of the thiol/disulfide oxidoreductase DsbA, which acts as a strong oxidizer of thiol groups, and DsbC, a protein disulfide isomerase, can enhance the production of active tPA, a protein with 17 disulfide bridges (Bessette et al. 1999). In another report, co-expression of DsbCD or DsbABCD improved production of the exported human nerve growth factor (NGF) by 80 % (Kurokawa et al. 2001).

Other chaperones, such as Skp/OmpH, which are involved in the folding of outer membrane proteins seem to have a broad substrate spectrum. Thus, overproduction of Skp improved the solubility of single-chain antibodies secreted into the periplasm (Hayhurst and Harris 1999). Some chaperones exhibit peptidyl-prolyl *cis/trans* isomerase activity (SurA, PpiD, FkpA, RotA/PipA). The co-expression of FkpA considerably improved the periplasmic solubility of a MalE mutant protein (Arie et al. 2001). There are about ten proteases in the periplasm, but probably only one of them – DegP – has a chaperone activity in addition, and is induced by the presence of misfolded proteins in the periplasm.

Overproduction of DegP has been found to improve the production of active penicillin amidase. High-level production of penicillin amidase leads to inclusion body formation in the periplasm, and leakage of active enzyme into the medium. Co-expression of DegP resulted in a significant decrease in the incidence of inclusion bodies, and the penicillin amidase was slightly more active (about 15 %). The main effect was the improvement of cell physiology. Whereas about 50 % of the enzyme was lost to the medium in the original strain, in the DegP-overproducing strain the outer membrane remained intact. It could be shown that this effect was due to the proteolytic activity of DegP. Overall, 80 % more active enzyme could be recovered from the cells (Pan et al. 2003).

2.6.5
Twin Arginine Translocation (TAT) of Folded Proteins

Several years ago it became clear that most bacteria have a second general protein export pathway, which is independent of Sec. This pathway was called TAT (*twin arginine translocation*), because its substrates are characterized by two consecutive arginine residues in the signal peptide sequence (reviewed by Berks et al. 2000). The most remarkable characteristic of the TAT system is that it transports into the periplasm those proteins which have already been folded in the cytoplasm, often together with a bound cofactor such as FeS centers or molybdopterin. Even heterodimeric proteins with a TAT signal in only one of the subunits are exported by this system (see also Chapter 4).

Signal peptides recognized by the Sec system are divided into three sections: a positively charged N-terminal sequence; a hydrophobic α-helical region; and a cleavage site. The TAT signal sequences are similar but exhibit some distinct features, most notably the conserved sequence (S/T-R-R-X-F-L-K) of the positively charged N-terminal region. Replacement of TAT by Sec signal sequences directs the TAT-exported protein exclusively to the Sec translocator, and vice versa, as has been shown for two *E. coli* proteins. Thus, it seems that sorting of the precursor proteins to the transport complex is determined only by the signal peptide. When the green fluorescent protein (GFP) was fused to a Sec signal peptide, it remained in the cytoplasm, whereas at least part of it was translocated into the periplasm when a TAT signal was used (DeLisa et al. 2002).

As with the Sec system, the addition of a TAT signal peptide to a large cytoplasmic protein such as β-galactosidase is not sufficient to ensure TAT-mediated translocation. Interestingly, penicillin amidase does not show the signal sequence typical of the TAT proteins (it has two arginine residues in the hydrophilic region, but these are separated by an asparagine), but it is nevertheless transported by the TAT translocator. When the original signal sequence was replaced by an OmpT signal (Sec), penicillin amidase was equally well exported (Ignatova et al. 2002). Since penicillin amidase folds predominantly in the periplasm, it seems that some proteins can be transported in a loosely folded state by the TAT system. Further investigations are necessary to understand what exactly is recognized by the TAT translocation system, before it can be routinely exploited for production of recombinant proteins.

2.6.6
Disulfide Bond Formation in the Cytoplasm

Disulfide bonds usually do not form in the cytoplasm, due to the presence of reducing components such as glutathione and thioredoxins and glutaredoxins. In addition, the cytoplasm lacks enzymes like the DsbA and DsbC found in the periplasm, which oxidize pairs of cysteines in proteins to form disulfide bonds, and function as a disulfide isomerase to reduce incorrectly formed disulfide bonds (Kadokura et al. 2003). The oxidized forms of the two thioredoxins and three glutaredoxins are able to oxidize proteins for disulfide formation, but they are kept in a reduced state by thioredoxin reductase (TrxB) and glutathione, which in turn is reduced by glutathione oxidoreductase (Gor).

Double mutants for *trxB* and *gor* have been isolated, which are viable but grow very slowly and only in the presence of exogenous reducing dithiothreitol (DTT). A fast-growing derivative has, however, been isolated which grows in the absence of DTT. Many proteins, such as *E. coli* alkaline phosphatase and the maltase MalS, as well as mouse urokinase and tissue plasminogen activator (t-PA), are only active when the correct disulfide bonds have been formed. Therefore, they must be exported into the periplasm to ensure activity. In the new mutant all of these proteins could be expressed intracellularly in an active conformation, and in higher amounts than in the periplasm. The amount of active protein could be further increased by intracellular production of the disulfide isomerase DsbC (Bessette et al. 1999). Hence, this strain seems to offer a suitable alternative to the secretion of proteins into the periplasm. Strains with the double mutation *trxB gor* are now being marketed commercially under the trade name Origami™.

2.6.7
Cell Surface Display and Secretion across the Outer Membrane

Display of peptides and proteins at the cell surface has a wide range of applications, including the development of live vaccines, screening of peptide libraries, adsorption and removal of hazardous chemicals, or construction of whole-cell biocatalysts and biosensors (reviewed by Lee et al. 2003). For this purpose, vehicles are needed which are targeted to the outer membrane, such as the *E. coli* Omp proteins OmpA and OmpC, the peptidoglycan-associated lipoprotein PAL, adhesin AIDA-I, TraT or LamB, as well as foreign proteins like the IgA1 protease from *Neisseria gonorrhea* or the ice nucleation protein (INP) of *Pseudomonas syringae*. Alternatively, proteins of cell appendages like pilins, fimbriae and flagella can be used. These so-called carrier proteins are fused at the N-terminal end, the C-terminal end or somewhere within the protein (sandwich technique) to a passenger peptide or protein. The site of fusion depends on where the carrier protein is anchored to the cell. Many carrier proteins allow the integration of small peptides only, otherwise the structure of the outer membrane or the appendages is disturbed and the protein no longer folds properly. The *Pseudomonas syringae* INP, which has been employed for the generation of a whole-cell biocatalyst, is the most promising example for displaying large proteins. INP is an outer membrane protein which can be easily produced in *E. coli*. It nucleates ice formation and is responsible for frost injuries in plants. Levansucrase, organophosphorus hydrolase, and carboxymethyl cellulase, as well as hepatitis B virus surface antigen (Kim and Yoo 1999) and single-chain antibodies, have been produced as fusions to the C-terminal end of the INP, and displayed in a biologically active form on the surface of the *E. coli* cell.

Sometimes it may be useful to secrete protein products through the cytoplasmic and outer membrane into the medium. In general, laboratory strains of *E. coli* do not seem to have the appropriate genetic equipment for this, in contrast to pathogenic *E. coli* and many other Gram-negative bacteria which segregate toxins, proteases, pullulanase and filamentous phages.

There are two principal ways to direct proteins into the medium (reviewed by Shokri et al. 2003). The first is to direct the target proteins into the periplasm, followed by subsequent passive passage through the outer membrane. A second approach involves the use of an active transport system from another bacterial strain or species. If the first option is chosen, methods can be applied to make the outer membrane leaky, such as osmotic shock, incubation with polyethylene glycol, EDTA or Triton X-100. Such procedures can only be carried out after the cells have been harvested, and there are no defined methods for particular proteins of interest. An alternative technique involves co-production of colicin lysis proteins, which permeabilize the outer membrane to deliver the toxin to the medium. An example is the colicin E1 lysis protein (Kil), which has been used to promote the passage of penicillin amidase into the medium. However, the damage caused to the outer membrane can rapidly lead to cell lysis and death.

E. coli mutants are known which are naturally more leaky than the wild-type, for example strains mutated in the Omp proteins. Finally mutants, so-called L-forms, which obviously have no outer membrane, have long been known (Lederberg and St. Clair 1958) but have largely been neglected. Although L-forms have been used successfully for single-chain antibody production on a laboratory scale (Rippmann et al. 1998), the very complex medium that is needed and the sensitivity of the cells prevents scale-up of this process for industrial purposes.

Many different systems have been described for the active transport of proteins, but all of them are very specific for a particular protein. An example is the transport of the hemolysin HlyA. This protein is translocated via a protein channel consisting of HlyB, HylD and TolC directly from the cytoplasm into the medium. The hemolysin is recognized by a signal sequence at its C-terminal end, which is not cleaved off. A two-stage process is observed in case of a *Klebsiella oxytoca*-derived pullulanase. First, the protein is transported into the periplasm, during which step the signal sequence at the N-terminus is cut off. Fourteen different gene products are then necessary to translocate the protein into the medium!

In summary, these and a series of other strategies have not yet led to efficient segregation of proteins; they are difficult to adapt to the transport of foreign proteins, and they are still not well understood.

2.7
Examples of Products and Processes

There are numerous examples of *E. coli*-based processes for the industrial production of proteins, including a range of pharmaceutical and therapeutic compounds. A selection of approved pharmaceutical products is listed in Table 2.3. In addition to the examples described here and in the previous sections, even very complex proteins such as new variants of tPA have been produced in high quality and large quantities (Mattes 2001). These thrombolytic (fibrinolytic) agents are used for the treatment of acute myocardial infarction. They are produced currently as inclusion bodies, and are purified with subsequent renaturation *in vitro* (Rudolph and Lilie 1996). This example is not unique, as several systems for the production of the original tPA and newly developed derivatives are available. Does *E. coli* perform as well as, or better than comparable animal cell systems? A unique comparative study performed some years ago provided some evidence for the superiority of the *E. coli*-based process in this particular development (Datar et al. 1993). In the meantime, there have been improvements which make the process even more efficient (Schaffner et al. 2001).

Tab. 2.3 Pharmaceutical products produced by *Escherichia coli* strains, approved in Germany (selection).

Class	Product	Brand name
Human insulin	insulin	Berlinsulin®
	insulin	Huminsulin®
	Insulin lispro (variant)	Humalog® 100
Pituitary hormones	somatotropin	Genotropin®
		Humatrope®
		Norditropin®
		Zomacton®
Cytokines	Aldesleukin, IL2	Proleukin®
	Interferon alfa-2a	Roferon®-A
	Interferon alfa-2b	Intron A®
	Interferon beta-1b	Betaferon®
	Interferon gamma-1b	Imukin®
	Molgramostim, rhGM-CSF	Leucomax®
	Filgrastim, r-metHuG-CSF	Neupogen®
Fibrinolytics	Reteplase (t-PA variant)	Rapilysin®

2.8
Conclusions and Future Perspectives

Many expression systems, ranging from microorganisms to animals and plants, have been developed or are under construction. Nevertheless, it is felt that *E. coli* is still the first choice as the host for heterologous gene expression, except in cases where special requirements such as protein modifications, glycosylation or unusual cofactors, must be met. However, even this may change in the near future, as by expanding the genetic code the incorporation of modified amino acids has been achieved (Wang et al. 2001; Döring et al. 2001). The "last" hurdle for prokaryotes – the lack of a glycosylation capability – may be solved by a new strategy and genome engineering, as depicted recently (Zhang et al. 2004). New methods of genome analysis (Gill et al. 2001) and efficient genomic engineering should therefore expand the potential of *E. coli* as an industrial producer of proteins and other valuable products (Causey et al. 2004).

Improvements are continuously being made – in the stability of expression vectors, in the tight regulation of promoters, the construction of new host strains for improved mRNA and protein stability, hosts which allow disulfide bond formation in the cytoplasm, or in the co-production of chaperones to facilitate correct protein folding. Other important developments are dealing with the economic production of proteins by improved feeding strategies during fed-batch fermentation and downstream processing.

Presumably, there will never be a single unique expression system which is applicable to all genes. However, we are confident that, with increasing knowledge and experience, the expression of even "difficult" genes can be successfully accomplished, and appropriate platforms and strategies will become available to overcome the problems discussed in this chapter.

Appendix

Tab. A2.1 Useful strains of *Escherichia coli* and their characteristics.

Strain	*Genotype*	*Relevant features/Remarks*	*Reference*
BL21(DE3)	*hsdS* (r_B^- m_B^-) *gal ompT* (λc_Its857 *ind*1 *Sam*7 *nin*5 *lac*UV5-T7 *gene 1*)	An *E. coli* B strain for expression vectors containing bacteriophage T7 promoter. Bacteriophage T7 RNA polymerase is carried on the bacteriophage λ DE3, which is integrated into the chromosome of BL21.	Studier and Moffatt (1986)
DH5α	*supE*44 Δ*lac*U169 (φ80 *lacZ*ΔM15) *hsdR*17 *recA*1 *endA*1 *gyrA*96 *thi*-1 *relA*1 *mcrA* Δ*mcrBC mrr*	A recombination deficient and suppressing strain used for plasmids and cosmids. The φ80 *lacZ*ΔM15 permits a complementation with the amino terminus of β-galactosidase encoded in pUC vectors.	Hanahan (1983)
HB101	*supE*44 *hsdS*20 (r_B^- m_B^-) *recA*13 *ara*-14 *proA*2 *lacY*1 *galK*2 *rpsL*20 *xyl*-5 *mtl*-1 Δ*mcrBC mrr*	A suppressing strain commonly used for large scale production of plasmids. It is an *E. coli* K12 × *E. coli* B hybrid that is highly transformable.	Boyer and Roulland-Dussoix (1969)
JM108	F^- *mcrA recA*1 *endA*1 *gyrA*96 *thi hsdR*17 (r_K^- m_K^+) *relA*1 *supE*44, Δ(*lac-proAB*)	A recombination deficient F^- strain that will support growth of vectors carrying amber mutations and will modify but not restrict transfected DNA.	Yanisch-Perron et al. (1985)
JM109	*mcrA recA*1 *supE*44 *endA*1 *hsdR*17 (r_K^- m_K^+) *gyrA*96 *relA*1 *thi* Δ(*lac-proAB*) F′ [*traD*36 *proAB*$^+$ *lacI*^{q} *lacZ*ΔM15]	A recombination deficient strain that will support growth of vectors carrying amber mutations and will modify but not restrict transfected DNA and contains F′. The F′ allows blue/white screening on X-gal and permits bacteriophage M13 superinfection.	Yanisch-Perron et al. (1985)
ED8654	*supE supF hsdR metB lacY gal trpR mcrA*	A suppressing strain commonly used to propagate bacteriophage λ vectors and their recombinants.	Borck et al. (1976)
W3110	inversion between *rrnD* and *rrnE*	Wild-type strain.	Hill and Harnish (1981)
RB791	W3110 *lacI*^{q} L8	A strain that makes high levels of lac repressor and is used for inducible expression of genes under the control of the *lac* and *tac* promoters.	Brent and Ptashne (1981)
XL1 Blue	*supE*44 *hsdR*17 *recA*1 *endA*1 *gyrA*46 *thi relA*1 *lac*$^-$ F′ [*proAB*$^+$ *lacI*^{q} *lacZ*ΔM15 Tn*10*(*tet*^{r})]	A recombination deficient strain that will support the growth of vectors carrying some amber mutations, but not those with the *Sam*100 mutation (e.g., λZAP). Transfected DNA is modified but not restricted. The F′ allows blue/white screening on X-gal and permits bacteriophage M13 superinfection.	Bullock et al. (1987)

Tab. A2.2 Useful vectors for *Escherichia coli* and their characteristics.

Plasmid	Resistance marker	Relevant features	Accession no./ Reference
pBR322	*bla* (Ap^r), *tet* (Tc^r)	Standard plasmid with 50–200 copies per cell	J01749
pKK223–3	*bla* (Ap^r)	expression vector, pBR322 derivative, *tacPO* (*trp/lac*UV5 hybrid promoter with *lac* operator), *rrnB* terminator	VB0070
pUC19	*bla* (Ap^r)	Cloning and expression vector (translational fusion to *lacZ* amino-terminal sequence), pBR322 derivative, *lac*UV5 promoter and *lacZα* coding region (blue-white screening with X-Gal); suitable with *lacZ*ΔM15 providing host cells, higher copy number due to loss of *rop*, contains 54-bp multiple cloning site polylinker within *lacZα*	M77789
pET11	*bla* (Ap^r)	Expression vector, pBR322 derivative, suitable for expression by T7 RNA polymerase provided by e.g. BL21(DE3), the T7 derived promoter is under control of the *lac* operator, *lacI* encodes the *lac* repressor	Dubendorff et al. (1991)
pJOE2702	*bla* (Ap^r)	Expression vector, pBR322 derivative, *rhaP* (promoter of the *rhaBAD* operon), *rrnB* terminator	Wilms et al. (2001a)
pACYC184	*tet* (Tc^r), *cat* (Cm^r)	p15A derivative, compatible to pBR322, lower copy number than pBR322	X06403
pACYC177	*bla* (Ap^r), *aphA* (Km^r)	p15A derivative, compatible to pBR322, lower copy number than pBR322	X06402
pUBS520	*aphA* (Km^r)	*argU* augmentation plasmid, pACYC177 derivative, providing also LacI from the *lacI*^{q} allele	Brinkmann et al. (1989)
pSC101	*tet* (Tc^r)	Low copy vector, compatible to pBR322 and p15A derivatives	NC_002056

References

Åkesson M, Hagander P, Axelsson JP (2001) Avoiding acetate accumulation in *Escherichia coli* cultures using feedback control of glucose feeding. Biotechnol Bioeng 73: 223–230

Amann E, Ochs B, Abel KJ (1988) Tightly regulated *tac* promoter vectors useful for the expression of unfused and fused proteins in *Escherichia coli*. Gene 69: 301–315

Andrews B, Adari H, Hannig G, Lahue E,

Gosselin M, Martin S, Ahmed A, Ford PJ, Hayman EG, Makrides SC (1996) A tightly regulated high level expression vector that utilizes a thermosensitive *lac* repressor: production of the human T cell receptor V beta 5.3 in *Escherichia coli*. Gene 182: 101–109

Arie JP, Sassoon N, Betton JM (2001) Chaperone function of FkpA, a heat shock prolyl isomerase, in the periplasm of *Escherichia coli*. Mol Microbiol 39: 199–210

ARISTIDOU AA, SAN KY, BENNETT GN (1999) Metabolic flux analysis of *Escherichia coli* expressing the *Bacillus subtilis* acetolactate synthase in batch and continuous cultures. Biotechnol Bioeng 63: 737–749

ARNOLD TE, YU J, BELASCO JG (1998) mRNA stabilization by the ompA 5′ untranslated region: two protective elements hinder distinct pathways for mRNA degradation. RNA 4: 319–330

ATLUNG T, NIELSEN A, RASMUSSEN LJ, NELLE-MANN LJ, HOLM F (1991) A versatile method for integration of genes and gene fusions into the lambda attachment site of *Escherichia coli*. Gene 107: 11–17

BACA AM, HOL WG (2000) Overcoming codon bias: a method for high-level overexpression of *Plasmodium* and other AT-rich parasite genes in *Escherichia coli*. Int J Parasitol 30: 113–118

BACHMANN BJ (1972) Pedigrees of some mutant strains of *Escherichia coli* K-12. Bacteriol Rev 36: 525–557

BADIA J, BALDOMA L, AGUILAR J, BORONAT A (1989) Identification of the *rhaA*, *rhaB* and *rhaD* gene products from *Escherichia coli* K-12. FEMS Microbiol Lett 53: 253–257

BAKER KE, MACKIE GA (2003) Ectopic RNase E sites promote bypass of 5′-end-dependent mRNA decay in *Escherichia coli*. Mol Microbiol 47: 75–88

BASS SH, YANSURA DG (2000) Application of the *E. coli trp* promoter. Mol Biotechnol 16: 253–260

BEN-BASSAT A (1991) Methods for removing N-terminal methionine from recombinant proteins. Bioprocess Technol 12: 147–159

BEN-BASSAT A, BAUER K, CHANG SY, MYAMBO K, BOOSMAN A, CHANG S (1987) Processing of the initiation methionine from proteins: properties of the *Escherichia coli* methionine aminopeptidase and its gene structure. J Bacteriol 169: 751–757

BERKS BC, SARGENT F, PALMER T (2000) The Tat protein export pathway. Mol Microbiol 35: 260–274

BERLYN MKB (1998) Linkage map of *Escherichia coli* K-12, edition 10: the traditional map. Microbiol Mol Biol Rev 62: 814–984

BESSETTE PH, ASLUND F, BECKWITH J, GEORGIOU G (1999) Efficient folding of proteins with multiple disulfide bonds in the *Escheri-chia coli* cytoplasm. Proc Natl Acad Sci USA 96: 13703–13708

BETLACH M, HERSHFIELD V, CHOW L, BROWN W, GOODMAN H, BOYER HW (1976) A restriction endonuclease analysis of the bacterial plasmid controlling the EcoRI restriction and modification of DNA. Fed Proc 35: 2037–2043

BLATTNER FR, PLUNKETT G, III, BLOCH CA, PERNA NT, BURLAND V, RILEY M, COLLADO-VIDES J, GLASNER JD, RODE CK, MAYHEW GF, GREGOR J, DAVIS NW, KIRKPATRICK HA, GOEDEN MA, ROSE DJ, MAU B, SHAO Y (1997) The complete genome sequence of *Escherichia coli* K-12. Science 277: 1453–1474

BÖCK A, FORCHHAMMER K, HEIDER J, LEIN-FELDER W, SAWERS G, VEPREK B, ZINONI F (1991) Selenocysteine: the 21st amino acid. Mol Microbiol 5: 515–520

BONEKAMP F, ANDERSEN HD, CHRISTENSEN T, JENSEN KF (1985) Codon-defined ribosomal pausing in *Escherichia coli* detected by using the *pyrE* attenuator to probe the coupling between transcription and translation. Nucleic Acids Res 13: 4113–4123

BORCK K, BEGGS JD, BRAMMAR WJ, HOPKINS AS, MURRAY NE (1976) The construction in vitro of transducing derivatives of phage lambda. Mol Gen Genet 146: 199–207

BOYER HW, ROULLAND-DUSSOIX D (1969) A complementation analysis of the restriction and modification of DNA in *Escherichia coli*. J Mol Biol 41: 459–472

BRENT R, PTASHNE M (1981) Mechanism of action of the *lexA* gene product. Proc Natl Acad Sci USA 78: 4204–4208

BRINKMANN U, MATTES RE, BUCKEL P (1989) High-level expression of recombinant genes in *Escherichia coli* is dependent on the availability of the *dnaY* gene product. Gene 85: 109–114

BROSIUS J, ERFLE M, STORELLA J (1985) Spacing of the −10 and −35 regions in the *tac* promoter. Effect on its *in vivo* activity. J Biol Chem 260: 3539–3541

BULLOCK WO, FERNANDEZ JM, SHORT JM (1987) XL1-Blue: A high efficiency plasmid transforming *recA Escherichia coli* strain with beta-galactosidase selection. Biotechniques 5: 376

CALDERONE TL, STEVENS RD, OAS TG (1996) High-level misincorporation of lysine for arginine at AGA codons in a fusion protein expressed in *Escherichia coli*. J Mol Biol 262: 407–412

Calos MP (1978) DNA sequence for a low-level promoter of the *lac* repressor gene and an 'up' promoter mutation. Nature 274: 762–765

Causey TB, Shanmugam KT, Yomano LP, Ingram LO (2004) Engineering *Escherichia coli* for efficient conversion of glucose to pyruvate. Proc Natl Acad Sci USA 101: 2235–2240

Chang AC, Cohen SN (1978) Construction and characterization of amplifiable multicopy DNA cloning vehicles derived from the P15A cryptic miniplasmid. J Bacteriol 134: 1141–1156

Chen GF, Inouye M (1990) Suppression of the negative effect of minor arginine codons on gene expression; preferential usage of minor codons within the first 25 codons of the *Escherichia coli* genes. Nucleic Acids Res 18: 1465–1473

Chou CH, Bennett GN, San KY (1994) Effect of modulated glucose uptake on high-level recombinant protein production in a dense *Escherichia coli* culture. Biotechnol Prog 10: 644–647

Cohen SN, McDowall KJ (1997) RNase E: still a wonderfully mysterious enzyme. Mol Microbiol 23: 1099–1106

Colloms SD, McCulloch R, Grant K, Neilson L, Sherratt DJ (1996) Xer-mediated site-specific recombination *in vitro*. EMBO J 15: 1172–1181

Datar RV, Cartwright T, Rosen CG (1993) Process economics of animal cell and bacterial fermentations: a case study analysis of tissue plasminogen activator. Biotechnology (NY) 11: 349–357

de Smit MH, van Duin J (1990) Secondary structure of the ribosome binding site determines translational efficiency: a quantitative analysis. Proc Natl Acad Sci USA 87: 7668–7672

deHaseth PL, Zupancic ML, Record MT, Jr. (1998) RNA polymerase-promoter interactions: the comings and goings of RNA polymerase. J Bacteriol 180: 3019–3025

Del Tito BJ, Jr., Ward JM, Hodgson J, Gershater CJ, Edwards H, Wysocki LA, Watson FA, Sathe G, Kane JF (1995) Effects of a minor isoleucyl tRNA on heterologous protein translation in *Escherichia coli*. J Bacteriol 177: 7086–7091

DeLisa MP, Samuelson P, Palmer T, Georgiou G (2002) Genetic analysis of the twin arginine translocator secretion pathway in bacteria. J Biol Chem 277: 29825–29831

Demerec M, Adelberg EA, Clark AJ, Hartman PE (1966) A proposal for a uniform nomenclature in bacterial genetics. Genetics 54: 61–76

Dong H, Nilsson L, Kurland CG (1996) Covariation of tRNA abundance and codon usage in *Escherichia coli* at different growth rates. J Mol Biol 260: 649–663

Döring V, Mootz HD, Nangle LA, Hendrickson TL, Crecy-Lagard V, Schimmel P, Marliere P (2001) Enlarging the amino acid set of *Escherichia coli* by infiltration of the valine coding pathway. Science 292: 501–504

Dubendorff JW, Studier FW (1991) Controlling basal expression in an inducible T7 expression system by blocking the target T7 promoter with *lac* repressor. J Mol Biol 219: 45–59

Etchegaray JP, Inouye M (1999) Translational enhancement by an element downstream of the initiation codon in *Escherichia coli*. J Biol Chem 274: 10079–10085

Forman MD, Stack RF, Masters PS, Hauer CR, Baxter SM (1998) High level, context dependent misincorporation of lysine for arginine in *Saccharomyces cerevisiae* a1 homeodomain expressed in *Escherichia coli*. Protein Sci 7: 500–503

Gallegos MT, Schleif R, Bairoch A, Hofmann K, Ramos JL (1997) AraC/XylS family of transcriptional regulators. Microbiol Mol Biol Rev 61: 393–410

Gill RT, DeLisa MP, Valdes JJ, Bentley WE (2001) Genomic analysis of high-cell-density recombinant *Escherichia coli* fermentation and 'cell conditioning' for improved recombinant protein yield. Biotechnol Bioeng 72: 85–95

Grabherr R, Nilsson E, Striedner G, Bayer K (2002) Stabilizing plasmid copy number to improve recombinant protein production. Biotechnol Bioeng 77: 142–147

Guzman LM, Belin D, Carson MJ, Beckwith J (1995) Tight regulation, modulation, and high-level expression by vectors containing the arabinose p_{BAD} promoter. J Bacteriol 177: 4121–4130

Haldimann A, Daniels LL, Wanner BL (1998) Use of new methods for construction of tightly regulated arabinose and rhamnose promoter fusions in studies of the *Escherichia coli* phosphate regulon. J Bacteriol 180: 1277–1286

Hanahan D (1983) Studies on transformation of *Escherichia coli* with plasmids. J Mol Biol 166: 557–580

Hasan N, Szybalski W (1987) Control of cloned gene expression by promoter inversion *in vivo*: construction of improved vectors with a multiple cloning site and the p$_{tac}$ promoter. Gene 56: 145–151

Hasan N, Szybalski W (1995) Construction of *lacI*^{ts} and *lacI*^{qts} expression plasmids and evaluation of the thermosensitive *lac* repressor. Gene 163: 35–40

Hayhurst A, Harris WJ (1999) *Escherichia coli* skp chaperone coexpression improves solubility and phage display of single-chain antibody fragments. Protein Expr Purif 15: 336–343

Hewitt CJ, Nebe-Von Caron G, Nienow AW, McFarlane CM (1999) Use of multi-staining flow cytometry to characterise the physiological state of *Escherichia coli* W3110 in high cell density fed-batch cultures. Biotechnol Bioeng 63: 705–711

Hill CW, Harnish BW (1981) Inversions between ribosomal RNA genes of *Escherichia coli*. Proc Natl Acad Sci USA 78: 7069–7072

Hirel PH, Schmitter MJ, Dessen P, Fayat G, Blanquet S (1989) Extent of N-terminal methionine excision from *Escherichia coli* proteins is governed by the side-chain length of the penultimate amino acid. Proc Natl Acad Sci USA 86: 8247–8251

Ignatova Z, Hornle C, Nurk A, Kasche V (2002) Unusual signal peptide directs penicillin amidase from *Escherichia coli* to the Tat translocation machinery. Biochem Biophys Res Commun 291: 146–149

Itoh T, Okayama T, Hashimoto H, Takeda J, Davis RW, Mori H, Gojobori T (1999) A low rate of nucleotide changes in *Escherichia coli* K-12 estimated from a comparison of the genome sequences between two different substrains. FEBS Lett 450: 72–76

Jones KL, Keasling JD (1998) Construction and characterization of F plasmid-based expression vectors. Biotechnol Bioeng 59: 659–665

Kadokura H, Katzen F, Beckwith J (2003) Protein disulfide bond formation in prokaryotes. Annu Rev Biochem 72: 111–135

Kaga N, Umitsuki G, Clark DP, Nagai K, Wachi M (2002) Extensive overproduction of the AdhE protein by *rng* mutations depends on mutations in the *cra* gene or in the Cra-box of the *adhE* promoter. Biochem Biophys Res Commun 295: 92–97

Kammerer W, Deuschle U, Gentz R, Bujard H (1986) Functional dissection of *Escherichia coli* promoters: information in the transcribed region is involved in late steps of the overall process. EMBO J 5: 2995–3000

Keiler KC, Waller PR, Sauer RT (1996) Role of a peptide tagging system in degradation of proteins synthesized from damaged messenger RNA. Science 271: 990–993

Kim EJ, Yoo SK (1999) Cell surface display of hepatitis B virus surface antigen by using *Pseudomonas syringae* ice nucleation protein. Lett Appl Microbiol 29: 292–297

Kleman GL, Strohl WR (1994) Acetate metabolism by *Escherichia coli* in high-cell-density fermentation. Appl Environ Microbiol 60: 3952–3958

Kolisnychenko V, Plunkett G, III, Herring CD, Feher T, Posfai J, Blattner FR, Posfai G (2002) Engineering a reduced *Escherichia coli* genome. Genome Res 12: 640–647

Kurokawa Y, Yanagi H, Yura T (2001) Overproduction of bacterial protein disulfide isomerase (DsbC) and its modulator (DsbD) markedly enhances periplasmic production of human nerve growth factor in *Escherichia coli*. J Biol Chem 276: 14393–14399

Kushner SR (2002) mRNA decay in *Escherichia coli* comes of age. J Bacteriol 184: 4658–4665

Lanzer M, Bujard H (1988) Promoters largely determine the efficiency of repressor action. Proc Natl Acad Sci USA 85: 8973–8977

LaVallie ER, McCoy JM (1995) Gene fusion expression systems in *Escherichia coli*. Curr Opin Biotechnol 6: 501–506

Lederberg J, St. Clair J (1958) Protoplasts and L-type growth of *Escherichia coli*. J Bacteriol 75: 143–160

Lee SY (1996) High cell-density culture of *Escherichia coli*. Trends Biotechnol 14: 98–105

Lee SY, Choi JH, Xu Z (2003) Microbial cell-surface display. Trends Biotechnol 21: 45–52

Lesley SA, Graziano J, Cho CY, Knuth MW, Klock HE (2002) Gene expression response to misfolded protein as a screen for soluble recombinant protein. Protein Eng 15: 153–160

Lilie H, Schwarz E, Rudolph R (1998) Advances in refolding of proteins produced in *E. coli*. Curr Opin Biotechnol 9: 497–501

Link AJ, Phillips D, Church GM (1997) Methods for generating precise deletions and insertions in the genome of wild-type *Escherichia coli*: application to open reading frame characterization. J Bacteriol 179: 6228–6237

Luli GW, Strohl WR (1990) Comparison of

growth, acetate production, and acetate inhibition of *Escherichia coli* strains in batch and fed-batch fermentations. Appl Environ Microbiol 56: 1004–1011

MACBEATH G, KAST P (1998) UGA read-through artifacts – when popular gene expression systems need a pATCH. BioTechniques 24: 789–794

MARTIN VJ, SMOLKE CD, KEASLING JD (2002) Redesigning cells for production of complex organic molecules. ASM News 68: 336–343

MATTES R (2001) The production of improved tissue-type plasminogen activator in *Escherichia coli*. Semin Thromb Hemost 27: 325–336

MOONEY RA, ARTSIMOVITCH I, LANDICK R (1998) Information processing by RNA polymerase: recognition of regulatory signals during RNA chain elongation. J Bacteriol 180: 3265–3275

MUTO A, USHIDA C, HIMENO H (1998) A bacterial RNA that functions as both a tRNA and an mRNA. Trends Biochem Sci 23: 25–29

NAGEL JH, GULTYAEV AP, GERDES K, PLEIJ CW (1999) Metastable structures and refolding kinetics in *hok* mRNA of plasmid R1. RNA 5: 1408–1418

NEIDHARDT FC, CURTISS R, INGRAHAM JL, LIN ECC, LOW KB, MAGASANIK B, REZNIKOFF WS, RILEY M, SCHAECHTER M, UMBARGER HE (1996) *Escherichia coli* and *Salmonella*: Cellular and molecular biology, 2nd edition. ASM Press, Washington, DC

NISHIHARA K, KANEMORI M, YANAGI H, YURA T (2000) Overexpression of trigger factor prevents aggregation of recombinant proteins in *Escherichia coli*. Appl Environ Microbiol 66: 884–889

O'HALLORAN T, WALSH C (1987) Metalloregulatory DNA-binding protein encoded by the *merR* gene: isolation and characterization. Science 235: 211–214

OEHLER S, AMOUYAL M, KOLKHOF P, WILCKEN-BERGMANN B, MÜLLER-HILL B (1994) Quality and position of the three *lac* operators of *E. coli* define efficiency of repression. EMBO J 13: 3348–3355

PAN KL, HSIAO HC, WENG CL, WU MS, CHOU CP (2003) Roles of DegP in prevention of protein misfolding in the periplasm upon overexpression of penicillin acylase in *Escherichia coli*. J Bacteriol 185: 3020–3030

PAN W, RAVOT E, TOLLE R, FRANK R, MOSBACH R, TURBACHOVA I, BUJARD H (1999) Vaccine candidate MSP-1 from *Plasmodium falciparum*: a

redesigned 4917 bp polynucleotide enables synthesis and isolation of full-length protein from *Escherichia coli* and mammalian cells. Nucleic Acids Res 27: 1094–1103

PETERS JE, THATE TE, CRAIG NL (2003) Definition of the *Escherichia coli* MC4100 genome by use of a DNA array. J Bacteriol 185: 2017–2021

POOLE ES, BROWN CM, TATE WP (1995) The identity of the base following the stop codon determines the efficiency of in vivo translational termination in *Escherichia coli*. EMBO J 14: 151–158

RAIVIO TL, SILHAVY TJ (2001) Periplasmic stress and ECF sigma factors. Annu Rev Microbiol 55: 591–624

REMAUT E, STANSSENS P, FIERS W (1981) Plasmid vectors for high-efficiency expression controlled by the P_L promoter of coliphage lambda. Gene 15: 81–93

RIESENBERG D, GUTHKE R (1999) High-cell-density cultivation of microorganisms. Appl Microbiol Biotechnol 51: 422–430

RILEY M, SERRES MH (2000) Interim report on genomics of *Escherichia coli*. Annu Rev Microbiol 54: 341–411

RIPPMANN JF, KLEIN M, HOISCHEN C, BROCKS B, RETTIG WJ, GUMPERT J, PFIZENMAIER K, MATTES R, MOOSMAYER D (1998) Procaryotic expression of single-chain variable-fragment (scFv) antibodies: secretion in L-form cells of *Proteus mirabilis* leads to active product and overcomes the limitations of periplasmic expression in *Escherichia coli*. Appl Environ Microbiol 64: 4862–4869

ROBINSON M, LILLEY R, LITTLE S, EMTAGE JS, YARRANTON G, STEPHENS P, MILLICAN A, EATON M, HUMPHREYS G (1984) Codon usage can affect efficiency of translation of genes in *Escherichia coli*. Nucleic Acids Res 12: 6663–6671

ROTHEN SA, SAUER M, SONNLEITNER B, WITHOLT B (1998) Growth characteristics of *Escherichia coli* HB101[pGEc47] on defined medium. Biotechnol Bioeng 58: 92–100

RUDD KE (1998) Linkage map of *Escherichia coli* K-12, edition 10: the physical map. Microbiol Mol Biol Rev 62: 985–1019

RUDOLPH R, LILIE H (1996) In vitro folding of inclusion body proteins. FASEB J 10: 49–56

SAGAWA H, OHSHIMA A, KATO I (1996) A tightly regulated expression system in *Escherichia coli* with SP6 RNA polymerase. Gene 168: 37–41

SANDMAN K, GRAYLING RA, REEVE JN (1995)

Improved N-terminal processing of recombinant proteins synthesized in *Escherichia coli*. Biotechnology (NY) 13: 504–506

SAWERS G, JARSCH M (1996) Alternative regulation principles for the production of recombinant proteins in *Escherichia coli*. Appl Microbiol Biotechnol 46: 1–9

SCHAFFNER J, WINTER J, RUDOLPH R, SCHWARZ E (2001) Cosecretion of chaperones and low-molecular-size medium additives increases the yield of recombinant disulfide-bridged proteins. Appl Environ Microbiol 67: 3994–4000

SCHLEIF R (1996) Two positively regulated systems, *ara* and *mal*. In: *Escherichia coli* and *Salmonella*: Cellular and molecular biology (Neidhardt FC, et al., Eds). ASM Press, Washington, DC, Vol 1, pp 1300–1309

SCHOLZ P, HARING V, WITTMANN-LIEBOLD B, ASHMAN K, BAGDASARIAN M, SCHERZINGER E (1989) Complete nucleotide sequence and gene organization of the broad-host-range plasmid RSF1010. Gene 75: 271–288

SCHONER BE, BELAGAJE RM, SCHONER RG (1990) Enhanced translational efficiency with two-cistron expression system. Methods Enzymol 185: 94–103

SCOTT HN, LAIBLE PD, HANSON DK (2003) Sequences of versatile broad-host-range vectors of the RK2 family. Plasmid 50: 74–79

SHARPE ME, CHATWIN HM, MACPHERSON C, WITHERS HL, SUMMERS DK (1999) Analysis of the CoIE1 stability determinant Rcd. Microbiology 145: 2135–2144

SHOKRI A, SANDEN AM, LARSSON G (2003) Cell and process design for targeting of recombinant protein into the culture medium of *Escherichia coli*. Appl Microbiol Biotechnol 60: 654–664

SIEGELE DA, HU JC (1997) Gene expression from plasmids containing the *araBAD* promoter at subsaturating inducer concentrations represents mixed populations. Proc Natl Acad Sci USA 94: 8168–8172

SORENSEN HP, SPERLING-PETERSEN HU, MORTENSEN KK (2003) Production of recombinant thermostable proteins expressed in *Escherichia coli*: completion of protein synthesis is the bottleneck. J Chromatogr B Analyt Technol Biomed Life Sci 786: 207–214

SPANJAARD RA, CHEN K, WALKER JR, VAN DUIN J (1990) Frameshift suppression at tandem AGA and AGG codons by cloned tRNA genes: assigning a codon to *argU* tRNA and T4 tRNA(Arg). Nucleic Acids Res 18: 5031–5036

SPEED MA, WANG DIC, KING J (1996) Specific aggregation of partially folded polypeptide chains: The molecular basis of inclusion body composition. Nat Biotechnol 14: 1283–1287

STUDIER FW, MOFFATT BA (1986) Use of bacteriophage T7 RNA polymerase to direct selective high-level expression of cloned genes. J Mol Biol 189: 113–130

SUMMERS D (1998) Timing, self-control and a sense of direction are the secrets of multicopy plasmid stability. Mol Microbiol 29: 1137–1145

SUNG WL, LUK CK, ZAHAB DM, BARBIER JR, LAFONTAINE M, WILLICK GE (1991) Internal ribosome-binding site directs expression of parathyroid hormone analogue (8–84) in *Escherichia coli*. Biochem Biophys Res Commun 181: 481–485

TAIT RC, BOYER HW (1978) Restriction endonuclease mapping of pSC101 and pMB9. Mol Gen Genet 164: 285–288

TOMIZAWA J (1990) Control of ColE1 plasmid replication. Interaction of Rom protein with an unstable complex formed by RNA I and RNA II. J Mol Biol 212: 695–708

TU GF, REID GE, ZHANG JG, MORITZ RL, SIMPSON RJ (1995) C-terminal extension of truncated recombinant proteins in *Escherichia coli* with a 10Sa RNA decapeptide. J Biol Chem 270: 9322–9326

VAN DE WALLE M, SHILOACH J (1998) Proposed mechanism of acetate accumulation in two recombinant *Escherichia coli* strains during high density fermentation. Biotechnol Bioeng 57: 71–78

VIA P, BADIA J, BALDOMA L, OBRADORS N, AGUILAR J (1996) Transcriptional regulation of the *Escherichia coli rhaT* gene. Microbiology 142: 1833–1840

WANG L, BROCK A, HERBERICH B, SCHULTZ PG (2001) Expanding the genetic code of *Escherichia coli*. Science 292: 498–500

WILMS B, HAUCK A, REUSS M, SYLDATK C, MATTES R, SIEMANN M, ALTENBUCHNER J (2001 a) High-cell-density fermentation for production of L-N-carbamoylase using an expression system based on the *Escherichia coli rhaBAD* promoter. Biotechnol Bioeng 73: 95–103

WILMS B, WIESE A, SYLDATK C, MATTES R, ALTENBUCHNER J (2001 b) Development of an *Escherichia coli* whole cell biocatalyst for the production of L-amino acids. J Biotechnol 86: 19–30

Wu X, Jornvall H, Berndt KD, Oppermann U (2004) Codon optimization reveals critical factors for high level expression of two rare codon genes in *Escherichia coli*: RNA stability and secondary structure but not tRNA abundance. Biochem Biophys Res Commun 313: 89–96

Yanisch-Perron C, Vieira J, Messing J (1985) Improved M13 phage cloning vectors and host strains: nucleotide sequences of the M13mp18 and pUC19 vectors. Gene 33: 103–119

Yavachev L, Ivanov I (1988) What does the homology between *E. coli* tRNAs and RNAs controlling ColE1 plasmid replication mean? J Theor Biol 131: 235–241

Yee L, Blanch HW (1992) Recombinant protein expression in high cell density fed-batch cultures of *Escherichia coli*. Biotechnology (NY) 10: 1550–1556

You J, Cohen RE, Pickart CM (1999) Construct for high-level expression and low misincorporation of lysine for arginine during expression of pET-encoded eukaryotic proteins in *Escherichia coli*. Biotechniques 27: 950–954

Zahn K (1996) Overexpression of an mRNA dependent on rare codons inhibits protein synthesis and cell growth. J Bacteriol 178: 2926–2933

Zdanovsky AG, Zdanovskaia MV (2000) Simple and efficient method for heterologous expression of clostridial proteins. Appl Environ Microbiol 66: 3166–3173

Zhang Y, Buchholz F, Muyrers JP, Stewart AF (1998) A new logic for DNA engineering using recombination in *Escherichia coli*. Nat Genet 20: 123–128

Zhang Z, Gildersleeve J, Yang YY, Xu R, Loo JA, Uryu S, Wong CH, Schultz PG (2004) A new strategy for the synthesis of glycoproteins. Science 303: 371–373

3
Pseudomonas fluorescens

Lawrence C. Chew, Tom M. Ramseier, Diane M. Retallack, Jane C. Schneider, Charles H. Squires, and Henry W. Talbot

List of Genes

Gene	Encoded gene product
rhGH	Recombinant human growth hormone (lacking the native secretion signal)
γ-IFN	Human gamma interferon (lacking the native secretion signal)
gal2	Single chain antibody Gal2
gal13	Single chain antibody Gal13
pelB	Pectate lyase
pyrF	Orotidine-5′-phosphate decarboxylase
proC	Pyrroline-5-carboxylate reductase
phoA	Alkaline phosphatase
pbp	Phosphate-binding protein
Pben	*benABCD* Promoter
benABCD	Benzoate degradation operon
Pant	*antABC* Promoter
antABC	Anthranilate degradation operon
antR	Transcriptional activator of the *antA* promoter
tetR	tetracycline resistance
Ptac	*tac* promoter
lacI	Repressor of *lac* promoter

3.1
Introduction

The rapid expansion of the biotechnology industry in recent years has necessitated the expression of a wide spectrum of recombinant proteins in different host systems for a wide variety of purposes. In some applications, a large number of proteins are needed in small quantities for screening applications or structural determinations.

Production of Recombinant Proteins. Novel Microbial and Eucaryotic Expression Systems. Edited by Gerd Gellissen
Copyright © 2005 WILEY-VCH Verlag GmbH & Co. KGaA, Weinheim
ISBN: 3-527-31036-3

In other cases, quantities approaching the metric tonne scale are needed for specific therapeutic applications. The majority of therapeutic proteins have been produced in either mammalian cell-culture systems, with Chinese hamster ovary (CHO) cells representing the most common mammalian cell system in use, or in microbial systems, *Escherichia coli* being the most common. A variety of alternative expression systems are also being used or developed and evaluated. These include other mammalian (human) cell lines, fungi, yeast, insect, and bacterial host cell systems. While each of these systems is being developed with very specific system advantages in mind, it is not clear which of them will ultimately be the most useful for therapeutic protein production. It is, in fact, likely that several of these systems will reach broad acceptance in particular niches in the protein production industry.

The efficient discovery of efficacious new protein pharmaceuticals is essential if medicine is to meet its challenge of treating diseases. However, the ability to manufacture pharmaceuticals economically has rapidly become as important as overall drug development costs continue to escalate. Production technologies that yield high titers of protein in active form are key. Also, proteins must often be post-translationally modified to achieve their intended performance as therapeutics. Many of the proteins in development today are monoclonal antibodies that present special production challenges in order to maximize recovery of active material for high dosage demands.

E. coli was the first heterologous host of any kind used to produce a recombinant DNA-based pharmaceutical, this being carried out in 1982 by the pharmaceutical company Eli Lilly to produce human insulin. Although *E. coli* has been the dominant host (in terms of economic value) for the production of protein pharmaceuticals (Swartz 2001), in recent years it has been overtaken by mammalian cell production. This has been driven by the need to produce more complex proteins, in particular, antibodies. In many cases (e.g., insulin), *E. coli* is unable to fold these products into their correct conformation, instead producing unfolded, inactive, intracellular protein masses called inclusion bodies. Although the recombinant insulin example suggests potential economic feasibility for protein production in *E. coli*, many producers are unwilling to implement an in-vitro protein folding process for these complex products. In addition, *E. coli* can neither synthesize nor attach mammalian glycosylation chains which may be required for functionality or pharmacokinetic properties, although for many products and applications glycosylation is not required, such as for antibodies in which only binding function is required.

Another bottleneck to protein pharmaceutical production can be the high-level expression of the target in a cell line or host organism. A *Pseudomonas fluorescens*-based manufacturing platform for high-yield production of protein pharmaceuticals has been developed based on *P. fluorescens* biovar I strain MB101 (Landry et al. 2003). The system's performance is due to the combination of a robust host strain and the availability of extensive molecular biology and bioinformatics tools. These tools include a range of stable plasmid vectors of various copy numbers, non-antibiotic-dependent plasmid maintenance, engineered host strains for stringent control of gene expression, as well as the capability of exporting proteins to the periplasmic space. A well-optimized, high-cell-density (HCD) fermentation process completes the utility of the system. Soluble and active protein yields in excess of 25 g L^{-1} with a variety of protein types ranging from enzymes for industrial use to proteins for pharmaceutical applications have been achieved. Production conditions are routinely optimized in 20-L fermentors, and have been successfully scaled up for the commercial production of a number of heterologous proteins.

3.2
Biology of *Pseudomonas fluorescens*

The species *P. fluorescens* is an abundant and natural component of the microbial flora of soil, water and plants, inhabiting plant rhizosphere and phyllosphere environments (OECD 1997). Members of the species *P. fluorescens* are aerobic, saprophytic, Gram-negative organisms that are not pathogenic (they do not cause a disease state) in plants (Bradbury 1986), mammals or immunocompetent humans (Palleroni 1982). They grow at neutral pH, and in the mesophilic temperature range (optimally between 25 and 30 °C), and do not grow above 40 °C or under acid conditions (pH <4.5) (Holt et al. 1994; OECD 1997). Although growth of *P. fluorescens* is restricted at higher temperatures, this organism has the ability to grow at 4 °C (Gilardi 1991).

Strains of *P. fluorescens* are commonly found on plant surfaces, as well as in decaying vegetation, soil, and water (Bradbury 1986). They can also grow on a wide range of very simple organic substrates and remain viable for long periods of time in a wide variety of habitats. The ubiquitous nature of *P. fluorescens* on the surface of plants typically grown for human consumption suggests that this bacterium has been widely consumed by humans for many years. The American Type Culture Collection (ATCC) has designated strains of *P. fluorescens* under Biosafety Level 1 (BSL-1), which is defined as " ... having no known potential to cause disease in humans or animals." (see Appendix A3.1).

P. fluorescens has been used in recent years in a variety of industrial applications, including for ice nucleation (for snow making and minimization of frost damage to plants) (Warren, 1987; Lindow and Panopoulos 1988; Wilson and Lindow 1993), to produce biological pesticides (Mycogen 1991; Herrera et al. 1994), and for the control of diseases in the phyllosphere of plants (Wilson and Lindow 1993). The ability of *P. fluorescens* to catabolize a wide variety of natural and synthetic compounds (e.g., chlorinated aliphatic hydrocarbons) and to produce a variety of enzymes has led to its wide use in bioremediation and biocatalysis (Wubbolts and Witholt 1998).

3.3
History and Taxonomy of *Pseudomonas fluorescens* Strain Biovar I MB101

Pseudomonas fluorescens strain MB101 was isolated in 1984 from the surface of a lettuce leaf located on a farm in San Diego County, California. Morphologic characteristics for MB101 are consistent with those identified in *Bergey's Manual* (Holt and Bergey 1984; Holt et al. 1994) for *P. fluorescens*:

- Straight or slightly curved rods, but not helical
- Motile by one or several polar flagella
- 0.5–1.0 µm in diameter by 1.5–5.0 µm in length
- The cells do not produce prosthecae, and are not surrounded by sheaths

The phylogeny of strain MB101 was established by thorough characterization of its morphological, biochemical, and genotypic properties. The results from a series of phenotypic tests confirmed that strain MB101 belongs to *P. fluorescens* biotype A – that is, biovar I. The tests included API rapid NFT (bio Merieux-Vitek, Hazelwood, MD USA), production of levan and phenazine, ability to denitrify and liquefy gelatin, utilization of trehalose and numerous other carbon sources, and analysis of fatty acid methyl esters (Roy 1988; Sasser 1990). Genotypic analyses were performed using the 16S rRNA genes of strain MB101 and other selected strains from the genus *Pseudomonas*. Analyses of the three hypervariable regions within the 16S rRNA genes (Moore et al. 1996) corroborated the findings of the phenotypic analyses that identify strain MB101 as *P. fluorescens* biovar I.

The taxonomic position of strain MB101 is therefore as follows:

Domain Bacteria
      Phylum BXII.   Proteobacteria
            Class III.   Gammaproteobacteria
                  Order VIII.   Pseudomonadales
                        Family I.   Pseudomonadaceae
                              Genus I.   *Pseudomonas*
                                    Species   *fluorescens*
                                      Biovar   I

P. fluorescens MB101 has been used as a host for production of several EPA-registered bioinsecticides (EPA 1991 a,b, 1995) and a generally-recognized-as-safe (GRAS) α-amylase preparation (FDA 2003; Landry et al. 2003). Therefore, the safety of the industrial application of *P. fluorescens* biovar I strain MB101 is well established.

3.4
Cultivation

While *P. fluorescens* can grow well in common laboratory media, such as Luria Broth and M9 minimal medium, its nutrient, growth temperature and oxygen requirements have to be taken into careful consideration in order to achieve high recombinant expression levels at high cell densities. As with other pseudomonads (Kim et al. 1997; Lee et al. 2000), *P. fluorescens* can be cultivated to high cell densities in bioreactors with simple but balanced defined mineral salts media supplemented with an inorganic nitrogen source such as ammonia, and a carbon source such as glucose. The optimal temperature for *P. fluorescens* growth is 32 °C. Being a strict aerobe, only adequate oxygen transfer to the culture will ensure better growth. Nevertheless, from our experience, maintaining dissolved oxygen above a certain level in the bioreactor (e.g., 20%) does not appear to be as critical in *P. fluorescens* as in *E. coli*. *P. fluorescens* MB101, being a prototrophic strain, obviates the need of any organic nitrogen supplementation, which decreases the risk of performance variability that is accompanied by dependence on crude hydrolysates. Unlike *E. coli*, *P. fluorescens* does not accumulate acetate during fermentation. Common to other pseudomonads, glucose uptake is preferentially through the oxidative rather than the phosphorylative pathway (Dawes et al. 1976; Temple et al. 1998). Using optimal fermentation conditions and carbon feeding, biomass levels of greater than 100 g L^{-1}, accompanied by recombinant protein expression levels of more than 30% total cell protein, can be achieved by *P. fluorescens* at production scales without the need for oxygen supplementation.

3.5
Genomics and Functional Genomics of *P. fluorescens* Strain MB101

The genome sequence of *P. fluorescens* strain MB214 (a derivative of MB101 with a chromosomal insertion of the *E. coli lac* operon) was determined to aide the development of MB214 into a production platform strain for the manufacture of proteins, peptides, and metabolites. Knowledge of the entire genome allows directed changes to individual genes or groups of genes to be performed quickly, thus enabling sophisticated pathway engineering and enhancement of specific gene expression in *P. fluorescens* (see below). *P. fluorescens* MB214 is also easy to manipulate genetically, thereby enhancing this capability.

The genome sequence was determined by random shotgun sequencing of three different DNA libraries, varying the type of vector and DNA insert size. The DNA sequences were assembled into overlapping reads using the Phred software (Ewing and Green 1998; Ewing et al. 1998). The genome of strain MB214 is 6.5 megabases, similar in size to other pseudomonads for which the genome sequences have already been established (Stover et al. 2000; Nelson et al. 2002; Buell et al. 2003). No indigenous plasmids were identified in strain MB214. The G+C content was calculated as 60%, which is slightly lower than the G+C content reported for *P. aeruginosa* PAO1 (Stover et al. 2000). Over 6200 open reading frames (ORFs) were identified using the gene finder tools Generation (Genomix Corporation, Oak Ridge, TN, USA), Critica (Badger and Olson 1999), and Glimmer (Salzberg et al. 1998) and the Glimmer post-processor, RBSfinder (Suzek 2001). Gene functions were assigned based on BLAST (Altschul et al. 1997) and FASTA (Pearson 2000) homology, as well as hits against COGs (Tatusov et al. 2000) and InterPro (Mulder et al. 2003).

Reconstruction of the metabolism based on the annotated genes reveals that metabolically, *P. fluorescens* is very versatile. Over 700 pathways were identified containing numerous, sometimes-redundant pathways for the same metabolic reaction (Table 3.1). Glucose uptake is achieved through two different routes, a phosphorylative and an oxidative pathway that are employed at different glucose concentrations. Unlike *E. coli*, catabolism of this inexpensive carbon source utilizes the Entner–Dou-

Tab. 3.1 Overview of number of pathways for each metabolic category.

Category	Pathways (n)
Amino acid metabolism	164
Aromatic hydrocarbons	27
Carbohydrate metabolism	139
Coenzymes and vitamins	38
Electron transport	21
Halide metabolism	1
Hydrocarbon metabolism	1
Hydrogen metabolism	1
Intracellular transport	1
Lipid metabolism	29
Membrane transport	124
Metabolism of oxygen/radicals	6
Nitrogen metabolism	3
Nucleic acids metabolism	28
One-carbon metabolism	9
Phosphate metabolism	7
Protein metabolism	50
Purine metabolism	42
Pyrimidine metabolism	36
Signal transduction	6
Sulfur metabolism	5

doroff pathway and not the Embden–Meyerhof–Parnas glycolytic pathway. This is due to an absence of the gene encoding phosphofructokinase. All the genes necessary for a fully functional Krebs cycle – including the glyoxylate shunt pathway – are present for efficient glucose catabolism. *P. fluorescens* has the "genomic power" to synthesize and degrade all 20 amino acids. Amino acid synthesis can occur via multiple routes for homoserine, methionine, glutamate, and homocysteine. *P. fluorescens* also contains two glutamate tRNA synthetases, in contrast to *E. coli* and *Bacillus subtilis*, both of which only have one corresponding enzyme.

Although there is a high level of genome conservation between *P. fluorescens* and the pathogenic *P. aeruginosa* PAO1, key virulence factors such as exotoxin A and PrpL proteinase are absent from the established genome sequence of *P. fluorescens* MB214. Other members of the species *P. fluorescens* whose genomes have been sequenced – for example, strains Pf0–1 (Joint Genome Institute 2004) and SBW25 (Sanger Centre 2004) – are also missing these factors.

Bioinformatics analyses of the MB214 genome to uncover protein secretion systems indicated that all but one known microbial protein export system (the type IV secretion system) are present in MB214. The absent secretion system is commonly plasmid-borne, and no plasmids have been found in MB214. For most of the present export systems, multiple paralogous protein exporters exist, which may have arisen because of the saprophytic lifestyle of this microbe. A comparison of the genomes of *P. aeruginosa* PAO1 and the two *P. fluorescens* strains MB214 and Pf0–1 revealed that the same protein secretion systems are present in all three pseudomonads, differing only in the number of paralogous exporters (Ma et al. 2003). The presence of multiple protein secretion systems opens avenues to genetically engineered production strains capable of secreting the desired protein products outside of the bacterial cells into the growth medium for easy and cost-effective protein recovery.

Knowledge of the genome sequence also allows establishment of functional genomics capabilities such as DNA microarray and proteomics for gene and protein expression analysis, respectively. These capabilities are useful to monitor cellular metabolic conditions during the growth and protein production phases of fermentation. For the gene expression capability, over 6200 50-mer oligodeoxyribonucleotides (oligos) were manufactured that represent each ORF of the *P. fluorescens* MB214 genome. The oligos were designed to exhibit consistent hybridization temperature, to lack strong secondary structure, and to show minimal cross-hybridization with other genes in the MB214 genome. All oligos are printed twice on epoxy-coated glass slides to increase the confidence of the obtained gene expression data. The printed microarrays are then hybridized with control and experimental labeled RNA samples, and their hybridization pattern is analyzed using software tools from BioDiscovery (El Segundo, CA, USA) and SpotFire Inc. (Somerville, MA, USA). The complementary proteomics capability involves separation of proteins using one- or two-dimensional SDS–polyacrylamide gel electrophoresis (SDS-PAGE), subsequent robotic protein spot cutting, automated proteolysis of the proteins to yield peptides, and matrix-assisted laser desorption/ionization time-of-flight (MALDI-TOF) mass spectrometric analysis of the obtained peptide to identify *P. fluorescens* proteins.

Both the transcriptional profiling and proteomics technologies are used as analytical tools to provide information about the metabolic state of the host cells, as well as the temporal profile of the gene transcripts and protein production during fermentation runs. The proteomics capability furthermore supplies information about the yield and stability of the overproduced protein as well as the processing sites of signal sequences that target proteins into the periplasm. Additionally, proteomics can be used to direct protein purification processes by identifying co-purifying proteins. This may lead to altering current purification processes or steering genetic engineering to eliminate interfering proteins. Together with other tools, these capabilities allow refined analysis and modification of both fermentation and protein expression in *P. fluorescens*.

Genome mining has also been used to discover new, useful genetic functions to enhance the performance of this organism. Novel promoter elements inducible by benzoate or anthranilate have been cloned and further engineered (see below). Furthermore, identification of essential metabolic genes within the genome led to the cloning of these genes into plasmids (coupled with the deletions of the same genes from the chromosome) replacing genes encoding antibiotic resistance, thus resulting in antibiotic-free stable plasmid maintenance during HCD production fermentation runs (see below). Computational analysis of the codon usage of strain MB214 also allows the construction of synthetic *P. fluorescens* expression-specific genes with improved protein production based on the preferred codons used in this organism.

3.6
Core Expression Platform for Heterologous Proteins

The current platform for recombinant protein production in *P. fluorescens* comprises a family of host strains derived from MB101 (Table A3.1) combined with stable, but nonconjugative expression plasmids of varying copy number (Table A3.2). The host strains are stable, amenable to genetic or molecular manipulations, and can be cultivated to high cell densities in fully defined mineral salts media in standard fermentors, without oxygen enrichment. The expression plasmid vectors are either derived from the medium copy number IncQ plasmid RSF1010 (Scholz et al. 1989) or the lower copy number *P. savastanoi* plasmid pPS10 (Nieto et al. 1990). These plasmids are compatible and, if needed, can be stably maintained concomitantly. Although broad host range IncP plasmids, such as RK2 derivatives, can replicate in *P. fluorescens*, they have been found to be unstable and not suitable for high-level heterologous gene expression. Expression of heterologous genes is driven by promoters of varying strengths, such as the *ben*, *ant*, *tac*, and *lacUV5* promoters, optimal translation initiation signals and strong transcription terminators. Derivatives of the *lac* promoter can be regulated by the introduction of a *lacI* gene in the host strain, permitting induction by isopropyl-thiogalactopyranoside (IPTG). Induction by lactose is possible in strains bearing the *lac* operon from *E. coli*, such as *P. fluorescens* MB214.

3.6.1
Antibiotic-free Plasmids using *pyrF* and *proC*

In traditional microbial heterologous protein production systems, antibiotic resistance genes are essential to maintain plasmids, but they cause a problem if they cannot be completely removed from the product. Regulatory agencies discourage residual antibiotic resistance-coding gene DNA in products due to the perceived risk of its transfer to the intestinal flora of humans. Additional, expensive processing steps are often needed to degrade or remove the DNA so that it does not contaminate the final product.

Removing the antibiotic resistance genes from the strain would save the cost of these steps. To achieve an antibiotic-resistance-gene-free system, antibiotic resistance genes on the plasmid were replaced by genes encoding essential proteins for a step in intermediary metabolism in *P. fluorescens*. Two genes were tested: the *pyrF* gene, which encodes orotidine 5'-decarboxylase, an essential step in the biosynthesis of uracil; and *proC*, a gene encoding pyrroline-5-carboxylate reductase, the last step in proline biosynthesis. Cells that lack *pyrF* can grow by supplementation with uracil, which is converted to uridine 5'-monophosphate (UMP) through a salvage pathway. Mutants of *pyrF* are no longer sensitive to 5'-fluoroorotic acid (FOA), which facilitated creation of further genomic changes (discussed further below). The *proC* gene was chosen because the ORF did not appear to be transcriptionally linked to adjacent ORFs; therefore, it was reasonable to assume that the endogenous promoter sequences were near the ORF. The ORFs for *pyrF* and *proC*, and a few hundred base pairs of flanking sequences, in order to include endogenous promoters and terminators, were PCR-amplified and cloned in place of the antibiotic-resistance genes on the two expression plasmids. The corresponding auxotrophic *Pseudomonas* strains were made by deleting the corresponding genes in the chromosome. The productivity of the strains carrying the auxotrophic markers in place of the antibiotic-resistance genes was identical (see below).

3.6.2
Gene Deletion Strategy and Re-usable Markers

A strain with a deletion of the *pyrF* ORF was created by PCR-amplifying the regions flanking the *pyrF* gene and joining them together on a plasmid that cannot replicate in *P. fluorescens* in order to carry out allele exchange (Stibitz 1994). Upon electroporation (Artiguenave et al. 1997), this plasmid crossed into the genomic region upstream or downstream of *pyrF*. Subsequent selection for growth in the presence of FOA identified strains in which the integrated plasmid had crossed out, resulting in a strain with a precise deletion of the *pyrF* ORF and no plasmid DNA.

This Δ*pyrF* strain is a platform for other precise genomic changes in the following manner. Any desired genomic change can be effected by PCR-amplifying about 500 bp of genomic DNA on each side of the change, then cloning it into a suicide vector carrying the *pyrF* gene (e. g., pDOW1261) (Figure 3.1). Allelic exchange is carried out as above, using the cloned *pyrF* as a counterselection marker, forming a strain with the desired change and restoring the Δ*pyrF* marker in the genome. In this manner, the *proC* gene was deleted from the production strain, and subsequently the *E. coli lacI*^{Q1} gene was inserted into a locus encoding levan sucrase.

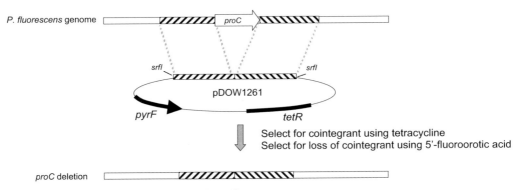

Fig. 3.1 Engineering of the *proC* deletion in the *P. fluorescens* genome. Regions flanking *proC* were PCR-amplified and cloned into the *Srf*I site of pDOW1261. Selection on tetracycline resulted in homologous recombination into one of the regions flanking *proC*. Subsequent growth in the absence of tetracycline and presence of proline allowed survival of strains that had recombined within the other flanking region. These strains were identified by selection on 5'-fluoroorotic acid, which is toxic to cells that are *pyrF+*.

3.6.3
Periplasmic Secretion and Use of Transposomes

In order to find signal sequences capable of efficiently secreting a heterologous protein to the periplasm, we designed a transposome (Goryshin and Reznikoff 1998) carrying near the insertion junction the *E. coli*-derived gene for alkaline phosphatase, *phoA*, from which the endogenous signal sequence had been removed. Since the *phoA* gene product is active only when localized to the periplasm (Manoil and Beckwith 1985), genomic *P. fluorescens* signal sequences were detected by electroporation of the transposome into cells, after which the transposome randomly integrated and created a signal sequence- *phoA* fusion. Of several signal sequences identified in this way, the signal from phosphate binding protein, in particular, was effective at efficiently transporting single-chain antibodies (see Section 3.7).

3.6.4
Alternative Expression Systems: Anthranilate and Benzoate-inducible Promoters

Native *P. fluorescens* promoters that provide an alternative to the traditional IPTG-inducible systems such as *lac* and *tac* promoters have been identified. Sodium benzoate and anthranilate are inexpensive chemicals that can be utilized as a carbon source by *P. fluorescens*, as well as by other *Pseudomonas* sp., *Acinetobacter* sp., and others (Neidle et al. 1991; Jeffrey et al. 1992; Bundy et al. 1998; Cowles et al. 2000; Eby et al. 2001). The promoters responsible for expression of the *P. fluorescens* benzoate (*benABCD*)-degradative genes (encoding catabolic pathway enzymes), which responds to sodium benzoate (*Pben*), and the promoter of the *P. fluorescens antABC*-degradative genes (*Pant*), which responds to anthranilate, were isolated and characterized. Transcriptional activators *benR* and *antR* have been identified directly upstream of the *benABCD* and *antABC* operons, respectively. Using the *E. coli*-derived *lacZ* gene as a reporter of promoter activity, a minimal promoter region that includes the activator binding site has been identified for both (Figure 3.2), which is active both at small (200 mL) and large (20 L) scales. Anthranilate-inducible activity is improved by adding the transcriptional activator *antR* in multi-copy. Inactivation of the benzoate and/or anthranilate metabolic pathways allows for use of these compounds as inducers without degradation of the inducer itself. By deleting gene(s) encoding the large and/or small subunits of the dioxygenases, we are able to mimic a gratuitous inducer system, such as IPTG induction of *lac*-based promoters, in that the inducer is added only once. Having a variety of promoters allows for greater flexibility in titrating gene expression, and allows for co-expression of genes using different induction profiles.

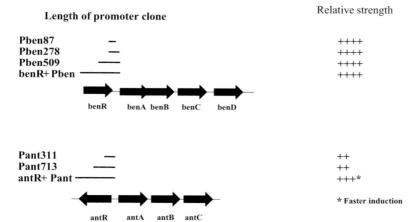

Fig. 3.2 Relative activity of benzoate and anthranilate-inducible promoters. Shown above are the arrangements of the benzoate (*benABCD*) and anthranilate (*antABC*) catabolic operons of *P. fluorescens* MB214. Promoter constructs tested are shown above each operon diagram. The relative strength of each promoter containing fragment, as measured by β-galactosidase expressed from *lacZ* fusion constructs, is depicted on the right with relative strength increasing from + to ++++.

3.7
Production of Heterologous Proteins in *P. fluorescens*

3.7.1
Pharmaceutical Proteins

The production of several therapeutic proteins in the *P. fluorescens* system was compared side-by-side with the *E. coli* T7 expression system (Studier 1991). In each case, *P. fluorescens* was equivalent to, or had some advantages over the commonly used T7 system in *E. coli*.

Commercial recombinant human growth hormone (rhGH) is currently produced in *E. coli* cytoplasmically as inclusion bodies that can be readily refolded to an active form (Patra et al. 2000). Secretion to the periplasm results in the production of soluble protein, but the yield is greatly decreased (Becker and Hsiung 1986; Uchida et al. 1997). The cDNA encoding rhGH, amplified from a human pituitary cDNA library with the native secretion signal sequence removed, was cloned and expressed cytoplasmically in both the classic *P. fluorescens* expression system and the *E. coli* T7 expression system, then evaluated at the 20-L fermentation scale. *P. fluorescens* produced 1.6-fold more rhGH per gram dry biomass than *E. coli*. Refolded rhGH isolated from *P. fluorescens* extracts was as active as commercially available rhGH, as determined by a cell proliferation assay using the rat lymphoma cell line Nb2–11 (Figure 3.3).

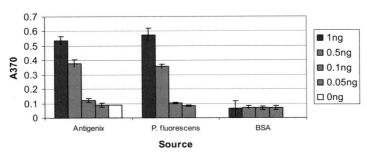

rhGH Activity Assay (Nb2-11 Proliferation)

Fig. 3.3 Analysis of recombinant human growth hormone (rhGH) activity. Activity of rhGH was measured essentially as described by Patra et al. (2000), except that proliferation of the rat lymphoma cell Nb2–11 was determined by BrdU incorporation, depicted on the Y-axis as A370. Proliferation induced by varying amounts of rhGH purified from *P. fluorescens* was compared to commercially available rhGH (Antigenix) purified from *E. coli*. Cells do not respond to bovine serum albumin (BSA).

Human gamma interferon (γ-IFN) represents another example of a therapeutic protein currently produced in *E. coli* as inclusion bodies (Simons et al. 1984; Perez et al. 1990; Haelewyn and DeLey 1995). Moreover, studies have found that γ-IFN is difficult to refold (Yip et al. 1981; Braude 1983; Braude 1984; Haelewyn and DeLey 1995). Mutants of γ-IFN that are produced as soluble protein, yet retain function, have been isolated and appear to overcome these problems (Wetzel et al. 1991). The comparative expression of γ-IFN in *P. fluorescens* versus *E. coli* revealed a significant advantage for recovery of the active cytokine in the *P. fluorescens* expression system. A human spleen cDNA library was used as a template to clone the γ-IFN cDNA, with the native signal sequence removed, into the *E. coli* T7 and *P. fluorescens* expression systems. The *E. coli* construct produced 2–4 g L^{-1} of insoluble protein at the 20-L scale (Figure 3.4). The *P. fluorescens* construct produced ~4 g L^{-1} of the cytokine during a typical 20-L fermentation. SDS-PAGE analysis of soluble and insoluble fractions shows that the majority of the protein (95%) is present in the soluble fraction when produced in *P. fluorescens*. This soluble protein was readily purified in a single ion-exchange chromatographic step to >95% purity, and was found in a viral inhibition assay to be as active as the commercially available compound (Meager 1987).

Monoclonal antibodies and antibody fragments represent a large percentage of therapeutic proteins currently in the development pipeline. Two anti-β-galactosidase single chain antibodies, gal13 and gal2 (Martineau et al. 1998; Martineau and Betton 1999) were produced in the cytoplasm or secreted to the periplasm of *P. fluorescens*. The Gal13 antibody, produced cytoplasmically, demonstrated an eightfold greater volumetric yield at the 20-L scale when produced in *P. fluorescens* as compared to expression in *E. coli*. Gal13 was produced in *P. fluorescens* primarily as soluble protein (96%), whereas only 48% of the gal13 was soluble in *E. coli* (Figure 3.5). Protein pur-

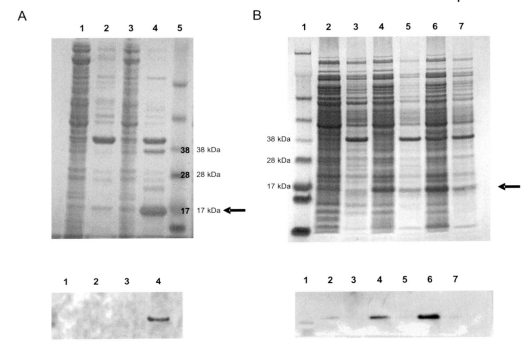

Fig. 3.4 Production of γ-IFN in *E. coli* versus *P. fluorescens*. γ-IFN produced by *E. coli* is shown in panel A. Lanes 1 and 2 show pre-induced samples, soluble and insoluble fractions, respectively; lanes 3 and 4 show 3-h-induced samples, soluble and insoluble fractions, respectively. Lane 5, See Blue Plus2 Marker. γ-IFN produced in *P. fluorescens* is shown in panel B. Lane 1, see Blue Plus2 Marker. Lanes 2 and 3 show pre-induced samples, soluble and insoluble fractions, respectively; lanes 4 and 5 show 24-h-induced samples, soluble and insoluble fractions, respectively. Lanes 6 and 7 show 48-h-induced samples, soluble and insoluble fractions, respectively. 5 μL samples normalized to OD_{575} 20 were loaded onto a 10% Bis-Tris NuPAGE run in 1 × MES. The arrow indicates the position of the recombinant protein. Western analyses for *E. coli* and *P. fluorescens* samples are depicted in the lower panels, showing 100% insoluble protein in *E. coli* and 95% soluble protein in *P. fluorescens*.

ified from both expression systems by Ni+ affinity chromatography was found to be active in an ELISA assay, as was previously shown for *E. coli*-derived material (Martineau et al. 1998).

The *pelB* secretion signal for *E. coli* expression/secretion system (Robert 2002), or the *pbp* (phosphate-binding protein signal sequence) secretion signal for the *P. fluorescens* expression/secretion system was fused to the N-terminus of the *gal2* ORF to test for Gal2 secretion to the periplasm and/or to the culture medium. The *pelB-gal2* fusion expressed in *E. coli* resulted in 1.6 g L^{-1} of protein of which about 54% was

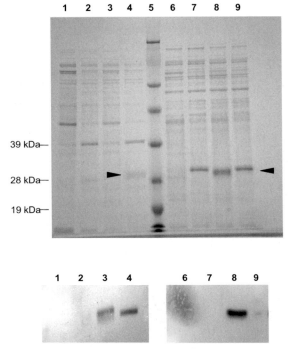

Fig. 3.5 Cytoplasmic production of Gal13. SDS-PAGE analysis of
E. coli and *P. fluorescens* soluble and insoluble fractions containing
Gal13. Lanes 1 and 2 show *E. coli* pre-induced samples, soluble and
insoluble fractions, respectively; lanes 3 and 4 show 3-h-induced
E. coli samples, soluble and insoluble fractions, respectively. Lane 5,
See Blue Plus2 Marker. Lanes 6 and 7 show *P. fluorescens* pre-induced samples, soluble and insoluble fractions, respectively; lanes 8
and 9 show *P. fluorescens* 24-h-induced samples, soluble and insoluble fractions, respectively. 5 μL samples normalized to OD_{575} 30
were loaded in each lane and run on a 10% NuPAGE gel (Invitrogen,
San Diego, CA, USA) in 1X MOPS buffer. Western analyses of these
samples are shown in the panels below. *E. coli* was found to have
48% soluble protein while *P. fluorescens* had 96% soluble protein.

processed by the cell, resulting in removal of the secretion signal and indicating that
the protein was secreted to the periplasm. Of that 54%, 11% was found in the soluble fraction. The *pbp-gal2* fusion expressed in *P. fluorescens* resulted in 9–10 g L^{-1} of
fully processed protein. The majority (96%) of the Gal2 protein produced in *P. fluorescens* was found in the insoluble fraction. Protein purified from both expression systems was found to be active in an anti-β-galactosidase ELISA. Although a greater percentage of the total protein was found in the soluble fraction for the *E. coli* construct,
as compared to the *P. fluorescens pbp:gal2* construct, the overall yield in *P. fluorescens*
of processed protein was significantly higher.

3.7.2
Industrial Enzymes

A wide range of industrial enzymes belonging to the glycosidase (EC 3.2.1.x), nitrilase (EC 3.5.5.x), and phosphatase (EC 3.1.3.x) families, both of mesophilic and hyperthermophilic origins, have been produced in *P. fluorescens* at high levels. Several native and laboratory-evolved α-amylases derived from environmental *Thermococcus* strains (Richardson et al. 2002) accumulated to amounts greater than 25% total cell protein in HCD cultures at pilot production scales of thousands of liters. Similarly, native and optimized nitrilase genes (DeSantis et al. 2003) have been expressed in *P. fluorescens* leading to yields greater than 50% total cell proteins in a soluble and active form (Figure 3.6). This represents a titer of >25 g L^{-1} when combined with a biomass of 100 g L^{-1} in a HCD fermentation process.

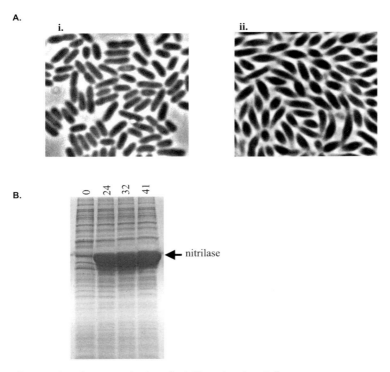

Fig. 3.6 Cytoplasmic production of soluble and active nitrilase. A) Phase-contrast micrographs of *P. fluorescens* cells: (i) before induction and (ii) producing nitrilase as nonrefractile inclusion bodies after induction. B) SDS-PAGE analysis of *P. fluorescens* producing nitrilase at 0, 24, 32, and 41 h post-IPTG induction time points.

3.7.3
Agricultural Proteins

The *P. fluorescens* system has been employed for many years to express a variety of genes encoding insecticidal proteins, ranging from 20 to 140 kDa in size, derived from *Bacillus thuringiensis*. *P. fluorescens* has proven to be a reliable tool to generate sufficient amounts of lead proteins for testing of specific insecticidal activities. The novel active proteins were then candidates for introduction into transgenic plants, or to be manufactured as spray-on bioinsecticides by fermentation. The Cry1 and Cry3 classes of *B. thuringiensis* δ-endotoxin insecticidal proteins of molecular weights 135 and 70 kDa, respectively, have been produced at the commercial scale of thousands of liters in the manufacture of M-Trak®, MVP®, MVPII®, and Mattch® biopesticides (EPA 1991 a,b, 1995). Expression of these insecticidal proteins was invariably greater than 20% of total cell protein in HCD fermentations of 100 g L^{-1} biomass. Fixation of the *P. fluorescens* during formulation afforded better stability over native *B. thuringiensis* products, permitting stable liquid concentrate formulations (Gaertner et al. 1993).

3.8
Conclusions

Although *P. fluorescens* has long been considered to be a metabolically diverse, versatile, and nonpathogenic organism, and has a history of successful and safe use for production of agricultural proteins and enzymes, until now it has not been seriously considered as a host for the production of therapeutic proteins. In this chapter, we describe results with a selected strain of *P. fluorescens* which shows great promise as an expression system for improved and lower-cost production of biotherapeutics. This strain, designated *P. fluorescens* MB101, has an excellent safety profile and is very well characterized. Extensive genomic and genetic information is now available for further strain development. At this point, a number of strong promoters are in place for the construction of expression vectors, the plasmids employed contain selection markers based on auxotrophic genes, and no antibiotic resistance markers are required. These plasmids are also stable, and are not mobilizable. A large and expanding set of molecular biology tools, and a functional genomics capability has been developed to enable rapid expression of a variety of therapeutic proteins.

As described in this chapter, several therapeutic proteins were produced in *P. fluorescens* MB101, and these results were compared to similar expression studies in *E. coli*. In most cases the *P. fluorescens* strain produced up to fivefold more protein (on a g L^{-1} basis) than the *E. coli* strain. An unexpected finding was the production of soluble and active proteins in *P. fluorescens*, whereas in *E. coli* the same proteins formed inclusion bodies. If a protein can be produced in soluble and active form, significant savings in purification costs can be achieved.

P. fluorescens has an impressive capability for producing heterologous proteins at high levels. The nitrilase described above was produced in a soluble and active form, amounting to yields >25 g L^{-1}, though this productivity level cannot be expected for all protein examples. MB101 was also shown to be an efficient producer of secreted proteins; yields in excess of 15 g L^{-1} have been achieved with single-chain antibody fragments secreted to the periplasm. Additional developmental work is likely to further improve these yields.

A rapid and efficient expression of therapeutics is desired, both now and in the future, in order to manage production costs and provide effective, affordable products, especially for higher volume applications. The *P. fluorescens* strain described in this chapter has many favorable properties in this regard. The organism is grown in a completely defined mineral salts medium with no added animal components or organic nitrogen of any kind. The fed-batch fermentation process is well-characterized, and scale up is predictable and rapid, up to thousands of liters. Cell densities in excess of 100 g L^{-1} dry cell weight are routinely obtained in standard fermentation vessels, without oxygen supplementation. The organism is unusually well suited for high level expression, and will tolerate a wide range of conditions. In addition, recovery and downstream purification procedures with *P. fluorescens* are standard, and consistent with those employed with *E. coli*.

A knowledge of the genome sequence has enabled strain engineering to improve gene expression in *P. fluorescens* and this has resulted in higher recombinant protein yields. Protein expression during fermentation runs can be monitored by functional genomic tools to gauge the metabolic state of host cells and to provide temporal production profiles of the desired gene transcript and protein. These data may then be used to further improve strain performance through directed genetic changes.

Extensive pathogenicity and toxicology studies with *P. fluorescens* have shown the organism to be a safe strain for the production of therapeutics. In this respect, investigations are continuing in order to add data which are relevant to regulatory submissions.

The combination of high volumetric and specific expression of a broad range of therapeutic proteins, coupled with a potential for soluble, active, and secreted products, makes *P. fluorescens* a compelling alternative for the microbial expression of biologicals for human health.

Appendix

Tab. A3.1 Selection of *P. fluorescens* strains.

Strain	Genotype	Phenotype	Reference
MB101	Wild-type	Biovar I	This report (Landry et al. 2003)
MB214	*Lac* operon insertion	Lac+	This report (Wilcox 1992)
DC206	MB101 *pyrF lac^{Q1}*	URA–	This report

Tab. A3.2 Selection of *P. fluorescens* plasmids.

Expression vectors	Expression cassette	Gene	Selection marker	Reference
pDOW2400	*tac* promoter and rrnBT1T2 terminator, RSF1010 origin	Human growth hormone without signal sequence (rhGH)	*tetR*	This report
pDOW1129	*tac* promoter and rrnBT1T2 terminator, RSF1010 origin	gamma interferon (γ-IFN) without native signal sequence	*tetR*	This report
pDOW1117	*tac* promoter and rrnBT1T2 terminator, RSF1010 origin	Single chain antibody Scfv gall 3	*tetR*	This report
pDOW1123	*tac* promoter and rrnBT1T2 terminator, RSF1010 origin, pbp secretion signal	pbp secretion signal fused to single chain antibody gal2	*tetR*	This report
pDOW2415	*tac* promoter and rrnBT1T2 terminator, RSF1010 origin	nitrilase	*pyrF*	This report (DeSantis et al. 2003)
Other plasmids				
pCN51*lacI*	pPS10 origin	*lacI*	*kanR*	(Nieto et al. 1990)
pDOW1261	ColEl origin does not replicate in *P. fluorescens*		*pyrF tetR*	This report

References

ALTSCHUL SF, MADDEN TL, SCHAFFER AA, ZHANG J, ZHANG Z, MILLER W, LIPMAN DJ (1997) Gapped BLAST and PSI-BLAST: a new generation of protein database search programs. Nucleic Acids Res 25: 3389–3402

ARTIGUENAVE F, VILAGINES R, DANGLOT C (1997) High-efficiency transposon mutagenesis by electroporation of a *Pseudomonas fluorescens* strain. FEMS Microbiol Lett 153: 363–369

BADGER JH, OLSEN GJ (1999) CRITICA: coding region identification tool invoking comparative analysis. Mol Biol Evol 16: 512–524

BECKER GW, HSIUNG HM (1986) Expression, secretion and folding of human growth hormone in *Escherichia coli*. Purification and characterization. FEBS Lett 204: 145–150

BRADBURY J (1986) Guide to Plant Pathogenic Bacteria, C.A.B. International Mycological Institute. Kew, Surrey, UK.

BRAUDE IA (1984) Purification of human gamma-interferon to essential homogeneity and its biochemical characterization. Biochemistry 23: 5603–5609

BRAUDE IA (1983) A simple and efficient method for the purification of human gamma interferon. Prep Biochem 13: 177–190

BUELL CR., JOARDAR V, LINDEBERG M, SELENGUT J, PAULSEN IT, GWINN ML, DODSON RJ, DEBOY RT, DURKIN AS, KOLONAY JF, MADUPU R, DAUGHERTY S, BRINKAC L, BEANAN MJ,

HAFT DH, NELSON WC, DAVIDSEN T, ZAFAR N, ZHOU L, LIU J, YUAN Q, KHOURI H, FEDOROVA N, TRAN B, RUSSELL D, BERRY K, UTTERBACK T, VAN AKEN SE, FELDBLYUM TV, D'ASCENZO M, DENG WL, RAMOS AR, ALFANO JR, CARTINHOUR S, CHATTERJEE AK, DELANEY TP, LAZAROWITZ SG, MARTIN GB, SCHNEIDER DJ, TANG X, BENDER CL, WHITE O, FRASER CM, COLLMER A (2003) The complete genome sequence of the *Arabidopsis* and tomato pathogen *Pseudomonas syringae* pv. tomato DC3000. Proc Natl Acad Sci USA 100: 10181–10186

BUNDY BM, CAMPBELL AL, NEIDLE EL (1998) Similarities between the *antABC*-encoded anthranilate dioxygenase and the *benABC*-encoded benzoate dioxygenase of *Acinetobacter* sp. strain ADP1. J Bacteriol 180: 4466–4474

COWLES CE, NICHOLS NN, HARWOOD CS (2000) *BenR*, a *XylS* homologue, regulates three different pathways of aromatic acid degradation in *Pseudomonas putida*. J Bacteriol 182: 6339–6346

DAWES EA, MIDGLEY M, WHITING PH (1976) Control of transport systems for glucose, gluconate and 2-oxo-gluconate, and of glucose metabolism in *Pseudomonas aeruginosa*. Presented at the Continuous Culture: Application of New Fields, [Plenary Lecture], 6th International Symposium of Continuous Culture of Micro-organisms

DESANTIS G, WONG K, FARWELL B, CHATMAN K, ZHU Z, TOMLINSON G, HUANG H, TAN X, BIBBS L, CHEN P, KRETZ K, BURK MJ (2003) Creation of a productive, highly enantioselective nitrilase through gene site saturation mutagenesis (GSSM). J Am Chem Soc 125: 11476–11477

EBY DM, BEHARRY ZM, COULTER ED, KURTZ DM, NEIDLE EL (2001) Characterization and evolution of anthranilate 1,2-dioxygenase from *Acinetobacter sp.* strain ADP1. J Bacteriol 183: 109–118

EPA (1991 a) M-One Plus Bioinsecticide; tolerance exemption. Federal Register 56: 28325–28356

EPA (1991 b) MVP Bioinsecticide; tolerance exemption. Federal Register 56: 28326–28328

EPA (1995) *CryIA*(c) and *CryIC* derived delta endotoxins of *Bacillus thuringiensis* encapsulated in killed *Pseudomonas fluorescens*; exemption from the requirement of a tolerance. Federal Register 60: 47487–47489

EWING B, GREEN P (1998) Base-calling of automated sequencer traces using Phred. II. Error probabilities. Genome Res 8: 186–194

EWING B, HILLIER L, WENDL MC, GREEN P (1998) Base-calling automated sequencer traces using Phred. I. Accuracy assessment. Genome Res 8: 175–185

FDA (2003) Agency Response Letter GRAS Notice No. GRN 000126 http://www.cfsan.fda.gov/~rdb/opa-g126.html. CFSAN/Office of Food Additive Safety.

GAERTNER FH, QUICK TC, THOMPSON MA (1993) CellCap: An encapsulation system for insecticidal biotoxin proteins. In: Advanced Engineered Pesticides (Kim L, Ed),. Marcel Dekker, New York, pp 73–83

GILARDI GL (1991) *Pseudomonas* and related genera. In: Manual of Clinical Microbiology, 5th edition (Balows A, Hausler WJ. Herrmann KL, Isenberg, HD, Shadomy HD, Eds). American Society for Microbiology, Washington, DC

GORYSHIN IY, REZNIKOFF WS (1998) Tn5 *in vitro* transposition. J Biol Chem 273: 7367–7374

HADDAD S, EBY DM, NEIDLE EL (2001) Cloning and expression of the benzoate dioxygenase genes from *Rhodococcus* sp. strain 19070. Appl Environ Microbiol 67: 2507–2514

HAELEWYN J, DE LEY M (1995) A rapid single-step purification method for human interferon-gamma from isolated *Escherichia coli* inclusion bodies. Biochem Mol Biol Int 37: 1163–1171

HERRERA GS, SNYMAN J, THOMSON JA (1994) Construction of a bioinsecticidal strain of *Pseudomonas fluorescens* active against the sugarcane borer, *Eldana saccharina*. Appl Environ Microbiol 60: 682–690

HOLT JG, KRIEG NR, SNEATH PHA, STALEY JT, WILLIAMS ST (1994) Bergey's Manual of Determinative Bacteriology, 9th edition. Lippincott, Williams & Wilkins, New York, pp 71–156

JEFFREY WH, CUSKEY SM, CHAPMAN PJ, RESNICK S, OLSEN RH (1992) Characterization of *Pseudomonas putida* mutants unable to catabolize benzoate: cloning and characterization of *Pseudomonas* genes involved in benzoate catabolism and isolation of a chromosomal DNA fragment able to substitute for *xylS* in activation of the *TOL* lower-pathway promoter. J Bacteriol 174: 4986–4996

Joint Genome Institute (2004) Pf0–1 genome sequence http://genome.jgi-psf.org/draft_microbes/psefl/psefl.home.html

KIM GJ, LEE IY, YOON SC, SHIN YC, PARK YH (1997) Enhanced yield and a high production of medium-chain-length poly(3-hydroxyalk-anoates) in a two-step fed-batch cultivation of *Pseudomonas putida* by combined use of glucose and octanoate. Enzyme Microbiol Technol 20: 500–505

LANDRY TD, CHEW L, DAVIS JW, FRAWLEY N, FOLEY HH, STELMAN SJ, THOMAS J, WOLT J, HANSELMAN DS (2003) Safety evaluation of an alpha-amylase enzyme preparation derived from the archaeal order *Thermococcales* as expressed in *Pseudomonas fluorescens* biovar I. Regul Toxicol Pharmacol 37: 149–168

LEE SY, WONG HH, CHOI JI, LEE SH, LEE SC, HAN CS (2000) Production of medium-chain-length polyhydroxyalkanoates by high-cell-density cultivation of *Pseudomonas putida* under phosphorus limitation. Biotechnol Bioeng 68: 466–470

LINDOW SE, PANOPOULOS NJ (1988) Field tests of recombinant Ice-*Pseudomonas syringae* for biological frost control in potato. In: Proceeding for the First International Conference on Release of Genetically Engineered Microorganisms (Sussman M, Collins CH, Skinner FA, Eds). Academic Press, London, pp 121–138

MA Q, ZHAI Y, SCHNEIDER JC, RAMSEIER, TM, SAIER MH. (2003) Protein secretion systems in *Pseudomonas aeruginosa* and *P. fluorescens*. Biochim Biophys Acta – Biomembranes 1611: 223–233

MANOIL C, BECKWITH J (1985) *TnphoA*: a transposon probe for protein export signals. Proc Natl Acad Sci USA 82: 8129–8133

MARTINEAU P, BETTON JM (1999) *In vitro* folding and thermodynamic stability of an antibody fragment selected in vivo for high expression levels in *Escherichia coli* cytoplasm. J Mol Biol 292: 921–929

MARTINEAU P, JONES P, WINTER G (1998) Expression of an antibody fragment at high levels in the bacterial cytoplasm. J Mol Biol 280: 117–127

MEAGER A (1987) Quantification of interferons by antiviral assays and their standardization. In: Lymphokines and Interferons, A Practical Approach (Clemens MJ, Morris AG, Gearing AJH, Eds). IRL Press, Oxford, pp 129–147

MOORE ERBM, ARNSCHEIDT A, BOETTGER EC, HUTSON RA, COLLINS MD, VAN DE PEER Y, DE WACHTER R, TIMMIS KN (1996) The determination and comparison of the 16S rRNA gene sequences of species of the genus *Pseudomonas* (sensu stricto) and estimation of the nat-

ural intrageneric relationships. System Appl Microbiol 19: 478–492

MULDER NJ, APWEILER R, T. ATTWOOD TK, BAIROCH A, BARRELL D, BATEMAN A, BINNS D, BISWAS M, BRADLEY P, BORK P, BUCHER P, COPLEY RR, COURCELLE E, DAS U, DURBIN R, FALQUET L, FLEISCHMANN W, GRIFFITHS-JONES S, HAFT D, HARTE N, HULO N, KAHN D, KANAPIN A, KRESTYANINOVA M, LOPEZ R, LETUNIC I, LONSDALE D, SILVENTOINEN V, ORCHARD SE, PAGNI M, PEYRUC D, PONTING CP, SELENGUT JD, SERVANT F, SIGRIST CJ, VAUGHAN R, ZDOBNOV EM (2003) The InterPro Database, 2003 brings increased coverage and new features. Nucleic Acids Res 31: 315–318

MYCOGEN (1991) Biotechnology PreManufacturing Notice submitted to The Office of Toxic Substances of USEPA to manufacture encapsulated *Bacillus thuringiensis* delta-endotoxin in *Pseudomonas fluorescens*.

NEIDLE EL, HARTNETT C, ORNSTON LN, BAIROCH A, REKIK M, HARAYAMA S (1991) Nucleotide sequences of the *Acinetobacter calcoaceticus benABC* genes for benzoate 1,2-dioxygenase reveal evolutionary relationships among multicomponent oxygenases. J Bacteriol 173: 5385–5395

NELSON KE, WEINEL C, PAULSEN IT, DODSON RJ, HILBERT H, MARTINS DOS SANTOS VA, FOUTS DE, GILL SR, POP M, HOLMES M, BRINKAC L, BEANAN M, DEBOY RT, DAUGHERTY S, KOLONAY J, MADUPU R, NELSON W, WHITE O, PETERSON J, KHOURI H, HANCE I, CHRIS LEE P, HOLTZAPPLE E, SCANLAN D, TRAN K, MOAZZEZ A, UTTERBACK T, RIZZO M, LEE K, KOSACK D, MOESTL D, WEDLER H, LAUBER J, STJEPANDIC D, HOHEISEL J, STRAETZ M, HEIM S, KIEWITZ C, EISEN JA, TIMMIS KN, DUSTERHOFT A, TUMMLER B, FRASER CM (2002) Complete genome sequence and comparative analysis of the metabolically versatile *Pseudomonas putida* KT2440. Environ Microbiol 4: 799–808

NIETO C, FERNANDEZ-TRESGUERRES E, SANCHEZ N, VICENTE M, DIAZ R (1990) Cloning vectors, derived from a naturally occurring plasmid of *Pseudomonas savastanoi*, specifically tailored for genetic manipulations in *Pseudomonas*. Gene 87: 145–149

OECD (1997) Consensus Document on Information Used in the Assessment of Environmental Applications Involving *Pseudomonas*.

Organisation for Economic Cooperation and Development, Paris, France

PALLERONI N (1992) Human and animal pathogenic *Pseudomonas*. In: The Prokaryotes, Vol. III 2nd edition (Balows A, Truper HG, Dworkin M, Harder W, Schleifer KH, Eds) Springer-Verlag, Berlin, New York, pp 3086–3103

PATRA AK, MUKHOPADHYAY R, MUKHIJA R, KRISHNAN A, GARG LC, PANDA AK (2000) Optimization of inclusion body solubilization and renaturation of recombinant human growth hormone from *Escherichia coli*. Protein Expr Purif 18: 182–192

PEARSON WR (2000) Flexible sequence similarity searching with the FASTA3 program package. Methods Mol Biol 132: 185–219

PEREZ L, VEGA J, CHUAY C, MENENDEZ A, UBIETA R, MONTERO M, PADRON G, SILVA A, SANTIZO C, BESADA V, HERRERA L (1990) Production and characterization of human gamma interferon from *Escherichia coli*. Appl Microbiol Biotechnol 33: 429–434

ROBERT RL (2002) Periplasmic expression and purification of recombinant Fabs. Methods Mol Biol 178: 343–348

RICHARDSON TH, TAN X, FREY G, CALLEN W, CABELL M, LAM D, MACOMBER J, SHORT JM, ROBERTSON DE, MILLER C (2002) A novel, high performance enzyme for starch liquefaction. Discovery and optimization of a low pH, thermostable alpha-amylase. J Biol Chem 277: 26501–26507

ROY MH (1988) Use of fatty acids for the identification of phytopathogenic bacteria. Plant Dis 72: 460

SALZBERG SL, DELCHER AL, KASIF S, WHITE O (1998) Microbial gene identification using interpolated Markov models. Nucleic Acids Res 26: 544–548

Sanger Centre (2004) SBW25: http://www.sanger.ac.uk/Projects/P_fluorescens/.

SASSER M (1990) Identification of bacteria through fatty acid analysis. In: Methods in Phytobacteriology (Klement KRZ, Rudolf K, Sands DC, Eds). Academiai Kiado, Budapest

SCHOLZ P, HARING V, WITTMANN-LIEBOLD B, ASHMAN K, BAGDASARIAN M, SCHERZINGER E (1989) Complete nucleotide sequence and gene organization of the broad-host-range plasmid RSF1010. Gene 75: 271–288

SIMONS G, REMAUT E, ALLET B, DEVOS R, FIERS W (1984) High-level expression of human interferon gamma in *Escherichia coli* under control of the *pL* promoter of bacteriophage lambda. Gene 28: 55–64

STIBITZ S (1994) Use of conditionally counterselectable suicide vectors for allelic exchange. Methods Enzymol, Vol 235, pp 458–465

STOVER CK, PHAM XQ, ERWIN AL, MIZOGUCHI SD, WARRENER P, HICKEY MJ, BRINKMAN FS, HUFNAGLE WO, KOWALIK DJ, LAGROU M, GARBER RL, GOLTRY L, TOLENTINO E, WESTBROCK-WADMAN S, YUAN Y, BRODY LL, COULTER SN, FOLGER KR, KAS A, LARBIG K, LIM R, SMITH K, SPENCER D, WONG GK, WU Z, PAULSEN IT, REIZER J, SAIER MH, HANCOCK RE, LORY S, OLSON MV (2000) Complete genome sequence of *Pseudomonas aeruginosa* PA01, an opportunistic pathogen. Nature 406: 959–964

STUDIER FW (1991) Use of bacteriophage T7 lysozyme to improve an inducible T7 expression system. J Mol Biol 219: 37–44

SUZEK, BE, ERMOLAEVA MD, SCHREIBER M, SALZBERG SL (2001) A probabilistic method for identifying start codons in bacterial genomes. Bioinformatics 17: 1123–1130. (ftp://ftp.tigr.org/pub/software/RBSfinder/. [Online.])

SWARTZ JR (2001) Advances in *Escherichia coli* production of therapeutic proteins. Curr Opin Biotechnol 12: 195–201

TATUSOV RL, GALPERIN MY, NATALE DA, KOONIN EV (2000) The COG database: a tool for genome-scale analysis of protein functions and evolution. Nucleic Acids Res 28: 33–36

TEMPLE LM, SAGE AE, SCHWEIZER HP, PHIBBS PV (1998) Carbohydrate catabolism in *Pseudomonas aeruginosa*. In: Pseudomonas, Vol. 10 (Montie TC, Ed). Plenum Press, New York, pp 35–72

UCHIDA H, NAITO N, ASADA N, WADA M, IKEDA M, KOBAYASHI H, ASANAGI M, MORI K, FUJITA Y, KONDA K, KUSUHARA N, KAMIOKA T, NAKASHIMA K, HONJO M (1997) Secretion of authentic 20-kDa human growth hormone (20K hGH) in *Escherichia coli* and properties of the purified product. J Biotechnol 55: 101–112

WARREN GJ (1987) Bacterial ice nucleation: molecular biology and applications. Biotechnol Genet Eng Rev 5: 107–135

WETZEL R, PERRY LJ VEILLEUX C (1991) Mutations in human interferon gamma affecting inclusion body formation identified by a general immunochemical screen. Biotechnology (NY) 9: 731–737

WILCOX ER (1992) Methods, vectors, and host cells for the control of expression of heterologous genes from lac operated promoters. US Patent 5,169,760

WILSON M, LINDOW SE (1993) Release of recombinant microorganisms. Annu Rev Microbiol 47: 913–944

WUBBOLTS MGW, WITHOLT B (1998) Selected industrial biotransformations. In: *Pseudomonas*, Vol. 10 (Montie TC, Ed). Plenum Press, New York, pp 271–329

YIP YK, PANG RH, URBAN C, VILCEK J (1981) Partial purification and characterization of human gamma (immune) interferon. Proc Natl Acad Sci USA 78: 1601–1605

4
Staphylococcus carnosus and other Gram-positive Bacteria
Roland Freudl

4.1
Introduction

Gram-positive bacteria are structurally very simple organisms. The cytosol is surrounded by the cytoplasmic membrane, which is covered by a thick cell wall composed mainly of peptidoglycan and negatively charged polymers such as teichoic or teichuronic acid. Since, in contrast to Gram-negative bacteria, this class of microorganisms does not possess an additional outer membrane, proteins which are exported across the cytoplasmic membrane can be released directly into the surrounding growth medium. Due to this fact, Gram-positive bacteria are considered especially interesting as host organisms for the secretory production of proteins, because the secretion of proteins offers considerable process advantages over intracellular deposition as insoluble aggregates (inclusion bodies). In fact, the enormous secretion potential of certain Gram-positive bacteria (e.g., *Bacillus* species) is one of the reasons for their extensive use in industry for the production of secretory proteins. A range of such microorganisms has been exploited as efficient sources for technical enzymes, such as proteases, lipases, amylases and other, mostly hydrolytic enzymes, with yields of several grams per liter of culture medium (Aunstrup 1979; Debabov 1982; Harwood 1992). Many of the bacteria that are presently used in industry for the production of technical enzymes have not been generated by molecular biological techniques, but have been identified by selection after classical mutagenesis. In most cases, the reasons for the improved productivities are therefore unknown.

Attempts to use Gram-positive bacteria for the secretory production of heterologous proteins have often failed or have led to disappointing results. The yields obtained for heterologous proteins were, in many cases, significantly lower than those observed for homologous enzymes (Simonen and Palva 1993). Several bottlenecks in the secretory pathway have been identified which, alone or in combination, can dramatically decrease the amount of the desired product in the culture supernatant (Figure 4.1). A detailed understanding of these bottlenecks at a molecular level will ease approaches for the selective improvement of Gram-positive bacteria as hosts for the secretory production of biotechnologically or pharmaceutically relevant heterologous proteins.

Production of Recombinant Proteins. Novel Microbial and Eucaryotic Expression Systems. Edited by Gerd Gellissen
Copyright © 2005 WILEY-VCH Verlag GmbH & Co. KGaA, Weinheim
ISBN: 3-527-31036-3

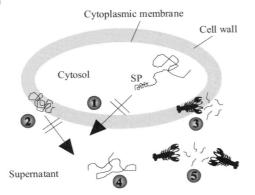

Fig. 4.1 Frequently observed problems during heterologous protein secretion in Gram-positive bacteria. (1) Inefficient translocation across the cytoplasmic membrane; (2) inefficient release into the supernatant; (3) degradation by membrane- and/or cell wall-associated proteases; (4) inefficient or wrong folding; (5) degradation by secreted proteases. SP: signal peptide.

4.2
Major Protein Export Routes in Gram-positive Bacteria

In eubacteria, the export of the vast majority of extra-cytosolic proteins is mediated by the general (Sec) secretion system. In addition, many bacteria possess a second protein export system (Tat) for the translocation of a certain subset of proteins (see also Chapter 2). One of the most remarkable differences between these two protein export pathways is the folding status of their respective substrate proteins during the actual translocation step. Sec-dependent proteins are translocated in more or less unfolded state. Subsequent folding takes place on the *trans*-side of the membrane after membrane translocation (Figure 4.2). In contrast, the Tat system translocates its substrates in a fully folded or even oligomeric form across the membrane (Figure 4.2).

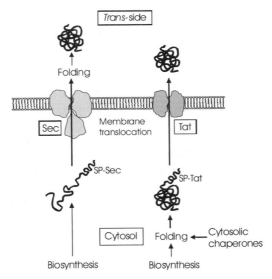

Fig. 4.2 Major protein export pathways in Gram-positive bacteria. The general Sec- and the alternative twin-arginine (Tat) protein export pathways differ dramatically with respect to the folding status of their substrates. Translocation via the Sec system requires that the export proteins are kept in an unfolded state. In contrast, fully folded proteins are exported by the Tat system. SP-Sec: Sec signal peptide, SP-Tat: Tat signal peptide.

4.2.1

The General Secretion (Sec) Pathway

The identification and characterization of genes in various Gram-positive bacteria, encoding homologues of the well-characterized Sec-proteins of the Gram-negative bacterium *Escherichia coli*, clearly has established that the mechanism of Sec-dependent protein translocation across the plasma membrane is highly conserved also in Gram-positive bacteria (van Wely et al. 2001): Proteins supposed to exit the cytosol are synthesized as larger precursors containing an N-terminal signal peptide. During or shortly after their synthesis, these precursor proteins are recognized by specific targeting factors (e.g., a bacterial signal recognition particle, b-SRP, consisting of Ffh and the Ffs-RNA, and its receptor FtsY) and subsequently delivered to the so-called translocase holoenzyme in the membrane, which consists of the subunits SecA, SecY, SecE, SecG, SecDF, and YajC (Figure 4.3). The translocation ATPase component SecA is the key component within the translocation scenario, and couples the energy of ATP-binding and ATP-hydrolysis to the movement of the translocating polypeptide chain across the membrane. This coupling is achieved by cycling of SecA between a membrane-peripheral and a membrane-integral state and, comparable to the needle of a sewing machine, SecA concomitantly transfers 20–30 amino acid residues of the translocating protein across the membrane during each cycle. In addition to ATP, the electrochemical membrane potential is required as an energy source for efficient protein translocation. The actual translocation across the membrane takes place at the SecY/SecE core translocase which forms the protein-conducting channel. SecG and the SecDF/YajC complex are thought to increase the efficiency of translocation at the SecY/SecE core translocase by facilitating the SecA cycle. During or shortly after translocation, the signal peptide is cleaved from the precursor proteins by specific signal peptidases (*Bacillus subtilis* has five of them) and the resulting mature protein is released from the translocase on the *trans*-side of the membrane. Assisted by specific folding factors, such as the PrsA lipoprotein (see also Section 4.3), the mature protein resumes its final folded structure and, after transport across the cell wall, is released into the supernatant.

Due to the fixed number of protein export sites, it is understandable that the capacity for heterologous protein secretion is limited. Overloading of the protein export machinery results in the intracellular accumulation of the corresponding heterologous precursor protein, which subsequently might become export-incompetent due to tight folding or due to aggregation. Furthermore, since ongoing functional protein translocation is essential for viability, blocking of the export sites by heterologous proteins is often toxic to the cells. The expression level required to overload the export machinery is dependent on the efficiency, by which the respective protein is translocated – that is, efficiently exported proteins require higher expression levels for saturating the translocase than those which are inefficiently translocated. Therefore, the application of moderate or low-level expression conditions can result in a significant improvement of the exportability and yield in the supernatant of the desired product (Meens et al. 1993). Likewise, the co-overproduction of limiting compo-

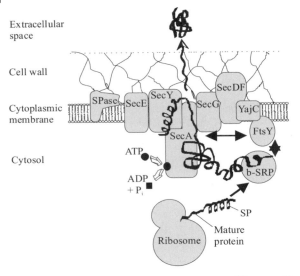

Extracellular
space

Cell wall

Cytoplasmic
membrane

Cytosol

Fig. 4.3 General (Sec) protein export apparatus of Gram-positive bacteria. Newly synthesized secretory precursor proteins are recognized by specific targeting factors (such as the bacterial signal recognition particle, b-SRP) and subsequently delivered to the translocase via the b-SRP receptor, FtsY. The translocation ATPase SecA pushes the secretory protein through a protein-conducting channel consisting of SecY, SecE and SecG. SecG and the SecDF-YajC complex increase the efficiency of translocation at the SecY-SecE core translocase. During translocation, the signal peptide (SP) is cleaved from the precursor by signal peptidase (SPase).

nents of the Sec-translocase with the protein of interest can be beneficial with respect to its secretion (Leloup et al. 1999).

Besides these quantitative aspects of substrate/translocase interactions, qualitative aspects of such interactions also play an important role. Successful secretion of a heterologous protein requires that the corresponding precursor protein can productively interact with all components of the secretion apparatus. Studies conducted in our laboratory have indicated that the formation of a productive translocation initiation complex, consisting of the foreign protein, SecA and the SecY/SecE core translocase, is a decisive step which determines whether or not the protein is channeled into the secretion pathway (M. Klein, D. Tippe, O. Köberling, R. Freudl, unpublished). The isolation of mutants with mutations in the translocase components that result in improved interactions between the heterologous protein and the translocase (Perez-Perez et al. 1996) can be considered as a promising strategy for enlarging the spectrum of heterologous proteins which can be accepted by the Sec protein export apparatus.

4.2.2
The Twin-Arginine Translocation (Tat) Pathway

In addition to Sec-mediated protein translocation, many bacteria possess a second protein export pathway for the translocation of a subset of precursor proteins. This Sec-independent mechanism has been designated Tat (for *twin-arginine translocation*) due to the fact that a characteristic amino acid motif including two consecutive arginine residues can be identified in the signal peptide of the respective precursor proteins. In marked contrast to the Sec system, which requires that its substrates are maintained in a more or less unfolded state, the Tat pathway translocates its protein substrates in a fully folded or even oligomeric form across the membrane. Many of the Tat substrates are proteins that have to recruit a cofactor in the cytosol and as a prerequisite have to acquire a folded status prior to export. In addition, some cofactor-less proteins are exported via this route, presumably because their rapid folding kinetics precludes transport via the Sec pathway (Müller et al. 2001; Berks et al. 2003).

The Tat export machinery consists of a surprisingly low number of components. In the ABC-type systems (found for example in *E. coli* and plant thylakoids), the three integral membrane proteins TatA, TatB, and TatC seem to be the minimal components of the Tat translocation pathway. These proteins are assumed to merge in the cytoplasmic membrane to form the Tat translocase. Although TatA and TatB show some weak sequence identity, they are thought to perform different functions. However, Tat systems of the AC type (found for example in *Bacillus subtilis* or *Staphylococcus carnosus*) even lack TatB, suggesting that in these systems TatA additionally takes over the TatB function. So far, very little is known about the mechanism of Tat-mediated protein transport. Experimental evidence from *E. coli* (Alami et al. 2003) and plant thylakoids (Mori and Cline 2002) suggests that Tat precursor proteins first bind to a receptor consisting of Tat(B)C in the cytoplasmic membrane. Subsequently, multiple copies of TatA are recruited to the Tat(B)C-precursor complex to form an active translocase, a step which is dependent on the presence of the transmembrane H^+ gradient (ΔpH). One attractive hypothesis for the formation of the actual translocation channel postulates that TatA molecules undergo a dynamic polymerization around the substrate molecule, such that the number of TatA proteins present in the pore matches the diameter of the respective substrate protein (which varies considerably between different known Tat substrates) within the pore (Berks et al. 2003). After transport of the substrate across the membrane plane, the substrate protein is released on the *trans*-side of the membrane and the pore component TatA re-dissociates from the Tat(B)C receptor complex, allowing the recycled components to catalyze further rounds of substrate binding and translocation.

Export of heterologous proteins via the Tat pathway can provide an attractive alternative to Sec-mediated export, especially in such cases where the heterologous protein does not fold properly in an extra-cytosolic environment. In *E. coli*, it has been shown that the fusion of the green fluorescent protein (GFP) from jellyfish to a Sec signal peptide results in an efficient Sec-dependent translocation of GFP into the periplasm. However, the exported GFP was completely inactive, most likely due to folding problems caused by the lack of chaperones in the periplasmic space (Feilmeier et al. 2000). In contrast, GFP could be transported into the periplasm in an active form, when the Tat transport route was used (Thomas et al. 2001). These results clearly demonstrate that pre-folding of a heterologous protein in the cytosol, assisted by the cytoplasmic Hsp60 and Hsp70 chaperone systems, and its subsequent translocation via the Tat pathway can be the superior to Sec-dependent secretion. The extent of Tat pathway-mediated secretion is also limited by the number of available export sites as described above for the Sec pathway. For *E. coli*, it has been shown that overproduction of GFP fused to a Tat signal peptide results in a rapid saturation of the Tat translocase and that the efficiency of export can significantly be increased by co-overexpression of the *tatABC* genes (Barrett et al. 2003).

To date, very few reports exist in the literature describing the use of the Tat pathway for the secretion of heterologous proteins in Gram-positive bacteria. Scherlaekens et al. (2001) demonstrated, that a tyrosinase from *Streptomyces antibioticus* (a copper-containing monooxygenase that catalyzes the formation of melanin pigment from tyrosine) could successfully be secreted in a Tat-dependent manner by *Streptomyces lividans*. Furthermore, it was shown by Pop et al. (2002) that the *B. subtilis* Tat system is capable of secreting the cytosolic *E. coli*-derived β-galactosidase when fused to a *B. subtilis* Tat signal peptide. Although further experimental evidence has to be provided, the available results strongly suggest that secretion of heterologous proteins by the Tat pathway of Gram-positive bacteria promises to be an attractive alternative to conventional approaches relying on Sec-dependent secretion.

4.2.3
Secretion Signals

In general, signal peptides for Sec-dependent protein export in bacteria and for import into the endoplasmic reticulum (ER) of eukaryotic cells as well as the membrane core of the respective translocases are the central features of protein export conserved during evolution. In many cases, it has been found that these signal peptides are functional across species boundaries. For example, the rat insulin signal peptide has been shown to mediate insulin export into the periplasm of *E. coli* (Talmadge et al. 1980). Likewise, it has been demonstrated that signal peptides of Gram-negative bacteria efficiently function in Gram-positive hosts (Meens et al. 1993) and vice versa (Malke and Ferretti 1984). Therefore, secretion of heterologous secretory proteins by Gram-positive bacteria can sometimes be achieved by maintaining their authentic signal peptide (Meens et al. 1993; Lao and Wilson 1996). However, it has also been observed in other cases that the natural signal peptide of a heterologous secretory protein functions only inefficiently in the chosen Gram-positive host and that its replacement by a host-derived signal peptide can significantly improve the secretion efficiency (Miller et al. 1987).

Secretion of a normally not-exported heterologous protein requires that the respective protein is attached to a signal peptide. The joint between the signal peptide and the foreign protein can be designed such that the protein is linked in an exact fusion located immediately to the signal peptide or a few amino acid residues downstream of it. The exact fusions are used to obtain an authentic NH_2-terminus for the secreted protein, the latter ones are used to preserve the steric environment for efficient removal of the signal peptide by the signal peptidase. However, it is still not predictable whether a certain combination of a signal peptide and a foreign protein will result in membrane translocation and processing of the respective hybrid precursor. For example, numerous attempts to secrete foreign proteins by *B. subtilis* have shown that certain signal peptides can support efficient export for a particular protein, but not for another one (Simonen and Palva 1993). When attempts to export a desired heterologous protein fail, use of alternative signal peptides derived from the chosen Gram-positive host might therefore be a promising strategy to succeed.

Tat signal peptides seem to function optimally only with their cognate translocases (Jongbloed et al. 2000; Blaudeck et al. 2001). For this reason, the fusion of heterologous proteins to a Tat signal peptide originating from the chosen Gram-positive host bacterium or a closely related species is highly recommended if the Tat-dependent secretion option is envisaged.

4.3
Extracytosolic Protein Folding

Folding of proteins that reside in the cytosol is assisted by various well-characterized general chaperones (Hsp60, Hsp70 proteins). These chaperones prevent the aggregation of newly synthesized proteins and assist their folding into the correct, biologically active conformation. In contrast, proteins exported via the Sec pathway have to resume their folded conformation at the *trans*-side of the membrane. Very little is known about the post-translocational mechanisms of folding of secreted proteins in Gram-positive bacteria; that is, it is still unclear to what extent cellular factors of the host bacterium (Figure 4.4) are involved in the respective folding events. In *B. subtilis*, the plasma membrane-anchored lipoprotein PrsA – which is homologous to the parvulin class of peptidyl prolyl cis/trans-isomerases – seems to be involved in the folding of various newly translocated proteins. Limiting the amount or the activity of PrsA results in the decreased folding of exported proteins and their subsequent degradation (see also Section 4.5) (Jacobs et al. 1993). Especially under high-level protein-secretion conditions, the amount of available PrsA is a serious bottleneck (Vitikainen et al. 2001). Consequently, the overproduction of the PrsA folding factor is a promising tool for improving the yields of correctly folded homologous and heterologous proteins. In fact, increasing the number of PrsA molecules resulted in an improved secretion of a *Bacillus amyloliquefaciens*-derived α-amylase (AmyQ) in *B. subtilis* (Vitikainen et al. 2001). Likewise, the amount of a fibrin-specific single-chain antibody fragment that was secreted into the culture supernatant of *B. subtilis* was significantly increased when the *prsA* gene was co-overexpressed (Wu et al. 2002).

Disulfide bonds are crucial for the activity and stability of many proteins of biotechnological or pharmaceutical interest. In principle, such bonds can form spontaneously by air oxidation. However this process is very slow and much less effective than the formation of disulfide bonds *in vivo*, catalyzed by specific enzymes, the so-

Extracellular
space

Cell wall

Plasma
membrane

Cytosol

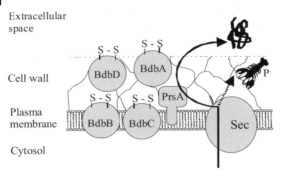

Fig. 4.4 Late stages of Sec-dependent protein secretion in Gram-positive bacteria. The lipoprotein PrsA and the thiol disulfide oxido-reductases BdbABCD are factors assisting protein folding on the *trans*-side of the plasma membrane. An inverse correlation exists between the efficiency of folding of newly translocated proteins into a protease-resistant conformation and their degradation by membrane- and/or cell wall-associated proteases (P).

called thiol disulfide oxido-reductases. In bacteria, disulfide bonds cannot be formed in the cytosol, due to its reducing conditions. Disulfide bond formation requires that the respective protein is exported into an extra-cytosolic compartment, such as the periplasm of Gram-negative bacteria. In *E. coli*, six proteins residing in the periplasm and the cytoplasmic membrane have been identified which are involved in disulfide bond formation and isomerization: DsbA, DsbB, DsbC, DsbD, DsbE, and DsbG (for details, see Kadokura et al. 2003). Very few extra-cytoplasmic proteins with disulfide bonds have been identified in Gram-positive bacteria; this raised the question whether this class of bacteria has catalytic systems for disulfide bond formation. However, recent genetic studies suggest the existence of such systems in *B. subtilis*. Here, genes encoding four proteins with similarities to thiol disulfide oxido-reductases (BdbA, BdbB, BdbC, and BdbD) have been identified and it has been shown that the natural substrates for these enzymes are a membrane protein involved in competence development (ComGC) and the lantibiotic sublancin 168, a peptidic antibiotic which is transported across the cytoplasmic membrane by an ABC transport system (Dorenbos et al. 2002; Meima et al. 2002). Moreover, it has been shown that *B. subtilis* can secrete active *E. coli*-derived alkaline phosphatase PhoA, which requires the correct formation of two disulfide bonds for activity. Mutations in the *bdbB* and *bdbC* genes led to a severe reduction in the production of the secreted PhoA with correct disulfide bonds (Bolhuis et al. 1999a). Secretion of active *E. coli*-derived TEM β-lactamase, which requires the presence of one disulfide bond for stability, only depends on BdbC (Bolhuis et al. 1999a). Active human interleukin-3, which again requires one disulfide bond for its biological activity, could be secreted by a recombinant *Bacillus licheniformis* strain, as shown in a previous study (van Leen et al. 1991). These results showed that disulfide bonds of heterologous proteins can be formed properly by Gram-positive bacteria and that thiol disulfide oxido-reductases are crucial for this process. On the other hand, secreted proteins of eukaryotes often contain several (i.e., more than one or two) disulfide bonds. Proteins containing multiple disulfide bridges such as the human serum albumin or the human pancreatic alpha-amylase were secreted only very poorly by *B. subtilis*, most likely due to impairments in disulfide bond formation (Saunders et al. 1987; Bolhuis et al. 1999b). Furthermore, not all Gram-positive bacteria seem to be equally capable in catalyzing the formation of disulfide bridges. In contrast to the situation in *B. subtilis*, the *E. coli*-derived alkaline phosphatase PhoA described before was found to be completely inactive when secreted from a *S. carnosus* host. This suggests either that this microorganism does not possess thiol disulfide oxido-reductases, or that these enzymes differ from their *B. subtilis* homologues with respect to their substrate specificity (M. Klein, R. Freudl, unpublished). Therefore, correct choice of the Gram-positive host organism is crucial if the desired heterologous protein requires disulfide bonds for activity.

4.4
The Cell Wall as a Barrier for the Secretion of Heterologous Proteins

Gram-positive bacteria possess a thick cell wall, a highly cross-linked semi-porous structure composed of peptidoglycan and negatively charged polymers such as teichoic or teichuroic acid. Proteins exiting the external surface of the cytoplasmic membrane first encounter the cell wall before being released into the culture medium. The cell wall therefore represents the final stage of protein secretion. Although some exo-proteins may be able to pass the cell wall by diffusion, the passage of larger proteins – or those extensively exposing positively charged amino acid residues at their surface – is likely to be affected by wall characteristics such as molecular-sieving and ion-exchanging properties. Accordingly, it has been found frequently that the wall can indeed represent a severe bottleneck in the secretion of heterologous proteins, but this is not observed in the export of autologous proteins. For example, no release of human serum albumin is observed unless the peptidoglycan layer of the cell wall is destroyed, despite the fact that the protein is translocated across the *B. subtilis* plasma membrane (Saunders et al. 1987). In such cases, testing different Gram-positive bacteria or mutant strains with slightly altered cell wall compositions (Thwaite et al. 2002) might be considered as a possible means of improving the cell wall passage of trapped or inefficiently released heterologous proteins. Moreover, the use of cell wall-less L-forms of Gram-positive bacteria (Gumpert and Hoischen 1998) might be an alternative approach for the secretory production of heterologous proteins with an otherwise impaired cell wall passage.

4.5
Degradation of Exported Proteins by Cell-associated and Secreted Proteases

Many Gram-positive bacteria secrete proteases into the culture supernatant. For example, at least eight different proteases efficiently degrade secreted, protease-sensitive heterologous proteins in *B. subtilis*. Mutants of *B. subtilis* have been generated which lack the corresponding protease genes and show less than 0.5% proteolytic activity compared to wild-type strains. In these mutants, an improved stability of secreted heterologous proteins can be observed (Murashima et al. 2002). In addition, strains of species secreting only low levels of proteases or no proteases at all (such as

S. carnosus; see the following sections) can be considered as alternative host organisms. In addition to soluble proteases in the supernatant, proteases localized on the outer surface of the plasma membrane and/or in the cell wall area of Gram-positive bacteria represent another serious bottleneck (Meens et al. 1997). An inverse correlation exists between the efficiency of folding of newly translocated proteins into a conformation resistant towards these cell-associated proteases and their degradation (see Figure 4.4). If the intrinsic folding of the target protein is inefficient, or if the folding is rendered inefficient by genetic manipulations (e.g., by depletion of the PrsA folding factor in *B. subtilis*), a rapid degradation of newly secreted proteins is observed, caused by the cell-associated proteases (Jacobs et al. 1993; Meens et al. 1997). So far, the respective proteases involved in the cell-associated degradation of exported heterologous proteins have not been characterized. The identification and subsequent inactivation of these proteases is a major challenge for the future which, in case of success, might result in substantial improvements of the performance of Gram-positive bacteria to secrete heterologous proteins.

4.6
Staphylococcus carnosus

The presence of secreted proteases in the culture supernatant of most Gram-positive bacteria is a major drawback that precludes their use for the secretory production of heterologous proteins. In contrast, the culture supernatant of *S. carnosus* TM300 does not contain any significant proteolytic activity. Mainly for this reason, this microorganism is considered to be a promising expression host for a variety of scientific and biotechnological applications (Götz 1986).

4.6.1
General Description

Gram-positive and catalase-positive cocci play an important role in the ripening process of dry sausages and meat. One of the predominant microorganisms in fermented meat is *S. carnosus*. In earlier reports, this bacterium has been named *Micrococcus* but, based on several criteria such as DNA sequence homology, peptidoglycan composition and biochemical properties, this microorganism was re-classified as a member of the genus *Staphylococcus*. The species name „*carnosus*" was chosen due to the fact that the bacterium occurs in, and actually was isolated from, meat (Schleifer and Fischer 1982). *S. carnosus* has now been in use for more than 50 years, either alone or in combination with other microorganisms (*Lactobacillus* sp., *Pediococcus* sp.), as a starter culture for the production of raw meat products. The use of starter cultures in meat fermentation processes offers a number of advantages over uncontrolled processes: (1) accelerated and controlled color formation, drop in pH and acquisition of the desired texture and flavor; and (2) control of food pathogens and spoilage organisms. The main characteristics of *S. carnosus* in this process is the reduction of nitrate, the development of a characteristic flavor, a moderate decrease in the pH value and the reduction of hydrogen peroxide that is produced by the cata-lase-negative lactobacilli (Liepe 1982).

To date, no *S. carnosus* strain is known to produce toxins, hemolysins, protein A, coagulase or clumping factors, all of which are typical markers for many pathogenic *Staphylococcus aureus* strains. Furthermore, *S. carnosus* has a long tradition of use in starter cultures, and no adverse effects on human health have ever been observed. Therefore, *S. carnosus* has obtained the GRAS status (generally recognized as safe), for its safe use in the food industry (Götz 1986).

4.6.2
Microbiological and Molecular Biological Tools

Among the various *S. carnosus* strains, the *S. carnosus* strain TM300 was identified as the most promising candidate as host for the expression, secretion and surface display of heterologous proteins, as this strain possesses hardly any extracellular proteolytic activity (Table A4.1). Due to the pioneering work of Friedrich Götz and his co-workers, methods allowing the introduction of DNA into *S. carnosus* by protoplast transforma-tion (Götz et al. 1983; Götz and Schumacher 1987) or electroporation (Augustin and Götz 1990) are available. Furthermore, a variety of vectors suitable for cloning and the expression of genes (Kreutz and Götz 1984; Keller et al. 1983; Wieland et al. 1995; Sandgathe et al. 2003), for gene replacement (Madsen et al. 2002), for promotor prob-ing (Wieland et al. 1995), and for secretion (Liebl and Götz 1986) or surface display (Samuelson et al. 1995) of heterologous proteins have been developed. In addition, a food-grade vector system has been described for *S. carnosus*, that is based on the genes that mediate sucrose uptake and hydrolysis in *Staphylococcus xylosus* (Brückner and Götz 1996) (Table A4.2).

4.6.3
S. carnosus as Host Organism for the Analysis of Staphylococcal-related
Pathogenicity Aspects

Due to its noninvasive, nonpathogenic characteristics, *S. carnosus* has been used suc-cessfully as a clean expression background to investigate the contribution of pathogeni-city factors in the virulence of pathogenic staphylococci. *S. aureus* invasion of mamma-lian cells, including epithelial, endothelial, and fibroblastic cells, critically depends on fibronectin bridging between *S. aureus* fibronectin-binding proteins (FnBPs) and the host fibronectin receptor integrin a5b1. To investigate whether this mechanism is suffi-cient for *S. aureus* invasion, the FnBPs of *S. aureus* were expressed in *S. carnosus*, which intrinsically lacks such FnBPs. Characterization of the corresponding recombinant strain revealed that the *S. aureus* FnBPs were expressed in a functional form at the cell surface. They conferred an invasiveness to *S. carnosus* comparable to that of the strongly invasive *S. aureus* strain Cowan 1, showing that the *S. aureus* FnBPs are *per se* sufficient to mediate fibronectin binding and invasion of host cells (Sinha et al. 2000).

In another study, the role of the intercellular adhesion (*ica*) gene products in the formation of biofilms was investigated using *S. carnosus* as an expression platform.

Nosocomial infections that result in the formation of biofilms on the surfaces of bio-medical implants are a leading cause of sepsis and are often associated with colonization of the implants by *Staphylococcus epidermidis*. Biofilm formation is thought to be a two-step process that requires the adhesion of bacteria to a substrate surface followed by cell–cell adhesion, forming the multiple layers of the biofilm. This latter process is triggered by the polysaccharide intercellular adhesin (PIA), which is composed of linear β-1,6-linked glucosaminylglycans. The intercellular adhesion (*ica*) locus was identified and shown to mediate cell-cell adhesion and PIA production in *S. epidermidis*. Transfer of the *ica* genes of *S. epidermidis* into *S. carnosus*, which intrinsically does not possess an intercellular adhesion phenotype, led to PIA expression and to the formation of a biofilm on a glass surface and of large cell aggregates, directly demonstrating the involvement of the *ica* gene products in biofilm formation (Heilmann et al. 1996).

4.6.4
Secretory Production of Heterologous Proteins by *S. carnosus*

4.6.4.1   The *Staphylococcus hyicus* Lipase: Secretory Signals and Heterologous Expression in *S. carnosus*

The secretory signals derived from a lipase of another staphylococcal species, *Staphylococcus hyicus*, have been successfully used in a variety of cases for the secretion of heterologous proteins by *S. carnosus*. They are among the most powerful signals that are presently available for this host system. In addition, analysis of the secretion mechanism of the *S. hyicus* lipase has provided some new insights into the secretion pathway of Gram-positive bacteria. These newly discovered features are of significance also with respect to the application of this class of microorganisms to the secretion of heterologous proteins. The *S. hyicus* lipase gene was one of the first heterologous genes that was cloned and expressed in *S. carnosus* (Götz et al. 1985). According to the gene sequence, the *S. hyicus* lipase consists of 641 amino acid residues (predicted molecular mass of 71 kDa) that contains a typical Gram-positive Sec signal peptide, 38 amino acid residues in length, at its amino-terminal end. Transformation of *S. carnosus* with a plasmid containing the lipase gene from *S. hyicus* resulted in the release of the active enzyme into the culture supernatant. The lipase is secreted in amounts of 40 mg protein L^{-1}, and accounts for about 15% of the total extracellular proteins in the recombinant *S. carnosus* strain. Comparison of the lipase proteins by sodium dodecyl sulfate-polyacrylamide gel electrophoresis (SDS-PAGE) revealed apparent masses of 46 kDa in the donor strain and 86 kDa in the recombinant *S. carnosus* strain, respectively (Götz et al. 1985). This remarkable size difference suggested that the lipase is processed proteolytically in *S. hyicus*, but not in *S.carnosus*. Amino-terminal sequencing of the two lipase variants showed that the 86-kDa form secreted by *S. carnosus* started with Asp^{39}, whereas the 46-kDa form found in the *S. hyicus* culture supernatant started with Val^{246}. These results confirmed that both forms were derived from the same precursor protein, and furthermore indicated that the 641-amino acid precursor is organized as a so-called pre-pro-protein possessing a 38-amino acid signal peptide (pre-peptide), a 207-amino acid

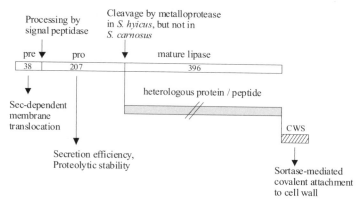

Fig. 4.5 The *Staphylococcus hyicus* pre-pro-lipase. The pre-pro part of the *S. hyicus* lipase is not only an efficient signal for the secretion of the cognate mature lipase, but also for heterologous proteins or peptides into the culture supernatant of *S. carnosus*. Attachment of a cell wall sorting signal (CWS) to the C-terminal end of a passenger protein or peptide by genetic engineering leads to covalent anchoring of the passenger to the peptidoglycan and to its display on the *S. carnosus* cell surface.

long pro-peptide followed by the 396-amino acid residues of the mature lipase (Wenzig et al. 1990) (Figure 4.5). Further characterization of the processing events showed that, in *S. hyicus* the pre-pro-lipase is first processed by signal peptidase to pro-lipase during or shortly after Sec-dependent membrane translocation. Then, the resulting 86-kDa pro-lipase is further processed to the 46-kDa form by a metalloprotease that is present in the culture supernatant of *S. hyicus* but not in *S. carnosus* (Ayora et al. 1994) (Figure 4.5).

The presence of a pro-peptide is not uncommon in proteins secreted from Gram-positive bacteria. In many examples, as in representatives of the well-characterized subtilisin protease family from various *Bacillus* species, the pro-peptide functions as an intramolecular chaperone that is strictly required for the folding of the attached mature protein. After the mature protease has adopted its active conformation, the pro-peptide is subsequently removed by autocatalytic proteolytic degradation (Shinde and Inouye 2000). In contrast, the pro-peptide of the *S. hyicus* lipase is obviously not essential for the folding of the mature part of the lipase, but it rather fulfils an additional important function in relation to its secretion. Deletion of parts of the pro-peptide or removal of the entire pro-peptide from the pre-pro-lipase precursor resulted in a dramatic reduction of the amount of lipase that is released into the supernatant of *S. carnosus* cultures. However, this secreted lipase was enzymatically active, showing that the pro-peptide *per se* is not absolutely required for adopting the correct conformation of the mature lipase. In contrast, the strongly reduced amounts of secreted lipase suggested that the presence of the pro-peptide is required for the protection of the mature lipase against proteolytic degradation at a certain step in the secretion pathway (Demleitner and Götz 1994). Furthermore, the secretion of the lipase pro-peptide *in trans* with a pro-peptide-deficient lipase precursor did not result in a stabilization of the mature lipase, showing that the protection against proteolytic degradation requires that the mature part of the lipase is covalently attached to the pro-peptide (Götz et al. 1998).

4.6.4.2 Use of the Pre-pro-part of the *S. hyicus* Lipase for the Secretion of Heterologous Proteins in *S. carnosus*

Subsequent studies showed that the pro-peptide promoted not only the efficient secretion of the cognate lipase, but also of totally unrelated heterologous proteins and peptides. The first study which demonstrated the beneficial effect of the pro-peptide on the secretion of foreign passenger proteins was performed by Liebl and Götz (1986) using the periplasmic β-lactamase of *E. coli* as a heterologous model protein. Fusion of the mature part of the *E. coli* β-lactamase to the pre-pro-part of the *S. hyicus* lipase resulted in the release of active pro-peptide-β-lactamase hybrid protein into the culture supernatant. The amounts obtained were close to those obtained with the native lipase. When the pro-peptide was omitted, hardly any secreted β-lactamase could be detected (Liebl and Götz 1986). More detailed information on the action of the pro-peptide was gained in studies aiming at the secretion of an *E. coli* outer membrane protein (OmpA) by *S. carnosus* (Meens et al. 1997). In its native environment, OmpA is a two-domain protein, consisting of an amino-terminal anti-parallel β-barrel outer membrane domain and a carboxy-terminal periplasmic domain. In contrast to the periplasmic domain, the folding properties of which can be considered similar to those of a soluble protein, the folding of the membrane domain into its native conformation requires the presence of lipopolysaccharide (LPS). Since Gram-positive bacteria do not contain LPS, the outer membrane domain of OmpA is not expected to fold into a stable conformation after secretion across the plasma membrane in *S. carnosus*. Therefore, it can be used as an excellent sensor for the detection of protease-related bottlenecks in the secretory pathway. Expression of native OmpA precursor protein containing its own signal peptide in *S. carnosus* resulted in the efficient secretion of two degradation products, 16 and 18 kDa in size. The amino-terminal sequence analysis revealed that the two fragments overlapped, and were both derived from the periplasmic domain of OmpA. Clearly, the membrane domain of OmpA was degraded subsequent to membrane translocation. This observation was surprising, since *S. carnosus* is known to secrete very small amounts of proteases, if any at all. The finding that purified OmpA, which was added externally to the culture medium of growing *S. carnosus* cells, remained intact indicated that newly synthesized and exported OmpA is degraded by one or more cell-associated proteases rather than by soluble proteases in the culture supernatant. Attachment of the mature OmpA to the pre-pro-part of the *S. hyicus* lipase led to an efficient secretion of the corresponding pro-peptide-OmpA hybrid protein into the supernatant of *S. carnosus*, and completely prevented the cell-associated proteolytic degradation of OmpA, despite the fact that the protease-sensitive membrane domain is present also in the fusion protein. Although the underlying molecular mechanism is still not understood, these results showed that a main function of the lipase pro-peptide is to prevent the proteolytic degradation of the attached passenger protein by proteases residing on the *trans*-side of the cytoplasmic membrane and/or in the cell wall area (Meens et al. 1997).

The beneficial effect of the lipase pro-peptide has also been used successfully for the secretion of mammalian proteins by *S. carnosus*. For example, pro-insulin (Götz et al. 1998), different antibody fragments (Pschorr et al. 1994; Schnappinger et al. 1995) and a precursor of the human peptide hormone calcitonin (see below; Section 4.6.4.3) could be secreted by employing vectors that had incorporated the pre-pro part of the *S. hyicus* lipase for secretion targeting. In all cases, it became evident that the pro-peptide is crucial for efficient secretion and proteolytic stability.

4.6.4.3 Process Development for the Secretory Production of a Human Calcitonin Precursor Fusion Protein by *S. carnosus*

Human calcitonin (hCT) is a peptide hormone of 32 amino acids which is effectively used for the treatment of diseases such as osteoporosis imperfect, menopausal osteoporosis and Paget's disease. To date, therapeutic hCT has been produced industrially by total chemical synthesis. Biologically active hCT requires the amidation of its C-terminal proline residue, a modification which is introduced enzymatically to the unmodified synthesized precursor peptide (Ishikawa and Tamaoki 1996).

Due to the high costs of the peptide synthesis and the increasing therapeutic demands for hCT, a biotechnological process for the large-scale (kg) production of hCT precursor peptide would be desirable. Previous attempts to establish such a process mostly used *E. coli* as an expression system. As small peptides such as hCT are rapidly degraded by intracellular proteases, hCT was produced as a fusion protein. However, the intracellular production of such fusion proteins resulted in the formation of inclusion bodies (Yabuta et al. 1995). Therefore, secretion of the hCT precursor peptide fused to the pre-pro part of the *S. hyicus* lipase by *S. carnosus* was considered as a promising approach to overcome problems of aggregation and proteolytic instability. Multiple repeats (2, 4, 6, 8, or 10 copies) of the hCT precursor peptide were fused to the pre-pro-part of the lipase. The hCT peptides were separated from each other and from the pro-peptide by linker sequences containing cleavage sites for the endopeptidase clostripain, allowing the enzymatic release of monomeric hCT precursor peptides subsequent to the isolation of the respective fusion proteins. The corresponding hybrid sequences were cloned into an expression vector fusing them to a xylose-inducible promoter. The respective plasmids were used to transform *S. carnosus*. In all cases, the lipase secretion signals allowed an efficient release of the respective pro-peptide-hCT fusion proteins. The yields obtained in laboratory shake flasks were close to those obtained with the native lipase under the same conditions (approx. 40–50 mg L^{-1}) (Sandgathe et al. 2003). In the next step, a fermentation process was established for the production of the pro-peptide-hCT fusion containing two hCT repeats. First, the recombinant strain was cultured in a pH-auxostatic fed-batch process in a 3-L stirred tank reactor using glycerol as carbon source, and with a pH-controlled addition of yeast extract. Following this process design, the production of 2000 mg L^{-1} of the pro-peptide-hCT fusion protein (equivalent to 420 mg L^{-1} hCT precursor peptide) was achieved within 14 hours. The process was then scaled up to 150 L, resulting in yields of 200 g of the pro-peptide-hCT fusion protein (Dilsen et al. 2000, 2001). Finally, a downstream procedure was developed based on an expanded bed adsorption (EBA) to the cation exchanger STREAMLINE SP matrix at pH 5. In this way, the pro-peptide-hCT fusion protein was recovered in a single step from the unclarified culture broth (Sandgathe et al. 2003). To summarize: the results obtained with hCT precursor peptide example clearly demonstrate the high potential of *S. carnosus* as an alternative expression system for the secretory production of heterologous proteins.

4.6.5
Surface Display on *S. carnosus*

The anchoring of heterologous proteins on the cell surface of food-grade bacteria, such as *S. carnosus*, has become an attractive strategy for various potential applications in biotechnology, microbiology, and biomedicine. Many of the major steps involved in the surface anchoring of proteins in Gram-positive bacteria have been elucidated in *S. aureus*, using protein A as the model protein (Mazmanian et al. 2001). For membrane translocation and subsequent anchoring to the cell wall, the proteins to be displayed on the surface must be equipped with an N-terminal secretion signal and a C-terminal cell wall sorting signal. The latter signal consists of: (1) a conserved penta-peptide motif, LPXTG; (2) a hydrophobic stretch of 15–22 amino acid residues; and (3) a short charged tail of six to seven amino acid residues. During secretion of proteins containing such a cell wall sorting signal, the hydrophobic region and the charged tail impede membrane translocation by the Sec system, allowing the recognition of the LPXTG motif by a membrane-associated sortase enzyme. In a following two-step transpeptidation reaction, sortase cleaves the LPXTG motif between the threonine and glycine residues and covalently attaches the threonine to the amino group of the penta-glycine cell wall cross-bridge, resulting in a cell wall-attached protein (Paterson and Mitchell 2004). Using the pre-pro part of the *S. hyicus* lipase as secretion signals and the cell wall sorting signal of either protein A or fibronectin-binding protein from *S. aureus* (see Figure 4.5), various heterologous proteins have successfully been displayed on the *S. carnosus* cell surface. The *S. hyicus*-derived lipase and the *E. coli*-derived β-lactamase are two examples of enzymes which have been tethered to the surface of *S. carnosus* in an active form. Approximately 10 000 enzyme molecules were found to be present on each cell, and it was suggested that, due to their rigid structure, Gram-positive bacteria would be particularly appropriate for applications such as whole-cell catalysts or biofilters (Strauss and Götz 1996). The surface display of chimeric proteins containing poly-histidyl peptides resulted in *S. carnosus* strains which have gained a significant Ni^{2+}- and Cd^{2+}-binding capacity, suggesting that such bacteria could find use as metal-binding bioadsorbents (Samuelson et al. 2000). However, most surface display approaches in *S. carnosus* are devoted to the exposition of an antigenic determinant on the cell surface. Use of the respective recombinant bacteria as live vaccine delivery vectors is considered to be an attractive vaccination alternative. For example, epitopes derived from streptococcal protein G (Cano et al. 1999) or the G glycoprotein of human respiratory syncytial virus (RSV) (Cano et al. 2000) were successfully displayed on the *S. carnosus* cell surface. Furthermore, intranasal or oral immunization of mice with the corresponding live recombinant staphylococci elicited significant serum IgG responses to the respective antigens. In addition, the challenge of mice, immunized with *S. carnosus* displaying RSV peptides on the cell surface, with 10^5 RSV tissue culture infectious doses$_{50}$ showed protection of the lung in approximately half of the mice in the two immunization groups. These results clearly demonstrate that immune protection can be elicited using the food-grade bacterium *S. carnosus* as a vaccine-delivery vehicle (Cano et al. 2000).

Appendix

Tab. A4.1 Selection of *S. carnosus* host strains.

Strain	Genotype	Reference
TM300	wild-type	Schleifer and Fischer (1982)
TM300 *recA*	*recA*	Sandgathe (2000)

Tab. A4.2 Selection of *S. carnosus* vectors.

Plasmid marker	Type	Resistance	Relevant features	Reference
pCA43	Cloning vector	cml	–	Kreutz and Götz (1984)
pCT20	Cloning vector	cml, tet	–	Keller et al. (1983)
pCE10	Cloning vector	cml, ery	–	Keller et al. (1983)
pSCRBA	Cloning vector	cml	food-grade vector system based on sucrose utilization	Brückner and Götz (1995)
pPS11	Promoter-probe vector	cml	promotor-less lipase gene of *S. hyicus* as reporter	Wieland et al. (1995)
pCX26Δ	Expression vector	cml	xylose-inducible promoter	Wieland et al. (1995)
pXR100	Expression vector	cml (Sc) amp (Ec)	xylose-inducible promoter (*E. coli/S. carnosus* shuttle vector)	Sandgathe et al. (2003)
pLipPS1	Secretion vector	cml	contains promoter and structural gene of *S. hyicus* pre-pro-lipase	Liebl and Götz (1986)

Tab. A4.2 (continued)

Plasmid marker	Type	Resistance	Relevant features	Reference
pSPP-mABPXM	Surface display vector	cml (Sc) amp (Ec)	contains lipase promoter and gene fragments encoding the pre-pro-part of *S. hyicus* lipase and the cell wall anchor of *S. aureus* protein A (*E. coli*/*S. carnosus* shuttle vector)	Samuelson et al. (1995)

References

ALAMI M, LÜKE I, DEITERMANN S, EISNER G KOCH HG, BRUNNER J, MÜLLER M (2003) Differential interaction between a twin-arginine signal peptide and its translocase in *Escherichia coli*. Mol Cell 12: 937–946

AUGUSTIN J, GÖTZ F (1990) Transformation of *Staphylococcus epidermidis* and other staphylococcal species with plasmid DNA by electroporation. FEMS Microbiol Lett 66: 203–208

AUNSTRUP K (1979) Production, isolation, and economics of extracellular enzymes. In: Applied Biochemistry and Bioengineering, Vol. 2 (Wingard LB, Katchalski-Kazir E, Goldstein L, Eds), Academic Press, New York, pp 27–69

AYORA S, LINDGREN PE, GÖTZ F (1994) Biochemical properties of a novel metalloprotease from *Staphylococcus hyicus* subsp. *hyicus* involved in extracellular lipase processing. J Bacteriol 176: 3218–3223

BARRETT CML, RAY N, THOMAS JD, ROBINSON C, BOLHUIS, A (2003) Quantitative export of a reporter protein, GFP, by the twin-arginine translocation pathway in *Escherichia coli*. Biochem Biophys Res Commun 304: 279–284

BERKS BC, PALMER T, SARGENT F (2003) The Tat protein translocation pathway and its role in microbial physiology. Adv Microbial Physiol 47: 187–254

BLAUDECK N, SPRENGER GA, FREUDL R, WIEGERT T (2001) Specificity of signal peptide recognition in Tat-dependent bacterial protein translocation. J Bacteriol 183: 604–610

BOLHUIS A, VENEMA G, QUAX WJ, BRON S, VAN DIJL JM (1999a) Functional analysis of paralogous thiol-disulfide oxidoreductases in *Bacillus subtilis*. J Biol Chem 274: 24531–24538

BOLHUIS A, TJALSMA H, SMITH HE, DE JONG A, MEIMA R, VENEMA G, BRON S, VAN DIJL JM (1999b) Evaluation of bottlenecks in the late stages of protein secretion in *Bacillus subtilis*. Appl Environ Microbiol 65: 2934–2941

BRÜCKNER R, GÖTZ F (1996) Development of a food-grade vector system for *Staphylococcus carnosus*. Syst Appl Microbiol 18: 510–516

CANO F, LILJEQVIST S, NGUYEN TN, SAMUELSON P, BONNEFOY JY, STAHL S, ROBERT A (1999) A surface-displayed cholera toxin B peptide improves antibody responses using food-grade staphylococci for mucosal subunit vaccine delivery. FEMS Immunol Med Microbiol 25: 289–298

CANO F, PLOTNICKY-GILQUIN H, NGUYEN TN, LILJEQVIST S, SAMUELSON P, BONNEFOY J, STAHL S, ROBERT A (2000) Partial protection to respiratory syncytial virus (RSV) elicited in mice by intranasal immunization using live staphylococci with surface-displayed RSV-peptides. Vaccine 18: 2743–2752

DEBABOV VS (1982) The industrial use of *Bacilli*. In: The Molecular Biology of the *Bacilli* (Dubnau D, Ed), Academic Press, New York, pp 331–370

DEMLEITNER G, GÖTZ F (1994) Evidence for importance of the *Staphylococcus hyicus* lipase pro-peptide in lipase secretion, stability and activity. FEMS Microbiol Lett 121: 189–198

DILSEN S, PAUL W, SANDGATHE A, TIPPE D, FREUDL R, THÖMMES J, KULA MR, TAKORS R,

WANDREY C, WEUSTER-BOTZ D (2000) Fed-batch production of recombinant human calcitonin precursor fusion protein using *Staphylococcus carnosus* as an expression-secretion system. Appl Microbiol Biotechnol 554: 361–369

DILSEN S, PAUL W, HERFORTH D, SANDGATHE A, ALTENBACH-REHM J, FREUDL R, WANDREY C, WEUSTER-BOTZ D (2001) Evaluation of parallel operated small-scale bubble columns for microbial process development using *Staphylococcus carnosus*. J Biotechnol 88: 77–84

DORENBOS R, STEIN T, KABEL J, BRUAND C, BOLHUIS A, BRON S, QUAX WJ, VAN DIJL JM (2002) Thiol-disulfide oxidoreductases are essential for the production of the lantibiotic sublancin 168. J Biol Chem 277: 16682–16688

FEILMEIER BJ, ISEMINGER G, SCHROEDER D, WEBBER H, PHILLIPS GJ (2000) Green fluorescent *protein* functions as a reporter for protein localisation in *Escherichia coli*. J Bacteriol 182: 4068–4076

GÖTZ F, KREUTZ B, SCHLEIFER KH (1983) Protoplast transformation of *Staphylococcus carnosus* by plasmid DNA. Mol Gen Genet 189: 340–342

GÖTZ F, POPP F, KORN E, SCHLEIFER KH (1985) Complete nucleotide sequence of the lipase from *Staphylococcus hyicus* cloned in *Staphylococcus carnosus*. Nucleic Acids Res 13: 5895–5906

GÖTZ F (1986) Ein neues Wirt-Vektor-System bei *Staphylococcus carnosus*. Umschau 10: 530–537

GÖTZ F, SCHUMACHER B (1987) Improvements of protoplast transformation in *Staphylococcus carnosus*. FEMS Microbiol Lett 40: 285–288

GÖTZ F, VERHEIJ HM, ROSENSTEIN R (1998) Staphylococcal lipases: molecular characterization, secretion, and processing. Chem Phys Lipids 93: 15–25

GUMPERT J, HOISCHEN C (1998) Use of cell wall-less bacteria (L-forms) for efficient expression and secretion of heterologous gene products. Curr Opin Biotechnol 9: 506–509

HARWOOD CR (1992) *Bacillus subtilis* and its relatives: molecular biological and industrial workhorses. Trends Biotechnol 10: 247–256

HEILMANN C, SCHWEITZER O, GERKE C, VANITTANAKOM N, MACK D, GÖTZ F (1996) Molecular basis of intercellular adhesion in the biofilm-forming *Staphylococcus epidermidis*. Mol Microbiol 20: 1083–1091

ISHIKAWA H, TAMAOKI H (1996) Production of human calcitonin in *Escherichia coli* from multimeric fusion protein. J Ferm Bioeng 82: 140–144

JACOBS M, ANDERSEN JB, KONTINEN V, SARVAS M (1993) *Bacillus subtilis* PrsA is required *in vivo* as an extracytoplasmic chaperone for secretion of active enzymes synthesized either with or without pro-sequences. Mol Microbiol 8: 957–966

JONGBLOED JDH, MARTIN U, ANTELMANN H, HECKER M, TJALSMA H, VENEMA G, BRON, S, VAN DIJL JM, MÜLLER J (2000) TatC is a specificity determinant for protein secretion via the twin-arginine translocation pathway. J Biol Chem 275: 41350–41357

KADOKURA H, KATZEN F, BECKWITH J (2003) Protein disulfide bond formation in prokaryotes. Annu Rev Biochem 72: 111–135

KELLER G, SCHLEIFER KH, GÖTZ F (1983) Construction and characterization of plasmid vectors for cloning in *Staphylococcus aureus* and *Staphylococcus carnosus*. Plasmid 10: 270–278

KREUTZ B, GÖTZ F (1984) Construction of *Staphylococcus* plasmid vector pCA43 conferring resistance to chloramphenicol, arsenate, arsenite and antimony. Gene 31: 301–304

LAO GF, WILSON DB (1996) Cloning, sequencing, and expression of a *Thermomonospora fusca* protease gene in *Streptomyces lividans*. Appl Environ Microbiol 62: 4256–4259

LELOUP L, DRIESSEN AJM, FREUDL R, CHAMBERT R, PETIT-GLATRON MF (1999) Differential dependence of levansucrase and α-amylase secretion on SecA (Div) during the exponential phase of growth of *Bacillus subtilis*. J Bacteriol 181: 1820–1826

LIEBL W, GÖTZ F (1986) Studies on lipase directed export of *Escherichia coli* β-lactamase in *Staphylococcus carnosus*. Mol Gen Genet 204: 166–173

LIEPE HU (1982). Bakterienkulturen und Rohwurst. Forum Mikrobiol 5: 10–15

MADSEN SM, BECK HC, RAVN P, VRANG A, HANSEN AM, ISRAELSEN H (2002) Cloning and inactivation of a branched-chain-amino-acid aminotransferase gene from *Stapyhlococcus carnosus* and characterization of the enzyme. Appl Environ Microbiol 68: 4007–4014

MALKE H, FERRETTI JJ (1984) Streptokinase: cloning, expression and excretion by *Escherichia coli*. Proc Natl Acad Sci USA 81: 3557–3561

MAZMANIAN SK, TON-THAT H, SCHNEEWIND O (2001) Sortase-catalyzed anchoring of surface proteins to the cell wall of *Staphylococcus aureus*. Mol Microbiol 40: 1049–1057

MEENS J, FRINGS E, KLOSE M, FREUDL R (1993) An outer membrane protein (OmpA) of *Escherichia coli* can be translocated across the cytoplasmic membrane of *Bacillus subtilis*. Mol Microbiol 9: 847–855

MEENS J, HERBORT M, KLEIN M, FREUDL R (1997) Use of the pre-pro part of *Staphylococcus hyicus* lipase as a carrier for secretion of *Escherichia coli* outer membrane protein A (OmpA) prevents proteolytic degradation of OmpA by cell-associated protease(s) in two different Gram-positive bacteria. Appl Environ Microbiol 63: 2814–2820

MEIMA R, ESCHEVINS C, FILLINGER S, BOLHUIS A, HAMOEN LW, DORENBOS R, QUAX WJ, VAN DIJL JM, PROVVEDI R, CHEN I, DUBNAU D, BRON S (2002) The *bdbDC* operon of *Bacillus subtilis* encodes thiol-disulfide oxidoreductases required for competence development. J Biol Chem 277: 6994–7001

MILLER JR, KOVACEVIC S, VEAL LE (1987) Secretion and processing of staphylococcal nuclease by *Bacillus subtilis*. J Bacteriol 169: 3508–3514

MORI H, CLINE K (2002) A twin arginine signal peptide and the pH gradient trigger reversible assembly of the thylakoid ΔpH/Tat translocase. J Cell Biol 157: 205–210

MÜLLER M, KOCH HG, BECK K, SCHÄFER U (2001) Protein traffic in bacteria: multiple routes from the ribosome to and across the membrane. Prog Nucleic Acid Res Mol Biol 66: 197–157

MURASHIMA K, CHEN CL, KOSUGI A, TAMARU Y, DOI RH, WONG SL (2002) Heterologous production of *Clostridium cellulovorans engB*, using protease-deficient *Bacillus subtilis*, and preparation of active recombinant cellulosomes. J Bacteriol 184: 76–81

PATERSON GK, MITCHELL TJ (2004) The biology of Gram-positive sortase enzymes. Trends Microbiol 12: 89–95

PEREZ-PEREZ J, BARBERO JL, MARQUEZ G, GUTIERREZ J (1996) Different PrlA proteins increase the efficiency of periplasmic production of human interleukin-6 in *Escherichia coli*. J Biotechnol 49: 245–247

POP O, MARTIN U, ABEL C, MÜLLER JP (2002) The twin-arginine signal peptide of PhoD and the TatAd/Cd proteins of *Bacillus subtilis* form an autonomous Tat translocation system. J Biol Chem 277: 3268–3273

PSCHORR J, BIESELER B, FRITZ HJ (1994) Production of the immunoglobulin variable domain REIv via fusion protein synthesized and secreted by *Staphylococcus carnosus*. Biol Chem Hoppe-Seyler 375: 271–280

SAMUELSON P, HANSSON M, AHLBORG N, ANDREONI C, GÖTZ F, BÄCHI T, NGUYEN TN, BINZ H, UHLEN M, STAHL S (1995) Cell surface display of recombinant proteins on *Staphylococcus carnosus*. J Bacteriol 177: 1470–1476

SAMUELSON P, WERNERUS H, SVEDBERG M, STAHL S (2000) *Staphylococcus* surface display of metal-binding polyhistidyl peptides. Appl Environ Microbiol 66: 1243–1248

SANDGATHE A (2000) Herstellung eines humanen Calcitonin-Vorläufers mit *Staphylococcus carnosus*: Sekretorische Expression und Aufarbeitung mittels Fließbettadsorption. Dissertation, Heinrich-Heine-Universität Düsseldorf

SANDGATHE A, TIPPE D, DILSEN S, MEENS J, HALFAR M, WEUSTER-BOTZ D, FREUDL R, THÖMMES J, KULA MR (2003) Production of a human calcitonin precursor with *Staphylococcus carnosus*: secretory expression and single-step recovery by expanded bed adsorption. Process Biochem 38: 1351–1363

SAUNDERS CW, SCHMIDT BJ, MALLONEE RL, GUYER MS (1987) Secretion of human serum albumin from *Bacillus subtilis*. J Bacteriol 169: 2917–2925

SCHAERLAEKENS K, SCHIEROVA M, LAMMERTYN E, GEUKENS N, ANNÉ J, VAN MELLAERT L (2001) Twin-arginine translocation pathway in *Streptomyces lividans*. J Bacteriol 183: 6727–6732

SCHLEIFER KH, FISCHER U (1982) Description of a new species of the genus *Staphylococcus*: *Staphylococcus carnosus*. Int J Syst Bacteriol 32: 153–156

SCHNAPPINGER D, GEISSDÖRFER W, SIZEMORE C, HILLEN W (1995) Extracellular expression of native anti-lysozyme fragments in *Staphylococcus carnosus*. FEMS Microbiol Lett 129: 121–128

SHINDE U, INOUYE M (2000) Intramolecular chaperones: polypeptide extensions that modulate protein folding. Semin Cell Dev Biol 11: 35–44

SIMONEN M, PALVA I (1993) Protein secretion in *Bacillus* species. Microbiol Rev 57: 109–137

SINHA B, FRANCOIS P, QUE YA, HUSSAIN M, HEILMANN C, MOREILLON P, LEW D, KRAUSE KH, PETERS G, HERRMANN M (2000)

Heterologously expressed *Staphylococcus aureus* fibronectin-binding proteins are sufficient for invasion of host cells. Infect Immun 68: 6871–6878

STRAUSS A, GÖTZ F (1996) *In vivo* immobilization of enzymatically active polypeptides on the cell surface of *Staphylococcus carnosus*. Mol Microbiol 21: 491–500

TALMADGE K, KAUFMAN J, GILBERT W (1980) Bacteria mature preproinsulin to proinsulin. Proc Natl Acad Sci USA 77: 3988–3992

THOMAS JD, DANIEL RA, ERRINGTON J, ROBINSON C (2001) Export of active green fluorescent protein to the periplasm by the twin-arginine translocase (Tat) pathway in *Escherichia coli*. Mol Microbiol 39: 47–53

THWAITE JE, BAILLIE LWJ, CARTER NM, STEPHENSON K, REES M, HARWOOD CR, EMMERSON PT (2002) Optimization of the cell wall microenvironment allows increased production of recombinant *Bacillus anthracis* protective antigen from *B. subtilis*. Appl Environ Microbiol 68: 227–234

VAN LEEN RW, BAKHUIS JG, VAN BECKHOVEN RRWC, BURGER H, DORSSERS LCJ, HOMMES RWJ, LEMSON PJ, NOORDAM B, PERSOON NLM, WAGEMAKER G (1991) Production of human interleukin-3 using industrial microorganisms. Bio/Technology 9: 47–52

VAN WELY KHM, SWAVING J, FREUDL R, DRIESSEN AJM (2001) Translocation of proteins across the cell envelope of Gram-positive bacteria. FEMS Microbiol Rev 25: 437–454

VITIKAINEN M, PUMMI T, AIRAKSINEN U, WAHLSTRÖM E, WU H, SARVAS M, KONTINEN VP (2001) Quantitation of the capacity of the secretion apparatus and requirement for PrsA in growth and secretion of α-amylase in *Bacillus subtilis*. J Bacteriol 183: 1881–1890

WENZIG E, LOTTSPEICH F, VERHEIJ B, DE HAAS GH, GÖTZ F (1990) Extracellular processing of the *Staphylococcus hyicus* lipase. Biochemistry (Life Sci Adv) 9: 47–55

WIELAND KP, WIELAND B, GÖTZ F (1995) A promoter-screening plasmid and xylose-inducible, glucose-repressible expression vectors for *Staphylococcus carnosus*. Gene 158: 91–96

WU SC, YEUNG JC, DUAN Y, YE R, SZARKA SJ, HABIBI HR, WONG SL (2002) Functional production and characterization of a fibrin-specific single-chain antibody fragment from *Bacillus subtilis*: Effects of molecular chaperones and a wall-bound protease on antibody fragment production. Appl Environ Microbiol 68: 3261–3269

YABUTA M, SUZUKI Y, OHSUYE K (1995) High expression of recombinant human calcitonin precursor peptide in *Escherichia coli*. Appl Microbiol Biotechnol 42: 703–708

5
Arxula adeninivorans
ERIK BÖER, GERD GELLISSEN, and GOTTHARD KUNZE

List of Genes

Gene	Encoded gene product
AEFG1	mitochondrial elongation factor G
AFET3	copper-dependent Fe(II) oxidase
AHSB4	histone H4
AHOG1	mitogen-activated protein kinase
AILV1	threonine deaminase
AINV	β-fructofuranoside fructohydrolase
ALEU2	β-isopropyl malate dehydrogenase
ALYS2	amino-adipate reductase
ARFC3	replication factor C component
GAA	glucoamylase
GFP	green fluorescent protein (*Aequorea victoria*)
HSA	human serum albumin
lacZ	β-galactosidase (*E. coli*)
MFα1	α mating factor (*S. cerevisiae*)
phbA	β-ketothiolase (*Ralstonia eutropha*)
phbB	acetoacetyl CoA reductase (*Ralstonia eutropha*)
phbC	PHA synthase (*Ralstonia eutropha*)
PHO5	acid phosphatase (*S. cerevisiae*)
TEF1	elongation factor 1α
XylE	catechol 2,3-dioxygenase (*Ps. putida*)
25S rDNA	25S ribosomal RNA

5.1
History of A. adeninivorans Research

Yeasts are simply organized ubiquitous unicellular eukaryotes that are able to adapt rapidly to alterations of environmental conditions. In addition to the traditional ba-

Production of Recombinant Proteins. Novel Microbial and Eucaryotic Expression Systems. Edited by Gerd Gellissen
Copyright © 2005 WILEY-VCH Verlag GmbH & Co. KGaA, Weinheim
ISBN: 3-527-31036-3

ker's yeast *Saccharomyces cerevisiae*, a wide range of nonconventional yeast species exist with attractive characteristics and growth properties. These species can be exploited for biotechnological applications as well as suitable model organisms for plant or animal research. They are either used as donor for genes encoding interesting gene products or employed as excellent hosts for the production of recombinant proteins. The range of yeast species that have been developed as platforms for heterologous gene expression includes *S. cerevisiae* and *Kluyveromyces lactis*; details of *Pichia pastoris*, *Yarrowia lipolytica*, and *Hansenula polymorpha* will be described in the following chapters. As a first yeast example, *Arxula adeninivorans* is described in more detail within this chapter (Wolf 1996; Wolf et al. 2003).

In 1984, Middelhoven et al. (1984) described a yeast species isolated from soil by enrichment culturing, named *Trichosporon adeninovorans*. The particular strain CBS 8244T displayed unusual biochemical activities. It was shown that it was able to assimilate a range of amines, adenine and several other purine compounds as sole energy and carbon source.

A second strain, LS3 (PAR-4) was isolated in Siberia (Kapultsevich, Institute of Genetics and Selection of Industrial Microorganisms, Moscow, Russia) from wood hydrolysates. As in the first case, this strain was found to use a very large spectrum of substances as carbon and nitrogen sources (Gienow et al. 1990).

In 1990, three additional *Tr. adeninovorans* strains were isolated from chopped maize herbage ensiled at 25 or 30 °C in The Netherlands, and yet another four strains were detected in humus-rich soil in South Africa (Van der Walt et al. 1990). A new genus name *Arxula* Van der Walt, M.T. Smith & Yamada (Candidaceae) was proposed for all these strains, which share properties such as nitrate assimilation and xerotolerance. All representatives of this newly proposed genus were ascomycetous, anamorphic, and arthroconidial (Van der Walt et al. 1990) (Figure 5.1).

The genus *Arxula* comprises two species, the type species of the genus *A. terrestre* (Van der Walt and Johanssen) Van der Walt, M.T. Smith & Yamada, nov. comb., and *A. adeninivorans* (Middelhoven, Hoogkamer Te-Niet and Kreger van Rij) Van der Walt, M.T. Smith and Yamada, nov. comb.

With 25S rDNA data available from a number of yeast species, the use of phylogenetic analysis led to a deeper insight into the *Arxula* evolution. After the basis of these sequences, the *Arxula* genus – including *A. adeninivorans* and *A. terrestre* – is an own subgroup which is closely related to a group comprising the genera *Saccharo-*

Fungi (Kingdom)
  Eumycota (Division)
    Ascomycota (Subdivision)
      Saccaromycetes (Class)
        Saccaromycetales (Order)
          Dipodascaceae (Family)
            Mitosporic Dipodascaceae (Sub-family)
              Arxula **(Genus)**
                Arxula adeninivorans **(Species)**

Fig. 5.1 Taxonomy of *Arxula adeninivorans* (after Kreger-van Rij 1984).

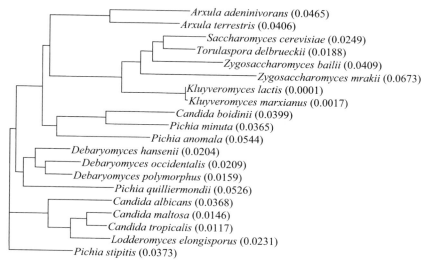

Arxula adeninivorans (0.0465)
Arxula terrestris (0.0406)
Saccharomyces cerevisiae (0.0249)
Torulaspora delbrueckii (0.0188)
Zygosaccharomyces bailii (0.0409)
Zygosaccharomyces mrakii (0.0673)
Kluyveromyces lactis (0.0001)
Kluyveromyces marxianus (0.0017)
Candida boidinii (0.0399)
Pichia minuta (0.0365)
Pichia anomala (0.0544)
Debaryomyces hansenii (0.0204)
Debaryomyces occidentalis (0.0209)
Debaryomyces polymorphus (0.0159)
Pichia quilliermondii (0.0526)
Candida albicans (0.0368)
Candida maltosa (0.0146)
Candida tropicalis (0.0117)
Lodderomyces elongisporus (0.0231)
Pichia stipitis (0.0373)

Fig. 5.2 Phylogenetic tree for 25 rDNA sequences. The tree is built using the Neighbor Joining method of Saitou and Nei (1987), which works on a matrix of distances between all pairs of sequence to be analyzed. These distances are related to the degree of divergence between the sequences (Vector NTI Advance User's Manual, Infor-Max-USA).

myces, *Zygosaccharomyces*, and *Kluyveromyces* and the species *Torulaspora delbrueckii*. In contrast, it seems to be evolutionarily more distant to the genus *Debaryomyces*, the *Candida* and *Pichia* species assessed to date, as well as to *Lodderomyces elongisporus* (Figure 5.2).

5.2
Physiology and Temperature-dependent Dimorphism

A detailed physiological description of the yeast was provided by Gienow et al. (1990) and Middelhoven et al. (1991, 1992). *A. adeninivorans* is able to assimilate nitrate in the same way as *Hansenula polymorpha*. It can utilize a range of compounds as sole energy and carbon sources, including adenine, uric acid, butylamine, pentylamine, putrescine, soluble starch, melibiose, melizitose, propylamine, or hexylamine (Middelhoven et al. 1984). It rapidly assimilates all the sugars, polyalcohols and organic acids used in the conventional carbon compound assimilation tests, except for L-rhamnose, inulin, lactose, lactate and methanol. Likewise, all conventionally used nitrogen compounds are suitable nitrogen sources, with the exception of creatine and creatinine. Several nitrogen compounds, including amino acids and purine derivatives, are metabolized as sole energy, carbon and nitrogen sources. This is also the case for many primary n-alkylamines and terminal diamines, while alcohols, dialcohols, carboxylic acids, dicarboxylic acids and other nitrogen-less analogous compounds, intermediates of the general metabolism are also assimilated. Furthermore, *A. adeninivorans* degrades some phenols and hydroxybenzoates.

A. adeninivorans produces several secretory enzymes including RNase, proteases, glucoamylase, some acid phosphatases, trehalase, some cellobiases, invertase, xylosidase and phytase. These secretory enzymes are listed in Table 5.1, together with a summary of their properties.

Special features of biotechnological impact are the thermotolerance and temperature-dependent dimorphism, which is especially pronounced in the strain LS3. This strain can grow at temperatures of up to 48 °C without previous adaptation to elevated temperatures, and is able to survive for some hours at a temperature of 55 °C (Böttcher et al. 1988; Wartmann et al. 1995). Strain LS3 exhibits a temperature-dependent morphological dimorphism. Temperatures above 42 °C induce a reversible transition from budding cells to mycelial forms. Budding is re-established when cultivation temperature is decreased below 42 °C.

Tab. 5.1 Properties of secretory enzymes of *A. adeninivorans* (Büttner et al. 1997, 1988, 1989 a, 1990 a, b, 1991 a, b, 1992 a, b; Büttner and Bode 1992; Kunze and Kunze 1994 a; Wartmann et al. 1995 a; Sano et al. 1999; Böer et al. 2004 a).

Enzyme	Optimum			Molecular mass (Da)
	Temperature	pH	k_m value	
Glucoamylase (1.4-α-D-glucan glucohydrolase, EC 3.2.1.3)	60–70 °C	4.0–5.0	1.2 g L^{-1} for starch 11.1 mM for maltose	225 000
Acid phosphatase I and II (ortho-phosphoric-monoester phosphohydrolase, EC 3.1.3.2)				
I	50–55 °C	5.2–5.5	3.5 mM for p-nitro-phenylphosphate	320 000
II	50–55 °C	5.2–5.5	5 mM for p-nitro-phenylphosphate	250 000
Trehalase (α,α-trehalose-gluco-hydrolase, EC 3.2.1.28) Cellobiase I and II (β-D-gluco-sidase, EC 3.2.1.21)	45–55 °C	4.5–4.9	0.8–1.0 mM for trehalose	250 000
I	60–63 °C	4.5	4.1 mM for cellobiose	570 000
II	60–63 °C	4.5	3.0 mM for cellobiose	525 000
Invertase (β-D-fructofuranoside fructohydrolase, EC 3.2.1.26)	50–60 °C	4.5	40–60 mM for sucrose 36 mM for raffinose	600 000
β-D-xylosidase (1,4-β-D-xylan xylohydrolase, EC 3.2.1.37)	60 °C	5.0	0.23–0.33 mM for p-nitro-phenyl-β-xylopyranoside	60 000
3-Phytase (*myo*-inositol hexakis phosphate 3-phosphohydrolase EC 3.1.3.8)	75 °C	4.5	0.23 mM for phytata	n.d.

Wartmann et al. (2000) selected mutants with altered dimorphism characteristics. These mutants grow already at 30 °C as mycelia, thus enabling a distinction between temperature-mediated and morphology-related effects on gene expression and protein accumulation. In analogy to other dimorphic yeasts, *A. adeninivorans* budding cells and mycelia differed in their contents of RNA, soluble protein and dry weight. During the middle as well as at the end of the exponential growth phase, mycelia were found to contain less RNA and protein. In contrast, the synthesis of secreted proteins is more pronounced in mycelia, resulting in a twofold higher extracellular accumulation of these proteins, including the enzymes glucoamylase and invertase. This indicates that morphology, rather than temperature, is the decisive factor in the analyzed features (Figure 5.3; Table 5.2).

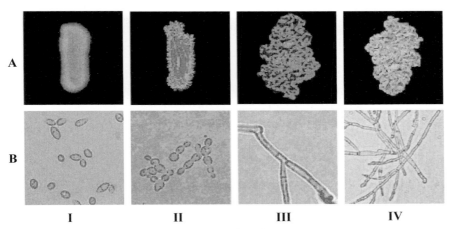

Fig. 5.3 A) Colony form and (B) cell morphology of *A. adeninivorans* LS3 grown at 30 °C (I), 42 °C (II) and 45 °C (III) as well as *A. adeninivorans* 135 grown at 30 °C (IV). The cells were cultured in YEPD medium for 18 h.

Tab. 5.2 DNA and RNA content, dry weight and amount of soluble protein of *A. adeninivorans* LS3 cultured at 30 °C (budding cell) and at 45 °C (mycelium), and of *A. adeninivorans* 135 cultured at 30 °C (mycelium) in yeast minimal medium (Tanaka et al. 1967) with 1% maltose as carbon source (Wartmann et al. 2000). The values are means ± SD from three separate cultures each with three batches in parallel.

	Budding cell LS3 (30 °C)	*Mycelium* LS3 (45 °C)	*Mycelium* 135 (30 °C)
Content (fg)			
DNA	25.3 ± 0.5	23.6 ± 2.4	24.8 ± 0.9
RNA (45 h)	118.0 ± 15.2	56.4 ± 9.1	44.7 ± 8.8
max. RNA	142.5 ± 18.0 (45 h)	73.0 ± 21.0 (36 h)	57.5 ± 9.6 (36 h)
Dry weight (pg, 45 h)	18.2 ± 0.2	22.4 ± 0.8	23.3 ± 0.3
Soluble protein (fg, 45 h)	169.2 ± 16.3	107.5 ± 20.2	76.9 ± 0.3
Max. soluble protein (fg)	234 ± 5.5 (60 h)	150.1 ± 17.9 (60 h)	125.8 ± 11.6 (60 h)

A strong correlation exists between the morphological status and iron uptake, transmitted by two transport systems that differ in iron affinity. In the presence of high Fe(II) concentrations (>2 μM), budding cells accumulate iron concentrations up to sevenfold higher than those observed in mycelia, while at low Fe(II) concentrations (<2 μM), both cell types accumulate similar amounts of iron. The copper-dependent Fe(II) oxidase (Afet3p) and the respective *AFET3* gene – both components of the high-affinity transport system – were analyzed in more detail. In this case, gene expression is strongly dependent on iron concentration but is independent of the morphological stage. However, cell morphology was found to influence the post-translational modifications of Afet3p, an observation of potential impact for heterologous gene expression. O-glycosylation was found in budding cells only, whereas N-glycosylation occurred in both cell types. The characteristic of differential O-glycosylation may provide an option to produce heterologous proteins in both, O-glycosylated and non-O-glycosylated forms, and to compare the impact of its presence on properties such as biological activity or immunological tolerance (Wartmann et al. 2002a).

In addition to temperature, the presence of compounds such as Cd^{2+}, tocopherol, NaCl and tunicamycin – as well as anaerobic cultivation conditions – cause alterations in the cell morphology, whereas dimorphism is not affected by Ca^{2+}, pH-value, carbon source, and substrate limitations (Table 5.3).

A further interesting property of *A. adeninivorans* is its osmotolerance (Figure 5.4). The organism can grow in minimal as well as rich media containing up to 3.32 osmol kg^{-1} H_2O in the presence of ionic (NaCl), osmotic (PEG400) and water stress (ethylene glycol). In strain LS3, the influence of NaCl on the growth characteristics was investigated in more detail. Supplementation with up to 3.4 M (10%) NaCl was of limited influence on growth only, but at NaCl concentrations higher than 3.4 M a decrease in the specific growth rate, a longer adaptation phase, and a lower cell count during the stationary growth phase was observed. Simultaneously, NaCl caused an induction of intracellular glycerol and erythritol accumulation in exponentially grow-

Tab. 5.3 Environmental factors and their influence on the dimorphism of *A. adeninivorans* LS3.

Transition budding cells → mycelia	⇒ temperature ≥ 43 °C
	⇒ 0.1 mM CdSO₄
	⇒ tocopherol
Transition mycelia → budding cells	⇒ temperature < 43 °C
	⇒ NaCl concentration ≥ 10%
	⇒ anaerobic conditions
	⇒ tunicamycin ≥ 8 μg mL^{-1}
Without influence	⇒ pH value
	⇒ cultivation media (YEPD, YMM)
	⇒ carbon source (glucose, fructose, maltose, sucrose, xylose, cellobiose, glycerol)
	⇒ sub- and emers cultivation
	⇒ substrate limitation
	⇒ Ca^{2+}

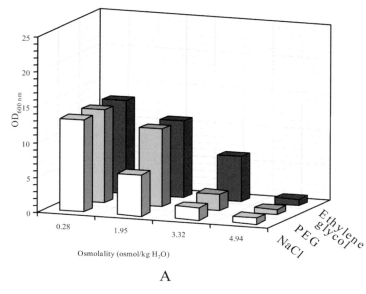

A

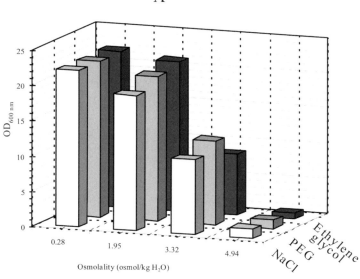

B

Fig. 5.4 Influence of osmolyte concentration on growth of
(A) logarithmical growth phase and (B) stationary growth phase
budding cells. *A. adeninivorans* LS3 cells were cultured in YMM
supplemented with 2 % glucose + NaCl, PEG400 or ethylene glycol
for (A) 40 h and (B) 70 h at 30 °C. The optical density was measured
at 600 nm.

ing cells, and a reduction of intracellular trehalose concentration in stationary cells and a decrease of the protein concentration in the cultivation medium (Yang et al. 2000).

5.3
Genetics and Molecular Biology

The DNA content of *A. adeninivorans* cells is similar to that of haploid cells of *S. cerevisiae* and other ascomycetous yeasts (Gienow et al. 1990; Wartmann et al. 2000; Samsonova et al. 1996). Furthermore, the fact that a relatively high frequency of auxotrophic mutants is obtained after nitrosoguanidine mutagenesis resulting in a broad range of phenotypes (Samsonova et al. 1989, 1996), as well as the quantity of chromosomal DNA and genome size (Gienow et al. 1990), sustain the conclusion for *A. adeninivorans* to be a haploid organism.

The complexity of the nuclear genome of *A. adeninivorans* was analyzed using DNA reassociation studies, and the karyotype was characterized by pulsed-field gel electrophoresis (PFGE) by Gienow et al. (1990) and Kunze and Kunze (1994b). Genome sizes of 16.1×10^9 and 16.9×10^9 Da were calculated from reassociation kinetics of chromosomal DNA from both, *A. adeninivorans* wild-type strains LS3 and 8244T, respectively. These are the highest values reported for yeasts so far, and much higher than the value of 9.2×10^9 Da determined for *S. cerevisiae* under identical conditions. The share of repetitive sequences was determined as 33.1% in LS3 and 35.9% in CBS 8244T – again, values higher than those reported for other yeasts. Karyotype polymorphisms were observed among the wild-type strains tested with four chromosomes ranging between 1.6 and 4.6 Mb in size.

Genetic studies were promoted by the isolation of mutants of this haploid species, and by the development of special techniques. After UV-induced mutagenesis or after treatment with nitrosoguanidine, a large number of auxotrophic mutants and of such mutants with altered catabolite repression that were resistant to 2-deoxy-D-glucose have been selected and characterized (Samsonova et al. 1989, 1996; Büttner et al. 1989b, 1990c) (Figure A5.1).

As no sexual cycle has been detected, mapping techniques based on parasexual mechanisms had to be applied to establish a genetic map of *A. adeninivorans*. After polyethylene glycol-induced fusion of spheroplasts, heterozygous diploids were obtained from auxotrophic mutants of strains LS3 and CBS8244T (Büttner et al. 1990c; Samsonova et al. 1996). Segregation of these diploids was achieved by treatment with benomyl (a drug known to induce haploidization), without affecting other mitotic recombination events (Böttcher and Samsonova 1983). This permitted the linkage analysis of various markers.

In this way, 32 genes could be assigned to four linkage groups, thus meeting the chromosome number of the *A. adeninivorans* genome. This was confirmed by relating the analyzed 32 auxotrophic mutations to particular chromosomes by PFGE and subsequent DNA hybridization with specific probes (Samsonova et al. 1996) (Figure 5.5).

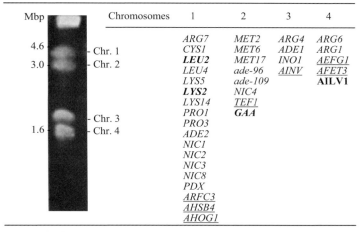

Mbp		Chromosomes	1	2	3	4
			ARG7	*MET2*	*ARG4*	*ARG6*
4.6		Chr. 1	*CYS1*	*MET6*	*ADE1*	*ARG1*
3.0		Chr. 2	**LEU2**	*MET17*	*INO1*	<u>*AEFG1*</u>
			LEU4	*ade-96*	<u>*AINV*</u>	<u>*AFET3*</u>
			LYS5	*ade-109*		**AILV1**
			LYS2	*NIC4*		
			LYS14	<u>*TEF1*</u>		
			PRO1	**GAA**		
		Chr. 3	*PRO3*			
1.6		Chr. 4	*ADE2*			
			NIC1			
			NIC2			
			NIC3			
			NIC8			
			PDX			
			<u>*ARFC3*</u>			
			<u>*AHSB4*</u>			
			<u>*AHOG1*</u>			

Data obtained by linkage group analyses are normal bold, by PFGE are underlined and by linkage group analyses as well as PFGE are extra bold

Fig. 5.5 *A. adeninivorans* chromosomes separated by PFGE and chromosomal gene localization by linkage group analyses and PFGE followed by DNA hybridization (Samsonova et al. 1996; Stoltenburg et al. 1999; Wartmann et al. 2001, 2002 a, 2003 a, b; Böer et al. 2004 a, b).

5.4
Arxula adeninivorans as a Gene Donor

A. adeninivorans genes were isolated by PCR amplification with specific consensus primer sequences. In this way, the genes *ALYS2, TEF1, ARFC3, AEFG1, AFET3, AINV, AHOG1,* and *AHSB4,* as well as the 25S rDNA, have been isolated from gene libraries containing either cDNA or chromosomal DNA from *A. adeninivorans* strain LS3 (Stoltenburg et al. 1999; Wartmann et al. 2001, 2002 a, 2003 b; Kunze and Kunze 1996; Rösel and Kunze 1995; Böer et al. 2004) (Table 5.4).

In contrast to *S. cerevisiae,* where introns are relatively unusual, their presence is more common in the *Arxula* genome. As such, *AHSB4, ARFC1,* and *AHOG1* are found to contain an intron. The comparison of 5′-splice site, branch site and 3′-splice site sequences demonstrate that the resulting consensus sequences are relatively similar to that of *S. cerevisiae* and filamentous fungi (Table 5.5).

Complementation of respective *E. coli* and *S. cerevisiae* mutants was used as an approach for the isolation of additional genes, namely the *ALYS2, AILV1,* and *ALEU2* genes which are suitable selection markers for the *Arxula*-based platform (Wartmann et al. 1998, 2003 a; Kunze and Kunze 1996).

The *GAA* gene encoding the biotechnologically important enzyme glucoamylase was identified from a cDNA library using an anti-glucoamylase antibody as probe for product detection. When heterologously expressed in *S. cerevisiae* and *Kluyveromyces*

Tab. 5.4 Isolated and sequenced genes of the yeast *A. adeninivorans* LS3.

Gene	Gene product	Accession no.	Reference
AILV1	Threonine deaminase	AJ222772	Wartmann et al. (1998)
ALYS2	Amino-adipate reductase	not sequenced	Kunze and Kunze (1996)
ALEU2	β-isopropylmalate dehydrogenase	AJ488496	Wartmann et al. (2003a)
GAA	Glucoamylase	Z46901	Bui et al. (1996a)
TEF1	Elongation factor 1α	Z47379	Rösel and Kunze (1995)
ARFC3	Replication factor C component	AJ007712	Stoltenburg et al. (1999)
AEFG1	Mitochondrial elongation factor G	AJ312230	Wartmann et al. (2001)
AHSB4	Histone H4	AJ535732	Wartmann et al. (2003b)
AFET3	Copper-dependent Fe(II) oxidase	AJ277833	Wartmann et al. (2002a)
AINV	β-fructofuranoside fructohydrolase	AJ580825	Böer et al. (2004a)
AHOG1	Mitogen-activated protein kinase	AJ626723	Böer et al. (2004b)
25S rDNA	25S rRNA	Z50840	Rösel and Kunze (1995)

Tab. 5.5 Consensus sequences at 5'-splice site, branch site and 3'-splice site.

Gene	5'-splice site	Branch site	3'-splice site	Reference
AHSB4	GG/GUAGGU	UACUAAC	AAG/A	Wartmann et al. (2003b)
ARFC1	UG/GUAAGU	AGCUAAC	CAG/G	Stoltenburg et al. (1999)
	UC/GUAAGU	CGCUAAC	UAG/A	Stoltenburg et al. (1999)
AHOG1	AG/GUAGGU	UGCUAAC	UAG/A	Böer et al. (2004b)
Cons. sequences				
A. adeninivorans	DS/GUARGU	HRCUAAC	HAG/R	
S. cerevisiae	NN/GUAUGU	UACUAAC	YAG/G	Tolstrup et al. (1997)
Filamentous fungi		YNCURAY		Turner (1993)

lactis, more than 90% of the synthesized glucoamylase was found to be secreted in both instances. The level of enzyme secretion was 20-fold higher in *Kl. lactis* than that observed in *S. cerevisiae* transformants when using a similar construct for transformation (Bui et al. 1996a, 1996b).

The *AINV* gene provides another example of an interesting enzyme gene. The encoded invertase preferentially hydrolyzes β-D-fructofuranosides, and could be applied to the hydrolysis of sugar cane molasses or sugar beet molasses on an industrial scale. The *AINV* gene was obtained by screening a cDNA and a chromosomal library with a PCR amplificate corresponding to a particular gene segment. The isolated gene was expressed in recombinant *A. adeninivorans* strains fusing the coding sequence to the strong constitutive *TEF1* promoter. The resulting transformants were found to secrete the enzyme in high concentrations independent of the carbon source used for cultivation (see Table 5.4).

5.5
The *A. adeninivorans*-based Platform

5.5.1
Transformation System

The first transformation system based on *A. adeninivorans* has been developed by Kunze et al. (1990) and Kunze and Kunze (1996), using the *LYS2* genes from *A. adeninivorans* and *S. cerevisiae* as selection markers. In these instances transformation vectors inconsistently either integrated into the chromosomal DNA or were of episomal fate displaying an altered restriction pattern.

Therefore, Rösel and Kunze (1998) developed a second transformation system based on a stable integration of heterologous DNA into the ribosomal DNA (rDNA). The generated *A. adeninivorans* plasmid pAL-HPH1 contains for selection the *E. coli*-derived *hph* gene inserted between the constitutive *A. adeninivorans*-derived *TEF*1 promoter and the *PHO5*-terminator from *S. cerevisiae* conferring resistance to hygromycin B. For rDNA targeting, it is equipped with a 25S rDNA fragment from *A. adeninivorans*. Transformation is performed with plasmids linearized at a unique restriction site within the 25S rDNA fragment. The resulting transformants were found to harbor 2–10 plasmid copies stably integrated into the ribosomal DNA (Rösel and Kunze 1998). Transformants could be obtained from both, the wild type strain and mutant strains (Figure 5.6A).

Employment of the undesired dominant marker gene results in a need for toxic compounds or antibiotics during the strain development. This can be avoided using auxotrophic strains and the respective gene sequence for complementation. As such, the *AILV1* and *ALEU2* genes isolated as described before and the respective auxotrophic strains were selected.

A. adeninivorans ailv1 or *aleu2* hosts were transformed with the plasmids pAL-AILV1 harboring the *AILV1* gene and pAL-ALEU2m harboring the *ALEU2* gene for complementation and the 25S rDNA for targeting (Figure 5.6B and C). Transformants generated in this way were found to harbor one to three copies of the heterologous DNA mitotically stable integrated into the 25S rDNA by homologous recombination (Wartmann et al. 1998, 2003a).

5.5.2
Heterologous Gene Expression

The construction of expression plasmids follows a two-step cloning strategy. Initially, the heterologous genes are inserted between the respective *A. adeninivorans*-derived promoter and fungal terminator such as the terminators *PHO5* from *S. cerevisiae* and *trpC* from *Aspergillus nidulans*, which are functional in *A. adeninivorans*. Between both structures multi-cloning sites are present. Subsequently, the resulting expression cassettes (*A. adeninivorans* promoter – heterologous gene – fungal terminator) are integrated into the respective *A. adeninivorans* expression plasmid. For this purpose the cassettes are flanked by unique restriction sites (*Apa*I – *Sal*I, *Apa*I

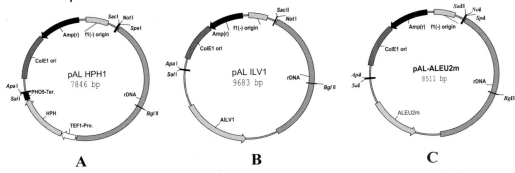

Fig. 5.6 Physical maps of vectors for the *A. adeninivorans*-based expression platform. A) The vector pAL-HPH1 contains the following elements: a 25S rDNA sequence (rDNA) for chromosomal targeting, expression cassette for the *E. coli*-derived *hph* gene in the order *A. adeninivorans*-derived *TEF1* promoter (TEF1-Pro.), the *hph*-coding sequence (HPH), *S. cerevisiae*-derived *PHO5* terminator (PHO5-Ter.). The vector further contains unique *Apa*I and *Sal*I restriction sites for the insertion of the expression cassettes for heterologous genes and a unique *Bgl*II site within the rDNA sequence for linearization. The vectors (B) pAL-AILV1 and (C) pAL-ALEU2m contains the selection marker *AILV1* (AILV1) or *ALEU2m* (ALEU2m) instead of the expression cassette for the *E. coli*-derived *hph* gene.

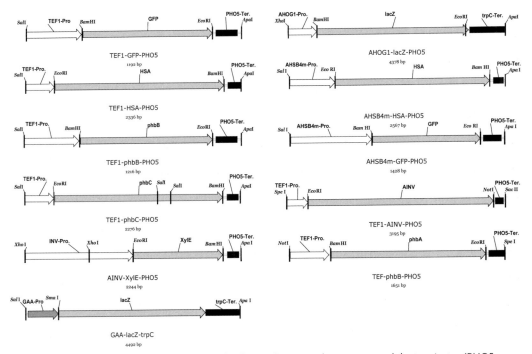

Fig. 5.7 Expression cassettes for the *Arxula*-based expression platform. The cassettes contain the following elements: a constitutive (TEF1-, AHSB4m-Pro.) or inducible promoter (AHOG1-, AINV-, GAA-Pro.), the respective heterologous gene and the terminator (PHO5-, trpC-Ter.). The flanked unique restriction sites allow a directional insertion of the cassettes into the respective basal plasmid.

– *Xho*I, *Spe*I – *Sac*II, *Spe*I – *Not*I) which allow a directional integration (Figures 5.6 and 5.7).

Several heterologous genes were assessed for expressibility in *A. adeninivorans*. As a first example, the *XylE* gene from *Pseudomonas putida* encoding the catechol-2,3-dioxygenase was fused to the *AILV1* promoter for expression control (Kunze et al. 1990; Kunze and Kunze 1996) (Table 5.6).

As further successful examples, *GFP* and *HSA* gene sequences were inserted into the vectors pAL-HPH1 and pAL-ALEU2m and used to transform *A. adeninivorans* wild-type and mutant strains. The resulting expression vectors were of similar design as described before, but now contained the coding sequences inserted between the strong constitutive *TEF1* promoter and the *PHO5* terminator. One to two linearized plasmid copies were integrated specifically into the chromosomal 25S rDNA re-

Tab. 5.6 Examples of heterologous genes expression in *A. adeninivorans* LS3.

Gene	Donor organism	Gene product	Promoter	Vector	Recombinant protein level	Reference
lacZ	*E. coli*	β-galactosidase	*GAA*	pAL-HPH1	350 kU mg^{-1}	Wartmann and Kunze (2000)
lacZ	*E. coli*	β-galactosidase	*AHOG1*	pAL-HPH1	350 U mg^{-1}	Böer et al. (2004 b)
GFP	*Aequorea victoria*	green fluorescent protein	*TEF1*	pAL-HPH1	n.d	Wartmann et al. (2002 b)
GFP	*Aequorea victoria*	green fluorescent protein	*TEF1*	pAL-ALEU2m	n.d	Wartmann et al. (2003 a)
GFP	*Aequorea victoria*	green fluorescent protein	*AHSB4*	pAL-ALEU2m	n.d	Wartmann et al. (2003 b)
HSA	*Homo sapiens*	human serum albumin	*TEF1*	pAL-HPH1	50 mg L^{-1}	Wartmann et al. (2002 b)
HSA	*Homo sapiens*	human serum albumin	*TEF1*	pAL-ALEU2m	50 mg L^{-1}	Wartmann et al. (2003 a)
HSA	*Homo sapiens*	human serum albumin	*AHSB4*	pAL-ALEU2m	50 mg L^{-1}	Wartmann et al. (2003 b)
XylE	*Ps. putida*	catechol 2,3-dioxygenase	*AILV1*	I1-ALYS2	0.4 pkat mg^{-1}	Kunze & Kunze (1996)
XylE	*Ps. putida*	catechol 2,3-dioxygenase	*AINV*	pAL-ALEU2m	4.5 pkat mg^{-1}	Böer et al. (2004 a)
AINV	*A. adeninivorans*	invertase	*TEF1*	pAL-ALEU2m	500 nkat mL^{-1}	Böer et al. (2004 a)
phbA	*R. eutropha*	β-ketothiolase	*TEF1*	pAL-HPH1	2.2 % PHA*	Terentiev et al. (2004 a)
phbB	*R. eutropha*	acetoacetyl CoA reductase	*TEF1*	pAL-HPH1	2.2 % PHA*	Terentiev et al. (2004 a)
phbC	*R. eutropha*	PHA synthase	*TEF1*	pAL-ALEU2m	2.2 % PHA*	Terentiev et al. (2004 a)
phbC	*R. eutropha*	PHA synthase	*TEF1*	pAL-HPH1	2.2 % PHA*	Terentiev et al. (2004 a)

*: per cent final product per dry weight.

gion. In the case of *GFP* expression, the recombinant protein was localized cytoplasmically, rendering the cells fluorescent. In the case of *HSA*, expression is based on an open reading frame (ORF) including the native signal sequence at the 5'-end. Accordingly, over 95% of the recombinant HSA was secreted into the culture medium. In initial fermentation trials of a single copy-transformant on a 200-mL shake flask scale, maximal HSA product levels of 50 mg L^{-1} were observed after 96 h of cultivation. Budding cells as well as mycelia secreted similar levels, demonstrating a morphology-independent productivity (Wartmann et al. 2003a, 2003b) (see Table 5.6).

In addition to the *TEF1* promoter, the strong constitutive *AHSB4* promoter was successfully tested for its suitability for the heterologous gene expression, and this resulted in similar expression levels. In order to facilitate the integration into the expression plasmids, a modified *AHSB4* promoter sequence (*AHSB4m* promoter) was generated by PCR and splicing by overlap extension (Wartmann et al. 2003b) (see Table 5.6).

To improve productivity, simultaneous transformation with two different expression plasmids based on pAL-HPH1 and pAL-ALEU2m was assessed using *HSA* as reporter gene. As a consequence, a twofold increase in productivity was observed in this double transformant (Wartmann et al. 2003b) (see Table 5.6; Figure 5.8).

For the construction of a recombinant biocatalyst, *A. adeninivorans* was equipped with the genes *phbA*, *phbB*, and *phbC* of the polyhydroxyalkanoate (PHA) biosynthetic pathway of *Ralstonia eutropha* encoding β-ketothiolase, NADPH-linked acetoacetyl-CoA reductase and PHA synthase, respectively. *A. adeninivorans* strains initially transformed with the PHA synthase gene (*phbC*) plasmids alone were able to produce PHA. However, the maximal content of the polymer detected in these strains was just 0.003% (w/w) poly-3-hydroxybutyrate (PHB) and 0.112% (w/w) poly-3-hydroxyvalerate (PHV). The expression of all three genes (*phbA*, *phbB*, *phbC*) resulted in small increases in only the PHA content. However, under controlled conditions, and using minimal medium and ethanol as the carbon source for cultivation, the recombinant yeast was able to accumulate up to 2.2% (w/w) PHV and 0.019% (w/w) PHB (Terentiev et al. 2004a) (Table 5.6; Figure 5.9).

In addition to the expression of biotechnologically interesting genes, the *A. adeninivorans* transformation/expression system was used for promoter assessment. For this purpose, the *lacZ* gene from *E. coli* and the *XylE* gene from *Ps. putida* were em-

---

Fig. 5.8 A) Transformation procedure based on simultaneous integration of the plasmids pAL-HPH-AHSB4m-HSA and pAL-ALEU2m-AHSB4m-HSA into the 25S rDNA of *A. adeninivorans* G1211 (*aleu2*). Both plasmids containing the expression cassettes with *AHSB4m*-promoter–*HSA* gene–*PHO5*-terminator are linearized by *Nar*I or *Esp*3I digestion. The resulting fragments flanked by 25S rDNA sequences are integrated together into the 25S rDNA of *Arxula* chromosomal DNA by homologous recombination. Transformants are selected either by resistance to hygromycin B (pAL-HPH-AHSB4m-HSA; Rösel and Kunze 1998) or the complementation of the *aleu2* mutation (plasmid pAL-ALEU2m-AHSB4m-HSA; Wartmann et al. 2003a). B) Gene dosage effect of *AHSB4m* promoter-driven HSA secretion. *A. adeninivorans* G1211/pAL-HPH-AHSB4m-HSA (▲), G1211/pAL-ALEU2m-AHSB4m-HSA (O) and G1211/pAL-HPH-AHSB4m-HSA – pAL-ALEU2m-AHSB4m-HSA (■) were cultured in YMM supplemented with 2% glucose for 24, 48, 72, 96, 120 and 144 h at 30 °C, harvested and measured on the secreted HSA concentration.

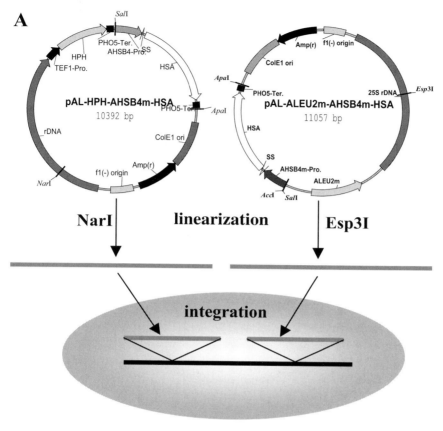

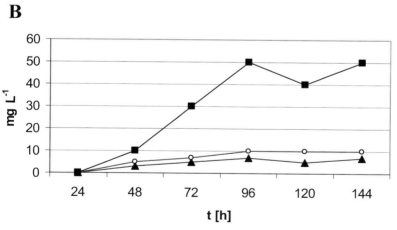

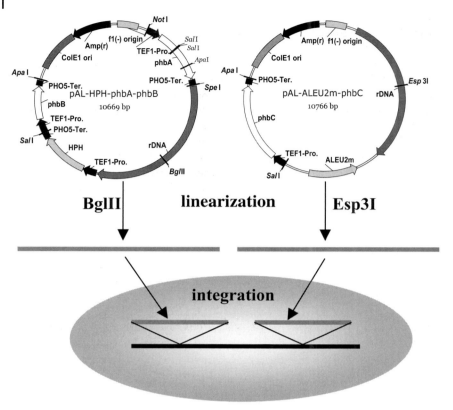

Fig. 5.9 Transformation procedure based on simultaneous integration of the plasmids pAL-HPH-phbA-phbB and pAL-ALEU2m-phbC into the 25S rDNA of *A. adeninivorans* G1211 (*aleu2*). The two plasmids pAL-HPH-phbA-phbB and pAL-ALEU2m-phbC containing the expression cassettes with *phbA*, *phbB* and *phbC* genes are linearized by *Bgl*II or *Esp*3I digestion, respectively. The resulting fragments flanked by 25S rDNA sequences are integrated together into the 25S rDNA of *Arxula* chromosomal DNA by homologous recombination. Transformants are selected either by resistance to hygromycin B (pAL-HPH-phbA-phbB; Rösel and Kunze 1998) or the complementation of the *aleu2* mutation (plasmid pAL-ALEU2m-phbC; Wartmann et al. 2003 a).

ployed as reporter genes. The expression cassettes containing the *GAA, AHOG1* as well as the *AINV* promoter-reporter gene-*PHO5* terminator were analyzed. Depending upon the carbon source, the osmolarity of the medium, and/or the morphological stage, the transformants successfully expressed the *lacZ/XylE* genes and synthesized the encoded enzymes β-galactosidase or catechol 2,3-dioxygenase (Böer et al. 2004 a,b; Wartmann and Kunze 2000; Kunze et al. 1990) (see Table 5.4; Figure A5.2).

5.6
Conclusions and Perspectives

A. adeninivorans is a haploid, dimorphic, nonpathogenic, ascomycetous, anamorphic, arthroconidial yeast of interest for both, basic and applied research. The very broad range of substances which can be used as carbon and/or nitrogen sources, the growth and secretion behavior, the thermo- and osmotolerance, and the temperature-dependent dimorphism make this yeast an attractive organism for biotechnological research and application. *A. adeninivorans* is an interesting host for the synthesis of special products because all essential prerequisites and components for heterologous gene expression are available. In addition, these exceptional properties make *A. adeninivorans* a potential gene donor to equip traditional biotechnologically applied organisms with new attractive capabilities.

Acknowledgments

The authors are grateful to Dr. I. Kunze and Prof. R. Bode for helpful discussion and critical reading of the manuscript. The experimental work was supported by grants from the "Ministry of Science and Research", Magdeburg, Sachsen-Anhalt, Germany (Grant No. 2067A/0025, 2463A/0086G, 2193A/0085G), Ministry of Economic, Nordrhein-Westfalen (TPW-9910v08), Deutsche Bundesstiftung Umwelt (AZ 13048) and by Funds of the Chemical Industry (GK).

Appendix

Tab. A5.1   Selection of *A. adeninivorans* host strains.

Strain	Genotype/Parental strains	Phenotype	Source
Parental strain			
LS3 (SBUG 724, PAR-4)	wild-type		Gienow et al. (1990)
CSIR 1136	wild-type		van der Walt et al. (1990)
CSIR 1138	wild-type		van der Walt et al. (1990)
CSIR 1147	wild-type		van der Walt et al. (1990)
CSIR 1148	wild-type		van der Walt et al. (1990)
CSIR 1149	wild-type		van der Walt et al. (1990)
CBS 8244T (CSIR 577)	wild-type		Middelhoven et al. (1984)
CBS 7370 (CSIR1117)	wild-type		van der Walt et al. (1990)
CBS 7377 (CSIR 1118)	wild-type		van der Walt et al. (1990)
CBS 8335	wild-type		Sano et al. (1999)

Tab. A5.1 (continued)

Strain	Genotype/Parental strains	Phenotype	Source
Mutant strains			
135	*adm1* (dimorphismus mutant)	mycelia-30 °C	Wartmann et al. (2000)
G648	*lys*	Lys⁻	Samsonova et al. (1989)
G698	*nic2-20*	Nic⁻	Samsonova et al. (1996)
G701	*cys1-9ade1-88*	Cys⁻ Ade⁻	Samsonova et al. (1996)
G704	*cys1-9 lys2-41*	Cys⁻ Lys⁻	Samsonova et al. (1996)
G709	*pro1-7 nic4-16*	Pro⁻ Nic⁻	Samsonova et al. (1996)
G835	*nic8 ade1*	Nic⁻ Ade⁻	Büttner et al. (1990c)
G836	*pro1 cys9*	Pro⁻ Cys⁻	Büttner et al. (1990c)
G841	*ino1-11*	Ino⁻	Samsonova et al. (1996)
G846	*ade-96*	Ade⁻	Samsonova et al. (1989)
G847	*arg-39*	Arg⁻	Samsonova et al. (1989)
G848	*lys5-38*	Lys⁻	Samsonova et al. (1996)
G849	*met17-32*	Met⁻	Samsonova et al. (1996)
G853	*pdx1-2*	Pyridoxine⁻	Samsonova et al. (1996)
G858	*arg4-34*	Arg⁻	Samsonova et al. (1996)
G859	*nic1-19*	Nic⁻	Samsonova et al. (1996)
G862	*pdx1-3*	Pyridoxine⁻	Samsonova et al. (1996)
G869	*ade-96 met6-25*	Ade⁻ Met⁻	Samsonova et al. (1996)
G873	*ade-96 leu4-24*	Ade⁻ Leu⁻	Samsonova et al. (1996)
G874	*ade-96 arg7-29 lys14-39*	Ade⁻ Arg⁻ Lys⁻	Samsonova et al. (1996)
G876	*ade-96 met2-26*	Ade⁻ Met⁻	Samsonova et al. (1996)
G877	*ade-96 ino1-15*	Ade⁻ Ino⁻	Samsonova et al. (1996)
G879	*lys5-38 ade2-106*	Lys⁻ Ade⁻	Samsonova et al. (1996)
G882	*lys5-38 nic3-17*	Lys⁻ Nic⁻	Samsonova et al. (1996)
G883	*lys5-38 ade1-107*	Lys⁻ Ade⁻	Samsonova et al. (1996)
G888	*ade9* (CBS 8244T)	Ade⁻	Büttner et al. (1990c)
G896	*arg2 ade8* (CBS 8244T)	Arg⁻ Ade⁻	Büttner et al. (1990c)
G897	*arg2 ade1* (CBS 8244T)	Arg⁻ Ade⁻	Büttner et al. (1990c)
G905	*lys*	Lys⁻	Samsonova et al. (1989)
G982	*lys*	Lys⁻	Samsonova et al. (1989)
G983	*ade lys*	Ade⁻ Lys⁻	Samsonova et al. (1989)
G984	*lys*	Lys⁻	Samsonova et al. (1989)
G985	*lys*	Lys⁻	Samsonova et al. (1989)
G1211	*leu2-29*	Leu⁻	Samsonova et al. (1996)
G1214	*arg1-41*	Arg⁻	Samsonova et al. (1996)
G1228	*met6-30 arg6-36*	Met⁻ Arg⁻	Samsonova et al. (1996)
G1229	*met6-30 ade-109*	Met⁻ Ade⁻	Samsonova et al. (1996)
G1296	*nic8-21 arg6-37*	Nic⁻ Arg⁻	Samsonova et al. (1996)
G1342	*lys5-38 ilv1-2*	Lys⁻ Ile⁻	Samsonova et al. (1996)
LS3-41	*dog*	2-deoxy-D-glucose resistance	Büttner et al. (1989b)
R208-12	*cys1-9 pro1-7*	Cys⁻ Pro⁻	Samsonova et al. (1996)
R208-83	*nic4-16 ade1-88*	Nic⁻ Ade⁻	Samsonova et al. (1996)
R212-2	*cys1-9 pro1-7 nic4-16 ade1-88*	Cys⁻ Pro⁻ Nic⁻ Ade⁻	Samsonova et al. (1996)
R218-1	*ade1-88 ino1-15*	Ade⁻ Ino⁻	Samsonova et al. (1996)
R251-2	*nic4-16 ino1-11*	Nic⁻ Ino⁻	Samsonova et al. (1996)

Tab. A5.1 (continued)

Strain	Genotype/Parental strains	Phenotype	Source
R251-181	*cys1-9 pro1-7 nic4-16 ino1-11*	Cys⁻ Pro⁻ Nic⁻ Ino⁻	Samsonova et al. (1996)
R252-1	*cys1-9 pro1-7 lys5-38 nic4-16 ade1-88*	Cys⁻ Pro⁻ Lys⁻ Nic⁻ Ade⁻	Samsonova et al. (1996)
R255-1	*pro1-7 leu4-24*	Pro⁻ Leu⁻	Samsonova et al. (1996)
R258-7	*cys1-9 pro1-7 lys14-39 arg7-29 ade-96*	Cys⁻ Pro⁻ Lys⁻ Arg⁻ Ade⁻	Samsonova et al. (1996)
R267-5	*cys1-9 lys2-41 ade-96 ino1-15*	Cys⁻ Lys⁻ Ade⁻ Ino⁻	Samsonova et al. (1996)
R268-1	*pdx1-3 ino1-11*	Pyridoxine⁻ Ino⁻	Samsonova et al. (1996)
R270-2	*nic2-20 ade-96 ino1-15*	Nic⁻ Ade⁻ Ino⁻	Samsonova et al. (1996)
R270-3	*cys1-9 lys2-41 nic2-20*	Cys⁻ Lys⁻ Nic⁻	Samsonova et al. (1996)
R279-1	*cys1-9 nic8-21 arg6-36*	Cys⁻ Nic⁻ Arg⁻	Samsonova et al. (1996)
R280-3	*lys5-38 pro3-8 leu4-24*	Lys⁻ Pro⁻ Leu⁻	Samsonova et al. (1996)
R283-3	*pdx1-3 met6-30 ino1-11*	Pyridoxin⁻ Met⁻ Ino⁻	Samsonova et al. (1996)
R319-1	*ade-96 ino1-15 ilv1-2*	Ade⁻ Ino Ile⁻	Samsonova et al. (1996)

Tab. A5.2 Selection of *A. adeninivorans* expression/integration vectors.

Plasmid	Expression cassette	Replication sequence	Selection marker	Integrated copy number	Reference
I1-X6	*AILV1* promoter; no terminator	–	*ALYS2*	1-2	Kunze and Kunze (1996)
pLYS2	–	–	*ALYS2*	1-2	Kunze and Kunze (1996)
pTALa2	–	–	*ALYS2*	1-2	Kunze and Kunze (1996)
pAL-HPH1	–	–	*hph*	2-10	Rösel et al. (1998)
pAL-AILV1	–	–	*AILV1*	1-3	Wartmann et al. (1998)
pAL-ALEU2m	–	–	*ALEU2m*	1-3	Wartmann et al. (2003 a)
pBS-TEF1 PHO5	*TEF1* promoter; *PHO* terminator	–	–	–	Wartmann et al. (2003 a)
pBS-AHSB4m PHO5	*AHSB4m* promoter; *PHO5* terminator	–	–	–	Wartmann et al. (2003 b)

References

Böer E, Wartmann T, Luther B, Manteuffel R, Bode R, Gellissen G, Kunze G (2004a) Characterization of the *AINV* gene and the encoded invertase from the dimorphic yeast *Arxula adeninivorans*. Antonie van Leeuwenhoek, 86: 121–134

Böer E, Wartmann T, Dlubatz K, Gellissen G, Kunze G (2004b) Characterization of the *Arxula adeninivorans AHOG1* gene and the encoded mitogen-activated protein kinase. Curr Genet (submitted)

Böttcher F, Samsonova IA (1983) Systemische Fungizide und antifungale Verbindungen. In: Systematische Fungizide und antifungale Verbindungen (Lyr H, Polter C, Eds). Akademie Verlag, Berlin, pp 255–258

Böttcher F, Klinner U, Köhler M, Samsonova IA, Kapultsevich J, Bliznik X (1988) Verfahren zur Futterhefeproduktion in zuckerhaltigen Medien. DD 278 354 A1

Bui DM, Kunze G, Förster S, Wartmann T, Horstmann C, Manteuffel R, Kunze G (1996a) Cloning and expression of an *Arxula adeninivorans* glucoamylase gene in *Saccharomyces cerevisiae*. Appl Microbiol Biotechnol 44: 610–619

Bui DM, Kunze G, Horstmann C, Schmidt T, Breunig KD, Kunze G (1996b) Expression of the *Arxula adeninivorans* glucoamylase gene in *Kluyveromyces lactis*. Appl Microbiol Biotechnol 45: 102–106

Büttner R, Bode R, Birnbaum D (1987) Purification and characterization of the extracellular glucoamylase from the yeast *Trichosporon adeninovorans*. J Basic Microbiol 27: 299–308

Büttner R, Bode R, Scheidt A, Birnbaum D (1988) Isolation and some properties of two extracellular β-glucosidases from *Trichosporon adeninovorans*. Acta Biotechnol 8: 517–525

Büttner R, Bode R, Birmbaum D (1989a) Purification and characterization of the extracellular glucoamylase from the yeast *Trichosporon adeninovorans*. J Basic Microbiol 30: 227–231

Büttner R, Scheit A, Bode R, Birnbaum D (1989b) Isolation and characterization of mutants of *Trichosporon adeninovorans* resistance to 2-deoxy-D-glucose. J Basic Microbiol 29: 67–72

Büttner R, Schubert U, Bode R, Birnbaum D (1990a) Purification and partial characterization of external and internal invertases

from two strains of *Trichosporon adeninovorans*. Acta Biotechnol 10: 361–370

Büttner R, Bode R, Birnbaum D (1990b) Extracellular enzymes from the yeast *Trichosporon adeninovorans*. Wiss Z Ernst-Moritz-Arndt Univ Greifswald, Math Naturwiss Reihe 39: 20–22

Büttner R, Bode R, Samsonova IA, Birnbaum D (1990c) Mapping of the glucoamylase gene of *Trichosporon adeninivorans* by mitotic haploidization using hybrids from protoplast fusions. J Basic Microbiol 30: 227–231

Büttner R, Bode R, Birnbaum D (1991a) Comparative study of external and internal β-glucosidases and glucoamylase of *Arxula adeninivorans*. J Basic Microbiol 31: 423–428

Büttner R, Bode R, Birnbaum D (1991b) Characterization of extracellular acid phosphatases from the yeast *Arxula adeninivorans*. Zbl Mikrobiol 146: 399–406

Büttner R, Bode R, Birnbaum D (1992a) Purification and characterization of trehalase from the yeast *Arxula adeninivorans*. Zbl Mikrobiol 147: 291–296

Büttner R, Bode R, Birnbaum D (1992b) Alcoholic fermentation of starch by *Arxula adeninivorans*. Zbl Mikrobiol 147: 237–242

Büttner R, Bode R (1992) Purification and characterization of ß-xylosidase activities from the yeast *Arxula adeninivorans*. J Basic Microbiol 32: 159–166

Gienow U, Kunze G, Schauer F, Bode R, Hofemeister J (1990) The yeast genus *Trichosporon* spec. LS3; Molecular characterization of genomic complexity. Zbl Mikrobiol 145: 3–12

Kreger-van Rij NJWN (1984) The yeasts. A taxonomic study, 3rd edition. Amsterdam: Elsevier Science Publisher, Amsterdam

Kunze G, Pich U, Lietz K, Barner A, Büttner R, Bode R, Conrad U, Samsonova IA, Schmidt H (1990) Wirts-Vektor-System und Verfahren zu seiner Herstellung. DD 298 821 A5

Kunze I, Kunze G (1994a) Comparative study of *Arxula adeninivorans* strains concerning morphological characteristics and activities of secretory invertase. J Eur Microbiol 212: 24–28

Kunze G, Kunze I (1994b) Characterization of *Arxula adeninivorans* from different habitats. Antonie van Leeuwenhoek 65: 29–34

Kunze G, Kunze I (1996) *Arxula adeninivorans*.

In: Nonconventional yeasts (Wolf K, Ed). Springer Verlag, Berlin-Heidelberg, pp 389–409

MIDDELHOVEN WJ, HOOGKAMER-TE NIET MC, and KREGER VAN RIJ NJW (1984) *Trichosporon adeninovorans* sp. nov., a yeast species utilizing adenine, xanthine, uric acid, putrescine and primary n-alkylamines as sole source of carbon. Nitrogen and energy. Antonie van Leeuwenhoek 50: 369–378

MIDDELHOVEN WJ, DE JONG IM, and DE WINTER M (1991) *Arxula adeninivorans*, a yeast assimilating many nitrogenous and aromatic compounds. Antonie van Leeuwenhoek 60: 129–137

MIDDELHOVEN WJ, COENEN A, KRAAKMAN B, SOLLEWIJN GELPKE MD (1992) Degradation of some phenols and hydroxybenzoates by the imperfect ascomycetous yeast *Candida parapsilosis* and *Arxula adeninivorans*: evidence for an operative gentisate pathway. Antonie van Leeuwenhoek 62: 181–187

RÖSEL H, KUNZE G (1995) Cloning and characterization of a *TEF1* gene for elongation factor 1α from the yeast *Arxula adeninivorans*. Curr Genet 28: 360–366

RÖSEL H, KUNZE G (1996) Identification of group-I intron within the 25S rDNA from the yeast *Arxula adeninivorans*. Yeast 12: 1201–1208

RÖSEL H, KUNZE G (1998) Integrative transformation of the dimorphic yeast *Arxula adeninivorans* based on hygromycin B resistance. Curr Genet 33: 157–163

SAMSONOVA IA, BÖTTCHER F, WERNER C, BODE R (1989) Auxotrophic mutants of the yeast *Trichosporon adeninovorans*. J Basic Microbiol 29: 675–683

SAMSONOVA IA, KUNZE G, BODE R, BÖTTCHER F (1996) A set of genetic markers for the chromosomes of the imperfect yeast *Arxula adeninivorans*. Yeast 12: 1209–1217

SANO K, FUKUHARA H, NAKAMURA Y (1999) Phytase of the yeast *Arxula adeninivorans*. Biotechnol Lett 21: 33–38

SAITOU N, NEI M (1987) The neighbor-joining method: a new method for reconstructing phylogenetic trees. Mol Biol Evol 4: 406–425

STOLTENBURG R, LÖSCHE O, KLAPPACH G, KUNZE G (1999) Molecular cloning and expression of the *ARFC3* gene, a component of the replication factor C from the salt tolerant, dimorphic yeast *Arxula adeninivorans* LS3. Curr Genet 35: 8–13

TANAKA A, OHNISHI N, FUKUI S (1967) Studies on the formation of vitamins and their function in hydrocarbon fermentation. Production of vitamin B6 by *Candida albicans* in hydrocarbon medium. J Ferment Technol 45: 617–623

TERENTIEV Y, BREUER U, BABEL W, KUNZE G (2004a) Non-conventional yeasts as producers of polyhydroxyalkanoates – Genetic engineering of *Arxula adeninivorans*. Appl Microbiol Biotechnol 64: 376–381

TOLSTRUP N, ROUZE P, BRUNAK S (1997) A branch point consensus from *Arabidopsis* found by non-circular analysis allows for better prediction of acceptor sites. Nucleic Acids Res 25: 3159–3163

TURNER G (1993) In: The Eukaryotic Genome (Borda P, Oliver SG, Sims PFG, Eds) Cambridge University Press, Cambridge, UK, pp 107–125

VAN DER WALT JP, SMITH MT, YAMADA Y (1990) *Arxula* gen. nov. (*Candidaceae*), a new anamorphic yeast genus. Antonie van Leeuwenhoek 57: 59–61

WARTMANN T, KUNZE I, BUI MD, MANTEUFFEL R, KUNZE G (1995a) Comparative biochemical, genetical and immunological studies of glucoamylase producing *Arxula adeninivorans* yeast strains. Microbiol Res 150: 113–120

WARTMANN T, KRÜGER A, ADLER K, BUI MD, KUNZE I, KUNZE G (1995b) Temperature dependent dimorphism of the yeast *Arxula adeninivorans* LS3. Antonie van Leeuwenhoek 68: 215–223

WARTMANN T, RÖSEL H, KUNZE I, BODE R, KUNZE G (1998) *AILV1* gene from the yeast *Arxula adeninivorans* LS3 – a new selective transformation marker. Yeast 14: 1017–1025

WARTMANN T, KUNZE G (2000) Genetic transformation and biotechnological application of the yeast *Arxula adeninivorans*. Appl Microbiol Biotechnol 54: 619–624

WARTMANN T, ERDMANN J, KUNZE I, KUNZE G (2000) Morphology-related effects on gene expression and protein accumulation of the yeast *Arxula adeninivorans* LS3. Arch Microbiol 173: 253–261

WARTMANN T, GELLISSEN G, KUNZE G (2001) Regulation of the *AEFG1* gene, a mitochondrial elongation factor G from the dimorphic yeast *Arxula* adeninivorans. Curr Genet 40: 172–178

WARTMANN T, STEPHAN UW, BUBE I, BÖER E, MELZER M, MANTEUFFEL R, STOLTENBURG R,

GUENGERICH L, GELLISSEN G, KUNZE G (2002 a) Post-translational modifications of the *AFET3* gene product – a component of the iron transport system in budding cells and mycelia of the yeast *Arxula adeninivorans*. Yeast 19: 849–862

WARTMANN T, BÖER E, HUARTO PICO A, SIEBER H, BARTELSEN O, GELLISSEN G, KUNZE G (2002 b) High-level production and secretion of recombinant proteins by the dimorphic yeast *Arxula adeninivorans*. FEMS Yeast Res 2: 363–369

WARTMANN T, STOLTENBURG R, BÖER E, SIEBER H, BARTELSEN O, GELLISSEN G, KUNZE G (2003 a) The *ALEU2* gene – a new component for an *Arxula adeninivorans*-based expression platform. FEMS Yeast Res 3: 223–232

WARTMANN T, BELLEBNA C, BÖER E, BARTELSEN O, GELLISSEN G, KUNZE G (2003 b) The constitutive *AHSB4* promoter – a novel component of the *Arxula adeninivorans*-based expression platform. Appl Microbiol Biotechnol 62: 528–535

WOLF K (1996) Nonconventional yeasts in biotechnology. Springer Verlag, Berlin-Heidelberg-New York

WOLF K, BREUNIG K, BARTH G (2003) Non-conventional yeasts in genetics, biochemistry and biotechnology. Springer Verlag, Berlin-Heidelberg-New York

YANG XX, WARTMANN T, STOLTENBURG R, KUNZE G (2000) Halotolerance of the yeast *Arxula adeninivorans* LS3. Antonie van Leeuwenhoek 77: 303–311

6
Hansenula polymorpha
Hyun Ah Kang and Gerd Gellissen

List of Genes

Gene	Encoded gene product
CAT	catalase
CNE1	calnexin
CTT1	catalase T (*S. cerevisiae*)
CMK2	calmodulin-dependent kinase
CWP1	Cell wall mannoprotein
DAK1	dihydroxyacetone kinase
DAS	dihydroxyacetone synthase
FLD1	formaldehyde dehydrogenase
FMD	formate dehydrogenase
GAM1	glucoamylase (*Schwanniomyces occidentalis*)
GAP	glyceraldehyde-3-phosphate dehydrogenase
GAS1	GPI-anchored surface glycoprotein
LEU2	β-isopropyl malate dehydrogenase
MFα1	α mating factor (*S. cerevisiae*)
MOX	methanol oxidase
PHO1	acid phosphatase
PMA1	plasma membrane ATPase
PMR1	P-type Ca^{2+} transport ATPase
SED1	Cell surface glycoprotein (Suppressor of Erd2 Deletion)
TIP1	Cell wall mannoprotein (Temperature shock-inducible protein)
TPS1	trehalose-6-phosphate synthase
TRP3	indole-3-glycerol-phosphate synthase
URA3 (ODC1)	ornithidine decarboxylase
YNT1	nitrate transporter
YNI1	nitrite reductase
YNR1	nitrate reductase

Production of Recombinant Proteins. Novel Microbial and Eucaryotic Expression Systems. Edited by Gerd Gellissen
Copyright © 2005 WILEY-VCH Verlag GmbH & Co. KGaA, Weinheim
ISBN: 3-527-31036-3

6.1
History, Phylogenetic Position, Basic Genetics and Biochemistry of *H. polymorpha*

A limited number of yeast species are able to utilize methanol as a sole energy and carbon source. The range of methylotrophic yeasts includes *Candida boidinii*, *Pichia methanolica*, *Pichia pastoris* (see Chapter 7) and *Hansenula polymorpha* (Gellissen 2000). The latter two are distinguished by a growing track record of application in heterologous gene expression. In particular, *H. polymorpha* has found successful application in the industrial production of heterologous proteins, as detailed in later sections of this chapter (Gellissen 2002; Guengerich et al. 2004). Since *H. polymorpha* is the more thermotolerant of the two yeasts, it might also be better suited as source and for the production of proteins considered for crystallographic studies. In basic research it is used as model organism for research in peroxisomal function and biogenesis, as well as nitrate assimilation (Gellissen and Veenhuis 2001; van der Klei and Veenhuis 2002; Siverio 2002; Gellissen 2002). Again, the presence of a nitrate assimilation pathway is a feature not shared by *P. pastoris*.

The first methylotrophic yeast described was *Kloeckera* sp. No 2201, later re-identified as *Candida boidinii* (Ogata et al. 1969). Subsequently, other species, including *H. polymorpha*, were identified as having methanol-assimilating capabilities (Hazeu et al. 1972). Three basic strains of this species with unclear relationships, different features, and independent origins are used in basic research and biotechnological applications: strain CBS4732 (CCY38–22-2; ATCC34438, NRRL-Y-5445) was initially isolated by Morais and Maia (1959) from soil irrigated with waste water from a distillery in Pernambuco, Brazil. Strain DL-1 (NRRL-Y-7560; ATCC26012) was isolated from soil by Levine and Cooney (1973). The strain named NCYC495 (CBS1976; ATAA14754, NRLL-Y-1798) is identical to a strain first isolated by Wickerham (1951) from spoiled concentrated orange juice in Florida and initially designated *Hansenula angusta*. Strains CBS4732 and NCYY495 can be mated, whereas strain DL-1 cannot be mated with the other two (K. Lahtchev, personal communication).

The genus *Hansenula* H. et P. Sydow includes ascosporogenic yeast species exhibiting spherical, spheroidal, ellipsoidal, oblong, cylindrical, or elongated cells. One to four ascospores are formed. Ascigenic cells are diploid arising from conjugation of haploid cells. The genus is predominantly heterothallic. *H. polymorpha* is probably homothallic, exhibiting an easy interconversion between the haploid and diploid states (Teunisson et al. 1960; Middelhoven 2002). Since the morphological characteristics of the *Hansenula* species are shared by species of the genus *Pichia* Hansen, Kurtzman (1984), after performing DNA/DNA reassociation studies, proposed to merge both genera and transfer *Hansenula* species with hat-shaped ascospores to *Pichia* Hansen emend Kurtzman, although *Hansenula* spp. can grow on nitrate and *Pichia* spp. cannot. Kurtzman and Robnett (1998) provided a phylogenetic tree in which nitrate-positive and nitrate-negative *Pichia* are clustered, demonstrating the unreliability of nitrate assimilation for prediction of kinship. The leading taxonomy monographs follow this proposal, re-naming *H. polymorpha* as *Pichia angusta* (Kurtzman and Fell 1998; Barnett et al. 2000). However, the merging of the genera is still criticized by some taxonomists, and there are arguments for maintain-

Fungi (Kingdom)
      Eumycota (Division)
            Ascomycotina (Subdivision)
                  (Hemoascomycetes (Class)
                        Endomycetales (Order)
                              Saccharomycetaceae (Family)
                                    (Saccharomycetoideae (Sub-family)
                                    Hansenula (Genus)

                                        Hansenula polymorpha (Species)

Fig. 6.1   Taxonomy of *Hansenula polymorpha* (after Kreger-van Rij 1984).

ing the established and popular name *Hansenula polymorpha* (Middelhoven 2002; Sudbery 2003).

Among a wealth of biochemical and physiological characteristics, some selected features are presented in the following sections; for a more comprehensive view, the reader is referred to the chapters of a recent monograph (Gellissen 2002). Some strains of *H. polymorpha* can tolerate temperatures of 49 °C and higher (Teunisson et al. 1960; Reinders et al. 1999). It was shown that trehalose synthesis is not required for growth at elevated temperatures, but that it is necessary for normal acquisition of thermotolerance (Reinders et al. 1999). The thermoprotective compound trehalose accumulates in large amounts in cells grown at high temperatures. The synthetic steps for trehalose synthesis have been detailed for *H. polymorpha*. The *TPS1* gene encoding trehalose 6-phosphate synthase is the key enzyme gene of this pathway (Romano 1998; Reinders et al. 1999). Transcripts of this gene were found to be present in high quantities in cells grown at normal temperatures, and to be especially abundant when grown at elevated temperatures (Reinders et al. 1999). Accordingly, the *TPS1*-derived promoter provides an attractive element to drive constitutive heterologous gene expression which can be further boosted at temperatures above 42 °C (Amuel et al. 2000; Suckow and Gellissen 2002; see also the following sections).

The capability of *H. polymorpha* to grow on methanol as a sole energy and carbon source is enabled by a methanol utilization pathway that is shared by all known methylotrophic yeasts (Tani 1984; Yurimoto et al. 2002; see also Chapter 7 on *P. pastoris*). Growth on methanol is accompanied by a massive proliferation of peroxisomes in which the initial enzymatic steps of this pathway take place (Figure 6.2) (Gellissen and Veenhuis 2001; van der Klei and Veenhuis 2002; Yurimoto et al. 2002).

The enzymatic steps of the compartmentalized methanol metabolism pathway are detailed in Figure 6.3. For more comprehensive information, the reader is referred to Yurimoto et al. (2002).

During growth on methanol, key enzymes of the methanol metabolism are present in high amounts. An especially high abundance can be observed for methanol

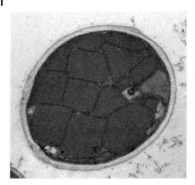

Fig. 6.2 *Hansenula polymorpha* cell. The cells grown in a methanol-limited chemostat are crowded with peroxisomes (Courtesy of M. Veenhuis).

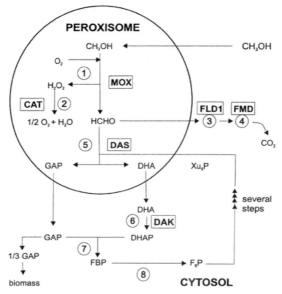

Fig. 6.3 The methanol utilization pathway and its compartmentalization in methylotrophic yeasts. (Modified after Gellissen 2000; Yurimoto et al. 2002.) *1* Methanol is oxidized by alcohol oxidase to generate formaldehyde and hydrogen peroxide. *2* The toxic hydrogen peroxide is decomposed by catalase to water and oxygen. *3, 4* Within a dissimilatory pathway the formaldehyde is oxidized by two subsequent dehydrogenase reactions to carbon dioxide, catalyzed by a formaldehyde dehydrogenase (FLD) and a formate dehydrogenase (FMD or FDH). *5* For assimilation, the formaldehyde reacts with xylulose 5-phosphate (Xu_5P) by the action of dihydroxyacetone synthase (DHAS) to generate the C_3 compounds glyceraldehyde 3-phosphate (GAP) and dihydroxyacetone (DHA). *6* DHA is phosphorylated by dihydroxyacetone kinase (DHAK) to dihydroxyacetone phosphate (DHAP). *7* GAP and DHAP yield in an aldolase reaction fructose 1,6-biphosphate (FBP). *8* In further steps of the pentose phosphate cycle, fructose 5-phosphate and xylulose 5-phosphate are finally generated. Identified and characterized genes of the *H. polymorpha* methanol utilization pathway are boxed and shown in the pathway position of the encoded enzymes. The genes are *MOX* (Ledeboer et al. 1985), *DAS* (Janowicz et al. 1985), *CAT* (Didion and Roggenkamp 1992), *DAK* (Tikhomirova et al. 1988), *FLD1* (Baerends et al. 2002), and *FMD* (Hollenberg and Janowicz 1988).

oxidase (MOX), formiate dehydrogenase (FMD), and dihydroxyacetone synthase (DHAS) (Gellissen et al. 1992a). The presence of these enzymes is regulated at the transcriptional level of the respective genes. Gene expression is subject to a carbon source-dependent repression/derepression/induction mechanism conferred by inherent properties of their promoters. Promoters are repressed by glucose, derepressed by glycerol, and induced by methanol. Again, these promoter elements – and in particular the elements derived from the *MOX* and the *FMD* genes – constitute attractive components for the control of heterologous gene expression that can be regulated by carbon source addition (see forthcoming sections). The possibility of eliciting high promoter activity with glycerol as sole carbon source and even with limited addition of glucose (glucose starvation) is unique among the methylotrophic yeasts. In the related species *C. boidinii*, *P. methanolica*, and *P. pastoris*, the active status of the promoter is strictly dependent on the presence of methanol or methanol derivatives (Gellissen 2000). However, this does not seem to be an inherent promoter characteristic; rather, it rather depends on the cellular environment of the specific host as upon transfer into *H. polymorpha* the *P. pastoris*-derived *AOX1* promoter is active under glycerol conditions (Raschke et al 1996; Rodriguez et al 1996).

6.2
Characteristics of the *H. polymorpha* Genome

As pointed out before, there are several *H. polymorpha* strains available. In the following section we focus on strains CBS4732 (CCY38–22-2; ATCC34438; NRRL-Y-5445) and DL-1 (NRRL-Y-7560; ATCC26012) which are the two ancestor strains of the preferred *H. polymorpha* hosts employed in heterologous gene expression. Data on karyotyping are available for both strains (Figure 6.4; Table 6.1).

Pulsed-field gel electrophoresis of *H. polymorpha* chromosomes revealed that both strains have six chromosomes, ranging from 0.9 to 1.9 Mbp, but the electrophoretic patterns of their chromosomes were quite different. The scientific and industrial significance of strain CBS4732 is now met by the recent characterization of its entire

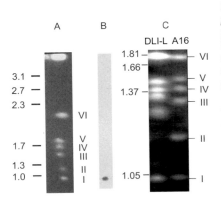

Fig. 6.4 Electrophoretic karyotype of *H. polymorpha* strains CBS4732 and DL-1. A) Chromosome pattern of *H. polymorpha* strain CBS4732 separated by pulsed-field gel electrophoresis (Schwarz and Cantor 1984) using the Pulsaphor apparatus (Pharmacia). Details of the separation conditions are provided elsewhere (Waschk et al. 2002). B) Chromo Blot. The separated chromosomes were transferred to a nylon membrane and hybridized to a labeled *URA3* probe. A signal was obtained exclusively with chromosome I. (Modified and supplemented according to Waschk et al. 2002.) C) Chromosome pattern of strain DL-1 (MJ Ok, HA Kang, unpublished results). Details of the separation conditions will be published elsewhere.

Tab. 6.1 Chromosomal localization of several cloned genes in strains CBS4732.

Cloned gene/sequence used as specific probe	Function	Reference	Chromosome no.
URA 3	Orotidine-5'-phosphate decarboxylase	Merckelbach et al. (1993)	I
CPY	Carboxypeptidase Y	Unpublished [1]	I
GAP	Glycerinaldehyde 3-phosphate dehydrogenase	Unpublished [2]	I
rDNA (5.8S, 18S, 26S)	Ribosomal DNA	Klabunde [3]	II
HARS1	Autonomously replicating sequence 1	Ledeboer et al. (1986)	III
TPS1	Trehalose 6-phosphate synthase	Reinders et al. (1999)	IV
HLEU2	β-Isopropylmalate dehydrogenase	Agaphonov et al. (1994)	IV
MOX	Methanol oxidase	Ledeboer et al. (1985)	V
FMD	Formiate dehydrogenase	Hollenberg and Janowicz (1989)	VI

1) Bae JH, Kim HY, Sohn JH, Choi ES, Rhee, SK. Accession number U67174.
2) Sohn JH, Choi ES, Rhee SK. Accession number U95625.
3) J Klabunde, personal communication.

genome (Ramezani-Rad et al. 2003). A few gene sequences are elucidated and can be compared for both strains (Table 6.2).

The sequence identity of the open reading frame (ORF) for the selected genes ranges between 94.5 and 97.2%, with an average value of 96.6%. The sequence differences are observed to be much more apparent at the 5'- and 3'-untranslated regions, which might be involved in controlling gene expression. This implies that

Tab. 6.2 Comparison of selected gene sequences from *H. polymorpha* strains CBS4732 (RB11) and DL-1.

Gene name	Amino acid identity (%)	Nucleic acid identity (%)	Accession No. in GenBank*	Reference
CST13	96.7	95.8	AF454544	Kim et al. (2002)
CPY	98.0	95.9	U67174	KRIBB
GSH2	96.0	94.5	AF435121	KRIBB
MNN9	96.3	95.5	AF264786	Kim et al. (2001)
PMI40	97.9	94.9	AF454544	Kim et al. (2002)
PMR1	98.5	95.2	U92083	Kang et al. (1998)
YPT1	99.5	97.2	AF454544	Kim et al. (2002)
Average	97.6	96.6		

* The sequences of genes isolated from the DL-1 strain were obtained from GenBank and compared with those from the RB11 strain (Ramezani-Rad et al. 2003).

two strains are closely related, but have distinct genetic and physiological characteristics.

Several groups worldwide initiated studies on the CBS4732 genome several years ago. Included in the comparative genome analysis on 13 hemiascomycetous yeasts, part of the *H. polymorpha* genome sequence was established using a partial random sequencing strategy with a coverage of 0.3 genome equivalents. Using this approach, about 3 Mbp of sequencing raw data of the *H. polymorpha* genome was yielded (Feldmann et al. 2000). The recently terminated genome analysis aimed at a higher coverage using a BAC-to-BAC approach and resulting in the comprehensive genome analysis of this organism (Ramezani-Rad et al. 2003). For sequencing of *H. polymorpha* strain RB11, an *odc1*-derivative of strain CBS4732, a BAC library with approximately $17 \times$ coverage was constructed in a pBACe3.6 vector according to Osoegawa et al. (1998, 1999). Details on base calling, assembly and editing are provided by Ramezani-Rad et al. (2003). Sequencing resulted in the characterization of 8.733 million base pairs assembled into 48 contigs. The derived sequence covers over 90% of the estimated total genome content of 9.5 Mbp located on six chromosomes which range in size from 0.9 to 2.2 Mbp (see Figure 6.4A). From the sequenced 8.73 Mb, a total of 5848 ORFs have been extracted for proteins longer than 80 amino acids (aa), and 389 ORFs smaller than 100 aa were identified. Likewise, 4771 ORFs have homologues to known proteins (81.6%). Calculation of the gene density and protein length, taking into account the gene numbers, showed an average gene density of one gene per 1.5 kb, and an average protein length of 440 amino acids. Ninety-one introns have been identified by homology to known proteins and confirmed by using GeneWise (Birney et al. 2000). Eighty tRNAs were identified, corresponding to all 20 amino acids. From approximately 50 rRNA clusters (Klabunde et al. 2002; Waschk et al. 2002; Klabunde et al. 2003), seven clusters have been fully sequenced. All clusters are completely identical and have a precise length of 5033 bp.

The main functional categories and their distribution in the gene set are manually predicted for energy, 4%; cellular communication, signal transduction mechanism, 3%; protein synthesis, 6%; cell rescue, defense and virulence, 4%; cellular transport and transport mechanisms, 9%; cell cycle and DNA processing, 9%; protein fate (folding, modification, destination), 17%; transcription, 13% and metabolism, 19%. A selection of the data obtained from the annotated sequence is provided in Tables 6.3 and 6.4.

Tab. 6.3 *Hansenula polymorpha* genome statistics.

Contigs:	48
Total length of contigs:	8 733 442 bp
Average contig length:	182 kb
No. of extracted ORFs:	5848
No. of ORFs <100 aa:	389
Average gene density:	1 gene/1.5 kb
Average gene size (start-stop):	1.3 kb (1320 nt)
Average protein length:	440 aa

Tab. 6.4 Functional categorization of genes.

Functional category	No. of ORFs	%
Metabolism	1114	19
Energy	231	4
Cell growth, cell division and DNA synthesis	518	9
Transcription	767	13
Protein synthesis	323	6
Protein destination	1014	17
Transport facilitation	423	7
Cellular transport and transport mechanisms	518	9
Control of cellular organization	417	7
Cellular communication/signal transduction	170	3
Cell rescue, defense, and virulence	260	4
Cell fate	282	5
Regulation of/interaction with cellular environment	184	3

6.3
N-linked glycosylation in *H. polymorpha*

The similarities between yeast and animal cell secretion pathways have made yeasts in general preferred microbial host systems for the production of human secretory proteins. A majority of human proteins with therapeutic potential are glycoproteins, and increasing evidence shows that oligosaccharides assembled on glycoproteins have profound effects on the properties of glycoproteins, such as antigenicity, specific activity, solubility, and stability. The initial processing of *N*-linked glycans on glycoproteins, which occurs in the endoplasmic reticulum (ER), is well conserved among eukaryotes and results in the core oligosaccharide, $Man_8GlcNAc_2$. However, further maturation of oligosaccharides in the Golgi apparatus is quite variable among organisms, even among yeast species (Gemmill and Trimble 1999). Yeasts elongate the core oligosaccharide mostly by addition of mannose, leading to the formation of core-sized structures ($Man_{<15}GlcNAc_2$) as well as hypermannose structures ($Man_{50-200}GlcNAc_2$) with extended poly-α-1,6 outer mannose chains, which are decorated with various carbohydrate side chains in a species-specific manner. In *Saccharomyces cerevisiae*, the linear backbone of the outer chain is often composed of 50 or more mannoses, highly branched by addition of α-1,2- linked mannoses and terminally capped with α-1,3-linked mannoses, generating heavily hypermannosylated glycoproteins. The outer chain also contains several mannosylphosphate residues, conferring the oligosaccharide with a negative net charge (Jigami and Odani 1999; Kim et al. 2004).

Compared to *S. cerevisiae*, the mannose outer chains of *N*-linked oligosaccharides generally appear to be much shorter in the methylotrophic yeasts *P. pastoris* and *H. polymorpha*, although extensive hyperglycosylation has also been reported in a few cases of recombinant glycoproteins produced in these yeasts (Scorer et al. 1993; Müller et al. 1998). Analyses of *N*-linked oligosaccharides added to native and recombinant glycoproteins from *P. pastoris* have indicated that the major oligosaccharide species in *P. pastoris* are $Man_{8-14}GlcNAc_2$ with short α-1,6 extensions. More significantly, *P. pastoris* oligosaccharides are reported to have no hyperimmunogenic terminal α-1,3 glycan linkages (Montesino et al. 1998; Bretthauer and Castellano 1999). Phosphomannose has been detected in both elongated and core oligosaccharides on some recombinant proteins of *P. pastoris* (Miele et al. 1997; Montesino et al. 1999). At present, very little information exists on the structural characteristics of N-linked oligosaccharides of *H. polymorpha*-derived glycoproteins. A comparative study on the glycosylation pattern of recombinant human $α_1$-antitrypsin produced in *S. cerevisiae*, *H. polymorpha*, and *P. pastoris* has suggested that the length of outer mannose chains attached to the recombinant protein in *H. polymorpha* was much shorter than in *S. cerevisiae*, but slightly longer than in *P. pastoris* (Kang et al. 1998b). A recent study on the structure of the oligosaccharides derived from the recombinant *Aspergillus niger* glucose oxidase (GOD) and the cell wall mannoproteins derived from *H. polymorpha* has revealed that most oligosaccharide species attached to the recombinant GOD have core-sized structures ($Man_{8-12}GlcNAc_2$) without terminal α-1,3-linked mannose residues (Kim et al. 2004). Therefore, the outer chain processing in the N-linked glycosylation pathway in *H. polymorpha* appears to be similar to that in *P. pastoris*, with the addition of short outer chains to the core and no terminal α-1,3-linked mannose addition (Figure 6.5).

In contrast to the yeast oligosaccharides composed solely of mannose, a variety of sugars including galactose, N-acetylgalactosamine and sialic acid, are added to oligosaccharides in mammals. The differences in glycan processing between yeasts and humans are a major limitation for yeasts to be used in the production of recombinant glycoproteins for therapeutic use. Glycoproteins derived from yeast expression systems contain nonhuman N-glycans of the high-mannose type, which are immunogenic in humans. Attempts have been made genetically to modify glycosylation processes in *S. cerevisiae* (Chiba et al. 1998) and *P. pastoris* (Callewaert et al. 2001), in order to trim the yeast N-glycans of the high-mannose type to the human glycans of the ($Man_5GlcNAc_2$) intermediate type. A more advanced achievement has been recently made genetically to re-engineer the glycosylation pathway of *P. pastoris* to produce the complex human N-glycan N-acetylglucosamine$_2$-mannose$_3$-N-acetylglucosamine$_2$ ($GlcNAc_2Man_3GlcNAc_2$) (Hamilton et al. 2003; see Chapter 7). Potentially, the development of *H. polymorpha* expression systems for proper glycosylation can be achieved as further understanding is gained of *H. polymorpha*-specific carbohydrate structure and processing sugar transferases. To date, only a few *H. polymorpha* genes and mutants related to protein glycosylation have been reported (Kang et al. 1998a; Agaphonov et al. 2001; Kim et al. 2001, 2002). Information from the *H. polymorpha* genome sequence (Blandin et al. 2000; Ramezani-Rad et al. 2003) will expedite the identification of genes that are predicted to be involved in the biosynthesis of sugar chains. The functional characterization of these genes should facilitate delineation of the *H. polymorpha*-specific N-glycosylation pathway, and this would provide valuable information for the development of glyco-engineering strategies in *H. polymorpha* to achieve the optimal glycosylation of recombinant proteins.

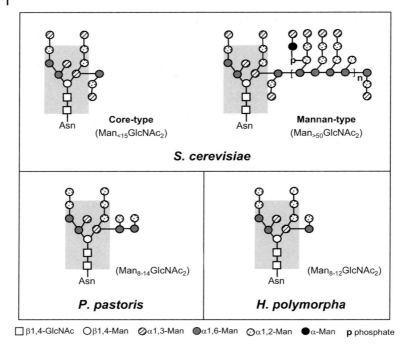

Fig. 6.5 The representative *N*-linked oligosaccharide structures assembled on the *S. cerevisiae*, *P. pastoris*, or *H. polymorpha*-derived glycoproteins. Information on the *S. cerevisiae* and *P. pastoris* oligo-saccharides is from Gemmill and Trimble (1999).

6.4
The *H. polymorpha*-based Expression Platform

6.4.1
Transformation

Recombinant *H. polymorpha* strain generation requires special tools. Application of the commonly used *S. cerevisiae* 2 µm sequence with its predominantly episomal fate is restricted to a limited set of host strains (Gellissen and Hollenberg 1997). A variety of gene replacement approaches have been described, but these require DNA fragments with termini comprising much longer target gene overlaps than those used in *S. cerevisiae* (Gonzalez et al. 1999; van Dijk et al. 2001). Typically, these overlaps must be in the magnitude of hundreds to thousands of base pairs. In addition, since the background caused by nonhomologous recombination is high, screening of more transformants than in *S. cerevisiae*-based systems is necessary to identify a strain with the desired integration/replacement (Kang et al. 2002). Plasmids harboring one of a set of several cloned sub-telomeric *ARS* sequences derived from the DL-1 strain have been described, which homologously integrate into a genomic counterpart, resulting in the recombinant strains harboring single or multiple tandemly repeated copies at the respective sub-telomeric genomic locus (Sohn et al. 1999a; Kim et al. 2003). A set of vectors has been described to target the heterologous DNA to the rDNA locus of *H. polymorpha* (Klabunde et al. 2002, 2003; Suckow and Gellissen 2002; see also Chapter 13). Most commonly, plasmids harboring *HARS1* (*Hansenula ARS1*) as a replication signal are used to generate recombinant *H. polymorpha* strains. The fate of these plasmids in *H. polymorpha* has been extensively analyzed for both RB11 (derived from CBS4732; see the following section) and DL-1 strains (Suckow and Gellissen 2002; Kang et al. 2002). The strain generation by use of *HARS1* plasmids will therefore be described in more detail.

After transformation, cells are plated on selective media according to the selection marker gene present on the *HARS1* plasmid. Macroscopic colonies typically appear after 4–5 days of incubation at 37 °C. In this early phase, all cells of a colony harbor the plasmid as an episome at a low copy number; plasmid loss can be induced by cultivation on a nonselective medium. Colonies are then transferred to liquid selective medium and incubated under vigorous shaking for 24–48 h at 37 °C. This procedure is called the "passaging step". An aliquot of the dense culture is then used to inoculate fresh selective medium, and the incubation is repeated. After three to eight subsequently applied passaging steps, cells grow with the initially episomal plasmid integrated into the genome. In a particular single strain developed for the production of a hepatitis B vaccine, a recombination within the *FMD* locus was observed (U Dahlems, unpublished results). In order to separate these cells from those still harboring the plasmid as an episome, one or two "stabilization steps" must be performed in nonselective liquid media. If aliquots of these cultures are plated onto selective media, the resulting single colonies will exclusively represent strains harboring one to multiple tandemly repeated copies of the *HARS1* plasmid. The various individual strains can vary significantly in the expression rates of the foreign gene, present on the plasmid. However, typically only a few strains display very high levels of target protein. For a high probability of obtaining a "high producer" in the first approach, parallel generation of up to 150 strains is recommended. Once a suitable strain is identified, a so-called "supertransformation" can be performed using a second *HARS1* plasmid with a different selection marker gene. This second plasmid may contain either the same heterologous expression cassette as the first plasmid, or a different one. In the first case, strains would result which might display higher production rates of the target protein than the basic strain; in the second case, strains would be obtained co-expressing two different heterologous genes at variable but individually fixed relative expression rates (Gellissen 2000).

To summarize this procedure, the generation of recombinant strains in *H. polymorpha* is clearly more laborious than in other yeasts. However, these additional difficulties are balanced by two positive features which are highly favorable in biotechnological applications. First, heterologous gene expression in *H. polymorpha* can be controlled by homologous promoters of extraordinary strength. While the carbon source-regulated *MOX* and *FMD* promoters are derived from genes of the methanol degradation pathway, the *TPS1* promoter, derived from the trehalose 6-phosphate synthase gene of *H. polymorpha*, is constitutive with regard to different carbon sources and can be influenced by different temperatures (Amuel et al. 2000). In combination with the obtained high copy numbers of the integrated *HARS1* plasmid (up to 100 copies per haploid genome may result from supertransformation), these strong promoters can provide very high expression rates of the heterologous gene in selected strains; indeed, for secreted phytase, product levels of up to 13.5 g L^{-1} have been obtained (Mayer et al. 1999). The second favorable feature of recombinant *H. polymorpha* is the unambiguous mitotic stability of the individual strains with regard to the copy numbers of the *HARS1* plasmids integrated, even upon cultivation on nonselective media for a long period of time. This stability has been well documented in several examples over periods of 800 generations. For plasmid examples, see Section 6.4.3 and Figures 6.7 and 6.8; for some product and process examples, see Section 6.5.

6.4.2
Strains

Starting with the three *H. polymorpha* parental strains, NCYC495, CBS4732, and DL-1 (see Section 6.1), some forty to fifty other strains have been derived. The DL-1 strain is not employed in classical genetic analyses, and no data are available regarding its ability to mate and to sporulate (Lahtchev 2002). The DL-1 strain has certain advantages in that it has a higher growth rate and adapts more quickly to culture media than the other parental strains; additionally, DL-1 strains have a higher frequency of homologous recombination than other strains (Kang et al. 2002; Lahtchev 2002). The inability of the DL-1 strain to copulate makes this strain inconvenient for classical genetic manipulation exploiting meiotic segregation. However, the relatively high frequency of homologous recombination in the DL-1 strain enables the application of several molecular genetic techniques developed in *S. cerevisiae* to be used in the organism. Several host strains suitable for heterologous protein expression, including auxotrophic mutants, protease-deficient strains, and mox-negative strains, have been constructed in the DL-1 strain, mostly using gene disruption techniques (see the list of strains in Appendix A6.1). A pop-out cassette using *HpURA3* as a selection marker has been constructed to recover the auxotrophic marker for the subsequent gene disruption or for subsequent transformation with expression vectors (Kang et al. 2002). Combined with fusion PCR and in-vivo recombination, the use of the *HpURA3* pop-out cassette becomes more simplified in constructing null mutant strains with disruption of the gene of interest (see Figure 6.6). This approach eliminates the time-consuming steps of ligation and sub-cloning, which are otherwise required for the construction of a gene deletion cassette.

Most classical genetic techniques have been performed using NCYC495, which shows both mating and sporulation (Lahtchev 2002). Unlike the other two parental strains, NCYC495 does not grow well on methanol-containing media and therefore does not have the strong methanol-induced promoters available to the other strains for gene expression. Instead, NCYC495 has other interesting applications, including its employment for the study of nitrate assimilation mentioned previously (Siverio 2002). Cells from CBS4732 grow well on methanol, and show strong mating and

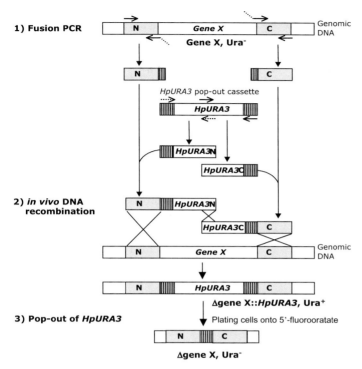

1) **Fusion PCR**

2) *in vivo* **DNA recombination**

3) **Pop-out of *HpURA3***

Fig. 6.6 Strategy of gene disruption using fusion PCR and in-vivo recombination in *H. polymorpha*. Step 1) Fusion-PCR. The N- and C-terminal fragments of "Gene X" are amplified and fused with the N- and C-fragments of the *HpURA3* pop-out cassette containing the overlapped internal sequence of *HpURA3* (100 bp) by PCR. Step 2) In-vivo DNA recombination. The two fusion-PCR products obtained are introduced into *H. polymorpha* cells and converted into one linear gene disruption cassette via in-vivo homologous recombination at the overlapped sequence. The double homologous crossover of the disruption cassette at the "Gene X" locus results in the disruption of "Gene X". Step 3) The *HpURA3* pop-out cassette can be removed to recover the auxotrophic marker for subsequent gene disruption or for subsequent transformation. A detailed procedure will be described elsewhere (MW Kim, HA Kang, unpublished results).

sporulation (Lahtchev 2002). Both CBS4732 (strain RB11 and its derivatives in particular) and DL-1 are employed in the production of recombinant products (Kang et al. 2001a; Gellissen 2000, 2002; Müller et al. 2002; Park et al. 2004; see also the forthcoming section of this chapter). In contrast to DL-1 strains, some sub-strains of CBS4732 are not easily applied in recombinatory methods, perhaps due to their high mitotic stability (Suckow and Gellissen 2002). For a selection of *H. polymorpha* strains and for protocols specific to parental strains, see Degelmann (2002) and the list in Appendix A6.1.

6.4.3
Plasmids and Available Elements

Expression and integration vectors in *H. polymorpha* are composed of prokaryotic and yeast DNA (Gellissen and Hollenberg 1997). Vectors are either supplied as circular plasmid or linearized and targeted to a specific genomic locus. Possible targets for homologous integration include the *MOX/TRP* locus (Agaphonov et al. 1995), an ARS sequence (Agaphonov et al. 1999; Sohn et al. 1996), the *URA3* gene (Brito et al. 1999), the *LEU2* gene (Agaphonov et al. 1999), the *GAP* promoter region (Heo et al. 2003), or the rDNA cluster (Klabunde et al. 2002, 2003). Clearly, the circular plasmids are not randomly integrated, but recombine with genomic sequences represented on the vector. This was recently shown with a particular vector harboring a *FMD* promoter/HBsAg fusion where recombination within the *FMD* gene was observed (U Dahlems, personal communication). It remains to be shown whether homologous recombination also takes place with vectors equipped with *MOX, TPS1* and other promoter elements.

Standard expression vectors and elements available for insertion into *H. polymorpha* is illustrated in Figures 6.7 and 6.8. (For detailed information on the various

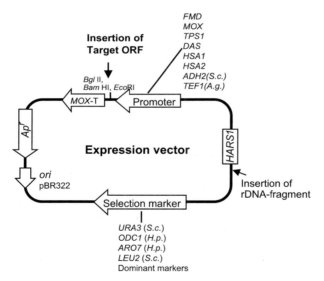

Fig. 6.7 General design of an expression vector for *H. polymorpha* RB11. A standard *H. polymorpha* expression vector contains the *ori* (origin of replication), a strong promoter, and a terminator (connected by a multiple cloning site for insertion of the foreign gene to be expressed), a selection marker from *H. polymorpha* or another yeast, and/or a selection marker for antibiotic resistance, a replication sequence (HARS), or alternatively a sequence for targeted integration into the genome sequence. For a list of elements available for insertion into a plasmid, see Table A6.2.

A. AMIp series

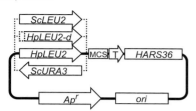

B. pGLG series

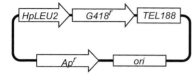

C. pHACT-HyL series

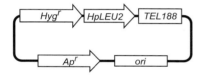

Fig. 6.8 Vectors designed for copy number-controlled gene integration in *H. polymorpha* DL-1 using auxotrophic selection markers (A) or antibiotic resistance markers (B and C). Refer to Kang et al. (2002) for more detailed information on these multiple integration vectors.

H. polymorpha expression platforms, see Suckow and Gellissen 2002; Kang et al. 2002; and Guengerich et al. 2004.) Plasmids that have been successfully developed for industrial use of RB11-based strains include pFPMT121 (for production of phytase) and a derivative of strain pMPT121 (for production of the anti-coagulant hirudin) (Suckow and Gellissen 2002) (Figure 6.7).

Plasmids AMpL1, AMIpLD1, and AMIpSU1 have been used as multiple integration systems based on the complementation of auxotrophic mutations, to elicit desired plasmid copy numbers in DL-1-derived recombinant strains (Figure 6.8A). When an appropriate mutant strain is transformed with one of these plasmids under selective conditions, transformants with plasmid integrated in low (1–2), moderate (6–9), or high (up to 100) copy numbers can be rapidly selected (Agaphonov et al. 1999). Another rapid and copy number-controlled selection system has been developed using antibiotic resistance markers (Figure 6.8B and C). The G418 and hygromycin B resistance cassettes were used as dominant selection markers, which allow selection of transformants on plates containing different concentrations of G418 (Sohn et al. 1999 b) or hygromycin B (Kang et al. 2001 b). Due to the strong correlation between antibiotic resistance and integration copy number, the selection of transformants with different copy numbers ranging from 1 to 50 can be easily achieved. For a selection of *H. polymorpha* expression/integration vectors, refer to Degelmann (2002) and to the list provided in Appendix A6.1.

Plasmids harboring *CEN/ARS* (automously replicating sequences) elements typically support autonomous replication of plasmids within host cells. In *H. polymorpha*, centromeres have not yet been isolated. The established plasmids contain *HARS* elements, but not a *CEN* region. These plasmids may integrate into the host DNA over a number of generations, resulting in mitotically stable strains with as many as 100 plasmids in tandem repeats (Gellissen 2000; Kang et al. 2002). Notably, a high number of integrated copies is not always a pre-requisite for high-level protein production, especially in the case of secretory production. In one case, four copies of a *HARS* vector were sufficient to obtain efficient expression of *Schwanniomyces occidentalis* glucoamylase in *H. polymorpha* (Gellissen et al. 1992b). Other good examples are the secretory expression of human urinary-type plasminogen activator (u-PA) and human serum albumin (HSA) in *H. polymorpha*. In these cases, a single- or two-copy integration of the expression vector resulted in the maximum levels of recombinant u-PA or HSA secreted into culture supernatants (Kang et al. 2002).

Signal sequences may be fused to the target ORF for direct release of synthesized proteins into the media, or into a preselected cell compartment, such as the peroxisome, the vacuole, the ER, the mitochondrion, or the cell surface. Available signal sequences include the peroxisomal targeting signals PTS1 and PTS2 (van Dijk et al. 2000), the repressible acid phosphatase (*PHO1*) secretion leader sequence (Phongdara et al. 1998), a *S. occidentalis*-derived *GAM1* (van Dijk et al. 2000; Weydemann et al. 1995), and the *S. cerevisiae*-derived *MFα1* (Gellissen 2000). Glycosylphosphatidylinositol (GPI)-anchoring motifs derived from the GPI-anchored cell surface proteins, such as *HpSED1*, *HpGAS1*, *HpTIP1*, and *HpCWP1*, have been recently exploited to develop a cell surface display system in *H. polymorpha*. When the recombinant glucose oxidase (GOD) was expressed as a fusion protein to these anchoring motifs, most enzyme activity was detected at the cell surface (Kim et al. 2002).

One of the main advantages of heterologous gene expression in *H. polymorpha* is that this yeast has unusually strong promoters, the most widely employed of which are derived from genes of the methanol utilization pathway (see Figure 6.3). These promoters include elements derived from the methanol oxidase (*MOX*), formate dehydrogenase (*FMD*), and dihydroxyacetone synthase (*DHAS*) gene (Gellissen et al. 2000; van Djik et al. 2000). Other available (but less frequently applied) regulative promoters are derived from inducible genes encoding enzymes involved in nitrate assimilation (e. g., *YNT1*, *YNI1*, *YNR1*, which can be induced by nitrate and repressed by ammonium) (Avila et al. 1998), or the enzyme acid phosphatase (the *PHO* promoter) (Phongdara et al. 1998; Baerends et al. 2002). Examples of constitutive promoters are *ACT* (Kang et al. 2001b), *GAP* (Heo et al. 2003), *PMA1* (Cox et al. 2000), and *TPS1* (Amuel et al. 2000). The *PMA1* promoter even competes with the outstanding *MOX* promoter in terms of high expression levels; *PMA1* is of interest in the co-expression of genes on industrial scale (Cox et al. 2000). The performance of the *TPS1* promoter is not linked to the use of a particular carbon source. In contrast to the constitutive promoters listed above, it can be applied at elevated temperatures, where its activity may be boosted even further (Amuel et al. 2000). The *H. polymorpha FLD1* gene encoding formaldehyde dehydrogenase has been characterized recently (Baerends et al. 2002). FLD1p is essential for the catabolism of methanol, and shows 82% sequence identity with the Fld1p protein from *P. pastoris* and 76% identity with Fld1p from *C. boidinii*. The *FLD1* promoter promises to be advantageous in that expression can be controlled at two levels: it is strongly induced under methylotrophic growth conditions, but shows moderate activity using primary amines as a nitrogen source. With these promising characteristics, the *FLD1* promoter is expected to augment the existing *H. polymorpha* promoters (Baerends et al. 2002). The *GAP* promoter also showed a higher specific production rate and required a much simpler fermentation process than the *MOX* promoter-based HSA production system, implying that the *GAP* promoter can be a practical alternative of the *MOX* promoter in the large-scale production of some recombinant proteins (Heo et al. 2003).

6.5
Product and Process Examples

We now provide a short summary of *H. polymorpha*-based processes. Here, a few industrially relevant examples are only briefly summarized; for a more detailed description of fermentation and purification procedures, the reader is referred to Jenzelewski (2002) and to Chapter 14 of this book.

Once stable recombinant integrants have been generated, production strain candidates are identified from a background of nonproducers or strains of low or impaired productivity. The subsequent design of a fermentation procedure greatly depends on the characteristics of the host cell, the intended routing of the recombinant gene product and, most importantly, on the promoter elements used. The commonly used culture media are based on simple synthetic components. They contain trace metal ions and adequate nitrogen sources, which are required for efficient gene expression and cell yield, but no proteins. The total fermentation time varies between 60 and 150 hours. Due to the inherent versatile characteristics of the two methanol-inducible promoters, fermentation modes vary for the most part in the supplemented carbon source: glycerol, methanol, glucose, and combinations thereof may be selected. The ability to achieve high yields of a recombinant product expressed from a methanol pathway promoter without the addition of methanol is a unique feature of the *H. polymorpha* system (see Section 6.1; Gellissen 2000; Suckow and Gellissen 2002). In contrast, activation of these promoters in the related yeast *P. pastoris* is strictly dependent on the presence of methanol (Cregg 1999).

In processes for secretory heterologous proteins, a "one-carbon source" mode is usually employed, supplementing the culture medium only with glycerol. A hirudin production process may serve as an example for this fermentation mode. In this process, a strain was employed that harbors 40 copies of an expression cassette for an MFα1 pre-pro-sequence/hirudin fusion gene under control of the *MOX* promoter (Weydemann et al. 1995; Avgerinos et al. 2001; Bartelsen et al. 2002). Hirudin production was promoted by reducing the initial glycerol concentration and maintaining it on a suitable level by a pO_2-controlled addition of the carbon source. The fermentation is started with 3% (w/v) glycerol. After consumption of the carbon source (after 25 h), the pO_2-controlled feeding mode is initiated and this results in a glycerol concentration of between 0.05 and 0.3% (w/v) (derepression of the *MOX* promoter). The fermentation run is terminated after 36 h of derepression (total fermentation time 72 h). The broth is then harvested and the secreted product purified from the supernatant by a sequence of ultrafiltration, ion exchange, and gel filtration steps.

In the case of HBsAg production, a "two carbon source" fermentation mode was used (see Chapter 15). The producer strain harbors high copy numbers of an expression cassette, with the coding sequence for the small surface antigen (S-antigen) under control of methanol pathway promoters. The selected strain is fermented on a 50-L scale. The product-containing cells are generated via a two-fermentor cascade, consisting of a 5-L seed inoculating the 50-L main fermentor. The initial steps of fermentation closely follow those described for the production of hirudin. Initially, cultivation is performed with a glycerol feeding in a fed-batch mode, to be followed by subsequent semi-continuous glycerol feeding controlled by the dissolved oxygen level in the culture broth. This derepression phase is then followed by batchwise feeding with methanol in the final fermentation mode. The product concentration increases to amounts in the multigram range. The product consists of a lipoprotein particle in which the recombinant HBsAg is inserted into host-derived membranes. As noted in Section 6.1, the addition of methanol also serves for the proliferation of organelles and consequently for the synthesis and proliferation of membranes. Methanol is thus needed in this case to provide a high yield and balanced co-production of both components of the particle (Schaefer et al. 2001, 2002; see also Chapter 15). For downstream processing, the harvested cells are disrupted and the particles purified in a multi-step procedure that includes adsorption of a debris-free extract to a matrix and the subsequent application of a sequence of ion exchange, ultra-filtration, gel filtration, and ultra-centrifugation steps as detailed by Schaefer et al. (2001, 2002), and in Chapter 15 for recombinant hepatitis B vaccines.

For the production of phytase, *H. polymorpha* has been used in a particularly efficient process (Mayer et al. 1999; Papendieck et al. 2002), a pre-requisite for an economically competitive production of a technical enzyme. In this development all steps and components of strain generation, fermentation, and purification are dictated by a rationale of efficiency and cost-effectiveness. This also applies to the definition of fermentation process using glucose as the main carbon source.

A strain was generated in which the phytase sequence is under control of the *FMD* promoter. Subsequent supertransformation yielded strains with up to 120 copies of the heterologous DNA, thus enabling a gene dosage-dependent high productivity. A fermentation procedure was then developed to achieve high levels of enzyme production. Significantly, it was found that the use of glycerol as the main carbon source was not required in this case, but could be substituted by low-cost glucose. The active status of the *FMD*-promoter was maintained by glucose starvation (fermentation with minimal levels of continuously fed glucose). At a 2000-L scale, fermentation with glucose as the sole carbon source led to high product yields and an 80% reduction in raw material costs compared to glycerol-based fermentations (Mayer et al. 1999; Papendieck et al. 2002). Strains were found to produce the recom-

Tab. 6.5 *H. polymorpha*-based products (selection).

Category	Product	Status	Brand name	Reference
Pharmaceuticals	HBsAg (*adr*)	Launched	HepaVax Gene	Schaefer et al. (2002)
	HbsAg (*adw*)	Launched	AgB	Schaefer et al. (2002)
	Insulin	Launched	Wosulin	
	IFNα-2a	Process transfer		Müller II et al. (2001)
	HSA	Pilot scale completed		Heo et al. (2003)
	EGF	Lab scale completed		Heo et al. (2002)
Food additive	hexose oxidase	Launched	Grindamyl-Surebake	Cook and Thygesen (2003)
Feed additive	phytase	Registration		Mayer et al. (1999)
Enzymes	Levansucrase	Lab scale completed		Park et al. (2004)

binant phytase at levels ranging up to 13.5 g L^{-1} (Mayer et al. 1999). The secreted product is purified through a series of steps, including flocculation centrifugation, dead-end filtration, and a final ultra-filtration yielding a high-quality, highly concentrated product at a recovery rate of up to 92 %.

A short outline of product recovery and downstream processing was provided for the previous examples of two secreted and one intracellular product. An individual procedure must be defined for each process developed. Especially in the case of secreted compounds, the fermentation and primary product recovery are intimately linked, and this interface of upstream and downstream processing is often the objective of successful integrated bioprocess development (Curvers et al. 2001; Thömmes et al. 2001). A typical example of this is the production of aprotinin variants in *H. polymorpha* (Zurek et al. 1996).

A selection of *H. polymorpha*-derived products is listed in Table 6.5.

6.6
Future Directions and Conclusion

6.6.1
Limitations of the *H. polymorpha*-based Expression Platform

Despite the most favorable characteristics of the *H. polymorpha*-based platform for application in heterologous gene expression, problems and limitations may be encountered in particular strain and product developments, as is similarly (and more frequently) seen in other yeast systems. These limitations include overglycosylation (Agaphonov et al. 2001), retention within the ER (Agaphonov et al. 2002), poor secre-

Tab. 6.6 Improvement of the expression performance of *H. polymorpha* production strains by the co-expression or deletion of other genes.

Gene expressed	Problem encountered	Co-expressed (+) or deleted (–) gene
IFNα-2a	Incorrect pre-pro-cleavage	*KEX2* (+)
Enzyme, IFNγ	Overglycosylation impaired secretion	*CMK2*; *CNE1* (+)
EGF	C-terminal truncation	*KEX1* (–)

tion, impaired processing (Müller II et al. 2001; Gellissen et al. 2002), and proteolytic degradation (Suckow and Gellissen 2002). A possible strategy to overcome these limitations is to identify genes and gene products that may, upon disruption or co-expression, positively influence the performance of respective strains. This has been applied successfully in several cases. For example, the deletion of the *KEX1* gene, coding for carboxypeptidase α, led to a significant improvement in the quality of recombinant human epidermal growth factor (hEGF) secreted from *H. polymorpha* by decreasing the generation of C-terminally truncated hEGF form (Heo et al. 2003). Among others, co-expression of the *S. cerevisiae*-derived *KEX2* gene provided a greatly improved processing of a IFNα-2a pre-pro-sequence in *H. polymorpha* in which production of predominantly N-terminally extended molecules had been observed previously (Müller II et al. 2001; Gellissen et al. 2002). In other examples, the co-expression of the *S. cerevisiae*-derived *CMK2* or the *H. polymorpha CNE1* (calnexin) gene led to an improved secretion and a reduction in overglycosylation of a secreted enzyme and a cytokine (Table 6.6).

6.6.2
Impact of Functional Genomics on Development of the *H. polymorpha* RB11-based Expression Platform

Several approaches have been initiated to identify *H. polymorpha* genes that may be manipulated to effect a positive influence on the performance of particular production strains. Examples include identification of the *PMR1* gene (Kang et al 1998) and glycosylation genes (Guengerich et al. 2004; Kim et al. 2002). With the completion of genome sequencing, transcriptome, proteome analysis and other related technologies are all now feasible and enable more systematic approaches to be introduced.

In a first approach, genes will be identified that are involved in methanol metabolism, peroxisome homeostasis, protein glycosylation, secretion, and cell wall integrity. These tasks are executed within a cooperative effort with partners in Russia, Ukraine, The Netherlands and Germany, and funded by INTAS (INTAS 2001–0583). For the identification of these genes, linear DNA fragments harboring reporter genes are used for random integration, thereby generating mutants. By using this random integration (RALF) approach (van Dijk et al. 2001), certain genes of potential impact for relevant gene expression functions may be disrupted and identified by sequencing the region adjoining the integration site and comparing the deduced sequence with the genome data. By applying a selection of suitable reporter proteins and a range of certain growth conditions, the generated strains can be screened for genes which might have an impact on the functions mentioned above.

For transcriptome analysis, *H. polymorpha* cDNA microarrays are being generated in cooperation with KRIBB (Korea Research Institute of Bioscience and Biotechnology, Daejeon, Korea), funded by the Korean Ministry of Science and Technology (Microbial Genomics and Applications R& D Program). As an initial trial, a partial genome cDNA chip spotted with 382 ORFs of *H. polymorpha* was constructed. Each ORF was PCR-amplified using gene-specific primer sets, of which the forward primers have a 5′-aminolink. The PCR products were printed in duplicate onto the aldehyde-coated slide glasses to link only the coding strands to the surface of the slide via covalent coupling between amine and aldehyde groups. The partial DNA chip was used to analyze differential expression profiling of *H. polymorpha* cells cultivated under defined environmental conditions (Oh et al. 2004). At present, the whole-genome cDNA microarrays of *H. polymorpha* are available, and these have been constructed using the same strategy as applied to fabrication of the partial cDNA chip. It is expected that these whole-genome DNA microarrays will provide a powerful tool for a genome-wide transcriptional profiling of *H. polymorpha* under defined genetic and physiological conditions. This will generate invaluable information for pathway engineering and process optimization in exploiting *H. polymorpha* as a cell factory.

A third project which is to be started soon is a comprehensive analysis of the proteome of recombinant *H. polymorpha* production strains in correlation to specific products, secretion efficiency, and other characteristics. The extraction of defined proteins from two-dimensional SDS gels and mass spectrometric analysis of proteolytic fragments will lead to the identification of proteins and their respective genes, with a potential impact on the performance of the *H. polymorpha* expression platform. The availability of the complete genome and various 'omic's approaches surely facilitates extensive exploration of interesting genes and strong promoters, and will thus further the development of expression systems to supplement the strong platform that already exists for *H. polymorpha*.

Yeasts have only come under intense molecular study during the past few decades, and yeast-derived recombinant products have been already developed, ranging from technical enzymes and anticoagulants (hirudin and saratin) (Avgerinos et al. 2001; Sohn et al. 2001; Barnes et al. 2001; Bartelsen et al. 2002) to vaccines such as hepatitis B (Schaefer et al. 2001; Schaefer et al. 2002). *S. cerevisiae* is well characterized, being the first organism in which recombinant vaccines were developed, and the first eukaryotic organism to have its genome entirely sequenced. However, *H. polymorpha* has several advantages over *S. cerevisiae*, including strong and tightly regulated promoters, the lack of hyperallergenic structures on target proteins, capabilities of dense growth on simple media, stability of expression plasmids, and high frequency of nonhomologous recombination. Moreover, *H. polymorpha* has the endogenous capacity for prolyl 4-hydroxylation, which is essential for the function and folding of certain recombinant proteins, such as gelatin (de Bruin et al. 2002). This post-translational modification is generally known to be absent in microbial eukaryotic systems. Consequently, *H. polymorpha* holds a secure place in representing the methylotrophic yeasts as an alternative system for heterologous gene expression.

Appendix

Tab. A6.1 Selection of *H. polymorpha* host strains.

Strain	Genotype	Phenotype	Source
Parental strain			
DL-1	wild-type (NRRL-Y-7560, ATCC26012)		Levine and Cooney (1973)
Auxotrophic strains			
DL-1-L	*leu2*	Leu⁻	Sohn et al. (1996)
uDL10	*leu2 ura3*	Leu⁻Ura⁻	KRIBB
DL-LdU	*leu2 Δura3::lacZ*	Leu⁻Ura⁻	KRIBB
DL1Δ-A	*leu2 Δade2*	Leu⁻Ade⁻	CRC
DL1Δ-L	*Δade2 Δleu2::ADE2*	Leu⁻	CRC
DL1Δ-U	*leu2 Δade2 Δura3::ADE2*	Leu⁻Ura⁻	CRC
Protease-deficient strains			
uDLB11	*leu2 ura3 Δpep4::lacZ*	Leu⁻Ura⁻ Pep4⁻	KRIBB
uDLB12	*leu2 ura3 Δprc1::lacZ*	Leu⁻Ura⁻ Prc1⁻	KRIBB
uDLB13	*leu2 ura3 Δkex1::lacZ*	Leu⁻Ura⁻ Kex1⁻	KRIBB
uDLB14	*leu2 ura3 Δpep4::lacZ Δprc1::lacZ*	Leu⁻Ura⁻Pep4⁻Prc1⁻	KRIBB
uDLB15	*leu2 ura3 Δpep4::lacZ Δkex1::lacZ*	Leu⁻Ura⁻Pep4⁻ Kex1⁻	KRIBB
uDLB16	*leu2 ura3 Δprc1::lacZ Δkex1::lacZ*	Leu⁻Ura⁻ Prc⁻ Kex1⁻	KRIBB
uDLB17	*leu2 ura3 Δpep4::lacZ Δprc1::lacZ Δkex1::lacZ*	Leu⁻Ura⁻Pep4⁻Prc1⁻ Kex1⁻	KRIBB
Methanol utilization-deficient strains			
DLT2	*leu2 Δmox-trp3::ScLEU2*	Mox⁻Trp⁻	CRC
DL1-LΔM	*leu2 Δmox*	Leu⁻Mox⁻	CRC
Parental strain			
NCYC495	wild-type (CBS1976, ATAA14754, NRRL-Y-1798, VKM-Y-1397)		Wickerham (1951)
L1	*leu1–1**	Leu⁻	Gleeson et al. (1986)
A11	*ade11–1*	Ade⁻	Parpinello et al. (1998)
M6	*met6–1*	Met⁻	Parpinello et al. (1998)
NCYC495	*leu1–1**	Leu⁻	Brito et al. (1999)
Nitrate assimilation-related strains			
NAG1995	*Δynr1::URA3, leu1–1**	Ynr1⁻ Leu⁻	Avila et al. (1995)
NAG1996	*Δyni1::URA3, leu1–1**	Yni1⁻ Leu⁻	Brito et al. (1996)
NAG997	*Δynt1::URA3, leu1–1**	Ynt1⁻ Leu⁻	Pérez et al. (1997)
NAG998	*Δyna1::URA3, leu1–1**	Yna1⁻ Leu⁻	Avila et al. (1998)
NAG2001	*yna2::URA3, leu1–1**	Yna2⁻	Avila and Siverio (unpublished)
Parental strain			
CBS4732	wild-type (CCY38–22-2, ATCC34438, NRRL-Y-5445)		Morais and Maia (1959)
LR9	*ura3–1 (odc1)*	Ura⁻	Roggenkamp et al. (1986)
RB11	*ura3–1*	Ura⁻	Weydemann et al. (1995)
RB12	*ura3 leu1–1**	Ura⁻Leu⁻	Rhein Biotech, unpublished
RB13	*ura3 leu1–1* mox*	Ura⁻Leu⁻Mox⁻	Rhein Biotech, unpublished

Tab. A6.1 (continued)

Strain	Genotype	Phenotype	Source
RB14	ura3 mox	Ura⁻Mox⁻	Rhein Biotech, unpublished
RB15	leu1–1* mox	Leu⁻Mox⁻	Rhein Biotech, unpublished
RB17	haro7	Tyr⁻	Krappmann et al. (2000)
RC296	ade	Ade⁻	Rhein Biotech, unpublished
A16	leu2	Leu⁻	
	trp3 mox	Trp⁻Mox⁻	Veale et al. (1992)
1B	ade2–88 leu2–2	Ade⁻Leu⁻	Bogdanova et al. (1998)
1-HP065	ade2–88 ura2–1 met 4–220	Ade⁻Leu⁻Met⁻	Mannazzu et al. (1997)
14C	leu2–2 cat1–14	Leu⁻Cat⁻	Lahtchev (2002)
5C-HP156	ade2–88	Ade⁻	Lahtchev (2002)
8V	leu2	Leu⁻	Agaphanov et al. (1995)

* *leu1–1* and *leu2* correspond to the same gene.

Tab. A6.2 Selection of *H. polymorpha* expression/integration vectors.

Plasmid	Expression cassette	Replication sequence	Selection marker	Integrated copy number	References/comments
DL-1 based plasmids					
AMIpL1	No promoter; Terminator from an unknown gene	HARS36	HpLEU2	1–2	Agaphonov et al. (1999) Multiple cloning sites for insertion of expression cassettes
AMIpLD1	No promoter; Terminator from an unknown gene	HARS36	HpLEU2-d	1–100	Agaphonov et al. (1999) Multiple cloning sites for insertion of expression cassettes
AMIpSL1	No promoter; Terminator from an unknown gene	HARS36	ScLEU2	6–9	Agaphonov et al. (1999) Multiple cloning sites for insertion of expression cassettes
AMIpSU1	No promoter; Terminator from an unknown gene	HARS36	ScURA3	30–50	Agaphonov et al. (1999) Multiple cloning sites for insertion of expression cassettes
pGLG61	No promoter; No terminator	TEL188	HpLEU2/G418ʳ	1–50	Sohn et al. (1999) *Not*I and a *Bam*HI sites for insertion of expression cassettes
pHACT90-HyL	No promoter; No terminator	TEL188	HpLEU2/Hygʳ	1–25	Kang et al. (2001) A *Not*I site for insertion of expression cassettes

Tab. A6.2 (continued)

Plasmid	Expression cassette	Replication sequence	Selection marker	Integrated copy number	References/comments
RB11-based plasmids					
pMPT121	*MOX*-promoter; *MOX*-terminator	*HARS1* (oppositely oriented than in pFPMT121 and pTPSPMT)	*ScURA3*	30–60	Gellissen and Hollenberg (1997) *Eco*RI, *Bam*HI, *Bgl*II, sites for insertion of ORFs
pFPMT121	*FMD*-promoter; *MOX*-terminator	*HARS1*	*ScURA3*	30–60	Zurek et al. (1996) *Eco*RI, *Bam*HI, *Bgl*II, sites for insertion of ORFs
pTPS1MT121	*TPS1*-promoter *MOX*-terminator	*HARS1*	*ScURA3*	30–60	Rhein Biotech, unpublished; *Eco*RI, *Bam*HI, *Bgl*II, sites for insertion of ORFs
pB14	*FMD*-promoter; *MOX*-terminator	*HARS1*	*ScURA3*	30–60	Rhein Biotech, unpublished; *Eco*RI, *Bam*HI, *Bgl*II, sites for insertion of ORFs; no ampR gene
pB14TPS1	*TPS1*-promoter; *MOX*-terminator	*HARS1*	*ScURA3*	30–60	Rhein Biotech, unpublished; *Eco*RI, *Bam*HI, *Bgl*II, sites for insertion of ORFs; no ampR
pB14-LEU2	*FMD*-promoter; *MOX*-terminator	*HARS1*	*ScLEU2*	30–60	Rhein Biotech, unpublished; *Eco*RI, *Bam*HI, *Bgl*II sites for insertion of ORFs; no ampR
pM1	No promoter *MOX*-terminator	*HARS1*	*ScURA3*	n.d.	Amuel et al. (2000) Multiple cloning sites for insertion of expression cassettes
pSK92	*FMD*-promoter; *MOX*-terminator	*HARS1*	*HARO7*	1–5	Krappmann et al. (2000) Rhein Biotech, unpublished; *Eco*RI, *Bam*HI, *Bgl*II, sites for insertion of ORFs

Tab. A6.2 (continued)

Plasmid	Expression cassette	Replication sequence	Selection marker	Integrated copy number	References/comments
NCYC495-based plasmids					
pHIPA4	AOX-promoter AMO-terminator	No replication sequence	HpPUR7/Amp	n.d.	Haan et al. (2002)
pHIPX4	AOX-promoter AMO-terminator	No replication sequence*	ScLEU2/Kan	n.d.	Gietl et al. (1994)
pHIPX5	AMO-promoter AMO-terminator	No replication sequence*	ScLEU2/Kan	n.d.	Kiel et al. (1995)
pHIPX6	PEX3-promoter AMO-terminator	No replication sequence*, **	ScLEU2/Kan	n.d.	Kiel et al. (1995)
pHIPX7	TEF1-promoter AMO-terminator	No replication sequence*	ScLEU2/Kan	n.d.	Baerends et al. (1997)
pHIPX8	TEF2-promoter AMO-terminator	No replication sequence*	ScLEU2/Kan	n.d.	M. Veenhuis, unpublished
pHIPZ4	AOX-promoter AMO-terminator	No replication sequence	Zeocin/Amp	n.d.	Salomons et al. (2000)
pREMI-Z	ScTEF1-promoter/ EM7 synthetic promoter/ ScCYC-terminator	No replication sequence	Zeocin	n.d.	Van Dijk et al. (2002)
pYT3	No promoter No terminator	C-ARS *	ScLEU2/Amp	n.d.	Tan et al. (1995)
pHS5	No promoter No terminator LacZα	No replication sequence*	ScLEU2/Amp	n.d.	M. Veenhuis, unpublished
pHS6	No promoter No terminator LacZα	HARS1*	ScLEU2/Amp	n.d.	M. Veenhuis, unpublished
pBSK-LEU2-Ca	No promoter No terminator	No replication sequence	CaLEU2/Amp	n.d.	M. Veenhuis, unpublished
pHI1	No promoter No terminator LacZα	No replication sequence	HpURA3/Amp	n.d.	Kiel et al. (1999)

Key: *Ca*, *Candida albicans*; *Hp*, *Hansenula polymorpha*; *Sc*, *Saccharomyces cerevisiae*.

 * These plasmids contain the *S. cerevisiae LEU2* gene, which harbors a cryptic HARS. As a consequence, pHIPX- and pHS-type plasmids replicate – albeit rather poorly – in *H. polymorpha* NCYC495. Addition of the C-ARS or HARS1 regions results in good replicating plasmids.

** pHIPX6 contains the *H. polymorpha PEX3* promoter. This fragment contains a HARS, and allows use as a replicating plasmid to express genes at rather low levels.

References

Agaphonov MO, Beburov MY, Ter-Avanesyan MD, Smirnov VN (1995) Disruption-displacement approach for the targeted integration of foreign genes in *Hansenula polymorpha*. Yeast 11: 1241–1247

Agaphonov MO, Deev AV, Kim S-Y, Sohn J-H, Choi E-S, Ter-Avanesyan MD (2003) A novel approach to isolation and functional characterization of genomic DNA sequences from the methylotrophic yeast *Hansenula polymorpha*. Mol Biol 37: 74–80

Agaphonov MO, Packeiser AN, Chechenova MB, Choi E-S, Ter-Avanesyan MD (2001) Mutation of the homologue of GDP-mannose pyrophosphorylase alters cell wall structure, protein glycosylation and secretion in *Hansenula polymorpha*. Yeast 18: 391–402

Agaphonov MO, Romanos NV, Trushina PM, Smirnov VN, Ter-Avanesyan MD (2002) Aggregation and retention of human urokinase type plasminogen activator in the yeast endoplasmic reticulum. BMC Mol Biol 3: 15–22

Agaphonov M, Trushkina PM, Sohn JS, Choi ES, Rhee SK, Ter-Avanesyan MD (1999) Vectors for rapid selection of integrants with different plasmid copy numbers in the yeast *Hansenula polymorpha* DL-1. Yeast 15: 541–551

Amuel C, Gellissen G, Hollenberg CP, Suckow M (2000) Analysis of heat shock promoters in *Hansenula polymorpha*: *TPS1*, a novel element for heterologous gene expression. Biotechnol Bioprocess Eng 5: 247–252

Avgerinos GC, Turner BG, Gorelick MD, Papendieck A, Weydemann U, Gellissen G (2001) Production and clinical development of a *Hansenula polymorpha*-derived PEGylated hirudin. Semin Thromb Hemost 27: 357–371

Avila J, González C, Brito N, Siverio JM (1998) Clustering of the *YNA1* gene encoding a ZN(II)$_2$Cys$_6$ transcriptional factor in the yeast *Hansenula polymorpha* with the nitrate assimilation genes *YNT1*, *YNI1* and *YNR1*, and its involvement in their transcriptional activation. Biochem J 335: 647–652

Avila J, Pérez MD, Brito N, González C, Siverio JM (1995) Cloning and disruption of the *YNR1* gene encoding the nitrate reductase apoenzyme of the yeast *Hansenula polymorpha* FEBS Lett 366: 137–142

Baerends RJS, Salomons FA, Kiel JAKW, van der Klei IJ, Veenhuis M (1997) Deviant Pex3p levels affect normal peroxisome formation in *Hansenula polymorpha*: a sharp increase of the protein level induces the proliferation of numerous, small protein import-competent peroxisomes. Yeast 13: 1449–1463

Baerends RJS, Sulter GJ, Jeffries TW, Cregg JM, Veenhuis M (2002) Molecular characterization of the *Hansenula polymorpha* *FLD1* gene encoding formaldehyde dehydrogenase. Yeast 19: 37–42

Barnes CS, Krafft B, Frech M, Hofmann UR, Papendieck A, Dahlems U, Gellissen G, Hoylaerts MF (2001) Production and characterization of saratin, an inhibitor of von Willebrand factor-dependent platelet adhesion to collagen. Semin Thromb Hemost 27: 337–347

Barnett JA, Payne RW, Yarrow D (2000) Yeasts: Characteristics and Identification, 3rd edition. Cambridge University Press, Cambridge, UK

Barr PJ, Brake AJ, Valenzuela P (Eds) Yeast genetic engineering. Butterworth, Boston, MA

Bartelsen O, Barnes CS, Gellissen G (2002) Production of anticoagulants in *Hansenula polymorpha*. In: Gellissen G (Ed) *Hansenula polymorpha* – biology and applications. Wiley-VCH, Weinheim, pp 211–228

Birney E, Durbin R (2000) Using GeneWise in the *Drosophila* annotation experiment. Genome Res 10: 547–548

Blandin G, Llorente B, Malpertuy A, Wincker P, Artiguenave F, Dujon B (2000) Genomic exploration of the hemiascomycetous yeasts: 13. *Pichia angusta*. FEBS Lett 487: 76–81

Bogdanova AI, Kustikova OS, Agaphonov MO, Ter-Avanesyan MD (1998) Sequences of *Saccharomyces cerevisiae* 2µm DNA improving plasmid partitioning in *Hansenula polymorpha*. Yeast 14: 1–9

Brito N, Avila J, Pérez MD, González C, Siverio JM (1996) The genes *YNI1* and *YNR1* encoding nitrite reductase and nitrate reductase respectively in the yeast *Hansenula polymorpha*, are clustered and co-ordinately regulated. Biochem J 317: 89–95

Brito N, Pérez MD, Perdomo G, González C, García-Lugo P, Siverio JM (1999) A set of *Hansenula polymorpha* integrative vectors to construct *lacZ* fusions. Appl Microbiol Biotechnol 53: 23–29

BRETTHAUER RK, CASTELLANO FJ (1999) Glycosylation of *Pichia pastoris*-derived proteins. Biotechnol Appl Biochem 30: 193–200

CALLEWAERT N, LAROY W, CADIRGI H, GEYSENS S, SAELNES X, JOU WM, CONTRERAS R (2001) Use of HDEL-tagged *Trichoderma reesei* mannosyl oligosaccharide 1,2-α-D-mannosidase for *N*-glycan engineering in *Pichia pastoris*. FEBS Lett 503: 173–178

CHIBA Y, SUZUKI M, YOSHIDA S, YOSHIDA A, IKENAGA H, TAKEUCHI M, JIGAMI Y, ICHISHIMA E (1998) Production of human compatible high mannose-type (Man₅GlcNAc₂) sugar chains in *Saccharomyces cerevisiae*. J Biol Chem 273: 26298–26304

COOK MW, THYGESEN HV (2003) Safety evaluation of a hexose oxidase expressed in *Hansenula polymorpha*. Food Chem Toxicol 41: 523–529

COX H, MEAD D, SUDBERY, ELAND M, EVANS L (2000) Constitutive expression of recombinant proteins in the methylotrophic yeast *Hansenula polymorpha* using the *PMA1* promoter. Yeast 16: 1191–1203

CREGG JM (1999) Expression in the methylotrophic yeast *Pichia pastoris*. In: Fernandez JM, Hoeffler JP (Eds) Gene expression systems: using nature for the art of expression. Academic Press, San Diego, pp 157–191

CURVERS S, BRIXIUS P, KLAUSER T, WEUSTER-BOTZ D, TAKORS R, WANDREY C (2001) Human chymotrypsinogen B production with *Pichia pastoris* by integrated development of fermentation and downstream processing Part I. Fermentation. Biotechnol Progr 17: 495–502

DE BRUIN EC, WERTEN MW, LAANE C, DE WOLF FA (2002) Endogenous propyl 4-hydroxylation in *Hansenula polymorpha* and its use for the production of hydroxylated recombinant gelatin. FEMS Yeast 1: 291–298

DEGELMANN A (2002) Methods. In: Gellissen G (Ed) *Hansenula polymorpha* – biology and applications. Wiley-VCH, Weinheim, pp 285–335

DEGELMANN A, MÜLLER F, SIEBER H, JENZELEWSKI V, SUCKOW M, STRASSER AWM, GELLISSEN G (2002) Strain and process development for the production of human cytokines in *Hansenula polymorpha*. FEMS Yeast Res 2: 349–361

DIDION T, ROGGENKAMP R (1992) Targeting signal of the peroxisomal catalase in the methylotrophic yeast *Hansenula polymorpha*. FEBS Lett 303: 113–116

FELDMANN, H (ED) (2000) Génolevures. Genomic exploration of the hemiascomycetous yeasts. FEBS Lett 487: 1–150

GELLISSEN G (2000) Heterologous protein production in methylotrophic yeasts. Appl Microbiol Biotechnol 54: 741–750

GELLISSEN G (2002) *Hansenula polymorpha* – biology and applications. Wiley-VCH, Weinheim

GELLISSEN G, HOLLENBERG CP (1997) Application of yeasts in gene expression studies: a comparison of *Saccharomyces cerevisiae*, *Hansenula polymorpha* and *Kluyveromyces lactis* – a review. Gene 190: 87–97

GELLISSEN G, VEENHUIS M (2001) The methylotrophic yeast *Hansenula polymorpha*: its use in fundamental research and as a cell factory. Yeast 18: i–iii

GELLISSEN G, JANOWICZ ZA, WEYDEMANN U, MELBER K, STRASSER AWM, HOLLENBERG CP (1992a) High-level expression of foreign genes in *Hansenula polymorpha*. Biotechnol Adv 10: 179–189

GELLISSEN G, MELBER K, JANOWICZ SA, DAHLEMS U, WEYDEMANN U, PIONTEK M, STRASSER AWM, HOLLENBERG CP (1992b) Heterologous protein expression in yeast. Antonie van Leeuwenhoek 62: 79–93

GELLISSEN G, MÜLLER F, SIEBER H, TIEKE A, JENZELEWSKI V, DEGELMANN A, STRASSER AWM (2002) Production of cytokines in *Hansenula polymorpha*. In: Gellissen G (Ed) *Hansenula polymorpha* – biology and applications. Wiley-VCH, Weinheim, pp 229–254

GELLISSEN G, PIONTEK M, DAHLEMS U, JENZELEWSKI V, GAVAGAN JE, DICOSIMO R, ANTON DL, JANOWICZ SA (1996) Recombinant *Hansenula polymorpha* as a biocatalyst: coexpression of the spinach glycolate oxidase (*GO*) and the *S. cerevisiae* catalase T (*CTT1*) gene. Appl Microbiol Biotechnol 46: 46–54

GEMMILL TR, TRIMBLE RB (1999) Overview of *N*-linked and *O*-linked oligosaccharide structures found in various yeast species. Biochim Biophys Acta 1426: 227–237

GIETL C, FABER KN, VAN DER KLEI IJ, VEENHUIS M (1994) Mutational analysis of the N-terminal topogenic signal of watermelon glyoxysomal malate dehydrogenase using the heterologous host *Hansenula polymorpha*. Proc Natl Acad Sci USA 91: 3151–3155

GLEESON MA, ORTORI GS, SUDBERY PE (1986) Transformation of the methylotrophic yeast *Hansenula polymorpha*. J Gen Microbiol 132: 3459–3465

Goffeau A (1996) 1996: A vintage year for Yeast and yeast. Yeast 2: 1603–1606

Gonzalez C, Perdomo G, Tejera P, Brito N, Siverio JM (1999) One-step, PCR-mediated, gene disruption in the yeast *Hansenula polymorpha*. Yeast 15: 1323–1329

Guengerich L, Kang HA, Behle B, Gellissen G, Suckow M (2004) A platform for heterologous gene expression based on the methylotrophic yeast *Hansenula polymorpha*. In: Kück U (Ed) The mycota II. Genetics and Biotechnology, 2nd edition, Springer-Verlag, Heidelberg, pp 273–287

Haan GJ, van Dijk R, Kiel JAKW, Veenhuis M (2002) Characterization of the *Hansenula polymorpha PUR7* gene and its use as selectable marker for targeted chromosomal integration. FEMS Yeast Res 2: 17–24

Hamilton SR, Bobrowicz P, Bobrowicz B, Davidson RC, Li H, Mitchell T, Nett JH, Rausch S, Stadheim TA, Wischenwsk H, Wildt S, Gerngross TU (2003) Production of complex human glycoproteins in yeast. Science 301: 1244–1246

Hazeu W, de Bruyn JC, Bos P (1972) Methanol assimilation by yeasts. Arch Mikrobiol 87: 185–188

Heo JH, Won HS, Kang HA, Rhee SK, Chung BH (2002) Purification of recombinant human epidermal growth factor secreted from the methylotrophic yeast *Hansenula polymorpha*. Protein Expr Purif 24: 117–122

Heo JH, Hong WK, Cho EY, Kim MW, Kim JY, Kim CH, Rhee SK, Kang HA (2003) Properties of the *Hansenula polymorpha*-derived constitutive *GAP* promoter, assessed using an *HSA* reporter gene. FEMS Yeast Res 4: 175–184

Hollenberg CP, Janowicz ZA (1988) DNA molecules coding for FMDH control regions and structured gene having FMDH activity and their uses. European Patent Application EPA 0299108

Janowicz ZA, Eckart MR, Drewke C, Roggenkamp RO, Hollenberg CP, Maat J, Ledeboer AM, Visser C, Verrips CT (1985) Cloning and characterization of the *DAS* gene encoding the major methanol assimilatory enzyme from the methylotrophic yeast *Hansenula polymorpha*. Nucleic Acids Res 13: 3043–306

Janowicz SA, Merckelbach A, Jacobs E, Harford N, Comerbach M, Hollenberg CP (1991) Simultaneous expression of the S and L surface antigens of hepatitis B, and formation of mixed particles in the methylotrophic yeast *Hansenula polymorpha*. Yeast 7: 431–443

Jenzelewski V (2002) Fermentation and primary product recovery. In: Gellissen G (Ed) *Hansenula polymorpha* – biology and applications. Wiley-VCH, Weinheim, pp 156–174

Jigami Y, Odani T (1999) Mannosylphosphate transfer to yeast mannan. Biochim Biophys Acta 1426: 335–345

Kang HA, Hong W-K, Sohn J-H, Choi E-S, Rhee SK (2001 b) Molecular characterization of the actin-encoding gene and the use of its promoter for a dominant selection system in the methylotrophic yeast *Hansenula polymorpha*. Appl Microbiol Biotechnol 55: 734–741

Kang HA, Kang W, Hong W-K, Kim MW, Kim J-Y, Sohn J-H, Choi E-S, Choe K-B, Rhee SK (2001 a) Development of expression systems for the production of recombinant human serum albumin using the *MOX* promoter in *Hansenula polymorpha* DL-1. Biotechnol Bioeng 76: 175–185

Kang HA, Kim J-Y, Ko S-M, Park CS, Ryu DY, Sohn J-H, Choi E-S, Rhee S-K (1998 a) Cloning and characterization of the *Hansenula polymorpha* homologue of the *Saccharomyces cerevisiae PMR1* gene. Yeast 14: 1233–1240

Kang HA, Sohn J-H, Agaphonov MO, Choi E-S, Ter-Avanesyan MD, Rhee SK (2002) Development of expression systems for the production of recombinant proteins in *Hansenula polymorpha* DL-1. In: Gellissen G (Ed) *Hansenula polymorpha* – biology and applications. Wiley-VCH, Weinheim, pp 124–146

Kang HA, Sohn J-H, Choi E-S, Chung B-H, Yu M-H, Rhee S-K (1998 b) Glycosylation of human α_1-antitrypsin in *Saccharomyces cerevisiae* and methylotrophic yeasts. Yeast 14: 371–381

Kiel JAKW, Hilbrands RE, van der Klei IJ, Rasmussen SW, Salomons FA, van der Heide M, Faber KN, Cregg JM, Veenhuis M (1999) *Hansenula polymorpha* Pex1p and Pex6p are peroxisome-associated AAA proteins that functionally and physically interact. Yeast 15: 1059–1078

Kiel JAKW, Keizer-Gunnink IK, Krause T, Komori M, Veenhuis M (1995) Heterologous complementation of peroxisome function in yeast: the *Saccharomyces cerevisiae PAS3* gene restores peroxisome biogenesis in a *Hansenula polymorpha per9* disruption mutant. FEBS Lett 377: 434–438

Kim MW, Agaphonov MO, Kim JY, Rhee SK,

KANG HA (2002) Sequencing and functional analysis of the *Hansenula polymorpha* genomic fragment containing the *YPT1* and *PMI40* genes. Yeast 19: 863–871

KIM MW, RHEE SK, KIM JY, SHIMMA YI, CHIBA Y, JIGAMI Y, KANG HA (2004) Characterization of N-linked oligosaccharides assembled on secretory recombinant glucose oxidase and cell wall mannoproteins from the methylotrophic yeast *Hansenula polymorpha*. Glycobiology 14: 243–251

KIM SY, SOHN S-H, KANG HA, CHOI E-S (2001) Cloning and characterization of the *Hansenula polymorpha* homologue of the *Saccharomyces cerevisiae MNN9* gene. Yeast 18: 455–461

KIM SY, SOHN JH, PYUN YR, CHOI ES (2002) A cell surface display system using novel GPI-anchored proteins in *Hansenula polymorpha*. Yeast 19: 1153–1163

KIM SY, SOHN JH, BAE JH, PYUN YR, AGAPHONOV MO, TER-AVANESYAN MD, CHOI ES (2003) Efficient library construction by *in vivo* recombination with a telomere-originated autonomously replicating sequence of *Hansenula polymorpha*. Appl Environ Microbiol 69: 4448–4454

KLABUNDE J, DIESEL A, WASCHK D, GELLISSEN G, HOLLENBERG CP, SUCKOW M (2002) Single step co-integration of multiple expressible heterologous genes into the ribosomal DNA of the methylotrophic yeast *Hansenula polymorpha*. Appl Microbiol Biotechnol 58: 797–805

KLABUNDE J, KUNZE G, GELLISSEN G, HOLLENBERG CP (2003) Integration of heterologous genes in several yeast species using vectors containing a *Hansenula polymorpha*-derived rDNA targeting element. FEMS Yeast Res 4: 185–193

KRAPPMANN S, PRIES R, GELLISSEN G, HILLER M, BRAUS GH (2000) *HARO7* encodes chorismate mutase of the methylotrophic yeast *Hansenula polymorpha* and is derepressed upon methanol utilization. J Bacteriol 182: 4188–4197

KREGER-VAN RIJ NJWN (1984) The yeasts. A taxonomic study, 3rd edition. Amsterdam: Elsevier Science Publisher, Amsterdam

KURTZMAN CP, FELL JW (1998) The yeasts, a taxonomic study, 4th edition. Elsevier, Amsterdam

KURTZMAN CP, ROBNETT CJ (1998) Identification and phylogeny of ascomycetous yeasts from analysis of nuclear large subunit (26S) ribosomal DNA partial sequences. Antonie van Leeuwenhoek 73: 331–371

KURTZMAN CP (1984) Synonymy of the yeast genera *Hansenula* and *Pichia* demonstrated through comparison of deoxyribonucleic acid relatedness. Antonie van Leeuwenhoek 50: 209–217

LAHTCHEV K (2002) Basic genetics of *Hansenula polymorpha*. In: Gellissen G (Ed) *Hansenula polymorpha* – biology and applications. Wiley-VCH, Weinheim, pp 8–20

LEDEBOER AM, EDENS L, MAAT J, VISSER C, BOS JW, VERRIPS CT, JANOWICZ ZA, ECKART M, ROGGENKAMP RO, HOLLENBERG CP (1985) Molecular cloning and characterization of a gene coding for methanol oxidase in *Hansenula polymorpha*. Nucleic Acids Res 13: 3063–3082

LEVINE DW, COONEY CL (1973) Isolation and characterization of a thermotolerant methanol-utilizing yeast. Appl Microbiol 26: 982–989

MANNAZZU I, GUERRA E, STRABBIOLI R, MASIA A, MAESTRALE GB, ZORODDU MA, FATICHENTI F. (1997) Vanadium affects vacuolation and phosphate metabolism in *Hansenula polymorpha*. FEMS Microbiol Lett. 147: 23–28

MAYER AF, HELLMUTH K, SCHLIEKER H, LOPEZ-ULIBARRI R, OERTEL S, DAHLEMS U, STRASSER AWM, VAN LOON APGM (1999) An expression system matures: a highly efficient and cost-effective process for phytase production by recombinant strains of *Hansenula polymorpha*. Biotechnol Bioeng 63: 373–381

MARRI L, ROSSOLINI GM, SATTA G (1993) Chromosome polymorphism among strains of *Hansenula polymorpha* (syn. *Pichia angusta*). Appl Environ Microbiol 59: 939–941

MEWES HW, FRISHMAN D, GÜLDENER U, MANNHAUPT G, MAYER K, MOKREJS M, MORGENSTERN B, MÜNSTERKOETTER M, RUDD S, WEIL B (2002) MIPS: a database for genomes and protein sequences. Nucleic Acids Res 30: 31–34

MIDDELHOVEN W (2002) History, habitat, variability, nomenclature and phylogenetic position of *Hansenula polymorpha*. In: Gellissen G (Ed) *Hansenula polymorpha* – biology and applications. Wiley-VCH, Weinheim, pp 1–7

MIELE RG, CASTELLINO FJ, BRETTHAUER RK (1997) Characterization of the acidic oligosaccharides assembled on the *Pichia pastoris*-expressed recombinant kringle 2 domain of tissue-type plasminogen activator. Biotechnol Appl Biochem 26: 79–83

MONTESINO R, NIMTZ M, QUINTERO O, GARCÍA R, FALCÓN V, CREMATA JA (1999) Characterization of the oligosaccharides assembled on the *Pichia pastoris*-expressed recombinant aspartic protease. Glycobiology 9: 1037–1043

MONTESINO R, GARCÍA R, QUINTERO O, CREMATA JA (1998) Variation in *N*-linked oligosaccharide structures on heterologous proteins secreted by the methylotrophic yeast *Pichia pastoris*. Protein Exp Purif 14: 197–207

MORAIS JOF DE, MAIA MHD (1959) Estudos de microorganismos encontrados em leitos de despejos de caldas de destilarias de Pernambuco. II. Una nova especie de *Hansenula: H. polymorpha*. Anais de Escola Superior de Quimica de Universidade do Recife 1: 15–20

MÜLLER S, SANDAL T, KAMP-HANSEN P, DALBØGE H (1998) Comparison of expression systems in the yeasts *Saccharomyces cerevisiae*, *Hansenula polymorpha*, *Kluyveromyces lactis*, *Schizosaccharomyces pombe* and *Yarrowia lipolytica*. Cloning of two novel promoters from *Yarrowia lipolytica*. Yeast 14: 1267–1283

MÜLLER F II, TIEKE A, WASCHK D, MÜHLE C, MÜLLER F I, SEIGELCHIFER M, PESCE A, JENZELEWSKI V, GELLISSEN G (2002) Production of IFNα-2a in *Hansenula polymorpha*. Proc Biochem 38: 15–25

OGATA K, NISHIKAWA H, OHSUGI M (1969) Yeast capable of utilizing methanol. Agric Biol Chem 33: 1519–1520

OH KS, KWON O, OH YW, SOHN MJ, JUNG S, KIM YK., KIM MG, RHEE SK, GELLISSEN G, KANG HA (2004) Fabrication of a partial genome microarray of the methylotrophic Yeast *Hansenula polymorpha*: optimization and evaluation for transcript profiling. J Microbiol Biotechnol 14: (in press)

OSOEGAWA K, WOON PY, ZHAO B, FRENGEN E, TATENO M, CATANESE JJ, DE JONG PJ (1998) An improved approach for construction of bacterial artificial chromosome libraries. Genomics 52: 1–8

OSOEGAWA K, DE JONG PJ, FRENGEN E, IOANNOU PA (1999) Construction of bacterial artificial chromosome (BAC/PAC) libraries. Curr Protocol Hum Genet 5.15.1–5.15.33

PAPENDIECK A, DAHLEMS U, GELLISSEN G (2002) Technical enzyme production and whole-cell biocatalysis: application of *Hanse-* *nula polymorpha*. In: Gellissen G (Ed) *Hansenula polymorpha* – biology and applications. Wiley-VCH, Weinheim, pp 255–271

PARK BS, ANANINE V, KIM CH, RHEE SK, KANG HA (2004) Secretory production of *Zymomonas mobilis* levansucrase by the methylotropic yeast *Hansenula polymorpha* Enzym Microbiol Tech 34: 132–138

PARPINELLO G, BERARDI F, STRABBIOLI R (1998) A regulatory mutant of *Hansenula polymorpha* exhibiting methanol utilization metabolism and peroxisome proliferation in glucose. J Bacteriol 180: 2958–2967

PHONGDARA A, MERCKELBACH A, KEUP P, GELLISSEN G, HOLLENBERG CP (1998) Cloning and characterization of the gene encoding a repressible acid phosphatase (*PHO1*) from the methylotrophic yeast *Hansenula polymorpha*. Appl Microbiol Biotechnol 50: 77–84

RAMEZANI-RAD M, HOLLENBERG CP, LAUBER J, WEDLER H, GRIESS E, WAGNER C, ALBERMANN K, HANI J, PIONTEK M, DAHLEMS U, GELLISSEN G (2003) The *Hansenula polymorpha* (strain CBS4732) genome sequencing and analysis. FEMS Yeast Res 4: 207–215

RASCHKE WC, NEIDITCH BR, HENDRICKS M, CREGG JM (1996) Inducible expression of a heterologous protein in *Hansenula polymorpha* using the alcohol oxidase 1 promoter of *Pichia pastoris*. Gene 177: 163–187

REHM H-J, REED G (EDS) (1993) Biotechnology, Vol 2, Genetic Fundamentals and genetic engineering. VCH, Weinheim

RODRIGUEZ L, NARCIANDI RE, ROCA H, CREMATA J, MONTESINOS R, RODRIGUEZ E, GRILLO JM, MUZIO V, HERRERA LS, DELGADO JM (1996) Invertase secretion in *Hansenula polymorpha* under the *AOX1* promoter from *Pichia pastoris*. Yeast 12: 815–822

ROGGENKAMP R, HANSEN H, ECKART M, JANOWICZ Z, HOLLENBERG, CP (1986) Transformation of the methylotrophic yeast *Hansenula polymorpha* by autonomous replication and integration vectors. Mol Gen Genet 202: 302–308

REINDERS A, ROMANO I, WIEMKEN A, DE VIRGILIO C (1999) The thermophilic yeast *Hansenula polymorpha* does not require trehalose synthesis for growth at high temperatures but does for normal acquisition of thermotolerance. J Bacteriol 181: 4665–4668

ROMANO I (1998) Untersuchungen zur Trehalose-6-phosphat Synthase und Klonierung und Deletion des *TPS1* Gens in der methylotrophen Hefe *Hansenula polymorpha*. PhD Thesis, University of Basel, Switzerland

SALOMONS FA, KIEL JAKW, FANER KN, VEENHUIS M, VAN DER KLEI IJ (2000) Overproduction of pex5p stimulates import of alcohol oxidase and dihydroxyacetone synthase in a *Hansenula polymorpha pex14* null mutant. J Biol Chem 276: 44570–44574

SCHAEFER S, PIONTEK M, AHN S-J, PAPENDIECK A, JANOWICZ SA, GELLISSEN G (2001) Recombinant hepatitis B vaccines – characterization of the viral disease and vaccine production in the methylotrophic yeast *Hansenula polymorpha*. In: Dembowsky K Stadler P (Eds), Novel therapeutic proteins: selected case studies. Wiley-VCH, Weinheim, pp 245–274

SCHAEFER S, PIONTEK M, AHN S-J, PAPENDIECK A, JANOWICZ ZA, TIMMERMANS I, GELLISSEN G (2002) Recombinant hepatitis B vaccines – disease characterization and vaccine production. In: Gellissen G (Ed) *Hansenula polymorpha* – biology and applications. Wiley-VCH, Weinheim, pp 175–210

SCORER CA, BUCKHOLZ RG, CLARE JJ, ROMANOS MA (1993) The intracellular production and secretion of HIV-1 envelop protein in the methylotrophic yeast *Pichia pastoris*. Gene 136: 111–119

SIVERIO JM (2002) Biochemistry and genetics of nitrate assimilation. In: Gellissen G (Ed) *Hansenula polymorpha* – biology and applications. Wiley-VCH, Weinheim, pp 21–40

SOHN J-H, CHOI E-S, KANG HA, RHEE J-S, RHEE S-K (1999a) A family of telomere-associated autonomously replicating sequences and their functions in targeted recombination in *Hansenula polymorpha* DL-1. J Bacteriol 181: 1005–1013

SOHN J-H, CHOI E-S, KANG HA, RHEE J-S, AGAPHONOV MO, TER-Avanesyan MD, RHEE S-K (1999b) A dominant selection system designed for copy number-controlled gene integration in *Hansenula polymorpha* DL-1. Appl Microbiol Biotechnol 51: 800–807

SOHN J-H, CHOI E-S, KIM C-H, AGAPHONOV, MO, TER-AVANESYAN MD, RHEE J-S, RHEE S-K (1996) A novel autonomously replicating sequence (ARS) for multiple integration in the yeast *Hansenula polymorpha* DL-1. J Bacteriol 178: 4420–4428

SOHN J-H, KANG HA, RAO KJ, KIM CH, CHOI E-S, CHUNG BH, RHEE S-K (2001) Current status of the anticoagulant hirudin: its bio-technological production and clinical practice. Appl Microbiol Biotechnol 57: 606–613

SPIER RE (2000) Yeast as a cell factory. Enzyme Microbiol Technol 26: 639

STRAHL-BOLSINGER S, GENTZSCH M, TANNER W (1999) Protein O-mannosylation. Biochim Biophys Acta 1426: 297–307

SUCKOW M, GELLISSEN G (2002) The expression platform based on *Hansenula polymorpha* strain RB11 and its derivatives-history, status and perspectives. In: Gellissen G (Ed) *Hansenula polymorpha* – biology and applications. Wiley-VCH, Weinheim, pp 105–123

SUDBERY P (2003) Book review: Gellissen G (Ed) *Hansenula polymorpha* – biology and Application. Yeast 20: 107–1308

TAN X, TITORENKO VI, VAN DER KLEI IJ, SULTER GJ, HAIMA P, WATERHAM HR, EVERS M, HARDER W, VEENHUIS M, CREGG JM (1995) Characterization of peroxisome-deficient mutants of *Hansenula polymorpha*. Curr Genet 28: 248–257

TANI Y (1984) Microbiology and biochemistry of the methylotrophic yeasts. In: Hou CT (ed) Methylotrophs: microbiology, biochemistry, and genetics. CRC Press, Boca Raton, FL, pp 55–86

TEUNISSON DJ, HALL HH, WICKERHAM LJ (1960) *Hansenula angusta*, an excellent species for demonstration of the coexistence of haploid and diploid cells in a homothallic yeast. Mycologia 52: 184–188

THÖMMES J, HALFAR M, GIEREN H, CURVERS S, TAKORS R, BRUNSCHIER R, KULA M-R (2001) Human chymotrypsinogen B production from *Pichia pastoris* by integrated development of fermentation and downstream processing Part 2. Protein recovery. Biotechnol Progr 17: 503–512

TIKHOMIROVA LP, IKONOMOVA RN, KUZNETSOVA EN, FODOR II, BYSTRYKH LV, AMINOVA LR, TROTSENKO YA (1988) Transformation of methylotrophic yeast *Hansenula polymorpha*: cloning and expression of genes. J Basic Microbiol 28: 343–351

VAN DIJK R, FABER KN, KIEL JAKW, VEENHUIS M, VAN DER KLEI IJ (2000) The methylotrophic yeast *Hansenula polymorpha*: a versatile cell factory. Enzyme Microbiol Technol 26: 793–800

VAN DIJK R, FABER KN, HAMMOND AT, GLICK BS, VEENHUIS M, KIEL JAKW (2001) Tagging *Hansenula polymorpha* genes by random integration of linear DNA fragments (RALF). Mol Genet Genomics 266: 646–656

VAN DER KLEI IJ, VEENHUIS M (2002) *Hansenula polymorpha*: a versatile model organism in peroxisome research. In: Gellissen G (Ed) *Hansenula polymorpha* – biology and applications. Wiley-VCH, Weinheim, pp 76–94

VEALE RA, GUISEPPIN ML, VAN EIJK HM, SUDBERY PE, VERRIPS CT (1992) Development of a strain of *Hansenula polymorpha* for the efficient expression of guar alpha-galactosidase. Yeast 8: 361–372

WASCHK D, KLABUNDE J, SUCKOW M, HOLLENBERG CP (2002) Characteristics of the *Hansenula polymorpha* genome. In: Gellissen G (Ed) *Hansenula polymorpha* – biology and applications. Wiley-VCH, Weinheim, pp 95–104

WEYDEMANN U, KEUP P, PIONTEK M, STRASSER AWM, SCHWEDEN J, GELLISSEN G, JANOWICZ SA (1995) High-level secretion of hirudin by *Hansenula polymorpha*-authentic processing of three different preprohirudins. Appl Microbiol Biotechnol 44: 377–385

WICKERHAM LJ (1951) Taxonomy of yeasts. Technical bulletin No 1029, US Dept Agric, Washington DC, pp 1–56

YURIMOTO H, SAKAI Y, KATO N (2002) Methanol metabolism. In: Gellissen G (Ed) *Hansenula polymorpha* – biology and applications. Wiley-VCH, Weinheim, pp 61–75

ZUREK C, KUBIS E, KEUP P, HÖRLEIN D, BEUNINK J, THÖMMES J, KULA M-R, HOLLENBERG CP, GELLISSEN G (1996) Production of two aprotinin variants in *Hansenula polymorpha*. Process Biochem 31: 679–689

7
Pichia pastoris
CHRISTINE ILGEN, JOAN LIN-CEREGHINO, and JAMES M. CREGG

List of Genes

Gene	Encoded gene product
ADE1	amidoimidazole succinocarboxamide synthase
ARG4	argininosuccinate lyase
AOX1	alcohol oxidase I
AOX2	alcohol oxidase II
FLD1	formaldehyde dehydrogenase
GAP	glyceraldehyde-3-phosphate dehydrogenase
HIS4	histidinol dehydrogenase
KEX1	carboxy-terminal protease
MFα1	α mating factor (S. cerevisiae)
OCH1	α-1,6-mannosyltransferase
PEP4	vacuolar protease
PEX8	peroxisomal biogenesis protein
PHO1	acid phosphatase
URA3	orotidine-5′-phosphate decarboxylase
URA5	orotate-phosphoribosyl transferase
YPT1	GTPase

7.1
Introduction

The methylotrophic yeast *Pichia pastoris* has developed into a highly successful system for the production of a variety of heterologous proteins. At present, over 500 reports describing its use have been published (http://faculty.kgi.edu/cregg/index.htm). The popularity of the system can be attributed to several factors, most importantly: (1) the presence, in expression vectors, of a strong, tightly regulated, and easily manipulated promoter derived from the *P. pastoris* alcohol oxidase I gene (*AOX1*) (Ellis et al. 1985; Cregg et al. 2000; Lin Cereghino and Cregg 2000); (2) a strong preference for respira-

Production of Recombinant Proteins. Novel Microbial and Eucaryotic Expression Systems. Edited by Gerd Gellissen
Copyright © 2005 WILEY-VCH Verlag GmbH & Co. KGaA, Weinheim
ISBN: 3-527-31036-3

tory versus fermentative growth; (3) the simplicity of techniques needed for the molecular genetic manipulation of *P. pastoris* (Higgins and Cregg 1998); (4) the capability of performing many eukaryotic post-translational modifications, such as glycosylation, disulfide bond formation and proteolytic processing; and (5) the availability of the expression system as a commercially available kit (www. invitrogen.com). This chapter is intended as a general overview of the system and how it works, plus a review of recent advances in modifying the structures of N-linked carbohydrates on glycoproteins secreted from the yeast.

The conceptual basis for the *P. pastoris* expression system stems from the ability of this yeast to grow on methanol as sole carbon and energy source (Ogata et al. 1969; Veenhuis et al. 1983; Wegner 1990) and the observation that some of the enzymes required for methanol metabolism are present at substantial levels only when cells are grown on methanol (Egli et al. 1980; Veenhuis et al. 1983). At least two of the methanol pathway enzymes, AOX and dihydroxyacetone synthase (DHAS), are present at high levels in cells grown on methanol, but are undetectable in cells grown on most other carbon sources. In cells fed methanol at growth-limiting rates in fermentor cultures, AOX levels are dramatically induced, constituting >30% of total soluble protein (Couderc and Baratti 1980).

The most commonly utilized promoter for controlling foreign gene expression in the system is derived from the *P. pastoris AOX1* gene (Tschopp et al. 1987 a). In fact, there are two genes that code for AOX in *P. pastoris*: *AOX1* and *AOX2*. Although both genes encode equally functional alcohol oxidase enzymes, *AOX1* is responsible for a vast majority of alcohol oxidase activity in the cell, due to the relative strength of its promoter (Ellis et al. 1985; Cregg et al. 1989). In methanol-grown cells, ~5% of polyA$^+$RNA is from *AOX1*; however, in cells grown on most other carbon sources, *AOX1* message is undetectable (Cregg and Madden, 1987). The regulation of the *AOX1* gene appears to involve two mechanisms: a repression/derepression mechanism; plus an induction mechanism, similar to the regulation of several of the *Saccharomyces cerevisiae GAL* genes. However, unlike *GAL*-pathway regulation, the absence of a repressing carbon source, such as glucose in the medium, does not result in substantial transcription of *AOX1*. The presence of methanol appears to be essential to induce high levels of *AOX1* transcription (Tschopp et al. 1987 a).

7.2
Construction of Expression Strains

Like most expression systems, the *P. pastoris* system is an *Escherichia coli*-based shuttle vector system. As a result of the efforts of numerous groups, a variety of expression vectors and *P. pastoris* host strains are available. Tables 7.1 and 7.2 provide lists of promoters and selectable marker genes that have been incorporated into *P. pastoris* expression vectors. Additional information on vectors and strains can be found in Tables A7.1 and A7.2, and elsewhere (http://faculty.kgi.edu/cregg/index.htm; Higgins and Cregg 1998; Cregg 1999; Cregg et al. 2000; Lin-Cereghino 2000).

Tab. 7.1 *P. pastoris* promoters and their properties.

Name[a]	Relative efficiency	Inducing compound	Reference
P_{AOX1}	High	Methanol	Ellis et al. (1985); Tschopp et al. (1987a)
P_{GAP}	High	Constitutive	Waterham et al. 1997)
P_{FLD1}	High	Methanol or methylamine	Shen et al. (1998)
P_{PEX8}	Moderate	Methanol or oleate	Liu et al. (1995)
P_{YPT1}	Low	Constitutive	Sears et al. (1998)

a) See Section 7.2.2 for definitions of abbreviations.

Tab. 7.2 Selectable marker genes for transformation of *P. pastoris*.

Biosynthetic markers	Useful property	Reference
HIS4	*P. pastoris* and *S. cerevisiae* genes complement *P. pastoris* his4 strains	Cregg et al. (1985)
ARG4	*P. pastoris* and *S. cerevisiae* genes complement *P. pastoris* arg4 strains	Cregg and Madden (1987)
ADE1	Pink color of *ade1* strains	Cereghino et al. (2001)
URA3	5-Fluoro-orotic acid selection for Ura⁻ strains	Cereghino et al. (2001)
URA5	5-Fluoro-orotic acid selection for Ura⁻ strains	Nett and Gerngross (2003)

Drug-resistance markers	Selective drug	Reference
Kan^R	G418/Geneticin	Scorer et al. (1994)
Zeo^R	Zeocin	Higgins et al. (1998)
Bsd^R	Blasticidin	www.invitrogen.com
FLD1	Formaldehyde	Sunga and Cregg (2004)

7.2.1
Expression Vector Components

Most *P. pastoris* expression vectors have an expression cassette composed of a 0.9-kb fragment from *AOX1* composed of the 5′ promoter sequences, and a second short *AOX1*-derived fragment with sequences required for transcription termination (Cregg et al. 1987; Tschopp et al. 1987a). Between the promoter and terminator sequences is a site or multiple cloning site (MCS) for insertion of the foreign coding sequence. Generally, the best expression results are obtained when the first ATG of the heterologous coding sequence is inserted as close as possible to the position of the corresponding *AOX1* ATG. This position coincides with the 5′-most restriction site in MCSs. In addition, for the secretion of foreign proteins, vectors are available where in-frame fusions of foreign proteins and the secretion signals of *P. pastoris* acid phosphatase (*PHO1*) or *S. cerevisiae* α-mating factor (α-MF) can be generated.

7.2.2
Alternative Promoters

Although the *AOX1* promoter has been successfully used to express numerous foreign genes, there are circumstances in which this promoter may not be suitable. For example, the use of methanol to induce gene expression may not be appropriate for the production of food products since methane – a petroleum-related compound – is one source of methanol. Also, methanol is a potential fire hazard, especially in quantities needed for large-scale fermentations. Therefore, promoters that are not induced by methanol are attractive for expression of certain genes. Alternative promoters to the *AOX1* promoter are the *P. pastoris GAP*, *FLD1*, *PEX8*, and *YPT1* promoters (see Table 7.1).

The *P. pastoris* glyceraldehyde 3-phosphate dehydrogenase (*GAP*) gene promoter provides strong constitutive expression on glucose at a level comparable to that seen with the *AOX1* promoter (Waterham et al. 1997). The advantage of using the *GAP* promoter is that methanol is not required for induction, nor is it necessary to shift cultures from one carbon source to another, making strain growth more straightforward. However, since the *GAP* promoter is constitutively expressed, it is not a good choice for the production of proteins that are toxic to the yeast.

The *FLD1* gene encodes glutathione-dependent formaldehyde dehydrogenase, a key enzyme required for the metabolism of certain methylated amines as nitrogen sources and methanol as a carbon source (Shen et al. 1998). The *FLD1* promoter can be induced with either methanol as a sole carbon source (and ammonium sulfate as a nitrogen source) or methylamine as a sole nitrogen source (and glucose as a carbon source). Thus, the *FLD1* promoter offers the flexibility to induce high levels of expression using either methanol or methylamine, an inexpensive nontoxic nitrogen source.

For some applications, the *AOX1*, *GAP*, and *FLD1* promoters may be too strong, expressing genes at too high a level. For these and other applications, moderately expressing promoters are desirable. Toward this end, the *P. pastoris PEX8* and *YPT1* promoters may be of use. The *PEX8* gene (formerly *PER3*) encodes a peroxisomal matrix protein that is essential for peroxisome biogenesis (Liu et al. 1995). It is expressed at a low but significant level on glucose and is induced modestly (three- to fivefold) when cells are shifted to methanol. The *YPT1* gene encodes a GTPase involved in secretion, and its promoter provides a low but constitutive level of expression in medium containing either glucose, methanol, or mannitol as carbon source (Sears et al. 1998).

7.2.3
Selectable Markers

The number of available vector marker genes for transformation of *P. pastoris* has increased significantly in recent years (see Table 7.2). The stable expression of human type III collagen illustrates the need for multiple selectable markers (Vuorela et al. 1997). The production of collagen required the co-expression of prolyl-4-hydroxylase, a central enzyme in the synthesis and assembly of trimeric collagen. Since prolyl-4-hydroxylase is an $\alpha_2\beta_2$ tetramer, the β subunit of which is protein disulfide isomerase (PDI), three markers – Arg, His, and Zeocin resistance – were used to co-express all three polypeptides in the same *P. pastoris* strain. Marker genes fall into two basic groups: the biosynthetic gene/auxotrophic mutant group; and the dominant drug resistance gene group.

Biosynthetic gene markers require a vector with such a gene along with a mutant host strain that is defective in that gene. For *P. pastoris* expression vectors, the *P. pastoris HIS4* (histidinol dehydrogenase), *ADE1* (PR-amidoimidazolesuccinocarboxamide synthase), *ARG4* (argininosuccinate lyase), *URA3* (orotidine 5'-phosphate decarboxylase), and *URA5* (orotate-phosphoribosyl transferase) genes have been developed (Cereghino et al. 2001; Nett and Gerngross 2003). Maps and sequences of vectors containing the first four marker genes are available (http://faculty.kgi.edu/cregg/index.htm). In addition, a series of host strains containing all possible combinations of *ade1*, *arg4*, *his4*, and *ura3* auxotrophies has been generated and made available (http://faculty.kgi.edu/cregg/index.htm; Cereghino et al. 2001).

Drug-resistant selectable markers are often more convenient to use as they usually do not require a specific mutation in the host strain for selection. Three such marker systems have been developed for *P. pastoris*. As shown in Table 7.2, these marker genes are the *E. coli* Tn903kan^R kanamycin/geneticin/G418 resistance gene (*Kan^R*) (Scorer et al., 1994), the *Streptoalloteichus hindustanus She ble* gene conferring resistance to the bleomycin-related drug Zeocin (*Zeo^R*) (Higgins et al. 1998), and the *Streptomyces griseochromogens* gene, conferring resistance to the nucleoside antibiotic Blasticidin S (*Bsd^R*) (www.invitrogen.com). At present, vectors containing the *Zeo^R* gene are most popular due to their small size, a consequence of the *Zeo^R* gene being the selectable marker for transformations of both *E. coli* and *P. pastoris* in these vectors (Figure 7.1). The *Bsd^R* vectors appear to be comparable to the *Zeo^R* vectors but are relatively new, with few reports of their use in the literature as yet.

As described in Section 7.2.6, all three systems offer the potential to enrich for strains with multiple integrated copies of a vector by selection for transformants with resistance to high levels of the appropriate drug. In addition, with each of these systems, it is only necessary to add drug during the initial selection of transformants (and perhaps once again to single colony purify strains). Once isolated, the strains are stable – provided one follows standard microbial storage and manipulation procedures – and do not need further drug selection, for example during growth of expression strains in shake flask or fermentor cultures.

A counterselective marker based on a corn mitochondrial gene *T-urf13* has been described (Soderholm et al. 2001). Expression of this gene in *P. pastoris* renders the

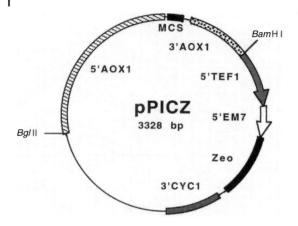

Fig. 7.1 Map of *P. pastoris* expression vector pPICZ.

cells sensitive to the insecticide methomyl, making this gene a useful counter marker for pop-in/pop-out gene replacement strategies in the yeast.

Finally, a marker system utilizing the *P. pastoris FLD1* gene as the selectable marker was recently reported (Sunga and Cregg 2004). Like biosynthetic gene systems, *FLD1*-based vectors require a mutant host (*fld1*) for initial selection of transformants for growth on methylamine as nitrogen source. However, like the drug resistance vectors, once the methylamine utilizing colonies are selected, one can enrich populations of transformants with multiple vector copies by selection for strains that are resistant to high levels of formaldehyde. Vectors containing the *FLD1* marker gene were constructed so that all non-yeast sequences (except the foreign gene itself) can be removed prior to transformation in *P. pastoris*. In particular, bacterial antibiotic resistance genes and origins of replication are not present in the final strains, substantially improving the biological safety of such strains over typical bacterial shuttle expression vectors.

7.2.4
Host Strains

All *P. pastoris* expression strains are derived from NRRL-Y 11430 (Northern Regional Research Laboratories, Peoria, IL, USA). Some have one or more auxotrophic mutations, which allow for selection of expression vectors containing the appropriate biosynthetic marker and of diploid products resulting from genetic crosses (Table A7.1). Prior to transformation, all of these strains grow on complex media but require supplementation with the appropriate nutrient(s) for growth on minimal media.

7.2.4.1   Methanol Utilization Phenotype
Most *P. pastoris* host strains grow on methanol at the wild-type rate (Mut$^+$, methanol utilization plus phenotype). However, two other types of host strains are available that vary with regard to their ability to utilize methanol because of deletions in one or both *AOX* genes. Strains with *AOX* mutations are sometimes better producers of

foreign proteins than wild-type strains (Cregg et al. 1987; Tschopp et al. 1987 a; Chiruvolu et al. 1997). Additionally, these strains require only small amounts of methanol for induction and, therefore, do not have the potential methanol feeding problems that large-scale fermentations of Mut$^+$ strains can experience (see Section 7.2.7). KM71 (*his4 arg4 aox1Δ::SARG4*) is a strain in which *AOX1* has been partially deleted and replaced with the *S. cerevisiae ARG4* gene (Cregg and Madden 1987). Since the strain must rely on the weaker *AOX2* for methanol metabolism, it grows slowly on this carbon source (Mut^s, methanol utilization slow phenotype). Another strain, MC100–3 (*his4 arg4 aox1Δ::SARG4 aox2Δ::Phis4*), is deleted for both *AOX* genes and is totally unable to grow on methanol (Mut$^-$, methanol utilization minus phenotype) (Cregg et al. 1989; Chiruvolu et al. 1997). All of these strains, including the Mut$^-$ strain, retain the ability to induce expression at high levels from the *AOX1* promoter.

7.2.4.2 Protease-deficient Host Strains

Several protease-deficient strains – SMD1163 (*his4 pep4 prb1*), SMD1165 (*his4 prb1*), and SMD1168 (*his4 pep4*) – have been shown to be effective in reducing the degradation of some foreign proteins (White et al. 1995; Brierley 1998). This is especially noticeable in fermentor cultures of secreted recombinant proteins, because the combination of high cell density and lysis of a small percentage of cells results in a relatively high concentration of these vacuolar proteases. Unfortunately, these protease-deficient cells are not as vigorous as wild-type strains with respect to *PEP4*. In addition to lower viability, they possess a slower growth rate and are more difficult to transform. Therefore, the use of protease-deficient strains is only recommended in situations where other measures to reduce proteolysis have yielded unsatisfactory results.

An additional protease-deficient strain, SMD1168 *Δkex1::SUC2* (*Δpep4::URA3 Δkex1::SUC2 his4 ura3*), was developed to inhibit proteolysis of murine and human endostatin (Boehm et al. 1999). The Kex1 protease cleaves carboxy-terminal lysines and arginines. Therefore, the *KEX1*-deleted strain was generated to inhibit carboxy-terminal proteolysis and proved to substantially reduce this type of degradation.

7.2.5
Construction of Expression Strains

Although autonomous episomal multicopy vectors exist for *P. pastoris* (Cregg et al. 1985), it was discovered that strains expressing a recombinant gene from a single integrated copy of an expression vector made more product than strains containing the identical expression cassette on a multicopy autonomous vector. Furthermore, integrated vectors are considerably more stable than autonomous vectors, requiring little or no continued selection for the vector after transformation and initial colony formation. Thus, virtually all *P. pastoris* expression vectors are designed to integrate into the genome of the yeast. *P. pastoris* expression vectors can be integrated into the genome in one of two ways. The simplest way is to restrict the vector at a unique site in either the marker gene (e.g., *HIS4*) or the *AOX1* promoter fragment and then transform it into the appropriate auxotrophic mutant. The free DNA termini stimulate homologous recombination events that result in single crossover-type integration events into these loci at high frequencies (50–80% of His$^+$ transformants). The remaining transformants have undergone gene conversion events in which only the marker gene from the vector has integrated into the mutant host locus without other vector sequences (Cregg et al. 1985).

Alternatively, certain *P. pastoris* expression vectors can be digested in such a way that the expression cassette and marker gene are released, flanked by 5' and 3' *AOX1* sequences (see for example Cregg et al. 1987; or Sunga and Cregg 2004). Approximately 10–20% of transformation events are the result of a gene replacement event in which the *AOX1* gene is deleted and replaced by the expression cassette and marker gene. This disruption of the *AOX1* gene forces these strains to rely on the transcriptionally weaker *AOX2* gene for growth on methanol (Cregg et al. 1987, 1989) and, as a result, these strains have a Mut^s phenotype. These gene replacement strains are easily identified among transformed colonies by replica-plating them to methanol and selecting those with reduced ability to grow on methanol (Mut^s phenotype). As mentioned in Section 7.2.4.1., the advantage of Mut^s strains is that they utilize less methanol, contain no bacterial antibiotic resistance genes or replication origins, and sometimes express higher levels of foreign protein than wild-type (Mut$^+$) strains, especially in shake-flask cultures.

7.2.6
Multicopy Strains

Most *P. pastoris* transformants contain a single copy of an expression vector. However, experience has shown that, despite the strength of the *AOX1* promoter, foreign gene expression is often transcriptionally limited. Thus, strains that contain multiple integrated copies of an expression cassette often, but not always (see for example Thill et al. 1990), yield more recombinant protein than single-copy strains (see for example Brierley 1998).

Two general approaches have been developed to construct multicopy expression strains in *P. pastoris*. The first approach involves constructing a vector with multiple head-to-tail copies of an expression cassette (Brierley, 1998). The key to generating this construction is a vector that has an expression cassette flanked by restriction sites that have complementary termini (e.g., *Bam*HI–*Bgl*II, *Sal*I–*Xho*I combinations). The process of repeated cleavage and reinsertion results in the generation of a series of vectors that contain increasing numbers of expression cassettes. A particular advantage to this approach, especially in the production of human pharmaceuticals, is that the precise number of expression cassettes is known and can be recovered for direct verification by DNA sequencing.

The second approach utilizes expression vectors that contain a drug resistance gene as the selectable marker. As described in Section 7.2.3 and Table 7.2, these genes include the bacterial *Kan*^{R}, *Zeo*^{R}, *Bsd*^{R} genes and the *P. pastoris FLD1* gene. With each of these genes, the level of drug resistance roughly correlates to vector copy number. However, with any of these four marker genes, vector copy number varies greatly, and most transformants still contain only a single vector copy, even at high levels of drug. Thus, a significant number (50–100) of transformants must be subjected to further analysis of copy number and expression level to identify one or two with high copy number. By this approach, strains carrying up to 30 copies of an expression cassette have been isolated (Scorer et al. 1994). Importantly, the multi-copy strains, once isolated, are stable with standard microbial handling procedures (e. g., stock of strain stored frozen at –80 °C, working stock kept on plate on non-inducing medium) and do not require continued drug selection on plates or in liquid medium.

7.2.7
Growth in Fermentor Cultures

P. pastoris is a poor fermentor, and this is a major advantage relative to *S. cerevisiae*. In high cell density (HCD) fermentor cultures, ethanol (the product of *S. cerevisiae* fermentation) rapidly builds to toxic levels, which limits further growth and foreign protein production. With its preference for respiratory growth, *P. pastoris* can be cultured at extremely high densities (150 g L^{-1} dry cell weight; 500 OD_{600} U mL^{-1}) in the controlled environment of the fermentor, with little risk of "pickling" itself. HCD growth is especially important for secreted proteins, as the concentration of product in the medium is roughly proportional to the concentration of cells in culture. For all cultures, high cell densities reduce the total volume of culture needed. Another positive aspect of growing *P. pastoris* in fermentor cultures is that the level of transcription initiated from the *AOX1* promoter can be three- to fivefold greater in cells fed methanol at growth-limiting rates compared to cells grown in excess methanol. Thus, even for intracellularly expressed proteins, product yields can be significantly higher from fermentor-cultured cells. Finally, methanol metabolism utilizes oxygen at a high rate, and expression of foreign genes is negatively affected by oxygen limitation. Only in the controlled environment of a fermentor is it feasible to monitor and adjust oxygen levels in the culture medium.

A hallmark of the *P. pastoris* system is the ease with which expression strains scale-up from shake-flask to high-density fermentor cultures. Although some foreign proteins have expressed well in shake-flask cultures, expression levels are typically low compared to fermentor cultures. Considerable effort has gone into the optimization of heterologous protein expression techniques, and detailed fed-batch and continuous culture protocols are available (Brierley 1998; Clare et al. 1998; Stratton et al. 1998). In general, P_{AOX1}-controlled expression strains are grown initially in a defined medium containing glycerol as its carbon source. During this time, biomass accumulates, but heterologous gene expression is fully repressed. Upon depletion of glycerol, a transition phase is initiated in which additional glycerol is fed to the culture at a growth-limiting rate. Finally, methanol or a mixture of glycerol and methanol is fed to the culture to induce expression. The concentration of foreign protein is monitored in the culture to determine time of harvest.

The growth medium for *P. pastoris* recombinant protein production fermentations is ideal for large-scale production because it is inexpensive and defined, consisting of pure carbon sources (glycerol and methanol), biotin, salts, trace elements, and water. This medium is free of undefined ingredients that can be sources of pyrogens or toxins and is, therefore, compatible with the production of human pharmaceuticals. Also, since *P. pastoris* is cultured in medium with a relatively low pH and methanol, it is less likely to become contaminated by most other microorganisms.

7.3
Post-translational Modification of Secreted Proteins

A major advantage of a fungal expression system over bacterial systems is that fungi are eukaryotes and, therefore, can perform many of the post-translational modifications typically associated with eukaryotes, such as processing of signal sequences, folding, disulfide bridge formation, certain types of lipid addition, and O- and N-linked glycosylation.

7.3.1
Secretion Signals

Foreign proteins expressed in *P. pastoris* can be produced either intracellularly or extracellularly. Because this yeast secretes only low levels of endogenous proteins, the secreted recombinant protein often constitutes the vast majority of total protein in the culture medium (Figure 7.2). Therefore, directing a heterologous protein to the culture medium can serve as a substantial first step in purification. However, due to protein stability and folding requirements, the option of secretion is usually reserved for foreign proteins that are normally secreted by their native hosts. Using the appropriate *P. pastoris* expression vector, researchers can clone a foreign gene with sequences encoding either the native signal of the foreign gene, the *S. cerevisiae* α-factor pre-pro peptide, or the *P. pastoris* acid phosphatase (*PHO1*) signal.

Although several different secretion signal sequences – including the native secretion signal present on heterologous proteins – have been used successfully, results have been variable. The *S. cerevisiae* α-factor pre-pro peptide has been used with the most success. This signal sequence consists of a 19-amino acid signal (pre) sequence followed by a 66-residue (pro) sequence containing three consensus N-linked glycosylation sites and a dibasic Kex2 endopeptidase processing site (Kurjan and Hersko-

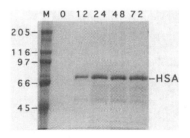

Fig. 7.2 Secreted expression of human serum albumin (HSA). 7.5% SDS-PAGE of 25-µL samples of culture supernatant from a *P. pastoris* strain expressing HSA. Cells were induced in BMMY (buffered methanol-complex medium) for 0, 12, 24, 48, and 72 h. Lane M contains molecular mass markers (kDa). (Reprinted with permission from Lin Cereghino and Cregg, 2000.)

witz 1982). The processing of this signal sequence involves three steps. The first is the removal of the pre signal by signal peptidase in the endoplasmic reticulum. Second, Kex2 endopeptidase cleaves between Arg–Lys of the pro leader sequence. This is rapidly followed by cleavage of Glu–Ala repeats by the Ste13 protein (Brake et al. 1984). The efficiency of this process can be affected by the surrounding amino acid sequence. For instance, the cleavage efficiencies of both Kex2 and Ste13 proteins can be affected by the close proximity of proline residues. In addition, the tertiary structure formed by a foreign protein may protect cleavage sites from their respective proteases.

The *S. cerevisiae* MFα1 pre-pro signal sequence is the most widely used signal for secretion of recombinant proteins from *P. pastoris*. In some cases, it is a better secretion signal for expression in *P. pastoris* than the leader sequence of the native heterologous protein. In a study concerning the expression of the industrial lipase Lip1 from *Candida rugosa*, the effect of heterologous leader sequences on expression and secretion was investigated (Brocca et al. 1998). It was found that the native Lip1p leader sequence allowed for secretion but somehow hampered expression. Either the α-factor pre or pre-pro signal was adequate for both secretion and expression, but the highest level of lipase secretion was from a clone with the full pre-pro sequence. This clone produced two species of secreted protein. A small percentage was correctly processed to the mature protein. However, a majority of the product contained four additional N-terminal amino acids. Variability in the amino terminus is commonly seen with heterologous proteins secreted by *P. pastoris* using the α-factor pre-pro leader.

In some cases, *MFα1* or *PHO1* secretion signals have not worked, so synthetic leaders have been created. Martinez-Ruiz et al. (1998) made mutations in the native leader to reconstruct a more efficient Kex2p recognition motif (Lys–Arg). This aided in secretion of the ribosome-inactivation protein α-sarcin from the mold *Aspergillus giganteus*. Another more drastic solution was to create an entirely synthetic pre-pro leader. For the expression of human insulin, a synthetic leader and spacer sequence was found to improve secretion and protein yield (Kjeldsen et al. 1999).

Recently, yet another signal peptide – PHA-E from the plant lectin *Phaseolus vulgaris* agglutinin – was found to be effective for the secreted expression of two plant lectins and green fluorescent protein (Raemaekers et al. 1999). Additionally, it was found that proteins fused to the PHA-E signal peptide were correctly processed at the amino termini, whereas the same proteins secreted using the *S. cerevisiae* -MFα1 signal had heterogeneous amino-terminal extensions, indicating improper processing of the MFα1 signal. It remains to be seen whether the PHA-E signal sequence works as well in the secretion and processing of other foreign proteins.

7.3.2
O-linked Glycosylation

P. pastoris is capable of adding both O- and N-linked carbohydrate moieties to secreted proteins (Tschopp et al. 1987b; Goochee et al. 1991; Trimble et al., 1991; Miele et al. 1997). Eukaroytic cells assemble O-linked saccharides onto the hydroxyl groups of serine and threonine. In mammals, O-linked oligosaccharides are composed of a variety of sugars, including N-acetylgalactosamine, galactose (Gal), and sialic acid (NeuAc). In contrast, lower eukaryotes such as *P. pastoris* add O-oligosaccharides composed solely of mannose (Man) residues. No consensus primary amino acid sequence for O-glycosylation appears to exist. Additionally, different hosts may add O-linked sugars on different residues in the same protein. Consequently, it should not be assumed that *P. pastoris* will not glycosylate a recombinant protein, even if that protein is not glycosylated by its native host. For instance, although insulin-like growth factor 1 (IGF-1) is not glycosylated in humans, *P. pastoris* was found to add O-linked mannose to 15% of expressed IGF-I product (Brierley 1998). It should also not be assumed that the specific Ser and Thr residues selected for O-glycosylation by *P. pastoris* will be the same as the original host.

Although there is little information concerning the mechanism and specificity of O-glycosylation in *P. pastoris*, the presence of O-glycosylation has been reported in some heterologous proteins, such as the *Aspergillus awamori* glucoamylase catalytic domain (Heimo et al. 1997), human IGF-1 (Brierley 1998), Barley α-amylases 1 and 2 (Juge et al. 1996), and human single-chain urokinase-type plasminogen activator (Tsujikawa et al. 1996).

Duman et al. (1998) used a variety of enzymatic and chromatographic procedures to study endogenous cellular proteins and recombinant human plasminogen produced in *P. pastoris*. The study revealed the presence of O-linked α-1,2 mannans containing dimeric, trimeric, tetrameric, and pentameric oligosaccharides. No α-1,3 linkages were detected. The majority of oligosaccharide was equally distributed between α-1,2-linked dimers and trimers.

7.3.3
N-linked Glycosylation

In all eukaryotes, N-glycosylation begins in the endoplasmic reticulum with the transfer of a lipid-linked oligosaccharide unit, $Glc_3Man_9GlcNAc_2$ (Glc = glucose; GlcNAc = N-acetylglucosamine), to asparagine at the recognition sequence Asn-X-Ser/Thr (Goochee et al. 1991). This oligosaccharide core is then trimmed to $Man_8GlcNAc_2$. At this point, glycosylation patterns of lower and higher eukaryotes begin to differ. The mammalian Golgi apparatus performs a series of trimming and addition reactions that generate oligosaccharides composed of $Man_{5-6}GlcNAc_2$ (high-mannose type), a mixture of several different sugars (complex type), or a combination of both (hybrid type). In *S. cerevisiae*, N-linked core units are elongated in the Golgi through the addition of mannose outer chains. Since these outer chains vary in length, endogenous and heterologous secreted proteins from *S. cerevisiae* are heterogeneous in size. These chains are typically 50–150 mannose residues in length – a condition referred to as hyperglycosylation.

Most foreign proteins, as well as native proteins, secreted in *P. pastoris* appear to be $Man_8GlcNAc_2$ or $Man_9GlcNAc_2$ with smaller amounts of carbohydrates with more Man units (Figure 7.3) (Trimble et al. 1991; Montesino et al. 1998). These fungal N-linked high-mannose oligosaccharides represent a significant problem in the

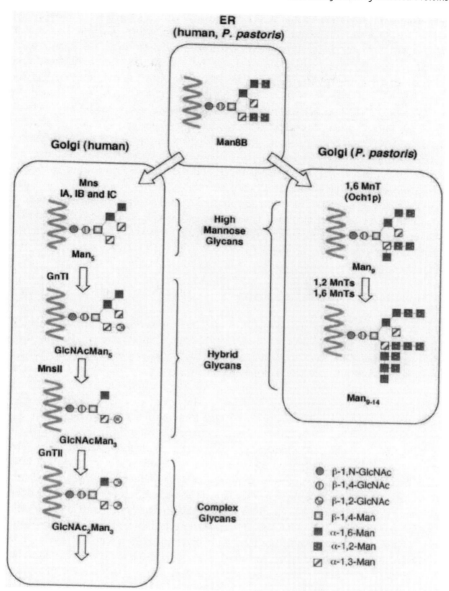

Fig. 7.3 N-linked glycosylation pathway in humans and *P. pastoris.* Mns, α-1,2-mannosidase; MnsII, mannosidase II; GnT1, β-1,2-*N*-acetylglucosaminyltransferase I; GnTII, β-1,2-*N*-acetylglucosaminyl-transferase II; MnT, mannosyltransferase. (Reprinted with permission from Hamilton et al. 2003.)

use of foreign-secreted proteins by the pharmaceutical industry. They can be exceedingly antigenic when introduced intravenously into mammals, and are rapidly cleared from the blood by the liver. An additional problem caused by the differences between yeast and mammalian N-linked glycosylation patterns is that the long outer chains may potentially interfere with the folding or function of a mammalian foreign protein.

Relative to the oligosaccharide structures on *S. cerevisiae*-secreted proteins, at least three differences are apparent in *P. pastoris*-produced proteins. First – and perhaps most importantly – is the frequent absence of hyperglycosylation. Another difference is the presence of α-1,6-linked mannose on core-related structures reported in *P. pastoris*-secreted invertase (Trimble et al. 1991), and the kringle-2 domain of tissue-type plasminogen activator (Miele et al. 1997) and other proteins. Finally, *P. pastoris* oligosaccharides appear not to have any terminal α-1,3-linked mannosylation (Romanos et al. 1992; Montesino et al. 1998). These linkages are unique to fungal glycoproteins and render recombinant proteins produced by fungi unsuitable for human pharmaceutical uses (Romanos et al. 1992).

7.3.4
"Humanization" of N-linked Carbohydrate

Recently, exciting results on the modification ("humanization") of N-linked carbohydrate structures on recombinant proteins secreted from *P. pastoris* have been described. First, Callewaert et al. (2001) showed that modification of carbohydrates on recombinant proteins secreted from *P. pastoris* was possible by co-expressing an endoplasmic reticulum-targeted *Trichoderma reesei* 1,2-α-D-mannosidase along with either of two foreign secreted glycoproteins. The result was recombinant glycoproteins that displayed a dramatic decrease in the number of α-1,2-linked mannose residues. For one of the two foreign proteins, a Trypanosome *trans*-sialidase, the major glycan was a near-human $Man_5GlcNAc_2$ structure. A second research group then pushed these results further by constructing a *P. pastoris* strain with the following modifications: (1) deletion of the *P. pastoris* OCH1 gene (encoding an α-1,6-mannosyltransferase); (2) addition and expression of two different α-1,2-mannosidase genes, and two different N-acetylglucosaminyl transferase genes; and (3) addition of a uridine 5'-diphosphate (UDP)-N-acetylglucosamine transporter gene (Choi et al. 2003; Hamilton et al. 2003). A novel method was utilized to search for mannosidase and transferase genes that performed the desired modifications and that were correctly localized in the secretory pathway: by expressing multiple genes for each of four enzyme types as fusions with a library of fungal type II membrane leader sequences. Each of the four combinatorial libraries (one for each enzymatic step) was sequentially introduced into *P. pastoris* and collections of transformed strains were screened for ones with the proper modifying activity. The result was a strain that predominantly added the human "high-mannose" N-glycan $GlcNAc_2Man_3GlcNAc_2$ to a model recombinant protein.

Further studies are needed to determine whether the humanized glycosylation strain will add this same structure to all recombinant proteins and to all sites on a recombinant protein. Furthermore, fully humanized glycans should also contain some amount of complex and/or hybrid structures, in particular, terminal sialic acid residues. It is not clear that *P. pastoris* will be able to perform these additions. Still, it is evident that great progress has been made in ameliorating this critical yeast post-translational problem.

These carbohydrate modification results should also be of interest to investigators studying protein structure/function. The eukaryotic secretory pathway typically produces glycoproteins with a mixture of glycan structures. This has been a problem for investigators since the resulting variations in carbohydrate structures influence the functional and structural properties of the protein. With these and other glycan-modifying strains, researchers will – for the first time – be able to produce recombinant glycoproteins with uniform carbohydrates of known structure.

7.4
Conclusions

The *P. pastoris* expression system has gained acceptance as an important host organism for the production of recombinant proteins, as illustrated by the number of investigations in which this system has been successfully used (http://faculty.kgi.edu/cregg/index.htm). Several proteins synthesized in *P. pastoris* are in use, or are being tested for use, as human pharmaceuticals in clinical trials. Human serum albumin (HSA), a serum replacement product marketed under the name ALBREC by Mitsubishi Pharmaceuticals, has passed clinical trials, and has received NDA approval in Japan (Ohtani et al. 1998; http://www.m-pharma.co.jp/e/ir/ar2002/pdfs/ar2002.pdf). The company is currently awaiting approval of its HSA manufacturing faculty. Another protein, hepatitis B surface antigen, is on the market as a subunit vaccine against the hepatitis B virus in South America. The angiogenesis inhibitors endostatin and angiostatin are in clinical trials (BKL Sim, unpublished results).

Acknowledgments

The preparation of this chapter was supported in part by a grant from US Department of Energy (DE-FG02-03ER15407).

Appendix

Tab. A7.1 Common *P. pastoris* expression host strains.

Strain name	Genotype	Phenotype	Reference
Y-11430		Wild-type	NRRL[a]
X-33		Wild-type	Invitrogen
GS115	*his4*	Mut⁺ His⁻	Cregg et al. (1985)
KM71	*aox1Δ::SARG4 his4 arg4*	Mutˢ His⁻	Cregg and Madden (1987)
MC100–3	*aox1Δ::SARG4 aox2Δ::Phis4 his4 arg4*	Mut⁻ His⁻	Cregg et al. (1989)
GS241	*fld1*	Mut⁻ Mta⁻	Sunga and Cregg (2004)
MS105	*fld1 his4*	Mut⁻ Mta⁻ His⁻	Sunga and Cregg (2004)
SMD1168	*pep4Δ his4*	Mut⁺ His⁻ Protease- deficient	Invitrogen Higgins and Cregg (1998)
SMD1165	*prb1 his4*	Mut⁺ His⁻ Protease deficient	Higgins and Cregg (1998)
SMD1163	*pep4 prb1 his4*	Mut⁺ His⁻ Protease deficient	Higgins and Cregg (1998)
JC220-JC308[b]	*ade1 arg4 his4 ura3*	Ade⁻ Arg⁻ His⁻ Ura⁻	Cereghino et al. (2001)

a) Northern Regional Research Laboratories, Peoria, IL, USA.
b) Strain set with all possible combinations of these four auxotrophic mutations.

Tab. A7.2 Common *P. pastoris* expression vectors.

Vector name	Selectable markers	Features	Reference
Intracellular pHIL-D2	*HIS4*	*Not*I sites for *AOX1* gene replacement	Invitrogen
pA0815	*HIS4*	Expression cassette bounded by *Bam*HI and *Bgl*II sites for generation of multicopy expression vector	Invitrogen Thill et al. (1990)
pPIC3.5K	*HIS4* and *kanʳ*	Multiple cloning sites for insertion of foreign genes; G418 selection	Invitrogen
pPICZ	*bleʳ*	Multiple cloning sites for insertion of foreign genes; Zeocin selection for multicopy strains; potential for fusion of foreign protein to His₆ and *myc* epitope tags	Invitrogen Miles et al. (1998)
pGAPZ	*bleʳ*	Expression controlled by constitutive *GAPp*; multiple cloning site or insertion of foreign genes; Zeocin selection for multicopy strains; potential for fusion of foreign protein *HIS₆* and *myc* epitope tags	Invitrogen Waterham et al. (1997)

Tab. A7.2 (continued)

Vector name	Selectable markers	Features	Reference
pJL-IX	*FLD1*	P_{AOX1} with *FLD1* gene as selectable marker	Sunga and Cregg (2004)
pBLHIS-IX pBLAR3-IX pBLADE-IX pBLURA-IX	*HIS4* *ARG4* *ADE1* URA3	P_{AOX1} vector series with one of four bio-synthetic gene markers (*HIS4, ARG4, ADE1* or *URA3*)	Cereghino et al. (2001) http://faculty.kgi.edu/cr
Secretion pHIL-S1	*HIS4*	P_{AOX1} fused to *PHO1 secretion signal*; *Xho*I, *Eco*RI, and *Bam*HI sites available for insertion of foreign genes	Invitrogen
pPIC9K	*HIS4* and *kan*^{r}	P_{AOX1} fused to α-MF pre-pro signal sequence; *Xho*I (not unique), *Eco*RI, *Not*I, *Sna*BI and *Avr*II sites available for insertion of foreign genes; G418 selection for multi-copy strains	Invitrogen Scorer et al. (1994)
pPICZα	*ble*^{r}	P_{AOX1} fused to α-MF pre-pro signal sequence; multiple cloning site for insertion of foreign genes; Zeocin selection for multi-copy strains; potential for fusion of foreign protein to HIS_6 and *myc* epitope tags	Invitrogen
pGAPZα	*ble*^{r}	Expression controlled by constitutive P_{GAP}; P_{GAP} fused to α-MF pre-pro signal sequence; multiple cloning site for insertion of foreign genes; Zeocin selection for multicopy strains; potential for fusion of foreign protein to HIS_6 and *myc* epitope tags	Invitrogen Waterham et al. (1997)
pJL-SX	*FLD1*	P_{AOX1} fused to α-MF pre-pro signal sequence; *FLD1* gene as selectable marker	Sunga and Gregg (2004)
pBLHIS-SX pBLARG-SX pBLADE-SX pBLURA-SX	*HIS4* *ARG4* *ADE1* URA3	Series of vectors with P_{AOX1} fused to α-MF pre-pro signal sequence and *H1S4, ARG4, ADE1,* or *URA3* genes as selectable marker	Cereghino et al. (2001); httpa/faculty.kqi.edu/ci

References

Boehm T, Pirie-Shepard S, Trinh LB, Shiloach J, Folkman, J (1999) Disruption of the *KEX1* gene in *Pichia pastoris* allows expression of full-length murine and human endostatin. Yeast 15: 563–567

Brake AJ, Merryweather JP, Coit DG, Heberlein UA, Masiarz FR, Mullenbach GT, Urdea MS, Valenzuela P, Barr PJ (1984) Alpha-factor-directed synthesis and secretion of mature foreign proteins in *Saccharomyces cerevisiae*. Proc Natl Acad Sci USA 81: 4642–4646

Brierley RA (1998) Secretion of recombinant human insulin-like growth factor I (IGF-1). Methods Mol Biol 103: 149–177

Brocca S, Schmidt-Dannert C, Lotti M, Alberghina L, Schmid RD (1998) Design, total synthesis, and functional overexpression of the *Candida rugosa lip1* gene coding for a major industrial lipase. Protein Sci 7: 1415–1422

Callewaert N, Laroy W, Cadirigi H, Geysens S, Saelens X, Min Jou W, Contreras R (2001) Use of HDEL-tagged *Trichoderma reesei* mannosyl oligosaccharide 1,2-α-D-mannosidase for *N*-glycan engineering in *Pichia pastoris*. FEBS Lett 503: 173–178

Cereghino GL, Lim M, Johnson MA, Cereghino JL, Sunga AJ, Raghavan D, Gleeson M, Cregg JM (2001) New selectable marker/auxotrophic host strain combinations for molecular genetic manipulation of *Pichia pastoris*. Gene 263: 159–169

Chiruvolu V, Cregg JM, Meagher MM (1997) Recombinant protein production in an alcohol oxidase-defective strain of *Pichia pastoris* in fed-batch fermentations. Enzyme Microb Technol 21: 277–283

Choi B-K, Bobrowicz P, Davidson RC, Hamilton SR, Kung DH, Li H, Miele RG, Nett JH, Wildt S, Gerngross TU (2003) Use of combinatorial genetic libraries to humanize N-linked glycosylation in the yeast *Pichia pastoris*. Proc Natl Acad Sci USA 100: 5022–5027

Clare J, Sreekrishna K, Romanos M (1998) Expression of tetanus toxin fragment C. Methods Mol Biol 103: 193–208

Couderc R and Baratti J (1980) Oxidation of methanol by the yeast *Pichia pastoris*: purification and properties of alcohol oxidase. Agric Biol Chem 44: 2279–2289

Cregg JM, Barringer KJ, Hessler AY, Madden KR (1985) *Pichia pastoris* as a host system for transformations. Mol Cell Biol 5: 3376–3385

Cregg JM, Madden KR (1987) Development of yeast transformation systems and construction of methanol-utilization-defective mutants of *Pichia pastoris* by gene disruption. In: Biological Research on Industrial Yeasts (Stewart GG, Russell I, Klein RD and Hiebsch RR, Eds), Vol. 2, CRC Press, Boca Raton, FL, pp 1–18

Cregg JM, Tschopp JF, Stillman C, Siegel R, Akong M, Craig WS, Buckholz RG, Madden KR, Kellaris PA, Davis GR, Smiley BL, Cruze J, Torregrossa R, Velicelebi G, Thill GP (1987) High level expression and efficient assembly of hepatitis B surface antigen in the methylotrophic yeast, *Pichia pastoris*. Bio/Technology 5: 479–485

Cregg JM, Madden KR, Barringer KJ, Thill GP, Stillman CA (1989) Functional characterization of the two alcohol oxidase genes from the yeast *Pichia pastoris*. Mol Cell Biol 9: 1316–1323

Cregg JM (1999) Expression in the methylotrophic yeast *Pichia pastoris*. In: Gene Expression Systems: Using Nature for the Art of Expression (Fernandez JM and Hoeffler JP, Eds), Academic Press, San Diego, CA, pp 157–191

Cregg J M, Lin Cereghino J, Shi J, Higgins D R (2000) Recombinant protein expression in *Pichia pastoris*. Mol Biotechnol 16: 23–52

Duman JG, Miele RG, Liang H, Grella DK, Sim KL, Castellino FJ, Bretthauer RK (1998) O-Mannosylation of *Pichia pastoris* cellular and recombinant proteins. Biotechnol Appl Biochem 28: 39–45

Egli T, van Dijken JP, Veenhuis M, Harder W, Fiechter A (1980) Methanol metabolism in yeasts: regulation of the synthesis of catabolic enzymes. Arch Microbiol 124: 115–121

Ellis SB, Brust PF, Koutz PJ, Waters AF, Harpold MM, Gingeras TR (1985) Isolation of alcohol oxidase and two other methanol regulatable genes from the yeast *Pichia pastoris*. Mol Cell Biol 5: 1111–1121

Goochee CF, Gramer MJ, Andersen DC, Bahr JB, Rasmussen JR (1991) The oligosaccharides of glycoproteins: bioprocess factors affecting oligosaccharide structure and their effect on glycoprotein properties. Bio/Technology 9: 1347–1355

HAMILTON SR, BOBROWICZ P, BOBROWICZ B, DAVIDSON RC, LI H, MITCHELL T, NETT JH, RAUSCH S, STADHEIM TA, WISCHNEWSKI H, WILDT S, GERNGROSS TU (2003) Production of complex human glycoproteins in yeast. Science 301: 1244–1246

HEIMO H, PALMU K, SUOMINEN I (1997) Expression in *Pichia pastoris* and purification of *Aspergillus awamori* glucoamylase catalytic domain. Protein Expr Purif 10: 70–79

HIGGINS DR, BUSSER K, COMISKEY J, WHITTIER PS, PURCELL TJ, HOEFFLER JP (1998) Small vectors for expression based on dominant drug resistance with direct multicopy selection. Methods Mol Biol 103: 41–53

HIGGINS DR, CREGG JM (1998) Methods in Molecular Biology: *Pichia* Protocols, Volume 103, Humana Press, Totowa, NJ

JUGE N, ANDERSEN JS, TULL D, ROEPSTORFF P, SVENSSON B (1996) Overexpression, purification, and characterization of recombinant barley alpha-amylases 1 and 2 secreted by the methylotrophic yeast *Pichia pastoris*. Protein Expr Purif 8: 204–214

KJELDSEN T, PETTERSSON AF, HACH M (1999) Secretory expression and characterization of insulin in *Pichia pastoris*. Biotechnol Appl Biochem 29: 79–86

KURJAN J, HERSKOWITZ I (1982) Structure of a yeast pheromone gene (MF alpha): a putative alpha-factor precursor contains four tandem copies of mature alpha-factor. Cell 30: 933–943

LIN CEREGHINO J, CREGG JM (2000) Heterologous protein expression in the methylotrophic yeast *Pichia pastoris*. FEMS Microbiol Rev 24: 45–66

LIU H, TAN X, RUSSELL KA, VEENHUIS M, CREGG JM (1995) *PER3*, a gene required for peroxisome biogenesis in *Pichia pastoris*, encodes a peroxisomal membrane protein involved in protein import. J Biol Chem 270: 10940–10951

MARTÍNEZ-RUIZ A, MARTÍNEZ DEL POZO A, LACADENA J, MANCHEÑO JM, OÑADERRA M, LÓPEZ-OTIN C, GAVILANES JG (1998) Secretion of recombinant pro- and mature fungal á-sarcin ribotoxin by the methylotrophic yeast *Pichia pastoris*: the Lys–Arg motif is required for maturation. Protein Expr Purif 12: 315–322

MIELE RG, CASTELLINO FJ, BRETTHAUER RK (1997) Characterization of the acidic oligosaccharides assembled on the *Pichia pastoris*-expressed recombinant kringle 2 domain of human tissue-type plasminogen activator. Biotechnol Appl Biochem 26: 79–83

MONTESINO R, GARCIA R, QUINTERO O and CREMATA JA (1998) Variation in N-linked oligosaccharide structures on heterologous proteins secreted by the methylotrophic yeast *Pichia pastoris*. Protein Expr Purif 14: 197–207

NETT JH, GERNGROSS TU (2003) Cloning and disruption of the *PpURA5* gene and construction of a set of integration vectors for the stable genetic modification of *Pichia pastoris*. Yeast 20: 1279–1290

OGATA K, NISHIKAWA H, OHSUGI M (1969) A yeast capable of utilizing methanol. Agric Biol Chem 33: 1519–1520

OHTANI W, NAWA Y, TAKESHIMA K, KAMURO H, KOBAYASHI K, OHMURA T (1998) Physico-chemical and immunochemical properties of recombinant human serum albumin from *Pichia pastoris*. Anal Biochem 256: 56–62

RAEMAEKERS RJM, DE MURO L, GATEHOUSE JA, FORDHAM-SKELTON AP (1999) Functional phytohaemagglutinin (PHA) and *Galanthus nivalis* agglutinin (GNA) expressed in *Pichia pastoris*: correct N-terminal processing and secretion of heterologous proteins expressed using the PHA-E signal peptide. Eur J Biochem 265: 394–403

ROMANOS MA, SCORER CA, CLARE JJ (1992) Foreign gene expression in yeast: a review. Yeast 8: 423–488

SCORER CA, CLARE JJ, MCCOMBIE WR, ROMANOS MA, SREEKRISHNA K (1994) Rapid selection using G418 of high copy number transformants of *Pichia pastoris* for high-level foreign gene expression. Bio/Technology 12: 181–184

SEARS IB, O'CONNOR J, ROSSANESE OW, GLICK BS (1998) A versatile set of vectors for constitutive and regulated gene expression in *Pichia pastoris*. Yeast 14: 783–790

SHEN S, SULTER G, JEFFRIES TW, CREGG JM (1998) A strong nitrogen source-regulated promoter for controlled expression of foreign genes in the yeast *Pichia pastoris*. Gene 216: 93–102

SODERHOLM J, BEVIS BJ, GLICK BS (2001) Vector for pop-in/pop-out gene replacement in *Pichia pastoris*. Biotechniques 31: 306–310, 312

STRATTON J, CHIRUVOLU V, MEAGHER M (1998) High cell-density fermentation. Methods Mol Biol 103: 107–120

Sunga AJ, Cregg JM (2004) The *Pichia pastoris* formaldehyde dehydrogenase gene (*FLD1*) as a marker for selection of multicopy expression strains of *P. pastoris*. Gene 330: 39–47

Thill GP, Davis GR, Stillman C, Holtz G, Brierley R, Engel M, Buckholtz R, Kenney J, Provow S, Vedvick T, Siegel RS (1990) Positive and negative effects of multicopy integrated expression vectors on protein expression in *Pichia pastoris*. In: Proceedings of the Sixth International Symposium on the Genetics of Microorganisms (Heslot H, Davies J, Florent J, Bobichon L, Durand G and Penasse L, Eds), Vol. 2, Société Française de Microbiologie, Paris, pp 477–490

Trimble RB, Atkinson PH, Tschopp JF, Townsend RR, Maley F (1991) Structure of oligosaccharides on *Saccharomyces SUC2* invertase secreted by the methylotrophic yeast *Pichia pastoris*. J Biol Chem 266: 22807–22817

Tschopp JF, Brust PF, Cregg JM, Stillman CA, Gingeras TR (1987 a) Expression of the *LacZ* gene from two methanol-regulated promoters in *Pichia pastoris*. Nucleic Acids Res 15: 3859–3876

Tschopp JF, Sverlow G, Kosson R, Craig W, Grinna L (1987 b) High level secretion of glycosylated invertase in the methylotrophic yeast, *Pichia pastoris*. Bio/Technology 5: 1305–1308

Tsujikawa M, Okabayashi K, Morita M, Tanabe T (1996) Secretion of a variant of human single-chain urokinase-type plasminogen activator without an N-glycosylation site in the methylotrophic yeast, *Pichia pastoris* and characterization of the secreted product. Yeast 12: 541–553

Veenhuis M, van Dijken JP, Harder W (1983) The significance of peroxisomes in the metabolism of one-carbon compounds in yeast. Adv Microb Physiol 24: 1–82

Vuorela A, Myllyharju J, Nissi R, Pihlajaniemi T, Kivirikko KI (1997) Assembly of human prolyl 4-hydroxylase and type III collagen in the yeast *Pichia pastoris*: formation of a stable enzyme tetramer requires coexpression with collagen and assembly of a stable collagen requires coexpression with prolyl 4-hydroxylase. EMBO J 16: 6702–6712

Waterham HR, Digan ME, Koutz PJ, Lair SV, Cregg JM (1997) Isolation of the *Pichia pastoris* glyceraldehyde-3-phosphate dehydrogenase gene and regulation and use of its promoter. Gene 186: 37–44

Wegner G (1990) Emerging applications of the methylotrophic yeasts. FEMS Microbiol Rev 7: 279–283

White CE, Hunter MJ, Meininger DP, White LR, Komives EA (1995) Large-scale expression, purification and characterization of small fragments of thrombomodulin: the roles of the sixth domain and of methionine 388. Protein Eng 8: 1177–1187

8
Yarrowia lipolytica
Catherine Madzak, Jean-Marc Nicaud, and Claude Gaillardin

List of Genes

Gene	Encoded gene product
AXP1	extracellular acidic protease (AXP)
GAP1	glyceraldehyde-3-phosphate dehydrogenase
ICL1	isocitrate lyase
KAR2	mammalian BiP (GPR78) homologue, endoplasmic reticulum
LEU2	β-isopropyl malate dehydrogenase
LIP2	extracellular lipase 2
POT1	peroxisomal 3-oxoacyl CoA thiolase
POX1	peroxisomal acyl-CoA oxidase 1
POX2	peroxisomal acyl-CoA oxidase 2
POX5	peroxisomal acyl-CoA oxidase 5
RPS7	ribosomal protein S7
SLS1	GTP exchange factor, endoplasmic reticulum
TEF1	translational elongation factor EF-1 alpha
URA3	orotidine-5'-phosphate decarboxylase
XPR2	extracellular alkaline protease (AEP)

8.1
History, Phylogenetic Position, Basic Genetics, and Biochemistry

8.1.1
Main Characteristics

The hemiascomycetous yeast *Yarrowia lipolytica* was formerly known as *Candida*, *Endomycopsis*, or *Saccharomycopsis lipolytica*. Its initial classification in the *Candida* genus was due to the ignorance of its sexual cycle, until identification of its perfect form in the late 1960s (Wickerham et al. 1970). *Y. lipolytica* is the only known species in its genus, but the asexual taxon *Candida deformans* appears phylogenetically clo-

Production of Recombinant Proteins. Novel Microbial and Eucaryotic Expression Systems. Edited by Gerd Gellissen
Copyright © 2005 WILEY-VCH Verlag GmbH & Co. KGaA, Weinheim
ISBN: 3-527-31036-3

sely related (Bigey et al. 2003). Nearly all natural isolates from *Y. lipolytica* are haploid, heterothallic, and belong either to the A or B mating type. Crossing of A and B strains results in the formation of a stable diploid, which can be induced to sporulate, forming one to four spored asci.

Y. lipolytica metabolizes only a few sugars (mainly glucose, but not sucrose), alcohols, acetate and hydrophobic substrates (such as alkanes, fatty acids and oils), as reviewed in Barth and Gaillardin (1997). The species is strictly aerobic, unlike most other hemiascomycetous yeasts. *Y. lipolytica* is considered as nonpathogenic, as the maximal growth temperature of most isolates does not exceed 32–34 °C. Moreover, several processes based on this yeast, mainly for the agro-food industry, have been classified as GRAS (generally *regarded as safe*) by the FDA (Food and Drug Administration). *Y. lipolytica* is a dimorphic organism which, depending on the growth conditions, is able to form predominantly either yeast cells, hyphae, and pseudohyphae.

Y. lipolytica diverges greatly from other ascomycetous yeasts in that it has: (i) a high GC content; (ii) a high frequency of introns (13% of the genes have one or occasionally more introns), often of a relatively large size (Bon et al. 2003); (iii) unusual structure of its rDNA genes; (iv) a low level of similarity of its genes with their counterparts from other yeasts (typically 30–60% at amino acid level); and (v) unusual types of transposable elements of the Ty3, LINE, or MULE families (Casarégola et al. 2002). On evolutionary trees based on sequences of "house-keeping" genes, *Y. lipolytica* appears isolated from *Schizosaccharomyces pombe* on the one hand, and from the group of other ascomycetous yeasts on the other hand (Barns et al. 1991; Ohkuma et al. 1993).

8.1.2
Historical Perspective on the Development of Studies

During the early 1940s, this yeast was noticed for its uncommon characteristics: it was isolated from lipids or proteinaceous substrates rather than from those containing sugars. In the late 1940s, *Y. lipolytica* cultures were shown to exhibit high extracellular protease and lipase activities. In the mid-1960s, *Y. lipolytica* attracted industrial interest for its ability to grow on n-paraffins as sole carbon source, and for its capacity to produce large amounts of organic acids (Tsugawa et al. 1969). *Y. lipolytica* has been used on a large scale for the production of single-cell proteins and of citric acid, the latter process still being operated in the USA (ADM company, Decatur, IL, USA) using rapeseed oil as a carbon source. This allowed the compilation of extensive data on its cultivation in bioreactors. In the 1970s and 1980s, several research groups performed studies on the genetics, biology, and biochemistry of this yeast (reviewed in Heslot 1990).

Early attempts to develop a genetic system for this species were hampered by poor fertility of the crosses, but inbreeding programs resulted in genetically tractable lines. Thereby, complementation and linkage studies – including tetrad dissection – became feasible (Barth and Gaillardin 1997).

An efficient integrative transformation system became available in the mid-1980s (Davidow et al. 1985; Gaillardin et al. 1985). This results from (generally) homologous recombination events between the vector and the chromosomal target sequences. Single-copy integration is the rule, permitting gene complementation and gene replacement or invalidation.

No natural episomal DNA was ever found in *Y. lipolytica*. For a long time, it was impossible to construct replicative vectors, but this problem was solved in the early 1990s when centromeric ARS sequences were discovered (Fournier et al. 1991). Indeed, both a replicative and a centromeric function are needed to maintain DNA under an episomal form in *Y. lipolytica* (Vernis et al. 2001). The replicative vectors based on the isolated ARSs behave like mini-chromosomes, segregate 2:2 at meiosis, and can be stably maintained as one to three copies per cell (Vernis et al. 1997; Madzak et al. 2000).

These tools, and the early identification of a highly expressed gene encoding an efficiently secreted protein (the alkaline extracellular protease, AEP, encoded by the *XPR2* gene), triggered the first attempts to express heterologous proteins in *Y. lipolytica* (see Gaillardin and Heslot 1988, for an early review). The processing pathway of the AEP was described in detail (Matoba et al. 1989), and the promoter, secretion signal, and terminator of the *XPR2* gene provided the first tools for heterologous protein production. This in turn elicited a strong interest in understanding the basic mechanisms that underlie protein secretion. The secretion pathway appeared to be closer to that of mammalian cell (co-translational secretion of secretory polypeptides) than to that of the baker's yeast *Saccharomyces cerevisiae* (post-translational secretion; see below and Beckerich et al. 1998).

Other aspects of the biology of *Y. lipolytica* were elucidated by various groups, and included: (i) the metabolism of alkanes and fatty acids (Wang et al. 1999; Mauersberger et al. 2001); (ii) dimorphic transition (reviewed in Dominguez et al. 2000; Perez-Campo and Dominguez 2001; Hurtado and Rachubinski 2002); (iii) peroxisome biogenesis (reviewed in Titorenko et al. 2000; Titorenko and Rachubinski 2001); and (iv) biogenesis of the mitochondrial respiratory complex I, which is absent from *S. cerevisiae* but is involved in several diseases in man (Kerscher et al. 2002). A comprehensive review of recent developments in these various fields has been published (Barth et al. 2003).

8.1.3
Secretion of Proteins

Y. lipolytica is distinguished by its capability to secrete a range of proteins, namely two proteases (AEP under alkaline conditions, and AXP under acidic conditions), several lipases and phosphatases, an RNase, and an esterase (reviewed in Barth and Gaillardin 1997). Under appropriate inducing conditions, *Y. lipolytica* secretes very large amounts of AEP (1–2 g L^{-1}; Tobe et al. 1976; Ogrydziak and Scharf 1982). The early steps of protein secretion – the import of the newly synthesized translation products into the endoplasmic reticulum (ER) – can follow either a co-translational or a post-translational translocation mechanism. Co-translation was shown to be predominant in *Y. lipolytica* (Boisramé et al. 1998), as is also the case in higher eukaryotes. This preference renders *Y. lipolytica* particularly competitive for the efficient production of heterologous proteins, in contrast to *S. cerevisiae*, in which the post-translational pathway is predominant. The availability of mutants with strong phenotypes affected in this pathway (in contrast to the situation in *S. cerevisiae*), permitted the identification of various genes involved in the quality control of secreted proteins. A particularly interesting gene product identified in this way is Sls1p that controls the activity of the secretory chaperone Kar2p. Sls1p was the first member of a new family of eukaryotic GEF-factors for Hsp70 proteins (Kabani et al. 2003).

8.1.4
Production of Heterologous Proteins and Glycosylation

Several non-*Saccharomyces* yeasts have been used for heterologous production, including the fission yeast *Saccharomyces pombe*, and several species combined under the somewhat misleading term of "non-conventional" yeasts: the methylotrophs *Hansenula polymorpha* and *Pichia pastoris* (described in Chapters 6 and 7, respectively), the dairy yeast *Kluyveromyces lactis*, the amylolytic yeast *Schwanniomyces occidentalis*, the alkane-utilizer *Yarrowia lipolytica*, and *Arxula adeninivorans*. A reliable comparison between these expression systems is difficult, due to the disparity of the available data and experimental set-ups. The comparative study of Müller et al. (1998) tested *S. cerevisiae*, *H. polymorpha*, *K. lactis*, *S. pombe*, and *Y. lipolytica* for the secretion of active forms of six selected fungal enzymes. The non-*Saccharomyces* yeasts were found to be the most efficient species, but the productivity varied greatly for the different heterologous proteins tested. The most attractive host in this study appeared to be *Y. lipolytica*, especially for the reproducibility of its performance. Since the mid-1980s, *Y. lipolytica* has been applied successfully to the production and secretion of heterologous proteins. However, the production levels obtained for heterologous proteins are in general much lower than those for secreted homologous proteins (e.g., for AEP). In the *Y. lipolytica* expression system, most heterologous proteins can be produced in shake-flasks at the milligram per liter level (often 5 to 20 mg L^{-1}; reviewed in Madzak et al. 2004). In scale-up fermentations in batch or fed-batch mode, a 10- to 20-fold increase in productivity can be obtained (Madzak et al. 2000; Park et al. 2000; Nicaud et al. 2002). Additionally, a further 10- to 20-fold increase can be obtained using multi-copy vectors (Juretzek et al. 2001; Nicaud et al. 2002). Thus, yields of hundreds of milligrams per liter, or even grams per liter can be achieved in processes when combining a multi-copy expression system with high cell density cultivation with optimal parameters (see below, and Nicaud et al. 2002).

Protein glycosylation represents an important protein modification, having a potential impact on stability, activity or immunogenicity (see also Chapters 6 and 7 on *H. polymorpha* and *P. pastoris*). Mammals generate three different types of N-linked glycosylation: high-mannose, complex, and hybrid types. Lower eukaryotes, including yeasts, are only able to perform *N*-glycosylation of the high-mannose type, adding only mannose residues to the outer chains. Moreover, the yeast mannose chains can be much longer than in mammals: in *S. cerevisiae*, they are typically of 50 to 150 mannose residues – a situation referred to as "hyperglycosylation". In some non-conventional yeasts, the extent of hyperglycosylation is less pronounced: *P. pastoris* and *H. polymorpha* produce 8 to 14 mannose chains (Grinna and Tschopp 1989; Gellissen and Hollenberg 1997; see also Chapters 6 and 7). Detailed data addressing this aspect of characterization in *Y. lipolytica* are restricted to a single recombinant protein of human origin where short oligosaccharide chains of about 8–10 mannose residues have been observed (A Franke, personal communication). Taking this example, it seems that the glycosylation pattern of *Y. lipolytica* is closer to the mammalian high-mannose type of glycosylation than to that of *S. cerevisiae.*

8.2
Characteristics of the *Y. lipolytica* Genome

Most natural isolates of *Y. lipolytica* that have been isolated are haploid and are poorly fertile when crossed under laboratory conditions. Hence, for a long time this species was considered as unfertile. Isolates tend to have a clonal evolution resulting in a high genetic diversity of individual strains. Although all strains seem to have six chromosomes, their size varies greatly and numerous cases of apparent gene transposition have been reported (Casarégola et al. 1997). Inbred lines were generated in different laboratories (Barth and Gaillardin 1997). These are genetically tractable and are individually distinguished by specific advantages such as suitability for tetrad dissection, efficient growth on alkanes, or efficient secretion. However, there exists no strain that combines all above advantages, and crossing of different inbred lines always results in poor fertility.

Special mention should be made of the *Y. lipolytica* transposon of the Ty3 family called Ylt1 (Schmid-Berger et al. 1994), as it was important for the development of multicopy expression systems (see Sections 8.3.3.2.2 and 8.3.3.2.3). The Ylt1 retrotransposon harbors large long terminal repeats (714 bp) called "zeta", which usually also exist as numerous dispersed solo copies in the genome. There are two classes of *Y. lipolytica* strains: those containing Ylt1 and zeta (so far, American isolates only), and those not containing them (so far, European isolates). Most laboratory strains result from crosses between American and European lines, and are thus Ylt1-positive: they contain at least 10 copies of Ylt1 together with at least 30–60 additional solo zeta long terminal repeats (LTRs) (Schmid-Berger et al. 1994; Juretzek et al. 2001).

Two types of genomic data are – or soon will be – available. The Genolevures I project assembled around 5000 readings from an industrial, Ylt1-free strain W29 (Casarégola et al. 2000). The Genolevures II project (Dujon et al. 2004) focused on one of the inbred, Ylt1-positive laboratory strains (E150; see Barth and Gaillardin 1997). A range of 182 cloned genes were used as probes on chromosome blots to constitute a scaffold of markers along the six chromosomes. Sequences were obtained from a combination of 3–5, 9–10, and 75–120 kb plasmid insert ends, and resulted in 231 161 reads or a $9 \times$ coverage. This yielded six super contigs representing the six chromosomes, all individual contigs being oriented and ordered along each chromosome. At present, there are less than 50 gaps in the assembly, most of them corresponding to repeated sequences (retrotransposons of the Ty3 and LINE families or

MULE elements). The total size of the assembled genome is around 20 Mb (without rDNA), which makes it the largest genome for a hemiascomycetous species, about twice that of *K. lactis*.

Gene content is surprisingly close to that of other yeast species with smaller genomes (around 6500 versus 6000), suggesting that extra DNA corresponds to noncoding regions such as introns, transposons and others (apparent gene density is around 46.3% versus 70.3% in *S. cerevisiae*).

Centromeres (five out of six were biologically identified previously; Vernis et al. 2001) are short (less than 250 nt), but are not conserved in sequence. Telomeric DNA is composed either of gene families or of rDNA (dispersed into at least six telomeric clusters), and terminate with 13-mer repeats (ACAGTGTATTGGG). A very large number of tRNA genes (510 versus 274 in *S. cerevisiae*) was observed. The significance of this large number remains unclear.

As stated previously, coding sequences of *Y. lipolytica* are distantly related to those of *S. cerevisiae*, making orthological relationships difficult to establish. Conclusive functional predictions are therefore restricted to a minority of genes, although many ORFs are similar to those of filamentous fungi (*A. nidulans* or *N. crassa*) or to those of *C. albicans*. The extent of gene redundancy is apparently similar to that observed in *S. cerevisiae*, but the underlying mechanisms seem to be quite different: instead of the long segmental duplications observed in *S. cerevisiae*, dispersed duplications are the rule in the *Y. lipolytica* genome. Several *Y. lipolytica* gene families have been identified encoding proteins with important functions like proteases, lipases, *POX* genes, cytochrome P450 families, and others.

8.3
Description of the Expression Platform

8.3.1
Host Strains

A study performed on *S. cerevisiae* showed that the genetic background of the host strain could change the efficiency of heterologous protein production by a factor of 100 (De Baetselier et al. 1991). However, no correlation of the performances with known genetic markers could be found. Thus, empirical methods prevail for the choice of a recipient strain.

Many *Y. lipolytica* strains have been used in the field of heterologous protein production (Table 8.1). A current trend is to resort to derivatives of sturdy, well-growing industrial strains such as W29 (Barth and Gaillardin 1997), including the widely used Po1 d strain and its derivatives, Po1 f, Po1 g, and Po1 h. These strains have been genetically engineered to carry nonreverting deletions of marker genes (*LEU2*, *URA3*) as well as additional features interesting for heterologous protein production: (i) the deletion of known extracellular proteases, which were potential threats for secreted heterologous proteins; (ii) the production of *S. cerevisiae* invertase, allowing utilization of sucrose as a new carbon source (Nicaud et al. 1989); and (iii) the integration of a "pBR322 docking platform", in order to facilitate further integration of pBR322-based vectors (Davidow et al. 1987; Blanchin et al. 1994). The production of *S. cerevisiae* invertase by some *Y. lipolytica* strains is particularly relevant for industrial applications, as it allows efficient growth on molasses, a cheap and abundant carbon source (Wojtatowicz et al. 1997). The recently developed strains Po1 f, Po1 g, and Po1 h have already been used for the production of several heterologous proteins (Madzak et al. 2000, 2001; Nicaud et al. 2002; Swennen et al. 2002; Laloi et al. 2002). The characteristics of these strains are listed in Table 8.1, and recapitulated in the Appendix (Table A8.1).

It should be stressed, however, that few published data are available to compare these strains and those derived from inbred lines (i.e., the Po1 g strain was described as 2.7-fold more efficient than the JM23SB strain for the production of bovine prorennin; Madzak et al. 2000).

Tab. 8.1 *Yarrowia lipolytica* strains used for heterologous expression.

Strain	Genotype	Deleted for Protease(s)[a]	Docking Platform[b]	Growth on Sucrose[c]	Reference
Strains issued from Pfizer (Groton, USA) inbreeding program (unpublished):					
PC 30869 (ATCC 20774)	*MatB, leu2-40, xpr2-1002, bio-6*	+	−	−	Davidow et al. (1987) Franke et al. (1988)
DL118 (ATCC 20794)	*MatB, leu2-40, xpr2-1002, bio-6::BIO*	+	+	−	Davidow et al. (1987) Franke et al. (1988)
Strain issued from Ogrydziak et al. (University of California, Davis, USA) inbreeding program (Barth and Gaillardin 1996):					
SMS397A	*MatA, ade1, ura3, xpr2*	+	−	−	Park et al. (1997 and 2000)
Strain issued from Barth et al. (University of Technology, Dresden, Germany) inbreeding program (Barth and Gaillardin 1996):					
B204-12A/213	*MatB, leu2-17, ura3-12*	−	−	−	Bauer et al. (1993)
Strains issued from Gaillardin et al. (INAPG, Grignon, France) inbreeding program (Barth and Gaillardin 1996):					
21501-4	*MatB, lys5-12, leu2-35, ade1, xpr2-34*	+	−	−	Hamsa et al. (1998)
21101-9	*MatA, lys2-5, leu2-35, lyc1-5, ade1, xpr2*	+	−	−	Gaillardin and Ribet (1987)
AM3	*MatA, lys11-23, leu2-270, ura3-302*	−	−	+	Cordero and Gaillardin (1996)
AM4	*MatB, his1, leu2-270, ura3-302*	−	−	+	Cordero and Gaillardin (1996)

Tab. 8.1 (continued)

Strain	Genotype	Deleted for Protease(s)[a]	Docking Platform[b]	Growth on Sucrose[c]	Reference
JM23	MatB, lys5-12, leu2-35, ura3-18, xpr2::LYS5	+	–	–	Nicaud et al. (1991)
JM23SB	MatB, lys5-12, leu2-35, ura3-18::URA3, xpr2::LYS5	+	+	–	Blanchin-Roland et al. (1994) Madzak et al. (1999)
INAG33122	MatB, lys2-5, leu2-35, ade1, xpr2	+	–	–	Nicaud et al. (1991)
E129	MatA, lys11-23, leu2-270, ura3-302, xpr2-322	+	–	+	Barth and Gaillardin (1996)
Po1d (CLIB[d] 139)	MatA, leu2-270, ura3-302, xpr2-322	+	–	+	Le Dall et al. (1994)
Po1f (CLIB 724)	MatA, leu2-270, ura3-302, xpr2-322, axp1-2	++	–	+	Madzak et al. (2000)
Po1g (CLIB 725)	MatA, leu2-270, ura3-302::URA3, xpr2-322, axp1-2	++	+	+	Madzak et al. (2000)
Po1h (CLIB 882)	MatA, ura3-302, xpr2-322, axp1-2	++	–	+	Madzak et al. (2003 and 2004)

a) In this column, "+" indicates that the strain was deleted for AEP (alkaline extracellular protease), and "++" that it was deleted for both extracellular proteases, AEP and AXP (acid extracellular protease).

b) In this column, "+" indicates that the strain carries a "pBR docking platform", provided by the integration of a pBR322 vector carrying a marker gene (*BIO* or *URA3*). This docking platform allows the ulterior integration of pBR-based vectors.

c) In this column, "+" indicates that the strain carries the allele *ura3-302*, corresponding to *ura3::pXPR2:SUC2*, namely to the disruption of *URA3* gene by the *SUC2* gene from *S. cerevisiae* under the control of the *XPR2* promoter. This allele confers the ability to grow on sucrose or mollasses (Nicaud et al. 1989).

d) CLIB: Collection de Levures d'Intérêt Biotechnologique (CBAI, INAPG, 78850 Thiverval-Grignon, France - http://www.inra.fr/clib/).

8.3.2
Selection Markers and Expression Signals

8.3.2.1 Selection Markers

Several genes have been used in *Y. lipolytica* as selective markers (Table 8.2). Few dominant selective markers have proved efficient for use. *Y. lipolytica* is naturally resistant to most commonly used antibiotics, with the exception of the bleomycin/phleomycin group, and of hygromycin B. Heterologous expression of genes conferring resistance to these antibiotics has been achieved (Gaillardin and Ribet 1987; Cordero Otero and Gaillardin 1996), but their use was initially impaired by a high frequency of spontaneous resistance (Barth and Gaillardin 1996). This was partially due to the inadequate strength of the promoter used to drive expression of the resis-

Tab. 8.2 Components used in *Yarrowia lipolytica* expression/secretion vectors.

Type	Components[a]	Characteristics[a]	References
Components for maintenance into yeast cells:			
Marker genes	LEU2, URA3, LYS5, ADE1	auxotrophy complementation	Barth and Gaillardin (1996)
	ura3d4	item + copy number amplification	Le Dall et al. (1994); Pignède et al. (2000); Juretzek et al. (2001); Nicaud et al. (2002)
	Phleo[R], hph (E. coli)	antibiotic resistance	Gaillardin and Ribet (1987); Cordero and Gaillardin (1996); Fickers et al. (2003)
	SUC2 (S. cerevisiae)	sucrose utilization	Nicaud et al. (1989)
Maintenance elements	ARS18, ARS68	Y. lipolytica autonomously replicative sequence	Nicaud et al. (1991); Fournier et al. (1993); Vernis et al. (1997); Madzak et al. (2000)
	homology to genome	homologous integration of linearized vectors	Barth and Gaillardin (1996)
	zeta (Ylt1 LTR)	non-homologous integration in Ylt1-devoid strains	Nicaud et al. (1998); Pignède et al. (2000); Juretzek et al. (2001); Nicaud et al. (2002)
Components from the expression cassette:			
Promoters	pLEU2	inducible by leucine precursor	Gaillardin and Ribet (1987); Franke et al. (1988)
	pXPR2	inducible by peptones	Davidow et al. (1987); Franke et al. (1988); Nicaud et al. (1989)
	pPOX2, pPOT1	inducible by fatty acids, their derivatives and alkanes	Juretzek et al. (2000) Nicaud et al. (2002)
	pICL1	inducible by fatty acids, their derivatives, alkanes, ethanol and acetate	Juretzek et al. (1995, 1998, 2000, 2001)
	pPOX1, pPOX5	weakly inducible by alkanes	Juretzek et al. (2000)
	pG3P	inducible by glycerol	Juretzek et al. (2000)
	pMTP	inducible by metallic salts	Dominguez et al. (1998)
	hp4d	growth-phase dependent	Madzak et al. (2000, 2003); Nicaud et al. (2002)
	pTEF, pRPS7	constitutive	Müller et al. (1998)
Secretion signals	heterologous	frequently efficient in Y. lipolytica	Müller et al. (1998); Park et al. (1997, 2000)
	XPR2 prepro	13 aa/dipeptide stretch/122 aa/cleavage site	Davidow et al. (1987); Franke et al. (1988); Nicaud et al. (1989)
	XPR2 pre + dipept.	13 aa/dipeptide stretch	Franke et al. (1988); Tharaud et al. (1992)
	XPR2 pre	13 aa/cleavage site	Swennen et al. (2002)
	LIP2 prepro	13 aa/dipeptide stretch/10 aa/cleavage site	Nicaud et al. (2002)
Terminators	XPR2t, LIP2t, PHO5t	respectively 430 bp, 150 bp and 320 bp fragments	Barth and Gaillardin (1996)
	minimal XPR2t	100 bp PCR-fragment with added restriction sites	Barth and Gaillardin (1996)

a) Abbreviations: p = promoter; t = terminator; aa = amino acids.

tance markers. This is no longer the case as the use of the recombinant defective hp4 d promoter (see below) now permits efficient selection of hygromycin B-resistant clones (S. Kerscher, personal communication, and unpublished data). This combination of promoter and dominant selection marker gene was also used in disruption cassettes, designed to allow rapid gene disruption and marker rescue in any *Y. lipolytica* strain (Fickers et al. 2003).

Wild-type *Y. lipolytica* strains are unable to use sucrose as a sole carbon source. Nicaud et al. (1989) expressed the *S. cerevisiae SUC2* gene in *Y. lipolytica*, and tested its use as a dominant selection marker. Unfortunately, selection was impaired by residual growth of *Y. lipolytica* due to impurities on the selective sucrose plates (Barth and Gaillardin 1996). Nevertheless, this approach permitted the design of invertase-producing strains such as Po1 d and its derivatives, that are able to grow on sucrose and molasses (see Section 8.3.1).

Auxotrophy markers constitute the most widely used selection method for *Y. lipolytica* transformants. Although several auxotrophy-complementing genes are currently available, the most commonly used are *LEU2* and *URA3*, as recipient strains carrying non-leaky and non-reverting *leu2* and *ura3* mutations have been constructed (see Section 8.3.1).

In order to select for multiple integration events, defective alleles of the *URA3* marker gene have been designed by sequential deletions in the *URA3* promoter (Le Dall et al. 1994). The *ura3d4* allele, with only 6 bp left upstream from the ATG sequence, constitutes the most commonly used defective allele: a single copy is unable to confer an Ura$^+$ phenotype, whereas around 10 copies permit almost normal growth in a uracil-free medium (Le Dall et al. 1994; see Section 8.3.3.2.2).

8.3.2.2 The Expression Cassette

Similarly to observations made previously in other yeasts, the expression of heterologous genes in *Y. lipolytica* was effective only when using "sandwich vectors", namely when the gene of interest was inserted between a yeast promoter and a yeast terminator (Franke et al. 1987). This set of sequences constitutes the "expression cassette", which can also include homologous or heterologous secretion signals. All of these elements are listed in Table 8.2.

8.3.2.2.1 Promoters

Historically, the *XPR2* promoter (p*XPR2*) has been of great importance in the development of molecular tools for *Y. lipolytica* research. The native form was initially used for the expression of heterologous proteins (Davidow et al. 1987; Nicaud et al. 1989, 1991; Hamsa and Chattoo 1994; Müller et al. 1998). However, industrial use was prevented by the complex regulation of this promoter: it is active only at pH above 6, on media lacking readily metabolizable carbon and nitrogen sources (e. g., glucose or glycerol and ammonia). Its full induction requires high levels of peptones in the culture medium (Ogrydziak et al. 1977). This situation stimulated the search for new promoters which were more adapted to the constraints of industrial production.

The functional dissection of p*XPR2* showed that one of its upstream activating sequences, UAS1, was poorly affected by environmental conditions (Blanchin-Roland et al. 1994; Madzak et al. 1999). Consequently, this element was used to construct a hybrid promoter, composed of four tandem copies of the UAS1, inserted upstream from a minimal *LEU2* promoter (reduced to its TATA box). This recombinant promoter, hp4d, is almost independent of environmental conditions such as pH, carbon and nitrogen sources, and the presence of peptones (Madzak et al. 1995, 2000). hp4d is able to drive a strong expression in virtually any medium, but retains unidentified elements that drive a growth-phase-dependent gene expression that is maximal at the beginning of stationary phase (Madzak et al. 2000; Nicaud et al. 2002). Fortunately, these characteristics are optimal for heterologous production, as the growth and expression phases are dissociated, thus minimizing the burden of over-expressing potentially toxic proteins during the growth phase. hp4d has been used successfully for the production of various heterologous proteins in *Y. lipolytica* (Madzak et al. 2000, 2001; Richard et al. 2001; Laloi et al. 2002; Nicaud et al. 2002; Swennen et al. 2002; also unpublished data from M. Chartier, C. Gysle, C. Jolivalt, N. Libessart, B.C. Sang and K. Uchida; see Table 8.3).

Two strong constitutive promoters have been isolated and described, namely those derived from the *Y. lipolytica TEF* and *RPS7* genes (Müller et al. 1998). These promoters are particularly suited to express genes from fungal sources. However, they are not generally recommended for heterologous production *per se* as the use of strong constitutive promoters results in high early expression of the heterologous protein that can be detrimental to culture growth.

For the above reason, an inducible promoter, the expression of which could be either completely repressed or strongly induced under easily controllable conditions, would be the optimal choice for an industrial production process. Unfortunately, such a promoter remains to be defined for *Y. lipolytica*. Nevertheless, a number of inducible promoters with interesting properties have been described. Dominguez et al. (1998) reported the use of the bi-directional metallothionein promoter, but the requirement of metal salts for induction renders it unattractive for most industrial purposes. In our laboratory, J.-M. Nicaud focused on the capacity of *Y. lipolytica* to grow on hydrophobic substrates, and isolated several promoters of genes encoding key enzymes of this pathway. Promoters from isocitrate lyase (*ICL1*), 3-oxo-acyl-CoA thiolase (*POT1*), and acyl-CoA oxidases (*POX1*, *POX2*, and *POX5*) genes were isolated and tested. They have been compared for strength and regulation by various carbon sources to the promoters of glycerol 3-phosphate dehydrogenase and alkaline protease genes, and to the recombinant promoter hp4d (Juretzek et al. 2000). The results of these studies showed that p*ICL1*, p*POT1*, and p*POX2* were the strongest inducible promoters available for *Y. lipolytica*. They are highly inducible by fatty acids and alkanes, and repressed by glucose and glycerol. Additionally, p*ICL1* is also strongly inducible by ethanol and acetate, but its repression by glucose and glycerol is incomplete. p*ICL1* and p*POX2* have been used successfully for heterologous protein production (Juretzek et al. 1995, 1998, 2001; Nicaud et al. 2002; Bhave and Chattoo 2003; also unpublished data from S. Mauersberger and B. Nthangeni B; see Table 8.3). However, industrial uses of these new inducible promoters still face some problems despite their high efficiency, namely the high basal level of expression of p*ICL1* on glucose, and the hydrophobic nature of p*POT1* and p*POX2* inducers, that may be incompatible with efficient protein production or purification.

Tab. 8.3   Heterologous protein expression in *Yarrowia lipolytica*.

Organism	Protein	Molecular mass (kDa)	Localiza-tion[a]	Source or Reference[b]
Viruses				
Hepatitis B virus	pre-HBs antigen	30	(−)	Hamsa and Chattoo (1994)
Bacteriophage P1	Cre recombinase	41	Int.	Richard et al. (2001); Fickers et al. (2003)
Eubacteria				
Escherichia coli	β-galactosidase	116	Int.	Gaillardin and Ribet (1987); Blanchin-Roland et al. (1994); Madzak et al. (2000); Juretzek et al. (2000, 2001)
E. coli	Tn5 phleomycin resistance gene	15	Int.	Gaillardin and Ribet (1987)
E. coli	β-glucuronidase	68	Int.	Bauer et al. (1993)
E. coli	hygromycin B resistance gene	41	Int.	Cordero and Gaillardin (1996); Fickers et al. (2003)
E. coli	XylE catechol dioxygenase	59	Int.	R. Cordero Otero, pers. commun.
E. coli	amylolytic enzyme	85	Sec.	N. Libessart, pers. commun.
Cyanobacteria				
Vitreoscilla stercoraria	single-chain hemoglobin VHb	18	Int.	Bhave and Chattoo (2003)
Fungi				
Saccharomyces cerevisiae	invertase	85	Sec.	Nicaud et al. (1989)
Aspergillus aculeatus	cellulase I	29	Sec.	Müller et al. (1998)
A. aculeatus	galactanase I	44	Sec.	Müller et al. (1998); D. Swennen, pers. commun.
A. aculeatus	polygalacturonase I	45	Sec.	Müller et al. (1998)
Humicola insolens	cellulase II	57	Sec.	Müller et al. (1998)
H. insolens	xylanase I	27	Sec.	Müller et al. (1998)
Thermomyces lanuginosus	lipase I	35	Sec.	Müller et al. (1998)
Trichoderma reesei	endoglucanase I	45	Sec.	Park et al. (2000)
Arxula adeninivorans	glucoamylase	90	Sec.	Swennen et al. (2002)
Aspergillus oryzae	leucine aminopeptidase	90	Sec.	Nicaud et al. (2002)
Alternaria alternata	recombinant Alta1p allergen	14	Sec.	Morin and Dominguez, TYLIM
Pycnoporus cinnabarinus	laccase I	54	Sec.	Madzak et al., SICRPP
Trametes versicolor	laccase IIIb	58	Sec.	Jolivalt et al. (2004)
Plants				
Oryza sativa	α-amylase	45	Sec.	Park et al. (1997)
Zea mays	cytokinin oxidase I	55	Sec.	Madzak et al. (2001)
Zea mays	cytokinin oxidase III	55	Sec.	Houba-Hérin et al., FMFSPP
Theobroma cacao	aspartic proteinase II	62	Sec.	Laloi et al. (2002)
Mammals				
Bos taurus	prochymosin	40	Sec.	Franke et al. (1988); Nicaud et al. (1991); Madzak et al. (2000)
Bos taurus	cytochrome P450 17α	61	Int.	Juretzek et al. (1995, 1998); Mauersberger et al. TYLIM

Tab. 8.3 (continued)

Organism	Protein	Molecular mass (kDa)	Localization[a]	Source or Reference[b]
Sus scrofa domestica	α1-interferon	23	Sec.	Nicaud et al. (1991)
Mus musculus	interleukin 6	20	Sec.	B.C. Sang, pers. commun.
Lama glama	anti-ACE VHH antibody	30	Sec.	M. Chartier, pers. commun.
Homo sapiens sapiens	anaphylatoxin C5a	9	Sec.	Davidow et al. (1987)
H. sapiens s.	blood coagulation factor XIIIa	80	Sec.	Tharaud et al. (1992)
H. sapiens s.	proinsulin analog	10	Sec.	James and Strick (1993)
H. sapiens s.	insulinotropin	4	Sec.	James and Strick (1993)
H. sapiens s.	epidermal growth factor	6	Sec.	Hamsa et al. (1998)
H. sapiens s.	tissue plasminogen activator	67	Sec.	A. Franke, pers. commun.
H. sapiens s.	α-foetoprotein	74	Sec.	K. Uchida, pers. commun.
H. sapiens s.	β2-microglobulin	12	Sec.	K. Uchida, pers. commun.
H. sapiens s.	soluble CD14 variant	48	Sec.	Gysler et al., TYLIM
H. sapiens s.	cytochrome P450 CYP1A1	61	Int.	Nthangeni et al., TYLIM
H. sapiens s.	anti-Ras single-chain antibody scFv	30	Sec.	Swennen et al. (2002)
H. sapiens s.	anti-estradiol scFv	30	Sec.	M. Chartier, pers. commun.

a) Int.: the protein accumulated intracellularly (no secretion signal was present). Sec.: the protein was secreted into the culture medium, due to the presence of either its native, or of a *Y. lipolytica* secretion signal. (–): the protein failed to be secreted despite the presence of a *Y. lipolytica* secretion signal (antigens assembled intracellularly, in the form of Dane particles).

b) pers. commun.: personal communication. TYLIM: Third *Yarrowia lipolytica* International Meeting, Dresden University of Technology, Dresden, July 2002. SICRPP: Second International Conference on Recombinant Protein Production, European Federation of Biotechnology, Cernobbio, Como, November 2002. FMFSPP: Fifth Meeting of the French Society for Plant Physiology, Plant Biology and the Challenge of Functional Genomics, Orsay, July 2003.

8.3.2.2.2 Secretion Targeting Signals

Both homologous and heterologous (native) secretion leader sequences have been used successfully in *Y. lipolytica*. The signal sequence which has been most widely used is the pre-pro-region from the *XPR2* gene. This sequence has been shown to target the early steps of protein secretion to the co-translational pathway of translocation (He et al. 1992; Yaver et al. 1992). The *XPR2* pro-region was shown to be required for the transit of the cognate AEP protein (Fabre et al. 1991, 1992). However, its presence does not seem to be important for the secretion of heterologous proteins, as the *XPR2* pre-region alone (with or without the following dipeptide stretch) was shown to be sufficient to drive efficient heterologous secretion (Franke et al. 1988; Tharaud et al. 1992; Swennen et al. 2002). More recently, other secretory signals have been successfully used: the pre-pro-region of the *Y. lipolytica LIP2* gene (Pignède et al. 2000), and a recombinant sequence derived from a fusion of *XPR2* and *LIP2* pre-pro-regions (Nicaud et al. 2002).

In several instances, the native secretory signals of heterologous proteins were found to drive efficient secretion in *Y. lipolytica*. As an example, Müller et al. (1998)

described the efficient secretion of six proteins from three different fungi using their native signal peptides. Other successful attempts have been reported not only for fungi (Park et al. 2000; Jolivalt et al. 2004; present authors' unpublished data), but also for plant proteins (rice: Park et al. 1997, and maize: unpublished data of N. Houba-Hérin; see Table 8.3).

In a few cases, several secretion sequences were tested for particular proteins including the preferred *XPR2* pre-pro- or *XPR2* pre-sequences. From this comparison no general rule could be deduced as the performance of a selected sequence differed depending on the heterologous protein tested (Franke et al. 1988; Tharaud et al. 1992; Park et al. 1997; Swennen et al. 2002; unpublished data). In a few cases, this variability was interpreted in terms of compatibility between the secretion sequence and the protein (Park et al. 1997): an abnormal processing was reported when the *XPR2* pre-pro-region was used to direct the secretion of rice α-amylase, possibly reflecting an unfavorable secondary structure at the dibasic cleavage site of the fusion protein.

8.3.3
Shuttle Vectors for Heterologous Protein Expression

The expression vectors used for transformation of *Y. lipolytica* are shuttle vectors, as in other yeast systems. Typically, they contain a bacterial part and, for the expression in *Y. lipolytica*, also: (i) a selection marker; (ii) the expression cassette; and (iii) elements for maintenance within yeast cells. Examples of the different types of vectors will be given hereafter, and in Appendix Table A8.2. Two major types of expression vectors can be used in *Y. lipolytica*, these differing by their mode of maintenance, namely episomal replicative and integrative vectors. These expression vectors can be introduced into *Y. lipolytica* host strains using either the lithium acetate method (see Section 8.5.3) or electroporation (Fournier et al. 1993). The lithium acetate method yields very high transformation efficiencies, but frequent recombination events between short repeated sequences are observed when using replicative plasmids (Barth and Gaillardin 1996). Thus, the lithium acetate method is recommended for integrative vectors, while electroporation is recommended for replicative ones.

8.3.3.1 Replicative Vectors
The need for both centromeric and replicative functions renders *Y. lipolytica* ARS elements unattractive for expression vector design, in that: (i) their copy number is limited to one to three copies per cell (Vernis et al. 1997; Madzak et al. 2000); (ii) as a consequence, the gene dosage-dependent expression is also limited (Nicaud et al. 1991; Madzak et al. 2000); and (iii) they require the maintenance of a selective pressure. Nevertheless, replicative vectors have been used successfully for the production of heterologous proteins (Nicaud et al. 1991; Müller et al. 1998).

8.3.3.2 Integrative Vectors
The integration of exogenous DNA into *Y. lipolytica* genome occurs mainly by homologous recombination, which is greatly stimulated by the linearization of the plasmid within the targeting region. This results in very high transformation frequencies (up

to 10^6 transformants per μg of DNA; Xuan et al. 1988). The integration of a linearized shuttle vector by single crossover can thus be precisely directed to a defined genomic site of the recipient strain (i. e., in the selection marker, in the terminator, or even in the bacterial moiety, for integration in strains carrying a corresponding docking platform). In more than 80% of the cases, a single complete copy of the vector will be integrated at the selected site (Barth and Gaillardin 1996). Integrated vectors exhibit a very high stability, as demonstrated by Hamsa and Chattoo (1994). The homologous integration of mono-copy vectors into *Y. lipolytica* offers several advantages over the situation in other yeast expression systems, such as *P. pastoris*: (i) a very high transformation efficiency; and (ii) a precise targeting of the copy to the genome. With these characteristics, the analysis of only a small number of transformants is needed to select a correct integrant. The performance of integrants bearing different constructs can be readily compared, as copy numbers and integration loci are well defined. These characteristics facilitate further genetic engineering via directed mutagenesis, DNA shuffling or in-vitro evolution for the improvement of constructs and of their products for industrial applications (Mougin et al. 2003).

8.3.3.2.1 Examples of Mono-copy Integrative Vectors

This series of vectors is based on the recombinant promoter hp4d and on the *LEU2* selection marker gene (see below). Their bacterial moiety is derived from the plasmid pBR322, and allows targeting, after linearization in this region, to a docking platform composed of an integrated pBR322 sequence into the genome of the Leu⁻ Po1g recipient strain. The series comprises: (i) an expression vector, pINA1269, without any secretion signal (allowing use of a native one); and (ii) two expression/secretion vectors, pINA1267 and pINA1296 (Figure 8.1), using a *Y. lipolytica* secretion signal (see Appendix, Table A8.2). These vectors were used successfully for the production of several heterologous proteins (see Table 8.3).

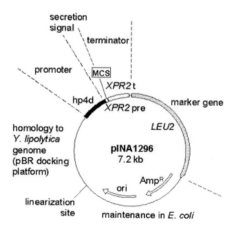

Fig. 8.1 Schematic map of the monocopy expression/secretion vector pINA1296. The bacterial moiety from the shuttle vector is represented by a thin line, with the bacterial replication origin and selection marker indicated as internal arrows. The yeast elements are represented as boxes (for details, see the text), and MCS indicates a "multiple cloning site" for the insertion of the heterologous gene of interest. The vector can be linearized at the corresponding site in order to target its integration to the pBR322 docking platform of the recipient strain genome.

8.3.3.2.2 Homologous Multiple Integrations

The first attempts to increase the copy number in *Y. lipolytica* used homologous multiple integrations into the ribosomal DNA (rDNA) cluster, together with a defective selection marker which does not allow growth under selective conditions when present as a single copy (James and Strick 1993; Le Dall et al. 1994). Le Dall et al. (1994) used an expression vector carrying a fragment of *Y. lipolytica* rDNA as the targeting region, the defective *ura3d4* allele to select for multicopy integrations, and *XPR2* as a reporter gene to quantify expression. These authors obtained transformants carrying up to 60 integrated copies of the expression vector. These copies contained mostly tandem repeats of the vector at one or two genomic sites, and also a few dispersed copies, and were fairly stable under non-inducing conditions. However, when the expression of the *XPR2* gene was induced, most copies were lost during subsequent cultivation, and the copy number eventually stabilized at about 10 copies per cell. This "de-amplification" phenomenon appeared to be linked to deleterious effects of overproducing the alkaline protease (AEP) encoded by the *XPR2* gene. For up to about 10 copies per genome, some correlation was observed between the copy number and AEP production.

Multiple integrations can be obtained even when the integration site is unique in the genome of *Y. lipolytica*. This was demonstrated by Juretzek et al. (2001), using vectors carrying the defective *ura3d4* allele (and the *lacZ* reporter gene under the control of an *ICL1* promoter). Vectors were integrated as tandem repeats at the single *XPR2* locus. As observed previously by Le Dall et al. (1994), the number of integrated copies tended to stabilize around 10–13 during cultivation under inducing conditions (Juretzek et al. 2001).

The above systems have two drawbacks: instability of tandem repeats; and a possible locus effect on gene expression. Although there is still no clear documentation on genome position effects in *Yarrowia*, assays were carried out using the (dispersed) long terminal repeats "zeta" derived from Ylt1 as target sequences (see Section 8.2).

The transformation frequency of a Ylt1-positive *Y. lipolytica* strain with a vector carrying both a zeta sequence and *ura3d4* is low, in the range of 10^2 transformants per µg (Juretzek et al. 2001). However, even this low efficiency remains comparable to those observed in other yeast expression systems, such as *P. pastoris* (Sreekrishna and Kropp 1996). In such a recipient, zeta-targeted vectors were found to integrate as expected as tandem repeats at a single, or sometimes at a few, zeta site(s). As each integration event affected a specific zeta locus, integrants differed in integration sites, copy numbers and, as a result, exhibited a great diversity in productivity. The identification of an optimal production strain requires therefore that a range of independent transformants be tested. This strategy has been used successfully to define strains over-expressing the gene for the *Y. lipolytica* extracellular lipase (Pignède et al. 2000) or other heterologous genes (Juretzek et al. 2001). The lipase over-producing transformants harbored about 10 integrated copies of the vector. This copy number was maintained after more than 120 generations, under both non-inducing and inducing production conditions (Pignède et al. 2000).

8.3.3.2.3 Non-homologous Multiple Integrations

A second advantage of zeta sequences was unexpectedly observed when zeta-free strains were used as hosts: these strains could be transformed with vectors carrying a zeta sequence, albeit at a low frequency, but much more efficiently than with non-zeta vectors devoid of genome homology (Nicaud et al. 1998). This was observed with all Ylt1-free strains tested, such as the German isolate H222 or the French isolate W29 and its derivatives (Casarégola et al. 2000; Juretzek et al. 2001).

Multi-copy zeta-based vectors (carrying the defective *ura3d4* allele) were shown to integrate non-homologously into a Ylt1-free strain with an efficiency that was 5- to 10-fold lower than into a Ylt1-positive strain (Pignède et al. 2000; Juretzek et al. 2001): transformation efficiency fell to about 10 transformants per µg (Juretzek et al. 2001). Despite this low transformation efficiency, these nonhomologous transformation events are highly interesting as in most cases they are derived from dispersed single-copy integrations, leading to a better stability of high-copy-number integrants (Nicaud et al. 1998). This method has already been used successfully to produce several heterologous proteins (Nicaud et al. 2002; Laloi et al. 2002).

8.3.3.2.4 Examples of Multicopy Integrative Vectors

This series of expression vectors, without any secretion signal, is based on p*ICL1*, an inducible promoter (see Section 8.3.2.2.1.). All carry the *ura3d4* defective selection marker. The series comprises: (i) the p64IP vector, carrying a *Y. lipolytica* rDNA region as targeting element, which allows it to transform any Ura⁻ strain (such as Po1 f or Po1 h); and (ii) the p67IP vector, carrying a zeta element for targeting, which allows it to integrate either homologously into any Ylt1-carrying Ura⁻ strain, or non-homologously into any Ylt1-free Ura⁻ strain (such as Po1 f or Po1 h). Mono-copy counterparts of these vectors (respectively p65IP and p66IP), in which *ura3d4* is replaced by a non-defective *URA3* allele, are also available (see Appendix Table A8.2).

8.3.3.2.5 Auto-cloning Vectors

During transformation of yeast with a standard integrative vector, the bacterial moiety of the shuttle vector integrates into the genome. The presence of this bacterial DNA (especially of the antibiotic resistance gene) constitutes a drawback for acceptance of such yeast production strain by regulatory authorities for commercial applications. With "auto-cloning" expression vectors, this problem can be avoided, as the resulting strains are devoid of bacterial DNA, including antibiotic resistance genes (Nicaud et al. 1998; Pignède et al. 2000). In these vectors, the bacterial moiety can be separated from the rest of the vector ("yeast cassette") using a restriction digestion followed by an agarose gel electrophoresis. The purified "yeast cassette" alone is then used for the transformation of the recipient strain. The transforming part of the auto-cloning vector corresponds to: (i) the (yeast) selection marker; (ii) the expression cassette (see Section 8.3.2.2); and (iii) a split zeta sequence constituting the borders of the DNA fragment. These zeta elements (long terminal repeats from *Y. lipolytica* Ylt1 retrotransposon) are able to promote either homologous or nonhomologous integration, depending on the strain used (see Sections 8.3.3.2.2 and 8.3.3.2.3).

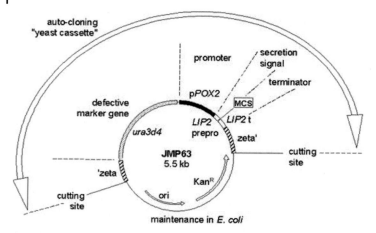

Fig. 8.2 Schematic map of the multicopy auto-cloning expression/ secretion vector JMP63. The bacterial moiety from the shuttle vector is represented as a thin line, with the bacterial replication origin and selection marker indicated as internal arrows. The yeast elements are represented as boxes (for details, see the text), and MCS indicates a "multiple cloning site" for the insertion of the heterologous gene of interest. The "yeast cassette" can be separated from the bacterial moiety by restriction digestion at the two "cutting sites", and then be purified and used to transform the recipient strain.

Two series of such vectors have been designed: one is based on the recombinant promoter hp4d, and the other is based on the inducible pPOX2 promoter.

Five hp4d-based multi-copy auto-cloning vectors are available: (i) two expression vectors, pINA1291 and pINA1292, without any secretion signal (allowing use of a native one); and (ii) three expression/secretion vectors, pINA1293, pINA1294, and pINA1297, using different Y. lipolytica secretion signals (see Appendix Table A8.2).

Four pPOX2-based multi-copy auto-cloning vectors are available: (i) two expression vectors, JMP64 and pYEG10, without any secretion signal; and (ii) two expression/secretion vectors, JMP63 (Figure 8.2) and pYSG10, using a Y. lipolytica secretion signal (see Appendix Table A8.2).

These vectors have been used successfully for the production of various heterologous proteins (Laloi et al. 2002; Nicaud et al. 2002; also unpublished data of C. Gysler and B. Nthangeni; see Table 8.3). Mono-copy counterparts of these vectors, in which ura3d4 is replaced by a non-defective URA3 allele, are also available (see Appendix Table A8.2).

8.4
Examples

To date, more than 40 heterologous proteins have been produced in *Y. lipolytica* (see Table 8.3). These proteins cover a large range of phylogenic origins, of sizes, and of divergent characteristics. A large majority of the producing strains were generated using integrative shuttle vectors, and mainly mono-copy vectors. Fairly good production results have been obtained using integrative mono-copy vectors, for example: (i) 100 mg L^{-1} for *Trichoderma reesei* endoglucanase I, using *XPR2* promoter for expression control (Park et al. 2000); (ii) 160 mg L^{-1} for bovine prochymosin, using hp4d promoter (Madzak et al. 2000); and (iii) 1 g L^{-1} for an *E. coli*-derived amylolytic enzyme, using hp4d promoter (N. Libessart, personal communication). When multi-copy vectors were used, they generally led to a 10- to 20-fold increase in heterologous protein production. For example, 500 mg L^{-1} was obtained in the case of human sCD14 milk protein, using hp4d promoter for expression control (C. Gysler, personal communication). In a few studies, batch or fed-batch cultivations have been performed, leading to a further 8- to 10-fold increase in heterologous protein production (Madzak et al. 2000; Park et al. 2000; Nicaud et al. 2002).

In *Y. lipolytica*, the codon usage differs to some extent to that of *S. cerevisiae*, and is more similar to that of the *Aspergillus* genus (Gaillardin and Heslot 1988; also unpublished data). In the expression studies reviewed here, the heterologous genes had not been adapted to the codon usage of *Y. lipolytica*. This could possibly account for the low production efficiencies (in the µg to the mg per liter range) obtained with some human genes, such as α-fetoprotein and β2-microglobulin (K. Uchida, personal communication).

A range of secretion leader sequences was successfully employed to direct foreign proteins to the secretory apparatus. As an exception, recombinant *S. cerevisiae* invertase was retained in the periplasm when using a *Y. lipolytica* secretion signal, but native invertase was shown to behave similarly in *S. cerevisiae* (Nicaud et al. 1989). The case of the human factor XIIIa is also remarkable: this protein is naturally secreted via an unknown signal-peptide-independent mechanism (Muesch et al. 1990). It was however secreted – albeit inefficiently – by the classical secretory pathway of *Y. lipolytica*, when using the *XPR2* pre-region followed by a dipeptide stretch, as secretion signal (Tharaud et al. 1992). This is one of the rare examples of a secretory protein lacking a classical signal peptide which was re-routed through a yeast secretory pathway (see also Baldari et al. 1987; Livi et al. 1991). In a few studies, the efficiency of the secretion has been determined more precisely: intracellular heterologous activity remained below detection level, when a high activity was detected in the culture supernatant (Franke et al. 1988; Nicaud et al. 1991; Madzak et al. 2000, 2001).

In some of the studies reviewed here, the overall pattern of glycosylation has been assessed using Western blotting and endoglycosidase treatment: no heavy hyperglycosylation has been observed. However, moderate levels of over-glycosylation of the heterologous protein (size increase in the range of 10 additional kDa) have been reported. Nevertheless, these modifications did not impair the activity of the recombinant protein (Nicaud et al. 1989; Tharaud et al. 1992; Park et al. 2000; Madzak et al. 2001; Jolivalt et al. 2004; L. Otterbein, personal communication).

8.5
Transformation Methods

The lithium acetate method was used to transform very efficiently *Y. lipolytica* host strains with integrative vectors (Xuan et al. 1988). A simplified one-step transformation protocol, avoiding laborious cultivation and centrifugation steps, has also been devised (Chen et al. 1997), and allowed to obtain, at least with some strains, transformation efficiencies higher or similar to those of multi-step methods. However, when using multi-copy auto-cloning vectors, the transformation efficiency of which is much lower, the above methods do not always allow enough transformants to be obtained. The following optimized protocol has been used successfully in our laboratory for transformation of *Y. lipolytica* with multicopy auto-cloning vectors, carrying the defective *ura3d4* allele as selection marker gene.

8.5.1
Materials

The following materials are required:

(i) Plates (15 g L^{-1} agar) of YPD (10 g L^{-1} yeast extract; 10 g L^{-1} bacto peptone; 10 g L^{-1} glucose) for recipient strain growth, and of YNB-N$_0$Glu (17 g L^{-1} yeast nitrogen base without amino acids and ammonium sulfate; 10 g L^{-1} glucose; 1 g L^{-1} sodium glutamate) for transformants selection. When transformants retain an auxotrophy, such as Leu$^-$ or Lys$^-$, casamino acids (1 g L^{-1}) must be added to selective plates. These plates should contain a volume of 50 mL of selective solid medium, in order to resist desiccation during long incubation times.
(ii) Liquid culture medium YPD-pH4 (as above + 50 mM citrate buffer at pH 4).
(iii) TE solution (10 mM Tris at pH 8, 1 mM EDTA).
(iv) LiAc solution (100 mM, adjusted to pH 6 with acetic acid).
(v) Carrier DNA: fish sperm DNA, with a mean fragment size of 500 bp (e.g., from ICN Biochemicals, France; or DNA sonicated to adjust mean fragment size to 500 bp), at 5 mg mL^{-1} in TE solution.
(vi) PEG solution (400 g L^{-1} PEG$_{4000}$, in LiAc solution), freshly prepared.

8.5.2
Preparation of Competent Cells

- On day 1, streak the *Y. lipolytica* recipient strain (from a glycerol stock at −80 °C) on a YPD plate. Incubate overnight at 28 °C.
- On day 2, in the morning, inoculate 4 mL of YPD-pH 4, in a 50-mL Erlenmeyer flask, with a loopful of yeast cells. Incubate at 28 °C with strong shaking. In the evening, count the cells (e.g., Malassez grid), and inoculate three overnight cultures of 20 mL of YPD-pH 4, into 500 mL Erlenmeyer flasks, at the following concentrations: 10^4 cells mL^{-1}, 3 × 10^4 cells mL^{-1}, and 10^5 cells mL^{-1}. Incubate overnight at 28 °C with strong shaking. Alternatively, the cell concentration can be as-

sessed by measuring OD_{600}, assuming that 1 unit corresponds to ~10^7 cells mL^{-1}, but this method is less precise than counting.

- On day 3, in the morning, count the cells and select an overnight culture, with a concentration between 8×10^7 cells mL^{-1} and 2×10^8 cells mL^{-1}, for the preparation of competent cells (the best level is ~10^8 cells mL^{-1}). Centrifuge the cells, at ambient temperature (5 min at $4000 \times g$). Gently resuspend the pellet in the same volume of TE solution. Centrifuge for 5 min at $2500 \times g$, then resuspend the pellet gently, at a concentration of 5×10^7 cells mL^{-1}, in LiAc solution. Transfer to a 500-mL Erlenmeyer flask and incubate at 28 °C with gentle shaking (120 r.p.m.) for 1 h. Centrifuge for 5 min at $1500 \times g$. Gently resuspend the pellet at a concentration of 5×10^8 cells mL^{-1} in LiAc solution.

8.5.3
Transformation

The transforming DNA corresponds to the "yeast cassette", liberated from the multi-copy auto-cloning vector using restriction digestion, and possibly purified. Starting from 1–5 µg of vector DNA, the transforming DNA should be recovered in a final volume of no more than 5–7 µL. In a 10-mL round-bottomed tube, add: (i) transforming DNA; (ii) 5 µL of carrier DNA; and (iii) 100 µL of the competent cells, as prepared above. Mix gently by pipetting. Incubate for 15 min in a water bath at 28 °C. Add 700 µL of PEG solution, mix gently and incubate at 28 °C with mild shaking (180 r.p.m.) for 1 h. Add 80 µL of DMSO and mix gently (omit this step when transforming with a vector integrating by homology, as DMSO is suspected to increase non-homologous integrations). Heat shock in a water bath at 39 °C for 10 min. Add 1.2 mL of LiAc solution, mix gently, and immediately pour aliquots of 250–300 µL onto YNB-N_0Glu selective plates, using a bent glass rod. Incubate the plates overnight at 28 °C. In the case of monocopy auto-cloning vectors, the transformant colonies appear after 2–3 days. In the case of multicopy auto-cloning vectors, in which multiple integration events must be selected for, transformant colonies can appear after one week, but the process can take as long as two to three weeks.

8.6
Conclusions and Future Trends

Several successful examples of protein production have now been published using *Y. lipolytica*. Some of these procedures have led to the analytical production of proteins, but to date none has led to any industrial application. Hopefully, this situation may change in the near future. Although over the years a wealth of data has been accumulated on *Y. lipolytica* biology – and specifically on protein secretion – there is at present no example of these results being used to improve any given process. A number of general tools have been designed (artificial promoters, protease-free strains, multi-copy vectors) which have resulted in highly significant increases in overall productivity. However, manipulation of the secretory pathway itself has not yet been much investigated in order to improve the production of any given protein. The selection of "super-secretors" for a given protein has been successfully performed in a few cases (A. Franke, personal communication), but no analysis of the resulting strains has been published, and the mutations leading to improvements have not been characterized. A better knowledge of the genes involved in quality control/protein degradation along the secretory pathway is clearly required in order to develop improved producers. In this regard, studies have recently been initiated to investigate protein glycosylation (Babour et al. 2003), and these may lead to the development of new strains exhibiting glycosylation patterns closer to that of mammals, and especially of man.

In addition to a better control of protein quality, an understanding of how yeast cells adapt the secretion of a foreign protein to nutrient availability and respond to an overloading of the secretory pathway is of primary importance. Refining culture modes in order to optimize heterologous protein production has shown much promise (Nicaud et al. 2002). Analyzing cell responses under precisely defined conditions may help to orient the engineering of diverse pathways. The recent availability of complete genomic data will foster studies on the global response of yeast strains, when challenged with the stress of overproducing foreign proteins, through transcriptome and proteome analysis. Likewise, they may also enrich and diversify our collection of strong regulatable promoters.

Acknowledgments

The authors wish to thank all colleagues and collaborators who provided their unpublished data, and particularly: Martine Chartier, Marie-Thérèse Le Dall, Ricardo Cordero Otero, Art Franke, Christof Gysler, Nicole Houba-Hérin, Claude Jolivalt, Nathalie Libessart, Christian Mougin, Ludovic Otterbein, Bi-Ching Sang, Dominique Swennen, Brigitte Tréton, and Kohji Uchida.

Appendix

Tab. A.8.1   Selection of *Y. lipolytica* host strains for heterologous proteins production.

Strain	Genotype	Phenotype	Source
Po1d	*MatA, leu2-270, ura3-302, xpr2-322*	Leu⁻, Ura⁻, ΔAEP, Suc⁺	Le Dall et al. (1994)
Po1f	*MatA, leu2-270, ura3-302, xpr2-322, axp1-2*	Leu⁻, Ura⁻, ΔAEP, ΔAXP, Suc⁺	Madzak et al. (2000)
Po1g	*MatA, leu2-270, ura3-302::URA3, xpr2-322, axp1-2*	Leu⁻, ΔAEP, ΔAXP, Suc⁺, pBR322	Madzak et al. (2000)
Po1h	*MatA, ura3-302, xpr2-322, axp1-2*	Ura⁻, ΔAEP, ΔAXP, Suc⁺	Madzak et al. (2003, 2004)

Tab. A8.2 Selection of *Y. lipolytica* integrative expression/secretion vectors.

Plasmid	Expression cassette prom./secretion signal/term.	Selection marker	Targeting sequence	Reference
Monocopy integrative vectors:				
pINA1269	hp4d/-/*XPR2*t	*LEU2*	pBR322	Madzak et al. (2000)
pINA1267	hp4d/*XPR2* prepro/*XPR2*t	*LEU2*	pBR322	Madzak et al. (2000)
pINA1296	hp4d/*XPR2* pre/*XPR2*t	*LEU2*	pBR322	Madzak et al. (2000)
p65IP	p*ICL1*/-/-	*URA3 (ura3d1)*	rDNA	Juretzek et al. (2001)
p66IP	p*ICL1*/-/-	*URA3 (ura3d1)*	zeta	Juretzek et al. (2001)
Multicopy integrative vectors:				
p64IP	p*ICL1*/-/-	*ura3d4*	rDNA	Juretzek et al. (2001)
p67IP	p*ICL1*/-/-	*ura3d4*	zeta	Juretzek et al. (2001)
Monocopy auto-cloning vectors:				
pINA1311	hp4d/-/*LIP2*t	*URA3 (ura3d1)*	zeta	Nicaud et al. (2002)
pINA1312	hp4d/-/*XPR2*t	*URA3 (ura3d1)*	zeta	Nicaud et al. (2002)
pINA1313	hp4d/*LIP2* prepro/*LIP2*t	*URA3 (ura3d1)*	zeta	Nicaud et al. (2002)
pINA1314	hp4d/*XPR2* prepro/*XPR2*t	*URA3 (ura3d1)*	zeta	Nicaud et al. (2002)
pINA1317	hp4d/*XPR2* pre/*XPR2*t	*URA3 (ura3d1)*	zeta	Nicaud et al. (2002)
JMP62	p*POX2*/-/*LIP2*t	*URA3 (ura3d1)*	zeta	Nicaud et al. (2002)
JMP61	p*POX2*/*LIP2* prepro/*LIP2*t	*URA3 (ura3d1)*	zeta	Nicaud et al. (2002)
pYEG1	p*POX2*/-/*XPR2*t	*URA3 (ura3d1)*	zeta	Nicaud et al. (2002)
pYSG1	p*POX2*/*LIP2* prepro/*XPR2*t	*URA3 (ura3d1)*	zeta	Nicaud et al. (2002))
Multicopy auto-cloning vectors:				
pINA1291	hp4d/-/*LIP2*t	*ura3d4*	zeta	Nicaud et al. (2002)
pINA1292	hp4d/-/*XPR2*t	*ura3d4*	zeta	Nicaud et al. (2002)
pINA1293	hp4d/*LIP2* prepro/*LIP2*t	*ura3d4*	zeta	Nicaud et al. (2002)
pINA1294	hp4d/*XPR2* prepro/*XPR2*t	*ura3d4*	zeta	Nicaud et al. (2002)
pINA1297	hp4d/*XPR2* pre/*XPR2*t	*ura3d4*	zeta	Nicaud et al. (2002)
JMP64	p*POX2*/-/*LIP2*t	*ura3d4*	zeta	Nicaud et al. (2002)
JMP63	p*POX2*/*LIP2* prepro/*LIP2*t	*ura3d4*	zeta	Nicaud et al. (2002)
pYEG10	p*POX2*/-/*XPR2*t	*ura3d4*	zeta	Nicaud et al. (2002)
pYSG10	p*POX2*/*LIP2* prepro/*XPR2*t	*ura3d4*	zeta	Nicaud et al. (2002)

References

BABOUR A, BECKERICH JM, GAILLARDIN C (2004) Identification of an UDP-Glc: glycoprotein glucosyltransferase in the yeast *Yarrowia lipolytica*. Yeast 21: 11–24

BALDARI C, MURRAY JAH, GHIARA P, CESARENI G, GALEOTTI CL (1987) A novel leader peptide which allows efficient secretion of a fragment of human interleukin 1β in *Saccharomyces cerevisiae*. EMBO J 6: 229–234

BARNS SM, LANE DJ, SOGIN ML. BIBEAU C, WEISBURG WG (1991) Evolutionary relationships among pathogenic *Candida* species and relatives. J Bacteriol 173: 2250–2255

BARTH G, GAILLARDIN C (1996) *Yarrowia lipolytica*. In: Nonconventional Yeasts in Biotechnology: a Handbook (Wolf K, Ed). Springer-Verlag, Heidelberg, pp 313–388

BARTH G, GAILLARDIN C (1997) Physiology and genetics of the dimorphic fungus *Yarrowia lipolytica*. FEMS Microbiol Rev 19: 219–237

BARTH G, BECKERICH JM, DOMINGUEZ A, KERSCHER S, OGRYDZIAK D, TITORENKO V, GAILLARDIN C (2003) Functional genetics of

Yarrowia lipolytica. In: Functional Genetics of industrial yeasts. Topics in Current Genetics, Vol. 1 (de-Winde H, Ed). Springer Verlag, Berlin, pp 227–271

BAUER R, PALTAUF F, KOHLWEIN SD (1993) Functional expression of bacterial β-glucuronidase and its use as a reporter system in the yeast *Yarrowia lipolytica*. Yeast 9: 71–75

BECKERICH JM, BOISRAMÉ-BAUDEVIN A, GAILLARDIN C (1998) *Yarrowia lipolytica*: a model organism for protein secretion studies. Internatl Microbiol 1: 123–130

BHAVE SL, CHATTOO BB (2003) Expression of vitreoscilla hemoglobin improves growth and levels of extracellular enzyme in *Yarrowia lipolytica*. Biotechnol Bioeng 84: 658–666

BIGEY, F, TUERY K, BOUGARD D, NICAUD JM, MOULIN G (2003) Identification of a triacylglycerol lipase gene family in *Candida deformans*: molecular cloning and functional expression. Yeast 20: 223–248

BLANCHIN-ROLAND S, CORDERO OTERO R, GAILLARDIN C (1994) Two upstream activation sequences control the expression of the *XPR2* gene in the yeast *Yarrowia lipolytica*. Mol Cell Biol 14: 327–338

BOISRAMÉ A, KABANI M, BECKERICH JM, HARTMANN E, GAILLARDIN C (1998) Interaction of Kar2p and Sls1p is required for efficient co-translational translocation of secreted proteins in the yeast *Yarrowia lipolytica*. J Biol Chem 273: 30903–30908

BON E, CASAREGOLA S, BLANDIN G, LLORENTE B, NEUVEGLISE C, MUNSTERKOTTER M, GULDENER U, MEWES HW, VAN HELDEN J, DUJON B, GAILLARDIN C (2003) molecular evolution of eukaryotic genomes: hemiascomycetous yeast spliceosomal introns. Nucleic Acids Res 31: 1121–1135

CASARÉGOLA S, FEYNEROL C, DIEZ M, FOURNIER P, GAILLARDIN C (1997) Genomic organization of the yeast *Yarrowia lipolytica*. Chromosoma 106: 380–390

CASARÉGOLA S, NEUVÉGLISE C, LÉPINGLE A, BON E, FEYNEROL C, ARTIGUENAVE F, WINCKLER P, GAILLARDIN C (2000) Genomic exploration of the hemiascomycetous yeasts: 17. *Yarrowia lipolytica*. FEBS Lett 487: 95–100

CHEN DC, BECKERICH JM, GAILLARDIN C (1997) One-step transformation of the dimorphic yeast *Yarrowia lipolytica*. Appl Microbiol Biotechnol 48: 232–235

CORDERO OTERO R, GAILLARDIN C (1996) Efficient selection of hygromycin-B-resistant *Yar-rowia lipolytica* transformants. Appl Microbiol Biotechnol 46: 143–148

DAVIDOW LS, APOSTOLAKOS D, O'DONNELL MM, PROCTOR AR, OGRYDZIAK DM, WING RA, STASKO I, DE ZEEUW JR (1985) Integrative transformation of the yeast *Yarrowia lipolytica*. Curr Genet 10: 39–48

DAVIDOW LS, FRANKE AE, DE ZEEUW JR (1987) New *Yarrowia lipolytica* transformants used for expression and secretion of heterologous proteins, especially prorennin and human anaphylatoxin C5a. European Patent Application EP86307839

DE BAETSELIER A, VASAVADA A, DOHET P, HA-THI V, DE BEUKELAER M, ERPICUM T, DE CLERCK L, HANOTIER J, ROSEMBERG S (1991) Fermentation of a yeast producing *A. niger* glucose oxidase: scale-up, purification and characterization of the recombinant enzyme. Bio/Technology 9: 559–561

DOMINGUEZ A, FERMINAN E, SANCHEZ M, GONZALEZ FJ, PEREZ-CAMPO FM, GARCIA S, HERRERO AB, SAN VICENTE A, CABELLO J, PRADO M, IGLESIAS FJ, CHOUPINA A, BURGUILLO FJ, FERNANDEZ-LAGO L, LOPEZ MC (1998) Non-conventional yeasts as hosts for heterologous protein production. Internatl Microbiol 1: 131–142

DOMINGUEZ A, FERMINAN E, GAILLARDIN C (2000) *Yarrowia lipolytica*: an organism amenable to genetic manipulation as a model for analyzing dimorphism in fungi. Contrib Microbiol 5: 151–172

DUJON B, SHERMAN D, FISCHER G, DURRENS P, CASAREGOLA S, LAFONTAINE I, DE MONTIGNY J, MARCK C, NEUVEGLISE C, TALLA E, GOFFARD N, FRANGEUL L, AIGLE M, ANTHOUARD V, BABOUR A, BARBE V, BARNAY S, BLANCHIN S, BECKERICH JM, BEYNE E, BLEYKASTEN C, BOISRAME A, BOYER J, CATTOLICO L, CONFANIOLERI F, DE DARUVAR A, DESPONS L, FABRE E, FAIRHEAD C, FERRY-DUMAZET H, GROPPI A, HANTRAYE F, HENNEQUIN C, JAUNIAUX N, JOYET P, KACHOURI R, KERREST A, KOSZUL R, LEMAIRE M, LESUR I, MA L, MULLER H, NICAUD JM, NIKOLSKI M, OZTAS S, OZIER-KALOGEROPOULOS O, PELLENZ S, POTIER S, RICHARD GF, STRAUB ML, SULEAU A, SWENNEN D, TEKAIA F, WESOLOWSKI-LOUVEL M, WESTHOF E, WIRTH B, ZENIOU-MEYER M, ZIVANOVIC I, BOLOTIN-FUKUHARA M, THIERRY A, BOUCHIER C, CAUDRON B, SCARPELLI C, GAILLARDIN C, WEISSENBACH J,

WINCKER P, SOUCIET JL. (2004) Genome evolution in yeasts. Nature 430: 35–44

FABRE E, NICAUD JM, LOPEZ MC, GAILLARDIN C (1991) Role of the proregion in the production and secretion of the *Yarrowia lipolytica* alkaline extracellular protease. J Biol Chem 266: 3782–3790

FABRE E, THARAUD C, GAILLARDIN C (1992) Intracellular transit of a yeast protease is rescued by trans-complementation with its pro-domain. J Biol Chem 267: 15049–15055

FICKERS P, LE DALL MT, GAILLARDIN C, THONART P, NICAUD JM (2003) New disruption cassettes for rapid gene disruption and marker rescue in the yeast *Yarrowia lipolytica*. J Microbiol Methods 55: 727–737

FOURNIER P, GUYANEUX L, CHASLES M, GAILLARDIN C (1991) Scarcity of ARS sequences isolated in a morphogenesis mutant of the yeast *Yarrowia lipolytica*. Yeast 7: 25–36

FOURNIER P, ABBAS A, CHASLES M, KUDLA B, OGRYDZIAK DM, YAVER D, XUAN JW, PEITO A, RIBET AM, FEYNEROL C, HE F, GAILLARDIN C (1993) Colocalization of centromeric and replicative functions on autonomously replicating sequences isolated from the yeast *Yarrowia lipolytica*. Proc Natl Acad Sci USA 90: 4912–4916

FRANKE AE, KACZMAREK FS, EISENHARD ME, GEOGHEGAN KF, DANLEY DE, DE ZEEUW JR, O'DONNELL MM, GOLLAHER MG, DAVIDOW LS (1988) Expression and secretion of bovine prochymosin in *Yarrowia lipolytica* In: Developments in Industrial Microbiology, Vol. 29 (Pierce G, Ed). Elsevier, Amsterdam, pp 43–57

GAILLARDIN C, RIBET AM, HESLOT H (1985) Integrative transformation of the yeast *Yarrowia lipolytica*. Curr Genet 10: 49–58

GAILLARDIN C, RIBET AM (1987) *LEU2* directed expression of β-galactosidase activity and phleomycin resistance in *Yarrowia lipolytica*. Curr Genet 11: 369–375

GAILLARDIN C, HESLOT H (1988) Genetic engineering in *Yarrowia lipolytica*. J. Basic Microbiol 28: 161–174

GELLISSEN G, HOLLENBERG CP (1997) Applications of yeast in gene expression studies: a comparison of *Saccharomyces cerevisiae*, *Hansenula polymorpha* and *Kluyveromyces lactis* – a review. Gene 190: 87–97

GRINNA LS, TSCHOPP JF (1989) Size distribution and general structural features of N-linked oligosaccharides from the methylotrophic yeast *Pichia pastoris*. Yeast 5: 107–115

HAMSA PV, CHATTOO BB (1994) Cloning and growth-regulated expression of the gene encoding the hepatitis B virus middle surface antigen in *Yarrowia lipolytica*. Gene 143: 165–170

HAMSA PV, KACHROO P, CHATTOO BB (1998) Production and secretion of biologically active human epidermal growth factor in *Yarrowia lipolytica*. Curr Genet 33: 231–237

HE F, BECKERICH JM, GAILLARDIN C (1992) A mutant of 7SL RNA in *Yarrowia lipolytica* affecting the synthesis of a secreted protein. J Biol Chem 267: 1932–1937

HESLOT H (1990) Genetics and genetic engineering of the industrial yeast *Yarrowia lipolytica*. Adv Biochem Eng Biotechnol 43: 43–73

HURTADO CA, RACHUBINSKI RA (2002) YlBMH1 encodes a 14–3-3 protein that promotes filamentous growth in the dimorphic yeast *Yarrowia lipolytica*. Microbiology 148: 3725–3735

JAMES LC, STRICK CA (1993) Multiple integrative vectors and *Yarrowia lipolytica* transformants. US Patent Application US08/117.375 (WO95/06739)

JOLIVALT C, MADZAK C, BRAULT A, CAMINADE E, MALOSSE C, MOUGIN C (2004) Expression of laccase IIIb from the white-rot fungus *Trametes versicolor* in the yeast *Yarrowia lipolytica* for environmental applications. Appl Microbiol Biotechnol in press

JURETZEK T, PRINZ A, SCHUNCK WH, BARTH G, MAUERSBERGER S (1995) Expressionskassetten zur heterologen Expression von Proteinen in der Hefe *Yarrowia lipolytica* unter Kontrolle des regulierbaren Promotors der Isocitratlyase. German patent DE19525282A1

JURETZEK T, MAUERSBERGER S, BARTH G (1998) Rekombinante haploide oder diploide *Yarrowia lipolytica* Zellen zur funktionellen heterologen Expression von Cytochrom P450 Systemen. German patent DE19932811.0

JURETZEK T, WANG HJ, NICAUD JM, MAUERSBERGER S, BARTH G (2000) Comparison of promoters suitable for regulated overexpression of β-galactosidase in the alkane-utilizing yeast *Yarrowia lipolytica*. Biotechnol Bioprocess Eng 5: 320–326

JURETZEK T, LE DALL M, MAUERSBERGER S, GAILLARDIN C, BARTH G, NICAUD JM (2001) Vectors for gene expression and amplification in the yeast *Yarrowia lipolytica*. Yeast 18: 97–113

KABANI M, BECKERICH JM, BRODSKY JL (2003) The yeast Sls1p proteins define a new family of Hsp70 nucleotide exchange factors. Curr Genomics 4: 465–473

KERSCHER S, DROSE S, ZWICKER K, ZICKERMANN V, BRANDT U (2002) *Yarrowia lipolytica*, a yeast genetic system to study mitochondrial complex I. Biochim Biophys Acta 1555: 83–91

LALOI M, MAC CARTHY J, MORANDI O, GYSLER C, BUCHELI P (2002) Molecular and biochemical characterisation of two aspartic proteinases TcAP1 and TcAP2 from *Theobroma cacao* seeds. Planta 215: 754–762

LE DALL MT, NICAUD JM, GAILLARDIN C (1994) Multiple-copy integration in the yeast *Yarrowia lipolytica*. Curr Genet 26: 38–44

LIVI GP, LILLQUIST JS, MILES LM, FERRARA A, SATHE GM, SIMON PL, MEYERS CA, GORMAN J, YOUNG PR (1991) Secretion of N-glycosylated interleukin-1β in *Saccharomyces cerevisiae* using a leader peptide from *Candida albicans*. J Biol Chem 266: 15348–15355

MADZAK C, BLANCHIN-ROLAND S, GAILLARDIN C (1995) Upstream activating sequences and recombinant promoter sequences functional in *Yarrowia* and vectors containing them. European Patent Application EP0747484A1

MADZAK C, BLANCHIN-ROLAND S, CORDERO OTERO R, GAILLARDIN C (1999) Functional analysis of upstream regulating regions from the *Yarrowia lipolytica XPR2* promoter. Microbiology 145: 75–87

MADZAK C, TRÉTON B, BLANCHIN-ROLAND S (2000) Strong hybrid promoters and integrative expression/secretion vectors for quasi-constitutive expression of heterologous proteins in the yeast *Yarrowia lipolytica*. J Mol Microbiol Biotechnol. 2: 207–216

MADZAK C, HOUBA-HÉRIN N, PETHE C, LALOUE M, GAILLARDIN C, BECKERICH JM (2001) An expression/secretion system for production of heterologous proteins in the non-conventional yeast *Yarrowia lipolytica*: the example of the cytokinin oxidase from *Zea mays*. Yeast 18: 297

MADZAK C (2003) New tools for heterologous protein production in the yeast *Yarrowia lipolytica*. In: Recent Research Developments in Microbiology, Vol. 7 (Pandalai SG, Ed). Research Signpost, Trivandrum, pp 453–479

MADZAK C, GAILLARDIN C, BECKERICH JM. (2004) Heterologous protein expression and secretion in the non-conventional yeast *Yarrowia lipolytica*: a review. J Biotechnol. 109(1–2): 63–81

MATOBA S, OGRYDZIAK DM (1989) A novel location for dipeptidylaminopeptidase processing sites in the alkaline extracellular protease of *Yarrowia lipolytica*. J Biol Chem 264: 6037–6043

MAUERSBERGER S, WANG HJ, GAILLARDIN C, BARTH G, NICAUD JM (2001) Insertional mutagenesis in the n-alkane-assimilating yeast *Yarrowia lipolytica*: generation of tagged mutations in genes involved in hydrophobic substrate utilization. J Bacteriol 183: 5102–5109

MOUGIN C, JOLIVALT C, BRIOZZO P, MADZAK C (2003) Fungal laccases: from structure–activity studies to environmental applications. Environ Chem Lett 1: 145–148

MUESCH A, HARTMANN E, ROHDE K, RUBARTELLI A, SITIA R, RAPOPORT TA (1990) A novel pathway for secretory proteins? Trends Biochem Sci 15: 86–88

MÜLLER S, SANDAL T, KAMP-HANSEN P, DALBOGE H (1998) Comparison of expression systems in the yeasts *Saccharomyces cerevisiae*, *Hansenula polymorpha*, *Kluyveromyces lactis*, *Schizosaccharomyces pombe* and *Yarrowia lipolytica*. Cloning of two novel promoters from *Yarrowia lipolytica*. Yeast 14: 1267–1283

NICAUD JM, FABRE E, GAILLARDIN C (1989) Expression of invertase activity in *Yarrowia lipolytica* and its use as a selective marker. Curr Genet 16: 253–260

NICAUD JM, FOURNIER P, LA BONNARDIERE C, CHASLES M, GAILLARDIN C (1991) Use of ars18 based vectors to increase protein production in *Yarrowia lipolytica*. J Biotechnol 19: 259–270

NICAUD JM, GAILLARDIN C, SEMAN M, PIGNÈDE G (1998) Process of non-homologous transformation of *Yarrowia lipolytica*. French Patent Application PCT/FR99/02079

NICAUD JM, MADZAK C, VAN DEN BROEK P, GYSLER C, DUBOC P, NIEDERBERGER P, GAILLARDIN C (2002) Protein expression and secretion in the yeast *Yarrowia lipolytica*. FEMS Yeast Res 2: 371–379

OGRYDZIAK DM, DEMAIN AL, TANNENBAUM SR (1977) Regulation of extracellular protease production in *Candida lipolytica*. Biochim Biophys Acta 497: 525–538

OGRYDZIAK DM, SCHARF SJ (1982) Alkaline extracellular protease produced by *Saccharomycopsis lipolytica* CX161–1B. J Gen Microbiol 128: 1225–1234

OHKUMA M, HWANG CW, MASUDA Y, NISHIDA H, SUGIYAMA J. OHTA A, TAKAGI M (1993) Evolutionary position of n-alkane-assimilating yeast *Candida maltosa* shown by nucleotide sequence of small-subunit ribosomal RNA gene. Biosci Biotechnol Biochem 57: 1793–1794

PARK CS, CHANG CC, KIM JY, OGRYDZIAK DM, RYU DDY (1997) Expression, secretion and processing of rice α-amylase in the yeast *Yarrowia lipolytica*. J Biol Chem 272: 6876–6881

PARK CS, CHANG CC, RYU DDY (2000) Expression and high-level secretion of *Trichoderma reesei* endoglucanase I in *Yarrowia lipolytica*. Appl Biochem Biotechnol 87: 1–15

PEREZ-CAMPO FM, DOMINGUEZ A (2001) Factors affecting the morphogenetic switch in *Yarrowia lipolytica*. Curr Microbiol 43: 429–433

PIGNÈDE G, WANG H, FUDALEJ F, SEMAN M, GAILLARDIN C, NICAUD JM (2000) Autocloning and amplification of *LIP2* in *Yarrowia lipolytica*. Appl Environ Microbiol 66: 3283–3289

RICHARD M, QUIJANO RR, BEZZATE S, BORDON-PALLIER F, GAILLARDIN C (2001) Tagging morphogenetic genes by insertional mutagenesis in the yeast *Yarrowia lipolytica*. J Bacteriol 183: 3098–3107

SCHMID-BERGER N, SCHMID B, BARTH G (1994) Ylt1, a highly repetitive retrotransposon in the genome of the dimorphic fungus *Yarrowia lipolytica*. J Bacteriol 176: 2477–2482

SREEKRISHNA K, KROPP K (1996) *Pichia pastoris*. In: Nonconventional yeasts in biotechnology, a handbook (Wolf K, Ed). Springer-Verlag, Berlin-Heidelberg, pp 203–252

SWENNEN D, PAUL MF. VERNIS L, BECKERICH JM, FOURNIER A, GAILLARDIN C (2002) Secretion of active anti-Ras single-chain Fv antibody by the yeasts *Yarrowia lipolytica* and *Kluyveromyces lactis*. Microbiology 148: 41–50

THARAUD C, RIBET AM, COSTES C, GAILLARDIN C (1992) Secretion of human blood coagulation factor XIIIa by the yeast *Yarrowia lipolytica*. Gene 121: 111–119

TITORENKO VI, SMITH JJ, SZILARD RK, RACHUBINSKI RA (2000) Peroxisome biogenesis in the yeast *Yarrowia lipolytica*. Cell Biochem Biophys 32: 21–26

TITORENKO VI, RACHUBINSKI RA (2001) Dynamics of peroxisome assembly and function. Trends Cell Biol 11: 22–29

TOBE S, TAKAMI T, IKEDA S, MITSUGI K (1976) Production and some enzymatic properties of alkaline protease of *Candida lipolytica*. Agric Biol Chem 40: 1087–1092

TSUGAWA R, NAKASE T, KOYABASHI T, YAMASHITA K, OKUMURA S (1969) Fermentation of n-paraffins by yeast. Part III. α-ketoglutarate productivity of various yeasts. Agr Biol Chem 33: 929–938

VERNIS L, ABBAS A, CHASLES M, GAILLARDIN C, BRUN C, HUBERMAN JA, FOURNIER P (1997) An origin of replication and a centromere are both needed to establish a replicative plasmid in the yeast *Yarrowia lipolytica*. Mol Cell Biol 17: 1995–2004

VERNIS L, POLJAK L, CHASLES M, UCHIDA K, CASARÉGOLA S, KAS E, MATSUOKA M, GAILLARDIN C, FOURNIER P (2001) Only centromeres can supply the partition system required for ARS function in the yeast *Yarrowia lipolytica*. J Mol Biol 305: 203–217

WANG HJ, LE DALL MT, WACH Y, LAROCHE C, BELIN JM, GAILLARDIN C, NICAUD JM (1999) Evaluation of the acyl coenzyme A oxidase (Aox) isozyme function in the n-alkane-assimilating yeast *Yarrowia lipolytica*. J Bacteriol 181: 5140–5148

WICKERHAM LJ, KURTZMAN CP, HERMAN AI (1970) Sexual reproduction in *Candida lipolytica*. Science 167: 1141

WOJTATOWICZ M, RYMOWICZ W, ROBAK M, ZAROWSKA B, NICAUD JM (1997) Kinetics of cell growth and citric acid production by *Yarrowia lipolytica* Suc+ transformants in sucrose media. Pol J Food Nutr Sci 6/47: 49–54

XUAN JW, FOURNIER P, GAILLARDIN C (1988) Cloning of the *LYS5* gene encoding saccharopine dehydrogenase from the yeast *Yarrowia lipolytica* by target integration. Curr Genet 14: 15–21

YAVER DS, MATOBA S, OGRYDZIAK DM (1992) A mutation in the signal recognition particle 7S RNA of the yeast *Yarrowia lipolytica* preferentially affects synthesis of the alkaline extracellular protease: *in vivo* evidence for translational arrest. J Cell Biol 116: 605–616

9
Aspergillus sojae
Margreet Heerikhuisen, Cees van den Hondel, and Peter Punt

List of Genes

Gene	Encoded gene product
ver-1	versicolorin A dehydrogenase (*Aspergillus* species in the section *Flavi*)
aflR	aflatoxin pathway-specific regulator (*Aspergillus* species in the section *Flavi*)
alpA	alkaline protease (*Aspergillus sojae*)
amdS	acetamidase (*Aspergillus nidulans*)
pyrG	orotidine-5-monophosphate decarboxylase (*Aspergillus niger*)
gpdA	glyceraldehyde-3-phosphate dehydrogenase (*Aspergillus nidulans*)
glaA	glucoamylase (*Aspergillus niger*)
alcA	alcohol dehydrogenase (*Aspergillus nidulans*)
cbh1	cellobiohydrolase (*Trichoderma reesei*)
GUS	glucuronidase (*Escherichia coli*)
IL-6	interleukin 6 (*Homo sapiens*)
GFP	green fluorescent protein (*Aequorea victoria*)
bipA	ER-chaperone from the HSP70 gene family (*Aspergillus niger*)
pclA	proprotein convertase (*Aspergillus niger, Aspergillus sojae*)
niaD	nitrate reductase (*Aspergillus sojae*)
trpC	trifunctional polypeptide; glutamine amido transferase, indole-3-glycerol phosphate synthase, and N-(5'-phosphoribosyl) anthranilate isomerase (*Aspergillus nidulans*)

9.1
Introduction

Among microbial production organisms, filamentous fungi hold a strong position in particular for the production of antimicrobial metabolites such as penicillins and cephalosporins (Elander 2003) and for the production of hydrolytic enzymes (Verdoes 1994). Bacterial hosts are frequently used for the production of primary and secondary metabolites such as amino acids and vitamins (Demain 2000). The use of fi-

Production of Recombinant Proteins. Novel Microbial and Eucaryotic Expression Systems. Edited by Gerd Gellissen
Copyright © 2005 WILEY-VCH Verlag GmbH & Co. KGaA, Weinheim
ISBN: 3-527-31036-3

lamentous fungi for protein production originates from a variety of traditional food processes involving these organisms. Already more than a century ago, fungi were identified as relevant microorganisms in cheese-making and a number of oriental food processes (e. g., soy sauce, tempeh, rice brewing). Based on this early biotechnological knowledge, fungi have been exploited as sources for enzymes beyond these original applications. Nowadays, a multitude of fungal enzymes are available commercially for numerous applications in the food and the feed industry, as well as being used in chemical processes related to detergent, paper and pulp generation and usage (see for example Sigoillot et al. 2004; Record et al. 2003).

With the availability of molecular tools for filamentous fungi, new developments emerged to improve yields in the production of established fungal proteins and to identify new protein products. The new protein products included both, modified proteins with altered or improved characteristics and hitherto poorly or non-accessible proteins (Gouka et al. 1997). Moreover, these molecular tools enabled the expression of heterologous proteins from completely unrelated species including mammals. The first reports on fungi-derived mammalian proteins in commercial yields date from more than a decade ago (for a review, see Gouka et al. 1997).

Only a relatively small number of fungal species is presently used as protein production hosts (Punt et al. 2002; see also Chapter 10 on *Sordaria macrospora*). Among these fungi, *Aspergillus* species play a dominant role. An important reason for this focus stems from the fact that *Aspergillus* species are highly represented among the strains used for traditional production processes. Consequently, one of the first fungal expression platforms is based on an *Aspergillus* species (Ballance et al. 1983; Kelly and Hynes 1985), thereby taking a clear lead to further developments based on fungal organisms in general. This position is further sustained by the fact that several *Aspergillus*-derived food additive products had already obtained a GRAS (generally *recognized as safe*) status from the regulatory authorities, thereby easing its recognition as a safe and reliable expression platform.

As with other industrial environments, an extensive intellectual property position was built up by a number of Biotech companies (e. g., DSM, NOVOzymes, Genencor) for particular *Aspergillus* species. As a consequence – and due to the obvious commercial attraction of fungal expression hosts – several parties set out to identify alternative species and to assess the possibilities of employing the newly identified fungal expression hosts for heterologous gene expression (see Punt et al. 2002). As a result of this search for new hosts, *Aspergillus sojae* was identified. This species is related to other *Aspergillus* species (in particular *A. oryzae*), and possibly allows the use of existing molecular tools. The products from this species have a history of safe use. In this chapter, the status of the protein production platform based on *A. sojae* will be described.

9.2
Taxonomy

Fungal taxonomy is a complex issue, which continues to be subject to changes. When deduced in particular from molecular data, the genus *Aspergillus* is classified into the *Ascomycetes*, now combining species with ascomycetous teleomorphs as well as species with no known teleomorphs which were previously classified into the *Deuteromycetes*.

By comparing complete 18S rDNA sequences, Tamura et al. found that the genus *Aspergillus* consists of three different evolutionary lines (Tamura et al. 2000). Based on monophyletic taxonomy, three subgenera were suggested that consist of 15 sections in total: *Aspergillus*, *Fumigati*, and *Nidulantes* (Peterson 2000). A phylogenetic tree of selected *Aspergillus* species is shown in Figure 9.1; these include species belonging to the section *Flavi*, which in turn can be divided into non-aflatoxigenic species such as *Aspergillus oryzae* and *A. sojae*, and aflatoxigenic species such as *Aspergillus flavus* and *A. parasiticus*. The non-aflatoxigenic species have been widely used in industry for food fermentation or for the production of enzymes. Discrimination between the non-aflatoxigenic species and the aflatoxigenic species is a very important characteristic with respect to their use in food applications, and represents a key criterion for their acceptance by regulatory authorities.

A. *sojae* strains can be distinguished from taxonomically closely related species, such as *A. oryzae* and *A. parasiticus*, in a number of ways. Traditionally, the identification of species belonging to the section *Flavi* was based on several morphological characteristics. However, these species are morphologically very similar, which impedes any clear distinction between them. Based on the high degree of DNA complementarity, Kurtzman et al. proposed that *A. flavus*, *A. oryzae*, *A. parasiticus*, and *A. sojae* are varieties of a single species. These authors found that the homology in total DNA hybridizations between *A. flavus* and *A. oryzae* is 100%, between *A. parasiticus* and *A. sojae* is 91%, and between these groups is 72% (Kurtzman et al. 1986). Others state that *Aspergillus sojae* and *A. oryzae* are domesticated variants of *A. parasiticus* and *A. flavus*, respectively.

During the past decade, molecular genetic techniques have been used for the classification of the species in the section *Flavi*. Reference is here made in particular to the PCR fragments derived from *ver-1*, *aflR* and rDNA sequences (Ushijima et al. 1981; Chang et al. 1995; Yuan et al. 1995; Kusomoto et al. 1998; Watson et al. 1999). For example, Yuan et al. (1995) have shown that the very closely related *A. parasiticus* and *A. sojae* species can be distinguished from each other by random amplification of polymorphic DNA (RAPD) analysis, or by the difference in resistance to bleomycin. In addition, it has been found that *Aspergillus oryzae* can be distinguished from *A. sojae* by comparison of the *alpA* sequences (Heerikhuisen et al. 2001). It has also been found that the *A. sojae* genome harbors an *Xmn*I restriction site at a specific location within the *alpA* gene, which is not present in *A. oryzae*. This provides an additional discrimination tool between the two fungal strains.

In our initial host screening, we have analyzed various *A. sojae* isolates available from the ATCC strain collection for the above-mentioned criteria.

Aspergillus sojae

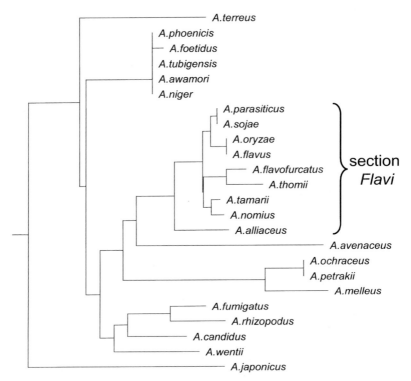

Fig. 9.1 Molecular phylogeny of selected *Aspergillus* species, including species belonging to the section *Flavi*, based on ITS sequence data.

9.3
The Expression Platform

9.3.1
Strain Selection

From a collection of 10 ATCC strains deposited as *A. sojae* (see Appendix Table A9.1), an initial selection was made for the most appropriate strain to be used as a host for protein production. We screened the strains for the previously described taxonomic criteria. A literature research resulted in the classification for nine out of the 10 strains to be *Aspergillus sojae*. On the basis of morphological parameters, *A. sojae* ATCC20235 strain did not meet the requirements to be classified as *A. sojae* (Ushijima et al. 1981). Based on the *alpA* restriction fragment length polymorphism (RFLP) which distin-

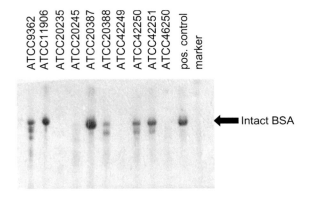

ATCC9362
ATCC11906
ATCC20235
ATCC20245
ATCC20387
ATCC20388
ATCC42249
ATCC42250
ATCC42251
ATCC46250
pos. control
marker

← Intact BSA

Fig. 9.2 Bovine serum albumin (BSA) degradation after overnight incubation with shake flask culture samples from various *A. sojae* strains grown in Mineral Medium (Bennett and Lasure 1991) containing 1% Trusoy (Soya flour; Spillers Premier Products Limited).

guishes *A. sojae* from *A. oryzae*, two strains (ATCC20235 and ATCC46250) were reclassified as *A. oryzae* strains, leaving eight *A. sojae* strains. Apparently, the deposition of fungal strains does not meet stringent taxonomic classification criteria.

An important property of a potentially useful fungal expression systems is the lack of (or a low level of) proteolytic degradation by the expression host. In order to analyze the protease activity of culture fluids of the various *A. sojae* strains, a milk-clearing assay was performed. The formation of a milk-clearing zone at the periphery of growing colonies – and thus the degradation of milk proteins – is a generally accepted criterion for protease activity (Mattern et al. 1992). In addition, shake flask culture samples were supplemented with different proteins (e.g., bovine serum albumin) and incubated. The extent of protein degradation was monitored to provide a valid acceptance criterion for suitability of the tested strains as expression hosts for a range of extracellular proteins (Figure 9.2). Two out of the eight *A. sojae* strains (ATCC11906 and ATCC20387) exhibited a low proteolytic profile in shake flask cultures. Under preliminary fed-batch fermentation conditions, the *A. sojae* ATCC11906 strain showed the lowest proteolytic activity and was thus selected for further research.

9.3.2
Transformation

9.3.2.1 Protoplasting
Two currently used protoplasting protocols – the modified OM-method (Yelton et al. 1984) and the NaCl method (Punt and van den Hondel 1992) – were tested for application to *A. sojae*. Transformation experiments with the selected *A. sojae* strains revealed that protoplasting efficiencies for *A. sojae* strains ATCC11906 (and also ATCC20387) were better using the NaCl method. Successful protoplasting was

achieved using various commercially available protoplasting enzyme preparations such as Novozym234 (NOVOzymes), Caylase C4 (Cayla), Glucanex (NOVOzymes).

9.3.2.2 Dominant Selection Markers

In general, fungi can be transformed with vectors either containing nutritional markers which complement an auxotrophy or dominant selection (mostly antibiotic resistance) markers. In particular, dominant selection markers can be very helpful as they do not require the availability of specific auxotrophic strains. Therefore, the sensitivity of *A. sojae* to various selective agents was tested on agar plates with the respective drug additions. The range of agents was restricted to those for which a corresponding fungal selection marker is available. From this analysis it was concluded that hygromycin B, phleomycin, bialaphos and acetamide selection cannot be used for *A. sojae*, whereas oligomycin C and acrylamide turned out to be useful as selective agents. Other known selective agents such as sulfonylurea and benomyl were not tested. Although *A. sojae* is able to grow on agar plates containing acetamide as a sole nitrogen source, the *amdS* gene of *A. nidulans* can also be used as a dominant selection marker for *A. sojae* on the basis of acrylamide utilization (Kelly and Hynes 1985). Unlike the other dominant markers, the *amdS* gene is not an antibiotic resistance marker but a dominant nutritional marker, encoding acetamidase, which is an enzyme that confers the ability to use acetamide and acrylamide as nitrogen and carbon sources.

For transformation of *A. sojae*, vector p3SR2 (carrying the *amdS* marker; Kelly and Hynes 1985) was used in combination with a derivative of the autonomously replicating *Aspergillus* vector Arp1 (Gems et al. 1991). In all *Aspergillus* species tested so far, this vector resulted in highly increased numbers of (instable) transformants when used as a co-transforming vector. Initial experiments revealed, that even on selective acrylamide plates, a considerable background from non-transformed protoplasts was observed. Selection of primary AmdS+ transformants required around three weeks, and many of the initially selected putative transformants turned out to be false positives, only showing background growth after transfer to fresh selective acrylamide plates. Further optimization of the acrylamide selection was obtained reducing background growth by omitting glucose from the acrylamide selective plates (Heerikhuisen et al. 2001).

9.3.2.3 Auxotrophic Selection Markers

As the transformation frequency was found to be too low for routine transformation when using the *amdS* marker for selection, an auxotrophic marker system was developed for *A. sojae*. In general, such a system provides a higher transformation frequency. A commonly used auxotrophic selection method for fungal transformants is based on the usage of orotidine 5-monophosphate decarboxylase (*pyrG*) mutants. Mattern et al. (1987) published the transformation of *A. oryzae* using the *A. niger pyrG* gene. The isolation of *pyrG* mutants is based on direct selection for resistance to fluoro-orotic acid (FOA). This approach has resulted in the isolation of numerous *pyrG* mutants for a variety of fungi to date. From our experience with a number of different filamentous fungi, the auxotrophic *pyrG*-based system has many favorable characteristics. Experiments were carried out to obtain *A. sojae pyrG* mutant strains, using a standard procedure based on direct selection for resistance to FOA on plates containing uridine to support growth of the mutant strain (van Hartingsveldt et al. 1987). However, use of this method did not result in *A. sojae pyrG* mutant strains, as all obtained FOA-resistant strains were able to grow without uridine. However, when using a modified selective FOA medium which contained uracil in addition to uridine, several FOA-resistant mutants were obtained, which were uracil- requiring (Heerikhuisen et al. 2001). Re-testing of these strains showed that they were unable to grow on uridine-supplemented minimal medium. Subsequent transformation experiments with some of the uracil-requiring strains showed that these mutants could indeed be complemented with a fungal *pyrG* gene (vector pAB4.1; see Figure 9.3). The inability of *pyrG* mutants to grow in minimal medium supplemented with uridine alone was not observed for the related species *A. nidulans, A. niger, A. oryzae*, and for various other fungal species.

9.3.2.4 Re-usable Selection Marker

Versatile genetic modification of *A. sojae* requires the possibility to modify, disrupt, and express a number of different genes in a single fungal strain. Therefore, the availability of a series of different selection markers is essential. However, the possibility of repeated use of the same marker in subsequent experiments can circumvent the need for multiple selection markers. Use of the *pyrG* gene is very suitable for this purpose, because it allows selection of the *pyrG* mutant (FOA selection) and the PyrG+ transformant (uracil-less medium). For such an approach, a *pyrG* marker gene was designed in which the complementing sequence was flanked by a direct repeat sequence originating from the 3′ flanking end of the *pyrG* gene. The construction of a vector containing this type of *pyrG* marker, pAB4-1rep, is detailed in Figure 9.3. Transformation of *A. sojae pyrG* mutants with this vector results in a similar number of PyrG+ transformants as with the vector pAB4-1.

In gene disruption experiments, the repeat-flanked *pyrG* fragment has been used for disruption of the gene of interest and subsequent removal of the *pyrG* selection marker (described in Section 9.5).

9.3.3
Promoter Elements

Promoter elements are essential components for the establishment of an expression platform. Therefore, we set out experiments for the identification and isolation of suitable elements for the *A. sojae* platform. As a result of gene expression studies, highly expressed fungal genes were identified and exploited as sources for efficient promoters. Currently, the most frequently used elements are derived from: (i) the constitutive *A. nidulans gpdA* gene, encoding glyceraldehyde 3-phosphate dehydrogenase; (ii) the *A. niger* glucoamylase-encoding gene *glaA*; and (iii) the *A. nidulans* alcohol dehydrogenase gene *alcA* and the *Trichoderma reesei* cellobiohydrolase (*cbh1*) gene (Verdoes 1994; Punt et al. 1991; Keranen and Penttila 1995). In contrast to the constitutive expression imposed by the *gpdA* promoter, a carbon source-dependent

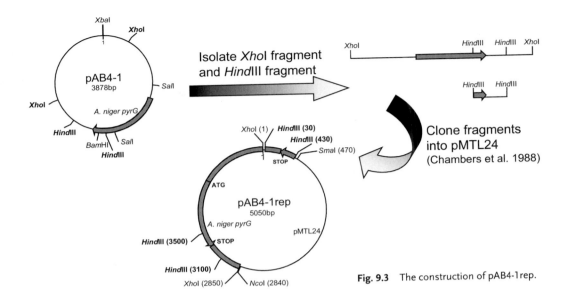

Fig. 9.3 The construction of pAB4-1rep.

gene expression profile is elicited by the others. Commonly used inducers for the latter three are starch/maltodextrin, ethanol, and cellulose/lactose, respectively.

The selected elements were assessed for heterologous gene expression in *A. sojae* by generating expression vectors using an *E. coli* glucuronidase (*GUS*) reporter gene (Roberts et al. 1989). Based on initial results, we focused on the *A. nidulans gpdA* and the *A. niger glaA* promoters (Roberts et al. 1989; Verdoes et al. 1994). In addition, a third *Aspergillus* promoter was evaluated for use in *A. sojae*. This promoter was obtained from the *A. niger*-derived *bipA* gene which exhibits a high and constitutive expression profile and encodes an endoplasmic reticulum (ER)-chaperone from the HSP70 gene family (van Gemeren et al. 1997). Interestingly, this promoter was induced through a stress-response mediated by the overproduction of homologous and heterologous proteins (van Gemeren et al. 1997; Punt et al. 1998). As a result of this, the *bipA* promoter is of potential interest for high-level induced gene expression. For assessment, the *bipA* promoter was included into a GUS expression vector designed identically to that described for the vectors harboring *gpdA* and *glaA* promoters (P.J. Punt et al., unpublished results).

The various expression vectors described above were applied to generate a number of *A. sojae* ATCC11906 transformants using the *amdS* gene for selection. GUS+ transformants were identified using X-glucuronide plate assays (Roberts et al. 1989). For each of the promoters, representative transformants were purified and analyzed for *GUS* expression in shake-flask cultures using nitrate-based mineral medium (Bennett and Lasure 1991) supplemented with a variety of carbon sources (Table 9.1). For comparison, results obtained with *A. niger* transformants were also included in this analysis (Verdoes et al. 1994; P.J. Punt et al., unpublished results). The *A. nidu-*

Tab. 9.1 Promoter strength in various *Aspergillus* transformants.

Transformant	Promoter	GUS activity (U mg⁻¹) in Minimal Medium		
		5% xylose	5% glucose	5% maltodextrin
Aspergillus sojae				
ATCC 11906 wt	–	0	0	0
ATCC11906[pGUS54]	*gpdA*	9141	6291	6667
ATCC11906[pGUS64]	*glaA*	33	50	25
ATCC11906[pBIPGUS]	*bipA*	2914	2849	1642
*Aspergillus niger**				
AB4.1 [pGUS54]	*gpdA*	ND	1992	2365
AB4.1[pGUS64]	*glaA*	ND	1966	4384
AB4.1 [pBIPGUS]	*bipA*	ND	785	1237

* *A. niger* transformants carrying a single copy of the expression vector at the *pyrG* locus were analyzed for GUS expression. For *A. sojae*, representative transformants were selected, based on a GUS plate assay.

lans gpdA promoter elicits the highest level of gene expression for each of the carbon sources, as shown in Table 9.1. In contrast, the *A. niger glaA* promoter is hardly active in *A. sojae* and shows no regulation by carbon source under the conditions tested. A previous analysis of the regulation of the *A. niger glaA* promoter suggested that this promoter is not only carbon source-, but also pH-regulated. Experiments carried out in various *A. niger* strains with different acidification characteristics revealed that the *glaA* promoter is not active at pH >6 (Withers et al. 1998; P.J. Punt et al., unpublished results). Since shake-flask culturing of *A. sojae* without pH adjustments results in the formation of pH values higher than 7, a lack of activity of the *glaA* promoter in this host is easily explained. In contrast, the *bipA* promoter is fully active in *A. sojae*, but of a strength lower than that of the *gpdA* promoter, as was also observed in *A. niger*. Furthermore, the *bipA* promoter is of decreased activity in *A. sojae* when culturing in maltodextrin-based media. This divergent gene expression profile was not further analyzed in detail.

Based on the results obtained with the *GUS* expression vectors, the *A. nidulans*-derived *gpdA* promoter was chosen as a preferred promoter for gene expression in *A. sojae*. Current research in our laboratory is focusing on the isolation of alternative promoters for similar high-level gene expression in *A. sojae*.

9.4
Aspergillus sojae as a Cell Factory for Foreign Proteins

Fungal cells are known to be efficient natural protein factories, as indicated above. This capability made them preferred target organisms for the production of recombinant proteins. Initial developments were aimed at increasing the productivity and the yields of fungal proteins by introducing additional gene copies (Verdoes 1994). Indeed, this approach resulted in largely increased levels of recombinant proteins

(Verdoes 1994). An additional encouragement was the recent substantial synthesis of fungal heme-peroxidases, the production of which in microbial hosts was reported to be extremely difficult (Conesa et al. 2002).

Based on these promising results, the production of non-fungal heterologous proteins was attempted, albeit with different levels of success. Based on a compilation of data from expression studies performed in other fungal hosts, the gene-fusion approach developed initially by researchers at Genencor has been particularly successful (reviewed in Gouka et al. 1997). In this method, the gene encoding the non-fungal protein of interest is fused to a gene encoding a well-secreted fungal protein, such as glucoamylase or cellobiohydrolase. This approach was applied successfully to the new system.

For recombinant protein production studies we have selected a number of divergent model proteins ranging from fungal hydrolases (glucoamylase, phytase), a fungal oxidase (laccase), and a mammalian protein (interleukin 6). As pointed out earlier, P*gpdA*-based expression vectors were employed and introduced into selected *A. sojae* host strains.

9.4.1
Production of Fungal Proteins

The various expression vectors used in this study are shown in the Appendix (Table A9.2). As shown in Figure 9.4, high levels of the fungal phytase and glucoamylase proteins were secreted from selected transformants in shake-flask cultures. In these transformants, the fraction of secreted recombinant proteins consisted of at least 80% of the total. The use of a glucoamylase-phytase gene-fusion did not result in higher levels of phytase. Apparently, in this case the gene fusion approach could not further improve the expression level of the fungal phytase. Since the copy number of the introduced expression cassette in the *A. sojae* transformants was low compared to that reported for *A. niger*, a cosmid approach was used to increase the copy number (Verdoes et al. 1993). This approach is based on the construction of a cosmid vector in which multiple copies of the expression cassette are cloned. Using a phytase expression vector carrying four copies of the expression cassette, the level of secreted phytase was increased at least twofold.

We were also successful in generating *A. sojae* transformants which secreted an active fungal blue copper-containing laccase (Record et al. 2002) at significant levels (Figure 9.5).

It should be noted that in all these cases only a limited set of transformants was screened. Screening of larger numbers of transformants using a platform for 96-well-based fungal cultures may result in the isolation of transformants with even higher levels of recombinant protein.

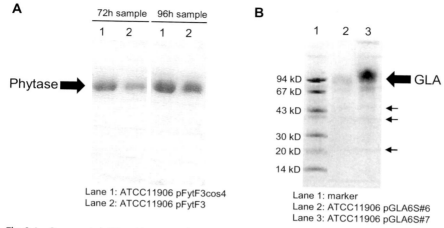

A

72h sample   96h sample

1   2   1   2

Phytase

B

1   2   3

94 kD
67 kD
43 kD
30 kD
20 kD
14 kD

GLA

Lane 1: ATCC11906 pFytF3cos4
Lane 2: ATCC11906 pFytF3

Lane 1: marker
Lane 2: ATCC11906 pGLA6S#6
Lane 3: ATCC11906 pGLA6S#7

Fig. 9.4 Coomassie brilliant blue-stained SDS-PAGE gels of medium samples from: A) phytase-secreting *A. sojae* transformants; and B) glucoamylase (GLA)-secreting *A. sojae* transformants. The small arrows indicate endogenous *A. sojae* proteins.

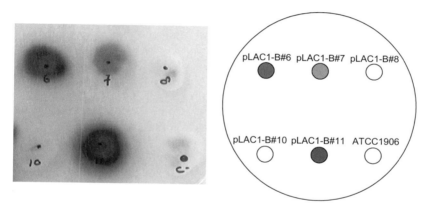

pLAC1-B#6 pLAC1-B#7 pLAC1-B#8

pLAC1-B#10 pLAC1-B#11 ATCC1906

Fig. 9.5 *A. sojae* pLAC1-B transformants in an ABTS plate assay (Record et al. 2002). The colored halo around the colonies indicate the secretion of active laccase. *A. sojae* strain ATCC11906 was included as a negative control.

9.4.2
Production of Non-fungal Proteins

As we are particularly interested in the production of non-fungal proteins, we have also expressed several genes for animal/human proteins in *A. sojae*. One of the proteins produced in *A. sojae* is human interleukin 6 (IL-6). For the production of IL-6, a glucoamylase-IL-6 fusion construct was used (pAN56–4; Broekhuijsen et al. 1993). In shake-flask cultures of the respective *A. sojae* transformants, reasonable levels of IL-6 could be detected by Western blot analysis using an IL-6 antiserum. The levels obtained in these cultures were clearly higher than those obtained initially in *A. niger* (Broekhuijsen et al. 1993). Attempts to produce IL-6 using vector constructs harboring a non-fused *IL-6* gene failed. From these results it can be concluded that fusion to a secreted fungal protein is a prerequisite to secrete non-fungal proteins in *A. sojae*, similar to the situation observed in other *Aspergillus* sp.

For the secretion of the intracellular green fluorescent protein (GFP), a glucoamylase-*GFP* fusion was constructed. Again, high-level secretion was achieved, and the recombinant *A. sojae* strains were found to be superior to *A. niger* reference strains. In contrast to the situation in *A. niger* (Gordon et al. 2000), significant amounts of GFP could be detected extracellularly, whereas in *A. niger* the secreted GFP was rapidly inactivated and/or degraded. For both IL-6 and GFP the acidification of culture medium in *A. niger* is suggested to be the cause for proteolytic degradation, and hence the low levels of active secreted protein. In shake-flask cultures with *A. sojae* no significant acidification occurs.

Based on these results it was concluded that the *A. sojae* expression system has good characteristics for the secretion of recombinant proteins. However, the level of protein secretion – and in particular of non-fungal proteins – had to be improved further in order to meet industrial requirements. To reach this goal, two approaches have been pursued, namely the development of improved production process conditions using controlled fermentation, and further strain improvement.

9.4.3
Controlled Fermentation

The transformant of the highest IL-6 productivity was selected for controlled batch fermentation experiments in fully controlled Bioflow fermentor systems (New Brunswick Scientific). Using a statistical experimental design, the effects of different process and culture parameters on IL-6 production were analyzed. An initial set of 16 batch fermentations was performed using Mineral Medium with two conditions for each of the four parameters (pH, temperature, C-source, and N-source). It transpired that pH was a critical parameter for culturing, and based on this finding a new series of fermentation runs was carried out at pH values ranging from 5 to 7.5.

Tab. 9.2 Production levels of glucoamylase and interleukin 6 (IL-6) in batch fermentations of an *A. sojae* pAN56-4 transformant at different pH values in Mineral Medium, 3% glucose, 10 mM NH_4Cl, 1% soya peptone, 33 °C. Glucoamylase levels were determined as described by Withers et al. (1998). IL-6 levels were determined using a BIACORE biosensor.

pH	Glucoamylase (mg L^{-1})	IL-6 (mg L^{-1})
5.0	180	0
5.5	350	173
6.0	500	181
6.5	370	106
7.0	440	54
7.5	290	54

As shown in Table 9.2, the highest levels of IL-6 were obtained at pH 6, at which point the relative levels of IL-6 were comparable to the glucoamylase levels, suggesting that no degradation occurred. In contrast, IL-6 levels were lower than glucoamylase levels at pH 5 and pH 7.5, most likely due to the degradation of IL-6 under these conditions. It was also observed that in most fermentation runs the IL-6 levels decreased at the end of fermentation (after 60 h).

9.5
Strain Development

9.5.1
Protease-deficient Strains

As described in Section 9.2, A. sojae strain ATCC11906 was selected for further development as an expression platform. Based on our previous studies with A. niger and the findings described earlier, it became clear that protease-deficient strains are required for the successful production of secreted recombinant proteins (Mattern et al. 1992; Broekhuijsen et al. 1993). Two approaches were pursued to obtain such protease-deficient strains. In a first approach, we carried out classical (ultra-violet) mutagenesis of spores from ATCC11906, followed by screening for a reduced milk-clearing zone and thus reduced proteolytic activity. In a second approach, the gene encoding alkaline protease was disrupted in strain ATCC11906.

9.5.1.1 Ultra-violet (UV) Mutagenesis
Freshly harvested spores from ATCC11906 were UV-mutagenized in a BioRad UV-chamber with a dose that resulted in 20–50% survival. Serial dilutions were plated onto skim-milk-containing plates (Mattern et al. 1992). Among 5000 surviving strains, four mutant strains with a considerably reduced milk-clearing zone were selected (Figure 9.6).

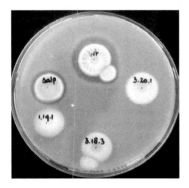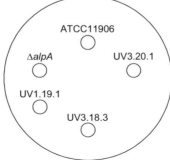

Fig. 9.6 Milk-clearing zones of the A. sojae protease-deficient strains compared to the wild-type strain on skim-milk-containing plates.

9.5.1.2 **Gene Disruption**

Protein analysis of *A. sojae* ATCC11906 fermentation media revealed a 35-kDa protein band on SDS-PAGE which was considered to represent an alkaline protease. Even though the proteolytic activity of this strain is relatively low, the putative 35-kDa protease band was relatively abundant. N-terminal sequence analysis confirmed that, indeed, the alkaline protease was present in this band. Primers were designed to clone the promoter region and the major part of the gene by PCR. The resulting PCR fragment was used as a probe for screening of an ATCC11906 cosmid library using the pAOpyrGcosarp1 vector (Hjort et al. 2000). A genomic clone containing the alkaline protease gene (*alpA*) was picked up. Two fragments (a 4-kb *Eco*RI fragment and a 2.5-kb *Hind*III fragment) from this clone were sub-cloned into a pUC19 vector and characterized by restriction enzyme digestion and sequence analysis. These sub-clones were used to construct a gene replacement vector replacing a 700-bp fragment of the coding region by the re-usable *pyrG* selection marker from pAB4−1rep (Figure 9.7).

An 8.7-kb *Eco*RI fragment from pAS1-Δalp, carrying the *alpA* disruption fragment, was used for transformation of the ATCC11906 *pyrG⁻* strain. A number of

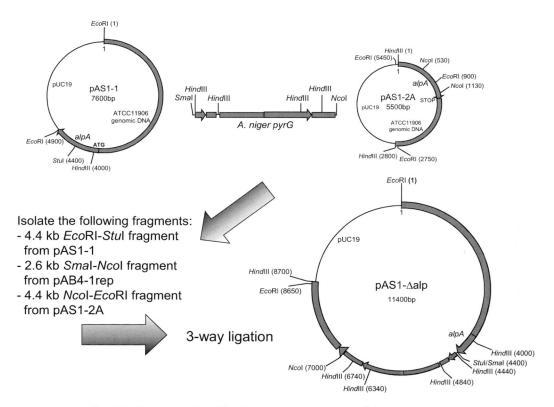

Fig. 9.7 The construction of the *alpA* disruption vector pAS1-Äalp.

transformants with a reduced milk halo were obtained. Southern analysis of these transformants revealed successful deletion of the alkaline protease gene. To allow subsequent use of the *pyrG* marker for transformation, 10^6 spores were plated on FOA-containing medium, selective for *pyrG* mutants. From strains with the correctly integrated disruption cassette a large number of FOA-resistant colonies were identified, whereas no spontaneous FOA-resistant mutants were obtained after plating 10^6 spores of a wild-type control strain. The FOA-resistant colonies derived from the disruption mutants were virtually all uracil- and uridine-requiring, and were shown again to be *pyrG⁻*. Southern analysis confirmed the removal of the *pyrG* gene at the *alpA* locus, leaving only a single copy of the 400-bp repeat at the site of disruption.

9.5.1.3 Controlled Batch Fermentations

Batch fermentations were carried out with the *A. sojae* ATCC11906 strain in comparison to the disrupted strain ATCC11906 Δ*alpA* and the UV-mutagenized strain ATCC11906 UV3.20.1. All fermentations were performed in a fully controlled Bioflow III fermentor (New Brunswick Scientific) using the following conditions: Mineral Medium at pH 6.5; 3% glucose; 10 mM NH_4Cl; 2% yeast extract; 30 °C.

When monitoring the proteolytic activity of the various culture samples, the disruptant strain showed, on completion of fermentation, about 40% reduction in proteolytic activity compared to that of the wild-type (Figure 9.8). The proteolytic activity of the UV-mutagenized strain was even further reduced (up to 70%) in comparison to the wild-type strain. Thus, we are confident that the newly generated protease-deficient strains will provide a considerable improvement for the use of *A. sojae* for protein production under controlled fermentation conditions.

Nevertheless, further improvement of the fermentation process is required. In particular, the culture viscosity was found to be very high due to the filamentous

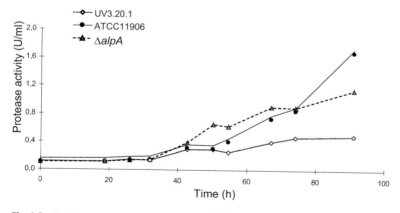

Fig. 9.8 Protease activity was measured at pH 7.8. The *A. sojae* protease-deficient strains were compared to the ATCC11906 strain. Universal Protease Substrate (Roche Applied Science, Cat. no. 1080733) was used to determine the activity.

growth of *A. sojae*. Consequently, oxygen transfer in the culture broth was insufficient, causing poor growth and poor biomass yields.

In order to improve oxygen transfer, the culture viscosity needs to be lowered. This can be achieved either by changing the process parameters, which could cause a difference in the growth characteristics of the fungus (Bhargava et al. 2003), or by changing the morphology of the fungus itself. Consequently, the latter option – to isolate morphological *A. sojae* mutants of reduced viscosity – was chosen.

9.5.2
Low-viscosity Mutants

As described previously, viscosity is of major concern during fermentations with *A. sojae* and many other fungal organisms (e.g., Withers et al. 1998; Bocking et al. 1999). The results of previous studies in our laboratory have shown that the disruption of a gene encoding the pro-protein convertase resulted in hyperbranching, low-viscosity mutants in *A. niger*. Based on this finding, the decision was taken to disrupt the corresponding *A. sojae* gene.

9.5.2.1 Gene Disruption

The pro-protein convertase-encoding gene (*pclA*) of *A. sojae* was isolated by functional complementation of an *A. niger pclA* disruption mutant with the ATCC11906 cosmid library. A genomic clone was identified to complement the hyperbranching phenotype of this mutant, resulting in wild-type growth characteristics. PCR and partial sequence analysis confirmed the presence of the *pclA* gene on this genomic clone. A 7.2-kb *Xba*I fragment contained on this clone was characterized by restriction analysis and sub-cloned into the pMTL24 cloning vector (Chambers et al. 1988). Subsequently, a 5.2-kb *Xba*I/*Bam*HI fragment was further sub-cloned, resulting in the vector pAS2-4 (Figure 9.9). The construction of a *pclA* disruption vector was achieved by cloning the repeat-flanked *pyrG* gene from pAB4-1rep as a *Sma*I fragment into the unique *Eco*RV site of the *pclA* gene (Figure 9.9).

The 8.2-kb *Not*I/*Asc*I fragment from pAS2-4Δpcl, containing the *pclA* disruption fragment, was used to transform strain ATCC11906 *pyrG⁻* and its protease-deficient derivatives. A number of transformants were obtained of which 5–10% showed a compact morphology. After purification of the compactly growing transformants, two types of compact morphology were observed; a poorly sporulating type, and another of normal sporulation (Figure 9.10). Microscopic analysis also revealed differences in morphology. The poorly sporulating transformants were hyperbranching, while the other transformants showed very short hyphae (Figure 9.11).

Southern analysis showed that the *pclA* gene was disrupted in the poorly sporulating, hyperbranching transformants. In the other type of transformants, the *pclA* disruption fragment was integrated in the genome elsewhere. Nevertheless, it was found that this morphologically aberrant type of transformants showed low-viscosity characteristics under shake-flask conditions, as did the *pclA* disruption strains. As the genotype of the non-*pclA* disrupted strains is still unknown, we refer to these strains as LFV (Low Fermentor Viscosity) strains.

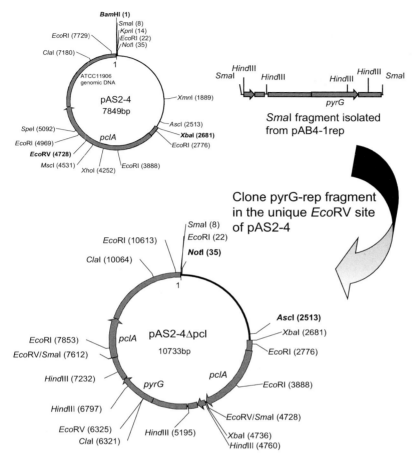

Fig. 9.9 The construction of the *pclA* disruption vector pAS2-4Δpcl.

9.5.2.2 Controlled Fermentations

The *pclA* disruption strains were grown in batch fermentations under the following conditions: Mineral Medium, 3% glucose; 10 mM NH₄Cl; 2% yeast extract; 33 °C; pH 5.5. The compact morphology of these strains in shake-flask cultures can be described as micropellets (pellet diameter <1 mm; pellet diameter of wild-type strains are >3 mm). Unexpectedly, in the batch fermentations rigid pellets were observed which tended to enlarge in time. From *A. niger* research it is known that pellet size decreases when the pH of the fermentation is increased (N. van Biezen, unpublished results). Therefore, the batch fermentations were repeated at pH 7.0, under which conditions micropellets of the *A. sojae pclA* disruption strains were observed, resulting in a lower viscosity. However biomass yields were not higher than those obtained with the wild-type strain. As the *A. sojae pclA* disruption strains poorly sporulate, further research is required on these to evaluate their potential as expression hosts.

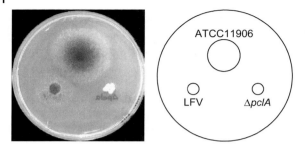

Fig. 9.10 Morphology of *A. sojae* wild-type and low-viscosity mutant strains grown on an agar plate.

A. sojae wild-type   *A. sojae* LFV   *A. sojae* ΔpclA

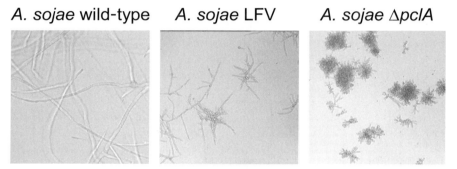

Fig. 9.11 Morphology of *A. sojae* wild-type and low-viscosity mutant strains grown in culture medium.

Controlled fed-batch fermentations were carried out with the *A. sojae* ATCC11906 strain and two LFV strains. The viscosity of whole fermentation broth (medium + cells) was determined in a Haake Viscotester (type VT500) using sensor system MV DIN (vessel number 7), operated at room temperature. A fresh fermentation sample (40 mL) was equilibrated for 1 min at a set spindle speed (5 r.p.m.). After this equilibration period, the viscosity was measured at 10 different spindle speeds. Viscosity ranges were determined for the *A. sojae* ATCC11906 and the two LFV strains and the viscosity of the whole fermentation broth of both LFV strains was significantly lower than that of the wild-type strain, while the biomass yields were comparable (Table 9.3). The LFV strain with the lowest viscosity (LFV#2) was selected for the development of high-density biomass fermentations.

High-density biomass fermentations were accomplished with the LFV#2 strain by optimizing the fed-batch fermentation conditions. Modifying the feed/batch ratios and higher amounts of yeast extract, phosphate and ammonia resulted in an up to fourfold higher biomass yield. The viscosity of these high-density biomass fermentations was still very low, and the macroscopic morphology of the LFV#2 strain could be described as "yeast-like". Initial protein production experiments with this strain showed good fermentation characteristics. Research investigations are currently un-

Tab. 9.3 Viscosity ranges of the various *A. sojae* strains determined on a Haake Viscotester (type VT500), using sensor system MV DIN (vessel number 7).

Strains	Viscosity (cP) at shear rate			Biomass (g L^{-1})
	~6.5 s^{-1}	83.2 s^{-1}	644.4 s^{-1}	
A. sojae ATCC11906	≫ 2000	1505	155	8.8 and 16.6
A. sojae LFV#2	1565	94	18	6.9 and 17.2
A. sojae LFV#4	2000	751	76	7.6 and 17.2

der way to determine the nature of the mutation in this strain. We are confident that the combination of protease-deficient and fermentor-adapted strains will further improve the *A. sojae* expression platform for recombinant protein production. An overview of strains is given in Appendix Table A9.1.

9.6
Outlook

A. sojae has been developed as a platform for recombinant protein production, in a similar manner as described previously for several other *Aspergillus* species. One aspect that makes *A. sojae* attractive is that it has a long history of (safe) use in the food industry. Several industrial processes are already in place applying both submerged and solid-state fermentations. However, additional experimental investigations are needed to define the conditions of controlled fungal fermentation for the production of non-fungal proteins when employing this organism, as well as using established fungal hosts such as *A. niger*. The established culture conditions defined for the production of endogenous proteins are often not suited to the production of heterologous non-fungal proteins.

This aspect is addressed within the department of Microbiology at TNO Nutrition and Food Research (www.voeding.tno.nl), as well as in a research program which was initiated in 2003 and funded by the Dutch Government: Kluyver Centre for Genomics of Industrial Fermentation (www.kluyvercentre.nl). In this program, integrative data-driven approaches such as transcriptomics, proteomics, and metabolomics will be used to address bottlenecks in the complex processes of fungal fermentation and protein production. This research will lead to the identification of new targets for process and strain optimization, which may subsequently be addressed with the molecular and process engineering approaches, as described in this chapter.

Acknowledgments

The authors thank Kurt Vogel and Markus Wyss for their contributions to the studies described in this chapter, and for their critical reading of the manuscript. DSM Nutritional Products is gratefully acknowledged for the financial support of these investigations.

Appendix

Tab. A9.1   Selection of *Aspergillus sojae* strains.

Strain	Genotype	Phenotype	Reference
Parental strains			
ATCC9362	Wild-type		Heerikhuisen et al. (2001)
ATCC11906	Wild-type		Heerikhuisen et al. (2001)
ATCC20235	Wild-type		Heerikhuisen et al. (2001)
ATCC20245	Wild-type		Heerikhuisen et al. (2001)
ATCC20387	Wild-type		Heerikhuisen et al. (2001)
ATCC20388	Wild-type		Heerikhuisen et al. (2001)
ATCC42249	Wild-type		Heerikhuisen et al. (2001)
ATCC42250	Wild-type		Heerikhuisen et al. (2001)
ATCC42251	Wild-type		Heerikhuisen et al. (2001)
ATCC46250	Wild-type		Heerikhuisen et al. (2001)
Protease-deficient strains			
ATCC11906 Δ*alpA*	*alpA*	AlpA⁻	Heerikhuisen et al. (2001)
UV1.19.1	*niaD*, unknown	NiaD⁻, Protease deficient	Heerikhuisen et al. (2001)
UV3.18.3	unknown	Protease deficient	Heerikhuisen et al. (2001)
UV3.20.1	*niaD*, unknown	NiaD⁻, Protease deficient	Heerikhuisen et al. (2001)
Low-viscosity mutants			
ATCC11906 Δ*pclA*	*pclA*	PclA⁻	Heerikhuisen et al. (2001)
ATCC11906 Δ*alpA* Δ*pclA*	*alpA, pclA*	AlpA⁻, PclA⁻	Heerikhuisen et al. (2001)
UV3.20.1 Δ*pclA*	*niaD, pclA*, unknown	NiaD⁻, PclA⁻, Protease deficient	Heerikhuisen et al. (2001)
ATCC11906 LFV	unknown	Low viscosity	Heerikhuisen et al. (2001)
Auxotrophic strains			
ATCC11906 *pyrG*	*pyrG*	PyrG⁻	Heerikhuisen et al. (2001)
ATCC11906 Δ*alpA pyrG*	*alpA, pyrG*	AlpA⁻, PyrG⁻	Heerikhuisen et al. (2001)

Tab. A9.1 (continued)

Strain	Genotype	Phenotype	Reference
ATCC11906 Δ*pclA pyrG*	*pclA, pyrG*	PclA⁻, PyrG⁻	Heerikhuisen et al. (2001)
ATCC11906 Δ*alpA* Δ*pclA pyrG*	*alpA, pclA, pyrG*	AlpA⁻, PclA⁻, PyrG⁻	Heerikhuisen et al. (2001)
UV1.19.1	*niaD*, unknown	NiaD⁻, Protease deficient	Heerikhuisen et al. (2001)
UV3.18.3 *pyrG*	*pyrG*, unknown	PyrG⁻, Protease deficient	Heerikhuisen et al. (2001)
UV3.20.1 *pyrG*	*niaD, pyrG,* unknown	NiaD⁻, PyrG⁻, Protease deficient	Heerikhuisen et al. (2001)
UV3.20.1 Δ*pclA pyrG*	*niaD, pclA, pyrG,* unknown	NiaD⁻, PclA⁻, PyrG⁻, Protease deficient	Heerikhuisen et al. (2001)

Tab. A9.2 Selection of vectors used in *Aspergillus sojae*.

Expression vectors	Expression cassette	Gene	Selection marker	Reference
pAN52-1Not	*A. nidulans gpdA* promoter and *trpC* terminator			Punt et al. (1991) (Accession #Z32524)
pAN56-1	*A. nidulans gpdA* promoter and *trpC* terminator	*A. niger glaA* (G2)		Punt et al. (1991) (Accession #Z32700)
pFytF3	*A. nidulans gpdA* promoter and *trpC* terminator	*A. fumigatus* phytase		Pasamontes et al. (1997)
pFytF3cos4	*A. nidulans gpdA* promoter and *trpC* terminator	*A. fumigatus* phytase		unpublished
pGLAFytF	*A. nidulans gpdA* promoter and *trpC* terminator	*A. niger glaA* (G2) – *A. fumigatus* phytase fusion		unpublished
pGLA6S	*A. nidulans gpdA* promoter and *trpC* terminator	*A. niger glaA* (G1)	*amdS*	Punt et al. (1991)
pLAC1-B	*A. nidulans gpdA* promoter and *trpC* terminator	*P. cinnabarius* laccase		Record et al. (2002)
pAN56-4	*A. nidulans gpdA* promoter and *trpC* terminator	*A. niger glaA* (G2) – human interleukin 6 fusion		Broekhuijsen et al. (1993)
Pgpd-GLA-GFP	*A. nidulans gpdA* promoter and *trpC* terminator	*A. niger glaA* (G2) – green fluorescent protein fusion		Gordon et al. (2000)

Tab. A9.2 (continued)

Selection vectors	Origin	Selection marker	Type	Reference
p3SR2	A. nidulans	amdS	Dominant	Kelly and Hynes (1985)
pAN4-cos1	A. nidulans	amdS	Dominant	Verdoes et al. (1993)
pAB4−1	A. niger	pyrG	Auxotrophic	van Hartingsveldt et al. (1987)
pAB4−1rep	A. niger	pyrG	Auxotrophic	Heerikhuisen et al. (2001)
pAOpyrGcosarp1	A. oryzae	pyrG	Auxotrophic	Hjort et al. (2000)

References

BALLANCE DJ, BUXTON FP, TURNER G (1983) Transformation of *Aspergillus nidulans* by the orotidine-5'-phosphate decarboxylase gene of *Neurospora crassa*. Biochem Biophys Res Commun 112: 284–289

BENNETT JW, LASURE LL (1991) Growth media. In: More Gene Manipulations in Fungi (Bennett JW, Lasure LL, Eds). Academic Press, San Diego, pp 441–457

BHARGAVA S, NANDAKUMAR MP, ROY A, WENGER KS, MARTEN MR (2003) Pulsed feeding during fed-batch fungal fermentation leads to reduced viscosity without detrimentally affecting protein expression. Biotechnol Bioeng 81: 341–347

BOCKING SP, WIEBE MG, ROBSON GD, HANSEN K, CHRISTIANSEN LH, TRINCI AP (1999) Effect of branch frequency in *Aspergillus oryzae* on protein secretion and culture viscosity. Biotechnol Bioeng 65: 638–648

BROEKHUIJSEN MP, MATTERN IE, CONTRERAS R, KINGHORN JR, VAN DEN HONDEL CA (1993) Secretion of heterologous proteins by *Aspergillus niger*: production of active human interleukin-6 in a protease-deficient mutant by KEX2-like processing of a glucoamylase-hIL6 fusion protein. J Biotechnol 31: 135–145

CHAMBERS SP, PRIOR SE, BARSTOW DA, MINTON NP (1988) The pMTL nic-cloning vectors. I. Improved pUC polylinker regions to facilitate the use of sonicated DNA for nucleotide sequencing. Gene 68: 139–149

CHANG PK, BHATNAGAR D, CLEVELAND TE, BENNETT JW (1995) Sequence variability in homologs of the aflatoxin pathway gene *aflR* distinguishes species in *Aspergillus* section *Flavi*. Appl Environm Microbiol 61: 40–43

CONESA A, PUNT PJ, VAN DEN HONDEL CA (2002) Fungal peroxidases: molecular aspects and applications. J Biotechnol 93: 143–158

DEMAIN AL (2000) Small bugs, big business: the economic power of the microbe. Biotechnol Adv 18: 499–514

ELANDER RP (2003) Industrial production of beta-lactam antibiotics. Appl Microbiol Biotechnol 61: 385–392

GEMS D, JOHNSTONE IL, CLUTTERBUCK AJ (1991) An autonomously replicating plasmid transforms *Aspergillus nidulans* at high frequency. Gene 98: 61–67

GORDON CL, ARCHER DB, JEENES DJ, DOONAN JH, WELLS B, TRINCI AP, ROBSON GD (2000) A glucoamylase:: GFP gene fusion to study protein secretion by individual hyphae of *Aspergillus niger*. J Microbiol Methods 42: 39–48

GOUKA RJ, PUNT PJ, VAN DEN HONDEL CA (1997) Efficient production of secreted proteins by *Aspergillus*: progress, limitations and prospects. Appl Microbiol Biotechnol 47: 1–11

HEERIKHUISEN M, VAN DEN HONDEL CA, PUNT PJ, VAN BIEZEN N, ALBERS A, VOGEL K (2001) Novel means of transformation of fungi and their use for heterologous protein production. World Patent Application WO 01/09352

HJORT C, VAN DEN HONDEL CA, PUNT PJ, SCHUREN FH (2000) Fungal transcriptional activator useful in methods for producing polypeptides. World Patent Application WO 00/20596

KELLY JM, HYNES MJ (1985) Transformation of

Aspergillus niger by the *amdS* gene of *Aspergillus nidulans*. EMBO J 4: 475–479

KERANEN S, PENTTILA M (1995) Production of recombinant proteins in the filamentous fungus *Trichoderma reesei*. Curr Opin Biotechnol 6: 534–537

KURTZMAN CP, SMILEY MJ, ROBERT CJ, WICKLOW DT (1986) DNA relatedness among wild and domesticated species in the *Aspergillus flavus* group. Mycologia 78: 955–959

KUSOMOTO KI, YABE K, NOGATA Y, OHTA H (1998) *Aspergillus oryzae* with and without a homolog of aflatoxin biosynthetic gene *ver-1*. Appl Microbiol Biotechnol 50: 98–104

MATTERN IE, UNKLES S, KINGHORN JR, POUWELS PH, VAN DEN HONDEL CA (1987) Transformation of *Aspergillus oryzae* using the *A. niger pyrG* gene. Mol Gen Genet 210: 460–461

MATTERN IE, VAN NOORT JM, VAN DEN BERG P, ARCHER DB, ROBERTS IN, VAN DEN HONDEL CA (1992) Isolation and characterization of mutants of *Aspergillus niger* deficient in extracellular proteases. Mol Gen Genet 234: 332–336

PASAMONTES L, HAIKER M, WYSS M, TESSIER M, VAN LOON AP (1997) Gene cloning, purification, and characterization of a heat-stable phytase from the fungus *Aspergillus fumigatus*. Appl Environ Microbiol 63: 1696–1700

PETERSON SW (2000) Phylogenetic relationships in *Aspergillus* based on rDNA sequence analysis. In: Integration of modern taxonomic methods for *Penicillium* and *Aspergillus* classification (Samson RA, Pitt JI, Eds). Singapore: Harwood Academic Publishers, pp 323–355

PUNT PJ, VAN BIEZEN N, CONESA A, ALBERS A, MANGNUS J, VAN DEN HONDEL CA (2002) Filamentous fungi as cell factories for heterologous protein production. Trends Biotechnol 20: 200–206

PUNT PJ, VAN DEN HONDEL CA (1992) Transformation of filamentous fungi based on hygromycin B and phleomycin resistance markers. Methods Enzymol 216: 447–457

PUNT PJ, VAN GEMEREN IA, DRINT-KUIJVEN-HOVEN J, HESSING JG, VAN MUIJLWIJK-HARTE-VELD GM, BEIJERSBERGEN A, VERRIPS CT, VAN DEN HONDEL CA (1998) Analysis of the role of the gene *bipA*, encoding the major endoplasmic reticulum chaperone protein in the secretion of homologous and heterologous proteins in black *Aspergillus*. Appl Microbiol Biotechnol 50: 447–454

PUNT PJ, ZEGERS ND, BUSSCHER M, POUWELS PH, VAN DEN HONDEL CA (1991) Intracellular and extracellular production of proteins in *Aspergillus* under the control of expression signals of the highly expressed *Aspergillus nidulans gpdA* gene. J Biotechnol 17: 19–33

RECORD E, ASTHER M, SIGOILLOT C, PAGES S, PUNT PJ, DELATTRE M, HAON M, VAN DEN HONDEL CA, SIGOILLOT JC, LESAGE-MEESSEN L, ASTHER M (2003) Overproduction of the *Aspergillus niger* feruloyl esterase for pulp bleaching application. Appl Microbiol Biotechnol 62: 349–355

RECORD E, PUNT PJ, CHAMKHA M, LABAT M, VAN DEN HONDEL CA, ASTHER M (2002) Expression of the *Pycnoporus cinnabarinus* laccase gene in *Aspergillus niger* and characterization of the recombinant enzyme. Eur J Biochem 269: 602–609

ROBERTS IN, OLIVER RP, PUNT PJ, VAN DEN HONDEL CA (1989) Expression of the *Escherichia coli* beta-glucuronidase gene in industrial and phytopathogenic filamentous fungi. Curr Genet 15: 177–180

SIGOILLOT C, RECORD E, BELLE V, ROBERT JL, LEVASSEUR A, PUNT PJ, VAN DEN HONDEL CA, FOURNEL A, SIGOILLOT JC, ASTHER M (2004) Natural and recombinant fungal laccases for paper pulp bleaching. Appl Microbiol Biotechnol 64: 346–352

TAMURA M, KAWAHARA K, SUGIYAMA J (2000) Molecular phylogeny of *Aspergillus* and associated teleomorphs in the trichocomaceae (Eurotiales). In: Integration of modern taxonomic methods for *Penicillium* and *Aspergillus* classification (Samson RA, Pitt JI, Eds). Singapore: Harwood Academic Publishers, pp 357–372

USHIJIMA S, HAYASHI K, MURAKAMI H (1982) The current taxonomic status of *Aspergillus sojae* used in *Shoyu* fermentation. Agric Biol Chem 46: 2365–2367

VAN GEMEREN IA, PUNT PJ, DRINT-KUYVEN-HOVEN A, BROEKHUIJSEN MP, VAN'T HOOG A, BEIJERSBERGEN A, VERRIPS CT, VAN DEN HONDEL CA (1997) The ER chaperone encoding *bipA* gene of black *Aspergilli* is induced by heat shock and unfolded proteins. Gene 198: 43–52

VAN HARTINGSVELDT W, MATTERN IE, VAN ZEIJL CM, POUWELS PH, VAN DEN HONDEL CA (1987) Development of a homologous transformation system for *Aspergillus niger* based on the *pyrG* gene. Mol Gen Genet 206: 71–75

VERDOES JC (1994) Molecular genetic studies of the overproduction of glucoamylase in *Aspergillus niger*. PhD thesis,Vrije Universiteit of Amsterdam

VERDOES JC, PUNT PJ, SCHRICKX JM, VAN VERSEVELD HW, STOUTHAMER AH, VAN DEN HONDEL CA (1993) Glucoamylase overexpression in *Aspergillus niger*: molecular genetic analysis of strains containing multiple copies of the *glaA* gene. Transgenic Res 2: 84–92

VERDOES JC, PUNT PJ, STOUTHAMER AH, VAN DEN HONDEL CA (1994) The effect of multiple copies of the upstream region on expression of the *Aspergillus niger* glucoamylase-encoding gene. Gene 145: 179–187

WATSON AJ, FULLER LJ, JEENES DJ, ARCHER DB (1999) Homologs of aflatoxin biosynthesis genes and sequence of *aflR* in *Aspergillus oryzae* and *Aspergillus sojae*. Appl Environ Microbiol 65: 307–310

WITHERS JM, SWIFT RJ, WIEBE MG, ROBSON GD, PUNT PJ, VAN DEN HONDEL CA, TRINCI AP (1998) Optimization and stability of glucoamylase production by recombinant strains of *Aspergillus niger* in chemostat culture. Biotechnol Bioeng 59: 407–418

YELTON MM, HAMER JE, TIMBERLAKE WE (1984) Transformation of *Aspergillus nidulans* by using a *trpC* plasmid. Proc Natl Acad Sci USA 81: 1470–1474

YUAN GF, LIU CS and CHEN CC (1995) Differentiation of *Aspergillus parasiticus* from *Aspergillus sojae* by Random Amplification of Polymorphic DNA. Appl Environm Microbiol 61: 2384–2387

10
Sordaria macrospora
Ulrich Kück and Stefanie Pöggeler

List of Genes

Gene	Encoded gene product
acl1	ATP citrate lyase
cpc2	Repressor of cross pathway control
egfp	enhanced green fluorescent protein
gfp	green fluorescent protein (*Aequorea victoria*)
gpd	glyceraldehyde-3-phosphate dehydrogenase
hph	hygromycin-B-phosphotransferase (*E. coli*)
leu1	β-isopropyl malate dehydrogenase
ndk1	nucleotide diphosphate kinase
ppg1	pheromone precursor 1
trpC	anthranilate synthase multifunctional protein (*Aspergillus nidulans*)
ura3	orotidine-5'-decarboxylase

10.1
Introduction

Filamentous fungi have been used for decades as major producers in the pharmaceutical, food, and food-processing industries. Their utilization led to a high technical standard in fermentation processes with large-scale fermentors. This, together with the fact that many filamentous fungi possess the GRAS (generally recognized as safe) status makes them ideal organisms for the production of recombinant proteins (Radzio and Kück 1997; Kück 1999). However, improvement attempts for recombinant strains are hampered by the fact that most fungi of biotechnical relevance lack any sexual cycle – the prerequisite for meiotic recombination. Instead, they propagate asexually by producing a vast amount of conidiospores, which may be disadvantageous when governmental regulations in production processes have to be met.

In this chapter, we present the filamentous fungus *Sordaria macrospora* as a potential host organism for the production of recombinant polypeptides. *S. macrospora* is closely

Production of Recombinant Proteins. Novel Microbial and Eucaryotic Expression Systems. Edited by Gerd Gellissen
Copyright © 2005 WILEY-VCH Verlag GmbH & Co. KGaA, Weinheim
ISBN: 3-527-31036-3

related to the genetic model organism *Neurospora crassa*, which belongs to the order of Sordariales within the ascomycetes. Both fungi propagate sexually, but in contrast to *N. crassa*, *S. macrospora* does not form any conidiospores. Furthermore, *S. macrospora* combines all benefits of filamentous fungi for expressing heterologous genes. For example, expression products can be secreted into the culture medium, which is desirable because downstream processing is made much easier. As in yeasts, many post-translational modifications of proteins occur during the secretory process, such as glycosylation, proteolytic processing, and disulfide bond formation. These processes – which are to some extent absent from bacteria, especially in Gram-negative species – can be essential for the stability and biological activity of recombinant proteins.

Here, we aim at presenting evidence for the generation of *S. macrospora* host strains, which fulfill all requirements of reliable producers for recombinant products. In addition, examples are provided for the genetic engineering and successful production of heterologous proteins in *S. macrospora*.

10.2
General Biology

Sordaria is a ubiquitous genus, which has been described in detail by Lundqvist (1972). From conventional phylogenetic studies it is well known that *Sordaria* species are closely related to different *Neurospora* species, and that both belong to the class of ascomycetes. Within higher fungi, this represents the largest class and includes the highest number of species.

The relationship between the genera *Neurospora* and *Sordaria* was recently elucidated for 17 different species by distance and parsimony trees. The construction of molecular phylogenetic trees was based on DNA sequences from mating type genes or from the *gpd* (glyceraldehyde 3-phosphate dehydrogenase) gene (Pöggeler 1999).

The natural habitat of *S. macrospora* is the dung of herbivorous animals. Thus, this fungus belongs to the group of coprophilous fungi that are of ecological importance in recycling nutrients from animal feces. In contrast, *N. crassa* is commonly found on bread and other carbohydrate-rich foodstuffs. In addition, it was also identified as the earliest colonizer of the remains of incinerated vegetation (Davis and Perkins 2002).

10.3
Morphological Characterization, Molecular Phylogeny and Life Cycle of *Sordaria macrospora*

In the laboratory, the life cycle of *S. macrospora* can be completed within seven days. *S. macrospora* propagates only sexually, and its fruiting bodies (perithecia) are prominent morphological structures with a size of about 200 μm.

The sexual life cycle of ascomycetes can be either heterothallic (self-sterile) or homothallic (self-fertile). Heterothallic fungi, including *N. crassa*, exist in two mating types,

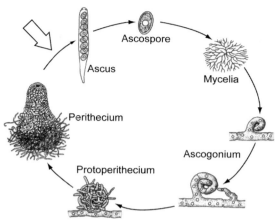

Fig. 10.1 Life cycle of the homo-thallic ascomycete *Sordaria macro-spora*. The cycle can be completed under laboratory conditions within seven days. The arrow indicates the site of a developmental block, which can be observed in the sterile mutant *inf*25 (see text for details).

A and *a*, and mating occurs only between sexual structures of opposite mating types. However, in the homothallic species *S. macrospora*, every strain is able to complete the sexual cycle without a mating partner. Detailed molecular analysis provided evidence that four mating type genes are located adjacent to each other on the chromosomal DNA in all *S. macrospora* strains (Pöggeler et al. 1997a; Pöggeler and Kück 2000). In heterothallic fungi, such as *N. crassa*, every mating type strain (A and *a*) has a physically different set of mating type genes (Pöggeler 2001). Mating-type proteins are thought to act as transcriptional regulators to control the expression of pheromone and phero-mone receptor genes. Pheromone genes as well as pheromone receptor genes were recently isolated from *S. macrospora* (Pöggeler 2000; Pöggeler and Kück 2001).

As illustrated in Figure 10.1, fruiting body development starts from vegetative my-celia with the formation of female gametangia, which are enveloped by sterile hy-phae to form closed spherical pre-fruiting bodies (protoperithecia). Subsequent cell differentiation gives rise to an outer-pigmented peridial tissue and inner ascus initi-als embedded in sterile paraphyses. The mature flask-like fruiting bodies are called perithecia. In ascomycetous fungi, fruiting bodies are generally termed ascomata which enclose sexual meiosporangia (asci) and meiospores (ascospores). In *S. macro-spora*, every perithecium contains about 100 asci, containing eight linearly ordered ascospores. After maturation, the ascospores are discharged from the fruiting bodies through a preformed hole, the ostiolum.

10.4
Generation of Sterile Mutants as Host Strains

It has been demonstrated for a number of ascomycetes that several genes control sexual development. To isolate sterile strains from *S. macrospora*, we used EMS (ethylmethane sulfonate) mutagenesis to generate mutants with defects in fruiting body formation (Masloff et al. 1999). Within our collection of mutants (Table A10.1),

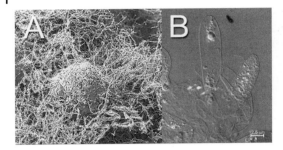

Fig. 10.2 Morphological characterization of a typical sterile "inf" strain from *Sordaria macrospora*. A) Electron scanning microscopy of the sterile mutant *inf*25; B) light microscopy of empty asci from a mature perithecium derived from mutant inf25.

we identified a class of strains, which generate fruiting bodies (perithecia), but are unable to produce ascospores. Sterile mutants of this type were designated with the prefix "inf" to indicate that they are "infertile" and as such are unable to produce any spores. As can be seen from Figure 10.2, these strains develop perithecia with asci lacking ascospores. As a consequence, these strains are sterile and thus fulfill high standard safety regulations when applied to production processes.

Albeit that they are sterile, every "inf" strain can be crossed in the laboratory with another non-allelic sterile strain. In the contact zone of both mycelia, fertile perithecia can be generated and used for manual ascospore manipulation. However, every "inf" strain can undergo meiotic recombination in spite of being sterile, and can be used as a convenient source for conventional strain improvement programs.

In this context, it seems worthwhile pointing out that several sterile mutants from *S. macrospora* have been used to isolate developmental genes for further molecular and functional characterization (Masloff et al. 1999, 2002; Nowrousian et al. 1999; van Heemst et al. 1999; Pöggeler and Kück 2004).

10.5
S. macrospora as a Safe Host for Heterologous Gene Expression

Production strains of recombinant proteins must meet several governmental requirements when used in biotechnical applications. Taking into account all data available from the literature, *S. macrospora* can be considered as a safe host for heterologous genes expression. This statement is based on the following facts:

1. The growth optimum lies between temperatures of 20–30 °C, and the ability to grow at 37 °C is sufficiently restricted.
2. Unlike most yeasts, *S. macrospora* is an obligate aerobe, thereby preventing growth in tissues or organs such as the gut or bladder, or systemically. Hence, it is very unlikely that *S. macrospora* should exercise adverse effects on either humans or animals.
3. *S. macrospora* does not have any conidiation, which distinguishes this fungus from *Penicillium camembertii*, *Penicillium roqueforti*, *Aspergillus oryzae* (= domesticated *Aspergillus flavus*) and *Fusarium venenatum*, which show severe conidiation.

Nevertheless, these filamentous fungi are considered safe for food according to the Food and Drug Administration (FDA) (see the FDA website [http://vm.cfsan.fda.gov/~dms/opa-micr.html]).

4. In the medical literature, *S. macrospora* is not mentioned for having allergenic properties. In addition, pathogenic effects of *S. macrospora* have never been reported from which, in summary, it can be deduced that *S. macrospora* has no allergenic or infectious potential.

5. No report could be found that describes *S. macrospora* as a mycotoxin producer, in contrast to *Penicillium camembertii*, *Penicillium roqueforti*, *Aspergillus oryzae* (= domesticated *Aspergillus flavus*) and *Fusarium venenatum*, which are still considered safe for food.

6. The available literature mentions the ascomycete *S. macrospora* mainly because of its significance as a genetic model organism that is frequently applied to basic research in cell and developmental biology.

In conclusion, *S. macrospora* can be considered as safe for in-vitro recombinant experiments. In the current research organism list of the German "Bundesgesundheits-blatt 40, 12 (1997)" (http://www.bba.de/gentech/orglist.htm), *S. macrospora* is therefore classified as bearing no risk for humans and the environment under "Risiko-gruppe 1 of Spender- und Empfängerorganismen für gentechnische Arbeiten zu Forschungszwecken" (i.e., risk group 1 of donor and recipient organisms for gene technological work for research purposes).

It is recommended, that *S. macrospora* should also be listed in "Spender- und Empfängerorganismen für gentechnische Arbeiten zu gewerblichen Zwecken" of the "Bundesgesundheitsblatt" in "Risikogruppe 1" (i.e., risk group 1 of donor and recipient organisms for gene technological work for commercial purposes). In this list, other closely related fungi already appear such as *Penicillium chrysogenum*, *Penicillium emmerssonii*, *Penicillium funiculosum*, and *Penicillium lilacinum*.

10.6
Molecular Genetic Techniques Developed for *S. macrospora*

The availability of suitable molecular tools is an essential prerequisite for the development of potential expression systems. In the case of *S. macrospora*, several technical improvements, together with conventional genetic methods, have made this fungus a versatile host for the expression of both homologous and heterologous genes.

A breakthrough in the genetic engineering of *S. macrospora* came, when reproducible DNA-mediated transformations became feasible using the bacterial hygromycin B phosphotransferase gene for selection (Walz and Kück 1995). Using strong fungal promoter sequences, the prokaryotic resistance gene can be used as a selection marker in transformation experiments. Following standard protocols, protoplasts derived from wild-type or mutant mycelia can be incubated together with isolated DNA. Transformants can then be selected on appropriate media. Usually, the transformation frequency is about 1 to 10 transformants per 10 µg of DNA.

In the meantime, other selection systems have been developed based on novel auxotrophic recipient strains. Using conventional mutagenesis, a leucine-auxotrophic strain was generated, which lacked the complete wild-type copy of the *leu1* gene, encoding β-isopropylmalate dehydrogenase (U. Kück, unpublished data).

An alternative strategy has been used to generate a uracil-auxotrophic host strain. The *ura3* (syn. *pyr4*) gene from *S. macrospora*, encoding orotidine 5′-phosphate decarboxylase, was isolated from genomic DNA (Nowrousian and Kück 1998), and the *hph* gene was inserted *in vitro* to disrupt the *ura3* open reading frame. The resulting recombinant plasmid pIG110 was employed in *S. macrospora* transformations to isolate a fungal derivative, in which the wild-type *ura3* copy was replaced by the chimeric *ura3-hph* gene through site-specific homologous recombination (U. Kück and S. Pöggeler, unpublished data). Both auxotrophic strains can be transformed with the corresponding wild-type gene from *S. macrospora*, and fungal transformants appear with frequencies as described above for the *hph* gene.

The availability of additional recipient strains for DNA-mediated transformations will be an alternative when, for example, a second recombinant gene has to be introduced into recombinant strains that already carry the frequently used *hph* resistance gene.

An additional important development for genetic engineering of *S. macrospora* is derived from the construction of various gene libraries with *S. macrospora* DNA. An indexed cosmid library, as well as a cDNA library, which is suitable for yeast two-hybrid screens should be mentioned here as examples (Pöggeler et al. 1997b; Pöggeler and Kück 2004).

Clones of indexed libraries are kept individually in micro-titer dishes to ensure an even representation of each clone. Several methods for the rapid screening of pooled cosmid DNA have been published, such as gene isolation by colony filter hybridization applied to cosmid clones, or rapid screening by PCR applying particular pairs of primers to pooled cosmid DNA. Using the indexed cosmid library, we successfully identified more than 20 genes from *S. macrospora*. Another application of indexed libraries is to clone genes by complementing mutant strains. Filamentous fungi with low transformation frequencies require specific cloning strategies, and with this approach several developmental genes from *S. macrospora* have been successfully isolated (Masloff et al. 1999; Nowrousian et al. 1999; Pöggeler and Kück 2004).

Using vectors of the yeast two-hybrid system, we have constructed a cDNA library from *S. macrospora* (unpublished data). This library facilitates a screen for the investigation of physical interactions between different polypeptides, which for example, can be components of a transcriptional regulatory network (Jacobsen et al. 2002).

Data from genome sequencing projects usually provide ample opportunities for advanced studies in regulatory mechanisms of gene expression at different levels of cellular complexity (genome, transcriptome, proteome, metabolome). The genome of *S. macrospora* consists of seven chromosomes (Pöggeler et al. 2000), but it has not yet been sequenced. However, the recent analysis of more than 85 sequences from this fungus indicated a high degree of sequence similarity and conservation of the overall genomic organization between *S. macrospora* and the completely sequenced genome of *N. crassa* (Galagan et al. 2003). The sequence information of the *N. crassa* genome can be used to identify functional elements in the *S. macrospora* genome (Pöggeler 1997). A comparative analysis of genome sequences from both fungi should further improve and simplify the annotation of the *N. crassa* genome (Nowrousian et al. 2004).

10.7
Isolation and Characterization of Strong Promoter Sequences from *S. macrospora*

In order to isolate promoter sequences, which mediate high transcriptional expression, we have characterized the *acl1* (ATP citrate lyase), *ndk1* (nucleotide diphosphate kinase), *cpc2* (repressor of cross pathway control), and *ppg1* (pheromone precursor gene) genes from *S. macrospora* by Northern hybridization.

In a first experimental step, appropriate probes were generated for hybridization experiments. While the *acl1* and *ppg1* genes were already characterized (Nowrousian et al. 1999; Pöggeler 2000) and the corresponding probes were available, probes for the two other genes had still to be generated. For this purpose, two pairs of primers were designed on the basis of the published *N. crassa* genomic sequence (Galagan et al. 2003). The oligonucleotide primers were used for PCR amplification with genomic DNA from *S. macrospora* as a template. The resulting amplificates had sizes of 655 bp (*cpc2*) and 594 bp (*ndk1*), respectively. The identity of the resulting fragments was verified by cloning and DNA sequencing.

In the next step, we used mRNA from the *S. macrospora* wild-type and from the sterile mutant strain inf25 for Northern hybridization. mRNA (5 μg) was separated by gel electrophoresis and transferred to nylon filters, after which individual hybridizations were performed with the four radiolabeled probes, as mentioned above. As shown in Figure 10.3, the *ppg1* probe results in a weak signal at 1.55 kb. Neverthe-

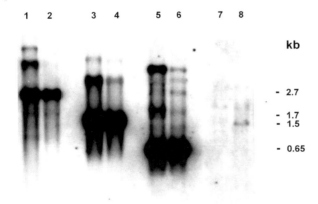

Fig. 10.3 Transcript analysis of mRNA from *S. macrospora* strains. mRNA (10 μg) from wild-type (odd lanes) or sterile (even lanes) strains were loaded on a denaturing RNA gel. RNA of individual lanes were hybridized to the following probes: *acl1* (lane 1,2), *ndk1* (lane 3,4), *cpc2* (lane 5,6), and *ppg1* (lane 7,8).

less, this signal is clearly stronger than those resulting from hybridizations with probes for house-keeping genes (e. g., β-tubulin gene, data not shown). Significantly stronger signals were obtained with the three other probes. In Figure 10.3, highly abundant transcripts can be identified representing the *acl1* (2.7 kb), *cpc2* (1.4 kb), and *ndk1* (0.65 kb) mRNAs. The size of these transcripts corresponds to the size of the respective cloned cDNAs deposited in the *N. crassa* cDNA library (http://www.genome.ou.edu/fungal.html).

10.8
Construction of Vectors for Efficient Gene Expression in *S. macrospora*

The results obtained with the Northern hybridization prompted us to isolate genomic fragments carrying the complete gene regions. These include promoter sequences, which were used in subsequent experiments for the construction of expression vectors (Table A10.2). The genomic sequences from the *acl1* and *ppg1* genes were already available from previous studies (Nowrousian et al. 1999; Pöggeler 2001). In case of the *cpc2* and *ndk1* genes, it was necessary to isolate cosmid clones carrying the corresponding gene regions. As described above, the available indexed cosmid library served as a source for the rapid isolation of the genomic clones. Subcloning and DNA sequencing finally led to the identification of the putative promoter sequences upstream of the ATG start codon. The sequences of the *cpc2* and *ndk1* promoters are provided in Figure 10.4.

According to cDNA sequences of the *N. crassa cpc2* and *ndk1* genes as deposited in the cDNA library (http://www.genome.ou.edu/fungal.html), a putative transcription start site can be assigned to position −67 for the *S. macrospora* in case of the *cpc2* gene and to position −56 in case of the *S. macrospora ndk1* gene. We investigated the nucleotide sequence upstream of the putative transcription start positions for motifs characteristic for fungal promoters (Meyer and Stahl 2004). Like other strongly transcribed genes in filamentous fungi, the *cpc2* and *ndk1* genes of *S. macrospora* lack a TATA-box in the promoter sequence. Instead, both genes contain pyrimidine-rich tracts immediately in front of the putative transcription start sites.

Typical *cis* elements in the putative promoter sequence of *cpc2* and *ndk1* are marked in Figure 10.4. These include CAAT-boxes and CCAAT-boxes located between 50 and 200 bp from the transcription start point. The CCAAT-box is recognized by a multimeric transcription factor (HAP complex) which is similarly found in all eukaryotes. In addition to elements defining a "core" promoter, several regulatory motifs have been found in the *cpc2* and *ndk1* promoters. Among these are binding sites for transcription factor CREA involved in carbon catabolite repression, and for transcription factor PACC, mediating pH regulation. In addition, we identified binding sites for GATA transcription factors and the zinc-finger transcription factors GCR1, as well as for the ACLR protein involved in ethanol utilization in *A. nidulans*, and for the transcriptions factors ABAA and BRLA. It is known from functional analysis of a variety of fungal promoters that these sequences function as binding sites of activating or repressing transcription factors.

cpc2

```
-1293  GCGGGGGAGTTGGCGGGGAAGTAGacattggcgggagggtacgagggagatggtacgagg
-1233  gcggtaatgatgaaggtggctacgagggaagaggtcatgttgaggaattatctctttcgg
-1173  tggctgaaggatgacgagttcaccatgtggaaatggtaaagatcaaaaatggctatgatc
-1113  atggccgtggtcttgcgaggaaattgcacatggaggCTGGAGagattggcatacgaagag
-1053  aggagtgtggtgtcagccttgaaagatggcaccgagctgtcattgcttggttcaaaaggc
 -993  atgaggagttagggaaggcgtatcattGATAgcattagcatttgtcgtctttctcaagtt
 -933  tacaagatacccatatgaaacatacaaataagacgagcagtggtctatatagcggatgttg
 -873  cttggtttcggtgttatgCAATgacttggtacattttgggtatcgtccttaagggggttca
 -813  cattaaggagaaacacgggactgaatttccaagaagctacAGAGGGAgcaaagagtcta
 -753  tggacgtccatgcgtgtgttgaagtCAATattgattattatgaacgactgagatgcgttt
 -693  tgcaaacagagctcttCAATaactcgaatctgccagtaaatgtcctagtaaaagggCAAT
 -633  ttgaagtcttgtcagttcgagtttaataccacctattggcaCCAATtgcataaacaaacc
 -573  aaacagaactgtcatgtgaaggtctgcatacttttgtgcacaggtacctactgtgtatga
 -513  gtgaagtcatcgtcgaacttggacgaatctgtttgtcatcgccgaggaagtgaagctcc
 -453  aaagtacacaactaaagctgccagcttttgccgggccgcacggcacagtgcacttgga
 -393  gtgacgacactgcagttccgggcaggCCTGGCcctCCAATttggtgaagtgtggcagaca
 -333  gtgtgCAATttgcgaccagccgtttgaggccacaggaggaaagcGCCAAGgcggcaacag
 -273  cccaaagagCCCCACcacgcaccacgcacccagcgaccgccgCCGCAgtgcctggctgtg
 -213  GCAGCTGTGGCAGGCGGGCTGCCCGCGTCGGCCGACCCTATTTTCCGCCGCACAAAAATT
 -153  tccggaattcctacctggtcccgacgCCCCAGtCTCCGCAttgtgccttttgacgatttt
  -93  ctgacttcccgcgcagagaggcaccgtGaaaaagctcaccaccccatttcccccattgat
                                    +1M   A   E   Q   L   I
  -33                    cctcccctgatcaaccagaaaggaactgcaaag   ATG GCT GAG CAA CTC ATC
```

ndk1

```
-1383  ctcgagtggtggtgcacctcgctgaggttagaactcgctaagaaagtcaaaagaaacacc
-1323  aagcaagaacagacaagcaacaacAAAGGGGactggttagaagctcgtctgggtcatca
-1263  agaacaacactcccgggtttgctaaggtatgtcgcagagTCCCTTGaagctgtcgtcgac
-1203  tgcagtataCTCCGCAgctgtgtaaggtaaggtaggtagaggtaatgtaagcggaagtca
-1143  cgagttgggttgacctgttgtcgtCAATgcCAATCAATgcctgccgtgttggcttcaaga
-1083  tggcagttgttagctggcagcatcgcaacaagtccacatctcggtggccgatggcgtcga
-1023  gtggccCCAATtcgcccCGTGGAAGCCAAGcagGCCAAGctgtcccctgccagctagcgc
 -963  accgCTTGGCatcgcgggattcgttagatgcatagtggctctagatGATAaaccGCGGGG
 -903  taagttttGATAtctgcatgacgacctcagcccaccgtctcagttctcagggcctgtagt
 -843  gtacaactgacttgggcgaacatatgttcctctaatcagttgacctggttgttcgtgatt
 -783  aatatggcattcgagcGGAATGacttcctttaaatattgaagagcattgggggttatctc
 -723  ggaaagaaaacctcagctgcacaggccgtacagattCAATggggggggagctctgcgagct
 -663  ttgctggcacccatcgtggcttctcgtcttccggcctcgttccgagacgttgcgcaac
 -603  cacttaaacagagacccataagaCCAATtgacggacgcgtgcctggtcgcttgctgttct
 -543  ctgctccatcaaaacattgccctaatttcttcaaacccgtcgcgtagagttgtcagctgt
 -483  ttgctgtcctgctgctgagtctggaccaaCATTCTtctaaaaatcagagctttgagctaa
 -423  gctttcatcacctcaccaactgaaatagctgcatCCAATaagaccatgtcgaacactgac
 -363  agctcggaacatctgcgacgagaatcagacggacaataccgctgCCGCAaactgaa
 -303  gtgcaacagaagcagcccgaatcgggaggcgCAATggcTGCGGgtgctgattctagtggt
 -243  gcCCAATctcgCCAATcgtcCCAATCGAGGGGaaggggggagCTCCAGaaggagacgagcc
 -183  aggtcacccaggggaacgggcccaCCAATttttttttccggccgcgtgtggactcctccta
 -123  gcagtcctccttttctatctcggccgctctcctcgtcttcctgctcgacttctccttct
  -63  ccatcatCaatcatcatcccatcacaaatctatactccctcacaggaaccacccataatc
                     +1M   S   N   Q   E   Q
   -3          aaa   ATG TCC AAC CAG GAG CAG
```

Fig. 10.4 Promoter sequences from the *S. macrospora cpc2* and *ndk1* genes. The beginning of the coding sequence of CPC2 and NDK1 is shown in capital letters. The putative start site of transcription according to cDNA clones of the *N. crassa* cDNA library at http://www.genome.ou.edu/fungal.html is marked with a boldface letter. The 5′-untranslated region is indicated in italics. Putative binding sites for fungal transcription factors are marked as follows: **CAAT**, CAAT-box; **GCCARG**, PACC; **CCAAT**, HAP2/3/4 complex; **CATTCY**, ABAA; **GATA**, GATA-factors; **CCGCA**, ACLR; **CGTGGAAGCC**, GCR1; **MRAGGGR**, BRLA; **SYGGRG**, CREA.

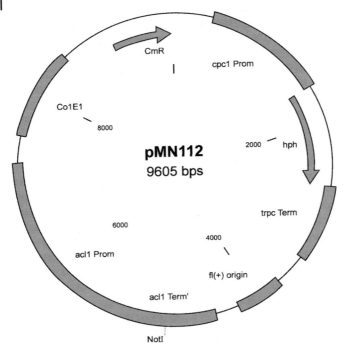

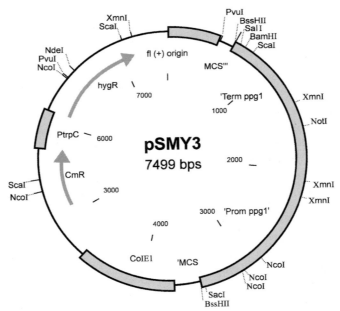

Fig. 10.5   Physical map of *S. macrospora* expression vectors. A) pMN112; B) pMY3.

For further vector constructions, the putative promoter sequences from the above-mentioned *S. macrospora* genes, *acl1, cpc2, ndk1,* and *ppg1* were selected. As examples, the vectors pMN112 and pSMY3 are exhibited in Figure 10.5A and B, which represent prototypes of *S. macrospora* expression vectors. In both plasmids, the promoter sequence and the transcription termination sequence of the *ppg1* and *acl1* gene, respectively, are separated by a *Not*I restriction site. This unique site can be utilized to insert any open reading frame, which will then be under the control of the *S. macrospora* promoter sequence.

10.9
Successful Expression of the Recombinant EGFP Reporter Gene in *S. macrospora*

We have constructed a series of reporter plasmids (Table A10.2) to express an *gfp* (*egfp*) reporter gene in *S. macrospora*. The *gfp* gene encodes the green fluorescent protein (GFP) from the jellyfish *Aequorea victoria*. This protein absorbs UV or blue light, and emits in the green region of the spectrum. Several applications are feasible and allow for example the cellular localization of recombinant gene products or the quantification of recombinant proteins. In filamentous fungi, the expression of the genuine *gfp* gene was initially hampered by an inefficient codon usage.

The generation of EGFP mutations, which contain chromophore mutations that increase fluorescence intensity on one hand and an adaptation of the codon usage to optimal gene expression in fungi on the other hand, enabled several applications. In a first series, several vectors containing the *egfp* gene have been constructed. As shown in Figure 10.6, the *egfp* gene has been derived from plasmid pEGFP/gfp/tel (Inglis et al. 1999), which was then used for the construction of several intermediates. Eventually, the *egfp* expression plasmid pSM2 was constructed. It harbors the *hph* gene for the selection of fungal transformants. The *egfp* reporter gene is flanked by the transcription termination sequence of the *A. nidulans trpC* gene. In front of the ATG codon of the *egfp* gene, plasmid pSM2 contains five unique recognition sites for commonly used restriction enzymes (*Pst*I, *Eco*RI, *Eco*RV, *Hind*II, and *Cla*I). These sites can be employed to integrate promoter sequences (Pöggeler et al. 2003).

Plasmid pSM2 was used for the insertion of promoter sequences from *S. macrospora*. For illustration, plasmid pSE43–2 is shown in Figure 10.7, which is equipped with the *efgp* gene under the control of the *cpc2* promoter. In addition, this plasmid carries the ampicillin and hygromycin B resistance gene for selection in both, bacteria and fungi, respectively.

Transformation of *S. macrospora* with various *egfp* reporter plasmids showed that the heterologous *egfp* gene under the control of the *S. macrospora* promoters integrates as multiple copies into the fungal transformants. In Figures 10.8A–D, the hyphae from a transgenic strain are shown with a green fluorescence uniformly distributed throughout the cytoplasm of the hyphal compartments. The fluorescence was less pronounced in the nuclei or other organelles than in the vacuoles, where it was present in high abundance. When this strain was selected for a cross with a

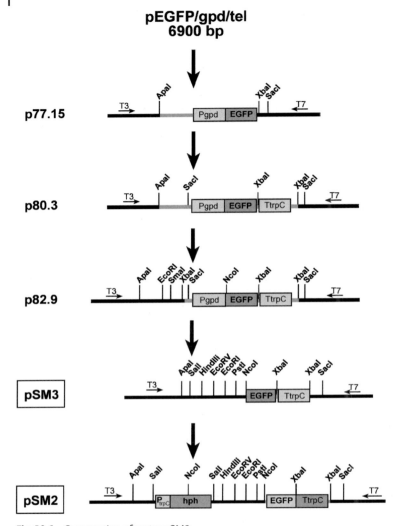

Fig. 10.6 Construction of vector pSM2.

non-transgenic spore color mutant, the ascospores of the derived strain showed a mendelian inheritance of green fluorescence. Comparing Figure 10.8E with 10.8F, the phenotype of eight ascospores can easily be distinguished in the linearly ordered asci. Four out of eight spores show the phenotype of one or the other parental strain. This is remarkable, as the expression of the transgenes is clearly not repressed by the RIP process. RIP stands for "repeat-induced point mutation", and occurs before karyogamy. It consists of a genome-wide scan for duplicated sequences, such as transposable elements. Initially, RIP was detected in *N. crassa* (Selker 2000), but later was also described for *P. anserina* (Hamann et al. 2000). One particular possible role

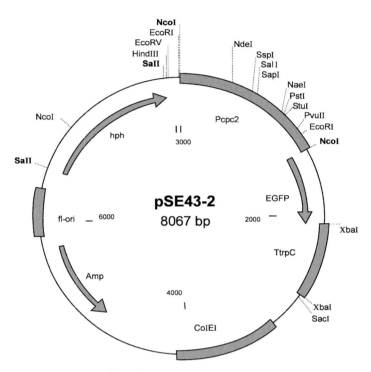

Fig. 10.7 Reporter plasmid pSE43-2.

of RIP is the inactivation of duplicated copies in the genome, which are inactivated by extensive introduction of C → T mutations.

10.10
Conclusions

The filamentous ascomycete *S. macrospora* provides a useful platform for the expression of heterologous genes. On the one hand, such usefulness is dependent upon the inherent characteristics of the wild-type strain and its mutant derivatives: *S. macrospora* has a simple life cycle, which allows genetic recombination using conventional genetic techniques. This homothallic organism does not produce conidiospores. Moreover, the availability of sterile strains makes this filamentous fungus a safe host for biotechnological production processes. On the other hand, an increasing knowledge relating to this fungal organism has led to the identification of strongly expressed genes that can be exploited as sources for strong promoter elements. The inclusion of such elements has resulted in useful vectors with high application potential for heterologous gene expression in *S. macrospora*.

Fig. 10.8 *S. macrospora*, expressing the gene encoding the green fluorescence protein from *Aequorea victoria*. Hyphae (A–D) and asci (E, F) are shown by light (A, C, E) and fluorescence (B, D, F) microscopy. Arrows indicate asci with eight linear ascospores.

Acknowledgments

The authors acknowledge the receipt of funding by the TPW project (TPW-9910v08), as well as by the Deutsche Forschungsgemeinschaft. They also thank Ingeborg Godehardt, Silke Nimtz, and Susanne Schlewinski for skilful technical assistance, and Dr. Minou Nowrousian and Dipl. Biol. Severine Mayrhofer for providing unpublished data.

Appendix

Tab. A10.1 Selection of S. *macrospora* host strains.

Strain	Genotype/ Parental strains	Phenotype	Source
Parental strains			
CBS 957.73	wild-type		CBS[1]
CBS396.69	wild-type		CBS[1]
DSM997	wild-type		DSM[2]
IMI 146.455	wild-type		IMI[3]
K(18)	wild-type		AMB[4]
K(3346)	wild-type		AMB[4]
K(1957)	wild-type		AMB[4]
Mutant strains			
T7	ura3⁻	Ura⁻	AMB[4]
pro4	leu1⁻	Leu⁻	AMB[4]
inf25	inf25-1	sterile	AMB[4]
inf40	inf40-1	sterile	AMB[4]
pro1	deltapro1	sterile	AMB[4]
pro11	pro11-1	sterile	AMB[4]
per5	acl1-1	sterile	AMB[4]

[1] Centraalbureau voor Schimmelcultures, Baarn, NL.
[2] Deutsche Stammsammlung für Mikroorganismen, Göttingen, GER.
[3] International Mycological Institute, London, GB.
[4] Lehrstuhl für Allgemeine und Molekulare Botanik, Bochum, GER.

Tab. A10.2 Selection of S. *macrospora* plasmids and expression/integration vectors.

Plasmid	Expression cassette	Replication sequence	Selection marker	Integrated copy number	Reference
pANsCos1	–	–	hph	1 to many	Osiewacz (1994)
pSM1	gpd promoter (A. nidulans) egfp gene trpC terminator (A. nidulans)	–	hph	1 to many	Pöggeler et al. (2003)
pSM2	promoter less egfp gene trpC terminator (A. nidulans)	–	hph	1 to many	Pöggeler et al. (2003)
pCB1003	–	–	hph	1 to many	Carroll et al. (1994)
pCB1004	–	–	hph	1 to many	Carroll et al. (1994)
pBChygro	–	–	hph	1 to many	Silar (1995)
pMN112	acl1 promoter, acl1 terminator	–	hph	1 to many	This reference
pSMY3	ppg1 promoter ppg1 terminator	–	hph	1 to many	This reference

Tab. A10.2 (continued)

Plasmid	Expression cassette	Replication sequence	Selection marker	Integrated copy number	Reference
pSE43-2	*cpc2* promoter *trpC* terminator (*A. nidulans*)	–	*hph*	1 to many	This reference
pIG1826-2	–	–	*leu1*	1 to many	U. Kück (unpublished)
p20.26	–	–	*ura3*	1 to many	U. Kück and S. Pöggeler (unpublished)

References

CARROLL AM, SWEIGARD JA, VALENT B (1994) Improved vectors for selecting resistance to hygromycin. Fungal Genet Newslett 41: 22

DAVIS RH, PERKINS DD (2002) *Neurospora*: a model of model microbes. Nat Rev Genet 3: 397–403

GALAGAN JE, CALVO SE, BORKOVICH KA, SELKER EU, READ ND, JAFFE D, FITZHUGH W, MA LJ, SMIRNOV S, PURCELL S, REHMAN B, ELKINS T, ENGELS R, WANG S, NIELSEN CB, BUTLER J, ENDRIZZI M, QUI D, IANAKIEV P, BELL-PEDERSEN D, NELSON MA, WERNER-WASHBURNE M, SELITRENNIKOFF CP, KINSEY JA, BRAUN EL, ZELTER A, SCHULTE U, KOTHE GO, JEDD G, MEWES W, STABEN C, MARCOTTE E, GREENBERG D, ROY A, FOLEY K, NAYLOR J, STANGE-THOMANN N, BARRETT R, GNERRE S, KAMAL M, KAMVYSSELIS M, MAUCELI E, BIELKE C, RUDD S, FRISHMAN D, KRYSTOFOVA S, RASMUSSEN C, METZENBERG RL, PERKINS DD, KROKEN S, COGONI C, MACINO G, CATCHESIDE D, LI W, PRATT RJ, OSMANI SA, DESOUZA CP, GLASS L, ORBACH MJ, BERGLUND JA, VOELKER R, YARDEN O, PLAMANN M, SEILER S, DUNLAP J, RADFORD A, ARAMAYO R, NATVIG DO, ALEX LA, MANNHAUPT G, EBBOLE DJ, FREITAG M, PAULSEN I, SACHS MS, LANDER ES, NUSBAUM C, BIRREN B (2003) The genome sequence of the filamentous fungus *Neurospora crassa*. Nature 422: 859–868

HAMANN A, FELLER F, OSIEWACZ HD (2000) The degenerate DNA transposon Pat and repeat-induced point mutation (RIP) in *Podospora anserina*. Mol Gen Genet 263: 1061–1069

INGLIS PW, QUEIROZ PR, VALADARES-INGLIS MC (1999) Transformation with green fluorescent protein of *Trichoderma harzianum* 1051, a strain with biocontrol activity against *Crinipellis perniciosa*, the agent of witches'-broom disease of cocoa. J Gen Appl Microbiol 45: 63–67

JACOBSEN S, WITTIG M, PÖGGELER S (2002) Interaction between mating-type proteins from the homothallic fungus *Sordaria macrospora*. Curr Genet 41: 150–158

KÜCK U (1999) Filamentous fungi. In: McGraw-Hill Yearbook of Science & Technology 2000, McGraw-Hill Book Co, New York, pp 163–166

LUNDQVIST N (1972) Nordic *Sordariaceae* s. lat. Symbolae Botanicae Upsalienses 20: 1–314

MASLOFF S, PÖGGELER S, KÜCK U (1999) The *pro1* gene from *Sordaria macrospora* encodes a C_6 zinc finger transcription factor required for fruiting body development. Genetics 152: 191–199

MASLOFF S, JACOBSEN S, PÖGGELER S, KÜCK U (2002) Functional analysis of the C_6 zinc finger gene pro1 involved in fungal sexual development. Fungal Genet Biol 36: 107–116

MEYER V, STAHL U (2004) Gene regulation in mycelial fungi. In: The Mycota II (Kück U, Ed). Springer Press, Berlin, Heidelberg, New York, pp 147–163

NOWROUSIAN M, KÜCK U (1998) Isolation and cloning of the *Sordaria macrospora ura3* gene and its heterologous expression in *Aspergillus niger*. Fungal Genet Newslett 45: 34–37

NOWROUSIAN M, MASLOFF S, PÖGGELER S, KÜCK U (1999) Cell differentiation during sexual development of the fungus *Sordaria macrospora* requires ATP citrate lyase activity. Mol Cell Biol 19: 450–460

Nowrousian M, Würtz C, Pöggeler S, Kück U (2004) Comparative sequence analysis of *Sordaria macrospora* and *Neurospora crassa* as a means to improve genome annotation. Fungal Genet Biol 41: 285–292

Osiewacz HD (1994) A versatile shuttle cosmid vector for the efficient construction of genomic libraries and for the cloning of fungal genes. Curr Genet 26: 87–90

Pöggeler S (1997) Sequence characteristics within nuclear genes from *Sordaria macrospora*. Fungal Genet Newslett 44: 41–44

Pöggeler S (1999) Phylogenetic relationships between mating-type sequences from homothallic and heterothallic ascomycetes. Curr Genet 36: 222–231

Pöggeler S (2000) Two pheromone precursor genes are transcriptionally expressed in the homothallic ascomycete *Sordaria macrospora*. Curr Genet 37: 403–411

Pöggeler S (2001) Mating type genes for classical strain improvements of ascomycetes. Appl Microbiol Biotechnol 56: 589–601

Pöggeler S, Kück U (2000) Comparative analysis of the mating-type loci from *Neurospora crassa* and *Sordaria macrospora*: identification of novel transcribed ORFs. Molec Gen Genet 263: 292–301

Pöggeler S, Kück U (2001) Identification of transcriptionally expressed pheromone receptor genes in filamentous ascomycetes. Gene 280: 9–17

Pöggeler S, Kück U (2004) A WD40-repeat protein regulates fungal cell differentiation and can be replaced functionally by the mammalian homologue striatin. Eukaryotic Cell 3: 232–240

Pöggeler S, Risch S, Kück U, Osiewacz HD (1997 a) Mating type genes from the homothallic fungus *Sordaria macrospora* are functionally expressed in a heterothallic ascomycete. Genetics 147: 567–580

Pöggeler S, Nowrousian M, Jacobsen S, Kück U (1997 b) An efficient procedure to isolate fungal genes from an indexed cosmid library. J Microbiol Methods 29: 49–61

Pöggeler S, Masloff S, Jacobsen S, Kück U (2000) Karyotype polymorphism correlates with intraspecific infertility in the homothallic ascomycete *Sordaria macrospora*. J Evol Biol 13/2: 281–289

Pöggeler S, Masloff S, Hoff B, Mayrhofer S, Kück U (2003) Versatile EGFP reporter plasmids for cellular localization of recombinant gene products in filamentous fungi. Curr Genet 43: 54–61

Radzio R, Kück U (1997) Synthesis of biotechnologically relevant heterologous proteins in filamentous fungi. Process Biochemistry 32: 529–539

Selker EU (2000) Premeitotic instability of repeated sequences in *Neurospora crassa*. Annu Rev Genet 24: 579–613

Silar P (1995) Two new easy to use vectors for transformations. Fungal Genet Newslett 42: 73

Van Heemst D, James F, Pöggeler S, Bertaux-Lecellier V, Zickler D (1999) Spo76p is a conserved chromosome morphogenesis protein that links the mitotic and meiotic programs. Cell 98: 261–271

Walz M, Kück U (1995) Transformation of *Sordaria macrospora* to hygromycin B resistance: characterization of transgenic strains by electrophoretic karyotyping and tetrad analysis. Curr Genet 29: 88–95

11
Mammalian Cells
VOLKER SANDIG, THOMAS ROSE, KARSTEN WINKLER, and RENE BRECHT

11.1
Why Use Mammalian Cells for Heterologous Gene Expression?

Mammalian cells constitute a demanding system for the production of heterologous proteins. The need for specialized media and sufficient oxygen supply, low cell densities and slow growth kinetics, and a high sensitivity of the cells to mechanical stress are obstacles which must be overcome in routine fermentation. Furthermore, mammalian cells are potential targets for adventitious viral agents, and processes based on such cells must be rigorously monitored. Despite these difficulties, mammalian cells are the preferred production system for the synthesis of glycoproteins intended for administration to humans. In 2004, mammalian cell-based therapeutic proteins are expected to reach a market share of 59%, followed by 27% for *E. coli*-based products (Source: Datamonitor).

The ease of production in bacterial systems must be counterbalanced against the need to dissolve and renature misfolded, aggregated, insoluble proteins. In contrast, the chaperone system in mammalian cells ensures that proteins are secreted in correctly folded form. Whereas eukaryotic microbial systems such as yeasts are also capable of modifying recombinant translation products by proteolytic processing of precursor proteins, formation of disulfide bridges and phosphorylation, only mammalian cells are able to glycosylate proteins in the patterns characteristic of higher eukaryotes, yielding products that are identical to their natural human counterparts. Although other modifications such as the attachment of lipids may have an impact on heterologous protein production in the near future, we focus in this chapter on authentic N- and O-linked glycosylation as the most important capability of mammalian cells that distinguishes them from other expression systems.

The significance of correct glycosylation is illustrated by the following examples: All antibodies are glycosylated at conserved positions in their constant regions, and the presence of these carbohydrate moieties is essential for pharmaceutical efficacy. The structure of the carbohydrate affects complement activation and binding to the Fc receptor (Jefferis and Lund 1997). In addition, glycosylation at or near the antigen binding site may have a dramatic influence on affinity. Human plasma proteins may be rendered antigenic upon exposure of epitopes that are normally masked by a car-

Production of Recombinant Proteins. Novel Microbial and Eucaryotic Expression Systems. Edited by Gerd Gellissen
Copyright © 2005 WILEY-VCH Verlag GmbH & Co. KGaA, Weinheim
ISBN: 3-527-31036-3

bohydrate chain. O-linked oligosaccharide chains (attached to serine and threonine hydroxyl groups) can present multivalent antigenic or functional determinants for antibody recognition and cell adhesion (Hounsell et al. 1996). Glycosylation patterns are not identical in all mammalian cells, but depend on the tissue type and species of origin, and on cell culture conditions (Wright and Morrison 1997). Again, these differences can be of functional significance. For example the alpha 1 → 3 gal epitope is absent from normal human cells but is present in human tumor cells, as well as in proteins derived from mice and yeast. This epitope is recognized by pre-existing antibodies, which mediate the rapid clearance of proteins that display it (Gollogly and Castronovo 1996).

11.2
Mammalian Cell Lines for Protein Production

One of the reasons why one might wish to produce proteins via heterologous gene expression is that a valuable protein of interest is normally secreted from its natural cellular source in very small amounts. Therefore, a recombinant approach is taken. A cDNA of the respective coding sequence is inserted between a strong promoter and a polyadenylation sequence contained on an expression vector, and transferred into a suitable cell line via transfection. Although almost any human or animal cell line would be able to generate the protein – provided that its growth properties meet fermentation requirements – therapeutic proteins are preferentially produced in CHO (Chinese hamster ovary), BHK (baby hamster kidney) and NS0 cells (mouse myeloma) and Sp2/0 (myeloma) cells (Table A11.1). More recently, special human designer cell lines made *in vitro* from primary cells by immortalization with defined oncogens, have become attractive candidates for glycoprotein production (Jones et al. 2003). HEK 293 derived from human embryonic kidney cells by transformation with E1 genes of Adenovirus 5 are used extensively for transient protein production due to their highly efficient DNA uptake system. The PER.C6 cell line – a better characterized line derived from human retinoblasts using the same transforming gene – is also intended for pharmaceutical application (Fallaux al. 1998). Extremely high cell densities can be achieved in batch processes (well over 10^7 cells mL^{-1}), allowing accumulation of a recombinant product to high titers. A second potential advantage of these cells is that they glycosylate target proteins in a precisely determined manner, which is characteristic of human glycoproteins. Other cell lines are currently under development that promise to be particularly suited for efficient protein secretion.

The vector and the cell line form an integrated system designed to produce maximal yields of a given protein. However, even when applying an optimal combination of vector and cell line, the productivity can vary greatly between individual clones. Furthermore, productivity depends crucially on the precise design of the fermentation process. Therefore, special emphasis must be placed on the careful screening of potential production lines and on optimization of the medium and growth conditions for the subsequent fermentation process.

11.3
Mammalian Expression Systems

11.3.1
Design of the Basic Expression Unit

To allow for transcription, the cDNA encoding the protein of interest must be linked to a promoter that controls its expression. Such an element is taken from a highly expressed viral or cellular gene. The "promoter" is a stretch of DNA immediately upstream of a transcription initiation site. The promoter binds factors of the transcription machinery and factors that recruit co-activators for the removal of core histones, thus giving the RNA polymerase access to the template and allowing it to bind. This process is supported by more distantly located enhancers which act in a distance- and orientation-independent manner (Berk 1999). From the wide range of promoter elements available, only a very few are routinely used: The SV40 early promoter, the Rous sarcoma virus (RSV) long terminal repeat, and the human cytomegalovirus (CMV) immediate early promoters are the most prominent virus-derived examples. Mammalian cell-derived promoters are often intrinsically less efficient, but on the other hand they may also be less susceptible to inactivation by DNA methylation, thus permitting stable long-term expression (Prosch al. 1996). In the majority of expression vectors currently in use, the immediate early CMV promoter/enhancer is employed, and several versions that differ in strength are available. Some versions included in commercial vectors lack upstream inhibitory elements such as the YY1 binding site (Liu et al. 1994), but also the strong activating downstream elements and even the natural transcription start site (pcDNA3; Invitrogen). Surprisingly, an extremely short version extending from −299 to 72 has exceptionally high activity (unpublished observation by the authors). Alternative promoters such as the mouse CMV promoter and the EF1 alpha promoter have been shown to be active in CHO cells when used alone (Rotondaro et al. 1996), or even more active when combined (Kim et al. 2002). The human ubiquitin C promoter, the human initiation factor 4A1 promoter, and the chicken beta actin promoter as a non-mammalian example provide additional alternatives (Quinn et al. 1999).

Promoter strength reflects the level of mRNA synthesis, but this not the only factor that determines the amount of protein obtained. Further determinants include the steady-state level of the mRNA, its transport to the cytoplasma, and its translatability. In cases where production is sub-optimal, the effects of re-introduction of introns into the cDNA can be assessed. In some cases this resulted in truncation of transcripts, since splicing commences from the 3'-end of a transcripts and a cryptic splice sequence within the cDNA could be used for processing instead of the expected donor. This problem can be eliminated by placing such introns 5' to the cDNA.

The definition of an appropriate polyadenylation signal is of similar importance for efficient gene expression. Sources for suitable elements are the SV40 late transcript, and the transcripts for bovine growth hormone and HSV tk genes (Denome and Cole 1988). In contrast, a large fraction of transcripts remains without a polyadenylated tail when the SV40 early poly(A) signal (frequently included in commercially available expression vectors) is used.

For efficient translation, the start codon of the gene to be expressed should be embedded within a nucleotide sequence that conforms to the Kozak rules (Kozak 1984). Although in a large number of genes the main start site is preceded by AUG codons (Peri and Pandey 2001), upstream AUGs – as well as hairpins close to the AUG – should be avoided by careful vector design. Their presence can result in a 100-fold reduction in protein yield.

11.3.2
Transient Expression and Episomal Vectors: Alternatives to Stable Integration

Typically, it takes between several months and a full year to establish a recombinant mammalian production cell line, and such developments require a large number of tedious and labor-intensive steps. However, it is desirable to be able to produce a limited amount of a protein of choice in a much shorter time frame when new product candidates are to be assessed, or a screening system has to be established. For required amounts of 10 mg to 1 g, mammalian cell systems offer the option of a shortcut, namely transient expression. Using especially efficient transfection techniques, almost 100% of the recipient cells of some cell lines (e.g., HEK293 or CHO) can be induced to take up vector DNA. Some techniques, such as a modified version of the traditional calcium phosphate co-precipitation technique (Lindell et al. 2004), or more recent approaches using polyethylenimine (PEI; Durocher et al. 2002) or dendrimers (Haensler and Szoka 1993), are suitable for suspension cells grown in bioreactors. The transfected DNA is transported to the nucleus where it is maintained for a few days. Unless it has any replication and nuclear retention signals, the foreign DNA is rapidly lost and expression consequently declines. The useful lifetime of such systems can be extended by using virus-based vectors (SV40, BPV, EBV). A large number of DNA copies per cell ensures maximal transcription, and protein processing then becomes the rate-limiting step in production. In the case of SV40, extrachromosomal replication can result in more than 100 000 DNA copies per cell. Due to their extremely high productivity, these cells eventually die because an appropriate energy balance cannot be maintained. In the case of BPV, attainable copy numbers are lower and expression persists for a longer time. All of the systems mentioned require the presence of a viral protein (SV40 large T antigen, BPV E1/E2, EBNA1, respectively) and a viral origin of replication. In the case of SV40 and BPV, replication origins are activated by the helicase activity of the respective viral protein, and replication is uncoupled from that of the chromosomal DNA. In contrast, EBNA1 appears to lack the enzymatic activities characteristic of ori-binding proteins. EBV-based plasmids replicate once per cell cycle, and expression is maintained for even longer. These proteins of all three viruses have one feature in common: they attach vector DNA to the nuclear matrix. This attachment is required for retention in the nucleus; the centromere functions that are lacking in a vector are substituted by association with host nuclear matrix (Calos 1998). For chromosomal DNA, attachment is mediated by *MAR* elements (AT-rich sequences of bent DNA). It is not surprising therefore that *MAR* elements can also elicit binding of viral sequences. Accordingly, it has been shown that vectors equipped with a specific assembly of replication origin, *MAR* element and properly terminated transcription units have all functional elements required for episomal replication over 100 passages (Piechaczek et al. 1999). This is not observed for EBV-based vectors, such as REP 4 and REP 8 (Invitrogen), which require permanent selection pressure for long-term stability. The recently described *MAR* element located adjacent to *oriP* in EBV may permit us to overcome these limitations (White et al. 2001). Such episomal vectors may in future represent the preferred option, not only for short-term transient expression but also for the generation of stable cell lines, provided that some open issues can be successfully addressed. Such vectors offer several obvious advantages. Almost all recipient cells will stably maintain the vector. There is no requirement for (inefficient) chromosomal integration relying on cellular repair mechanisms. Moreover, episomal DNA is assumed to be not affected by cellular defense mechanisms that would silence expression, as it is often observed with integrated DNA.

For the time being, traditional episomal vectors grown in HEK293 cells carrying the respective genes (293EBNA, 293T) will retain their importance as short- and medium-term mammalian expression systems.

Quite different DNA-RNA replication systems for intermediate and long-term expression have been defined which do not require DNA integration. These systems are based on RNA viruses such as alphaviruses. DNA vectors are introduced into the cell and transcribed as described for the previous vectors. However, in addition to the open reading frame (ORF) for the protein of interest, such a vector harbors the gene for an RNA-dependent RNA polymerase. This polymerase first allows for replication of the transcript via a minus strand and a sub-genomic plus strand RNA. The plus strand serves as a template for translation of the gene of interest. This system provides a stable level of RNA that can also be regulated via temperature shifts and is compatible with the physiology of the host cell (Boorsma al. 2000). This contrasts with the previous systems, in which the newly introduced virus can completely take over the host cell's metabolism and eventually kill it.

11.3.3
"Stable" Integration into the Host Genome

The most typical approach for protein production in mammalian cells is the generation of so-called stable cell lines. This term is used to describe an individual cell clone which carries the expression unit embedded at a single or at multiple sites in host chromosome(s). The substrate for integration is a linear DNA molecule with free ends, irrespective of whether the vector is provided as a supercoiled or linearized molecule. These free ends are recognized by the double-strand break repair system and joined to cellular DNA. This process represents a survival strategy that is vital to the maintenance of permanent cell lines because it prevents the loss of essential genes telomeric to a breakpoint. Since integration usually occurs in non-essential DNA – which represents the majority of the mammalian genome – this is unlikely to be detrimental to the cell. Depending on the plasmid load introduced into a cell,

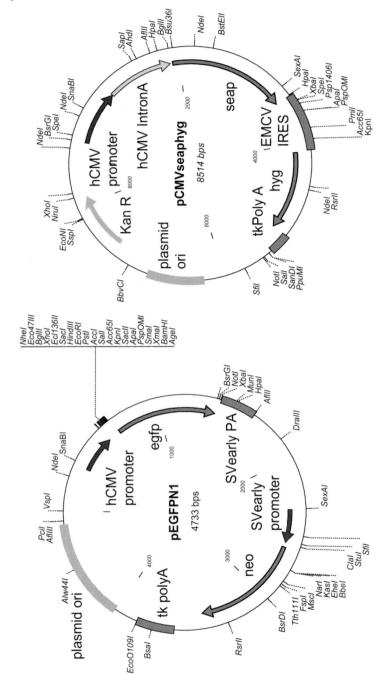

Fig. 11.1 Representative vectors for stable expression. *Left*: plasmid pEGFPN1 (Clonetech): the plasmid contains separate expression units for GFP (with hCMV promoter and SV40 early Poly(A) signal) and neomycin phosphotransferase as selection marker (with Sv40 early promoter and HSVtk Poly(A) signal). The neomycin gene serves as prokaryotic selection marker.

Right: pCMVseaphyg (ProBioGen) The plasmid contains a bi-cistronic expression unit controlled by the human CMV promoter. CMV intron A is introduced to facilitate RNA transport and to include downstream promoter elements. The EMCV IRES is positioned between *seap* and *hyg* genes to allow expression of *hygR*. Kan R serves as a prokaryotic marker only.

integration occurs as a single copy or in the form of concatamers. Calcium phosphate transfection is known to favor the integration of multimeric molecules. The resulting increase in gene dosage may lead to higher expression levels. Hence, the method has traditionally been preferred for the generation of high-level producer cell lines.

Alternatively, multiple expression cassettes can be placed on a single vector, and plasmids with up to 12 copies of a complete gene and a single selection marker gene can easily be constructed. The insertion of multiple cassettes at a single position may even be enforced, if the expression cassettes are integrated into a single plasmid using restriction endonucleases (such as *Sfi*I) that recognize interrupted palindromes, and cosmid-based vectors are employed (Monaco et al. 1994). In contrast to the products of random concatamerization, the expression units on such plasmids are present in head-to-tail array after transfection, thereby precluding the formation of antisense RNA. However, the increase in expression is never proportional to the copy number, and expression levels can be further reduced by epigenetic modifications. Thus, after extended passaging the expression level may actually be inferior to that attainable with individual inserts. Epigenetic inactivation is induced as a defense mechanism against retroviruses, and is based on recognition of identical DNA repeats of several kilobases in length (Monaco et al. 1994; McBurney et al. 2002). Therefore, transfection techniques that result in the preferential integration of a single copy, such as electroporation and lipofection, are now preferred.

11.3.4
Selection Strategies for Mammalian Cells

Although illegitimate recombination (end joining without homology among the participating sequences) is strongly enhanced in permanent cell lines, it still occurs in only a minor fraction of transfected cells. This is not surprising because pre-existing double-strand breaks in the genome are required for this process. The identification of the few cells with an insertion requires the co-integration of marker genes and drug selection of the transfected cell pool.

Initially, the gene of interest and the selection marker were introduced on separate plasmids provided in a molar ratio of at least 10:1 in favor of the plasmid bearing the product gene. In more recent approaches, both genes are delivered on the same plasmid (Figure 11.1). This strategy still does not ensure co-expression of both genes because the gene of interest or its expression signals may be inactivated by random double-strand breaks introduced before integration. This is rendered much less likely when both genes are provided as a single expression unit. In this case, both are transcribed from a single promoter and constitute a single RNA species with only one polyadenylation signal. Whereas translation of the upstream gene in this bi-cistronic transcript follows conventional rules (e.g., dependence on a m7G-cap structure at the 5'-end of the RNA and ribosome scanning to the first ATG; Kozak 2002), the second ORF is inactive unless an IRES element (internal ribosome entry site) is incorporated (Figure 11.1). IRES elements have been isolated mainly from picornaviruses, wherein the positive-strand RNA genome lacks a cap structure, and translation is dependent upon a non-coding 450-bp fragment of complex secondary structure.

However, class II IRES elements, such as those derived from the encephalomycarditis virus (EMCV) and foot and mouth disease virus (FMDV), are generally more efficient and easier to use compared to those of class I (e.g., poliovirus), because they do not require exact positioning of the ATG relative to the element. More recently, a number of elements of similar function have been identified in cellular genes (Nanbru et al. 1997; Miller et al. 1998) that are normally expressed under conditions when cap-dependent translation is inhibited. Even when the more efficient elements are used, the expression level of the downstream gene is approximately one order of magnitude lower than that of the upstream gene. Therefore, the gene encoding the selection marker is placed to the downstream position. The tight linkage of the two genes provides a tool for selecting high-level producer clones by increasing the selection pressure. However, the selection conditions must be re-established for each individual construct, because the efficiency of translation of the marker sequence strongly depends on the structure of the upstream gene. Inactivation of the gene of interest is highly unlikely when an IRES is used. However, under more stringent conditions, we have frequently observed cases in which the marker alone is expressed. Indeed, such events can occur as often with such coupled expression plasmids as when selection and product units are located on the same plasmid.

Even more important than the linkage between the two genes is the definition of an appropriate selection marker. The *npt* gene conferring resistance against G418 is an established selection marker for a wide range of eukaryotic cells. However, even very few molecules of its product neomycin phosphotransferase render the cell resistant to G418. Therefore selection of high-level producers is impaired since more stringent conditions cannot be imposed.

Indeed, selection approaches based on factors that confer resistance to toxic antibiotics generally suffer from a number of disadvantages:

1. The resistance gene usually acts via an indirect mechanism: it inactivates an antibiotic that typically blocks protein synthesis in treated cells.
2. Inactivated antibiotics must be replenished to maintain the selective pressure, thereby leading to a variable selective concentration.
3. In some cases, temporary selection pressure results in permanent changes as a harmful effect of partial inactivation of antibiotics.
4. Often, a subfraction of untransfected cells can survive under selective conditions in acquiring resistance via an alternative mechanism such as activation of the multidrug resistance gene. The latter phenomenon is very pronounced when puromycin is used for selection of recombinant HEK 293 cells.

These disadvantages have therefore led to the use of auxotrophic markers as a more promising alternative.

11.3.5
Auxotrophic Selection Markers and Gene Amplification

Two auxotrophic markers have emerged as preferred options for the selection of high-level producers, namely the genes for dihydrofolate reductase (*dhfr*) and gluta-mine synthetase (*gs*). The use of the *dhfr* gene is restricted to CHO cells, as cell var-iants have been engineered that lack the corresponding enzyme. These cells depend on low levels of exogenous enzyme for nucleotide synthesis if cultured in the ab-sence of thymidine and hypoxanthine. Two such cell lines were constructed by Chasin and colleagues during the 1980s. Whereas in CHO DXB11 cells (Urlaub and Chasin 1980) one allele is still present but inactivated by point mutation, thus allow-ing reversion to the dhfr+ phenotype by homologous recombination, in the DG44 mutant (Urlaub et al. 1986) *dhfr* sequences have been completely deleted. Both cell lines are frequently used to generate high-level producer cell lines.

In contrast to *dhfr*, the gene encoding glutamine synthetase (Bebbington et al. 1992) acts as a dominant selection marker, because the enzyme content in most mammalian cell lines is too low to ensure an adequate supply of glutamine. Accord-ingly, the system was initially developed for myeloma cells, such as the NS0, in which endogenous enzyme levels are negligible.

In addition to efficient selection, both systems offer an even more important ad-vantage: selective pressure can be sequentially increased using the specific enzyme inhibitors methotrexate (MTX) for DHFR and methionine sulphoximine (MSX) for GS. The single-copy *dhfr* or *gs* gene cannot provide sufficient amounts of the enzyme to counteract the effects of high inhibitor concentrations. This situation selects for mutants that increase gene dosage by intrachromosomal amplification of the selec-tion marker and adjacent sequences. This process is well known from tumor cells, which can acquire resistance to chemotherapeutics by a similar mechanism. Accord-ing to the breakage-fusion-bridge model (Coquelle et al. 1997), this requires the ex-posure of fragile sites close to the marker gene – double-strand breaks which must be repaired by sister chromatids, and secondary breakage of covalently linked chro-matids. In agreement with this model, MTX selection strongly reduces the size of cellular nucleotide pool, resulting in DNA breakage. MTX selection thus results in the accumulation of several hundred copies of the gene, in contrast to MSX selection where lower copy numbers are observed. This extended amplification promotes higher expression levels, but is also a potential cause of instability. Depending on the site of the first breakage, a telomeric fragment may be amplified that is unequally distributed to the daughter cells. As a consequence, the amplified region is rapidly lost in the absence of selection.

Therefore, amplification is considered highly unpredictable, and cell lines gener-ated without amplification are currently to be preferred.

11.3.6
The Integration Locus: a Major Determinant of Expression Level

Despite the presence of identical plasmids in comparable copy numbers, expression levels can vary by two to three orders of magnitude between individual clones. This phenomenon can be attributed to variations in the accessibility of the integrated gene to the transcriptional machinery. Only if the gene sequence is in the open chromatin conformation does it becomes available for transcription. This conformation, in turn, is determined by types of sequence elements, such as matrix attachment sites or chromatin opening elements, and mediated via histone acetylation and the lack of DNA methylation. Loci that display high expression efficiency are extremely rare in the genome (Figure 11.2). Hence, intensive screening (coupled with stringent selection and tight linkage between the marker and the gene of interest) is required to identify clones that show high productivity.

Furthermore, the selection must be repeated for every individual cell line development. A high frequency of illegitimate recombination in permanent mammalian cell lines further impedes the identification of lines resulting from these rare homologous recombination events, and makes targeting as labor-intensive as random screening. The isolation of site-specific recombinases from heterologous sources, and their introduction into mammalian cells, can potentially overcome this problem. Recombinases from bacteriophage P1 (cre) and *Saccharomyces cerevisiae* (flp) elicit intermolecular strand exchange between the short target sequences (*lox* and *FRT*, respectively) in their natural host. For site-specific integration an intermolecular reaction is required between a circular plasmid and a chromosomal target site. However, the thermodynamically preferred excision of a fragment flanked by recombinase sites is more often applied for the generation of recombinant mammalian cells and transgenic animals (Figure 11.3) (Ray et al. 2000; Rodriguez et al. 2000).

Generation of CHO-Receptor Cell Lines

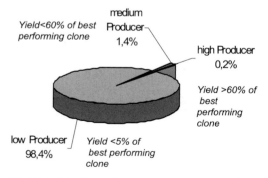

Fig. 11.2 Distribution between low producers, medium producers and high producers within a pool of resistant cell clones after transfection and selection.

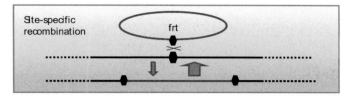

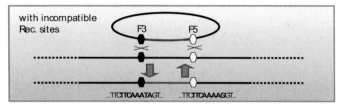

Fig. 11.3 Site-specific recombination using a single site versus two heterospecific sites. Different core sequences within the frt recognition sequences (bold) of frt F5 and F3 (Schlake and Bode 1994) prevent recombination within the chromosome.

In order to achieve efficient integration, three options are available:

1. The use of heterospecific recombinase sites, in which the core sequences are modified to prevent a reaction *between* these sites (see Figure 11.3).
2. Restriction of recombinase activity to a short time interval immediately after introduction of the template DNA, when the template concentration is still high (regulated flp expression, protein transfer, RNA transfection).
3. The introduction of an incomplete promoter-less selection marker into the target site, and activation of this marker via the subsequent assembly of a functional promoter as a direct result of the recombination event (promoter trapping).

We have successfully applied techniques (1) and (3) to achieve highly efficient targeting, using, respectively, heterospecific *FLP* sites (Schlake and Bode 1994) and repair of a defective *neo* gene lacking a start codon by a promoter/ATG combination (Figure 11.4).

In a large-scale screening effort, we identified recipient CHO cell clones that show stable high-level expression of a model gene (Figure 11.5). While the model gene could be reproducibly replaced by the gene of interest in these clones, we nevertheless observed variations in expression level among clones originating from a single recipient cell. All of these clones harbored a single copy of the gene of interest at the pre-determined locus. The heterogeneity of expression can possibly be attributed to perturbance of the architecture of a previously stable locus caused by epigenetic phenomena. After re-cloning, we were able to identify clones that secreted up to 40 pg of the desired of a recombinant glycoprotein per cell per day.

In addition to the integration loci identified by random screening, we also have used the human IgH locus of a human-mouse heterohybridoma for targeted integra-

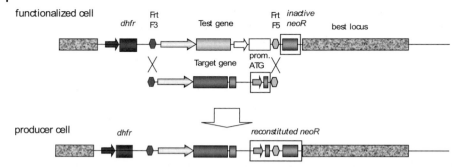

Fig. 11.4 Position-mediated expression enhancement (PMEE). The functionalized cell is a pre-selected high producer for a test protein. This construct contains a conventional selection marker (white box) and the *dhfr* gene allowing for gene amplification at a later stage. In addition, it contains an incomplete *neo* gene lacking a promoter. The cassette to be exchanged is flanked by heterospecific *frt* sites (hexagon). To generate the producer cell the test gene is exchanged for the gene of interest via recombination at the *frt* sites mediated by co-transfection of an flp-expressing plasmid. Recombination reconstitutes a functional *neo* gene. The procedure ensures rapid and reliable insertion into an optimized locus.

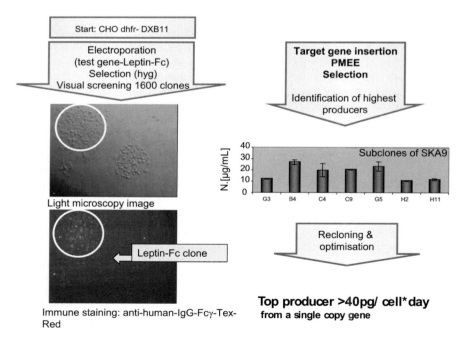

Fig. 11.5 Generation of high producers for multiple genes using a pre-selected clone expressing the gene for a test protein suitable for targeting.

tion. The locus was made accessible by insertion of *FRT* sites via homologous recombination and targeted by Flp-mediated exchange between heterospecific *FRT* sites. The best clones secreted 45 pg per cell per day of an IgM antibody. Insertion of multiple genes into this locus typically resulted in cell lines yielding typically between 10 and 25 pg per cell per day.

Loci with superior expression characteristics can also be created by specific vector design: thus, specific genomic sequences that locate several kilobases away from enhancers and display features of *MAR* elements can be inserted into a vector. The inclusion of such elements can significantly reduce the screening effort required for the development of particular lines (Zahn-Zabal et al. 2001).

Finally, it should be stressed that optimal molecular design and selection of the producer cell line are not the only factors that determine whether an economic cell-based process can be established. The subsequent definition and fine-tuning of media composition and of fermentation parameters can have a comparable impact on the economics of a manufacturing process.

11.4
Mammalian Cell-based Fermentation Processes

Mammalian cells are the preferred option to produce therapeutics that require authentic glycosylation for activity or for improved distribution and elimination properties *in vivo*, as pointed out before. The lower productivity per volume of culture space in comparison to bacterial and fungal strains is the major challenge for process development and process design (Molowa and Mazanet 2003).

In addition, reproducibility, long-term stability and compliance with the regulatory requirements are of concern when initiating the development for a mammalian cell-based production process.

11.4.1
Batch and Fed-batch Fermentation

Fermentation processes in stirred tank reactors (STR) carried out in a batch mode (Nelson and Geyer 1991) can be scaled up to a 15 000-L scale. Due to the limitations in nutrient supply inherent in this mode of fermentation, cells grow logarithmically and die shortly after reaching maximal density (up to 5×10^6 cells mL^{-1}). At a defined end-point, the content of the bioreactor is harvested and the total harvest is subjected to a purification (downstream) process. For the manufacturing of approved biopharmaceuticals, extensive cleaning in place (CIP) and sterilization in place (SIP) procedures are required to prepare the bioreactor for the next production run.

As the total protein yield is proportional to the number of viable cells over the time (see area below the gray line in Figure 11.6), great efforts have been made to extend the effective production time by supplementing the medium with nutrient solution during fermentation. Cell line-specific feeding strategies were developed result-

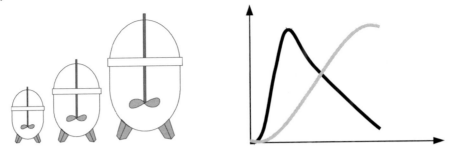

Fig. 11.6 Production train of stirred tank reactors; the smaller tanks are required to inoculate the production reactor. Only the reactor with the highest volume is used for production. The schematic diagram on the right illustrates the typical profile of viable cell number (black line) and the amount of protein produced (gray line) over the culture time (x-axis).

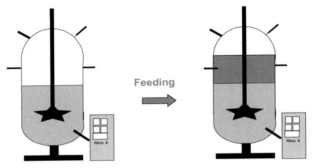

Fig. 11.7 Schematic illustration of fed-batch production process with increasing culture volume.

ing in highly efficient "fed-batch" processes. In the fed-batch mode, nutrients such as glucose or amino acids are added to the bioreactor at defined culture times after initiating the culture (Figure 11.7) (Nelson and Geyer 1991). Higher cell densities of up to 10×10^7 cells mL^{-1} are now feasible, and the productivity of the substrate is much higher (up to fivefold) in comparison to the batch fermentation mode.

11.4.2
Continuous Perfusion Fermentation

More recently, variants of continuous production processes have been assessed. These processes require a continuous flow of culture medium through the bioreactor, in addition to cell retention (Figure 11.8). Cell retention enables higher cell densities (up to 5×10^7 cells mL^{-1}) and, when combined with extended culture periods, a substantial increase of volumetric productivity per bioreactor run can be observed.

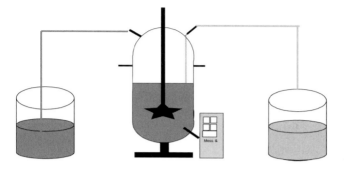

Fig. 11.8 Schematic illustration of perfusion fermentation (media tank on the left side, harvest tank on the right side).

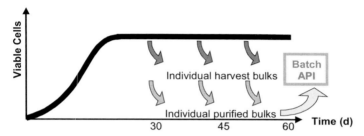

Fig. 11.9 Scheme of batch production with flexible downstream processes in continuous fermentation. API = active pharmaceutical ingredient. The individual purified bulks can be pooled together to get one batch per reactor run.

Cell viability can be stably maintained during the complete fermentation run, thereby achieving consistent product quality during the entire fermentation process (Lullau et al. 2003; Voisard et al. 2003).

The product is also removed from the culture via the continuous flow of medium. This reduces the potential problems are caused by product inhibition, or by limited stability of the product. It furthermore allows a down-sizing of the downstream process and an overall reduction of investments (Figure 11.9).

Despite these advantages, continuous fermentation processes are rarely used in biopharmaceutical production at present. The equipment is more complex and more difficult to maintain in comparison to batch or fed-batch fermentation. The need for long-term sterility and additional reservoirs and equipment for the media and harvest must also be considered. Likewise, the process development and process validation required are more time-consuming.

Several different cell-retention systems are available which exploit the following methods: filtration, acoustic cell retention, sedimentation, centrifugation, and fluidized bed retention (Castilho and Medronho 2002).

11.4.3
Continuous Production with Hollow-fiber Bioreactors

Hollow-fiber bioreactors represent special continuous perfusion systems, characterized by independent medium and harvest streams. The system allows retention of
both cells and product in the culture chamber. The separation is realized by using
hollow-fiber cartridges with intra- and extracapillary spaces. The system has a cycling
pathway that provides a continuous flow of medium through the hollow fibers and
the oxygenator (medium circulation), and an inoculation and harvest pathway, that
runs through the extracapillary space, also termed the culture space. The mass transfer through the membrane is diffusion-controlled.

The use of older types of these bioreactors is associated with problems involving
oxygen supply, membrane fouling, homogeneous product harvest, and removal of
dead cells and debris. In attempt to avoid these problems, a specific second-generation type of bioreactor was developed. Reactors of this type are equipped with an expansion chamber for each pathway, in which gas pressure can be selectively increased (up to 100 mmHg) (Figure 11.10). This leads to a controlled trans-membrane
pressure, and provides a much larger mass transfer than that obtained via diffusion.
An advantageous mixing effect occurs when this trans-membrane pressure is applied and, as a result, the fluid flow is reversed. The trans-membrane flow can be adjusted as high as sixfold that of the cell culture volume per hour. Because harvesting
is independent of the feeding flow, the product concentration is also adjustable.

Another advantage is the use of disposable culture devices (equipment that is in
contact with cells and medium/harvest flow), and therefore, no CIP and SIP procedures need to be applied. This in turn minimizes the risk of process variation and of
cross-contamination between different reactor runs.

Very high cell densities ($>10^8$ cells mL^{-1}) are reached in the culture space. As
shown in Figure 11.11, the cells are arranged in a single hollow fiber, and densities
are comparable to those in the tissues. Because it is impossible to determine the
count of viable cells or to isolate large cell amounts from the bioreactor, the provision
of proof of batch-to-batch consistency is a major challenge. As an example, three different production runs for a monoclonal antibody are shown in Figure 11.12. The

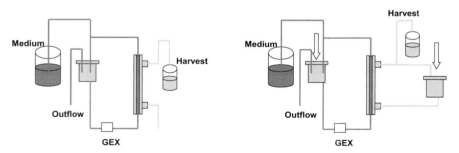

Fig. 11.10 Schematic flow chart of traditional (left) and recent
(right) hollow-fiber bioreactor systems. Gas pressure is marked with
arrows in the more recent system. (GEX = gas exchange).

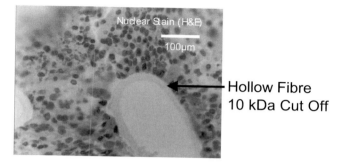

Fig. 11.11 Histological preparation of a hollow-fiber cartridge. A homogeneous cell distribution was found throughout the vast majority of the extracapillary space in the modules.

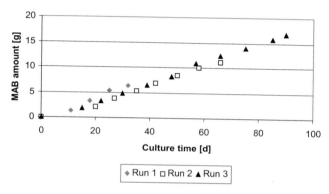

Fig. 11.12 Three different production runs for a monoclonal antibody with varying process times (◆, 1 month; □, 2 months; ▲, 3 months). The runs were performed by using an ASM-Maximizer 500 (100 mL of culture space).

production rate was shown to be constant for up to three months, and to be independent of the culture time (Brecht et al. 2003).

Hollow-fiber bioreactors have many advantages over traditional bioreactors. Their only limitation is that the capacity of commercially available bioreactors is so far limited to a culture space of 2000 mL. Therefore, their use is restricted to the production of low-volume biopharmaceuticals and in-vivo diagnostics (Chu and Robinson 2001).

11.5
Conclusions

Mammalian cells are the system of choice for the production of therapeutic glycoproteins, and in particular of full-length therapeutic antibodies. To make the system efficient, careful screening and selection of the producer cell line and a careful process design are required.

Highly versatile vector systems have been defined that allow the production of single or multiple candidate proteins in stable producer lines on small and medium scales. Despite continuous advances being made in basic research, the mechanisms of expression and secretion remain poorly understood. Therefore, any success in identifying a line with superior characteristics remains unpredictable. Currently, new approaches are being taken for standardized cell lines and process development, thereby reducing the time and effort required to establish mammalian cell-based production processes.

Acknowledgments

The authors thank Prof. Gellissen and Michele Großstück for carefully reading the manuscript, and for their suggestions and modifications.

Appendix

Tab. A11.1 Mammalian cell lines used for glycoprotein production.

Cell line	Genotype	Selection	Source	Reference
CHO K1	parental/no defect	GS	DSMZ (ACC110)	Puck (1965)
		Antibiotic resistance markers (no dhfr)	ATCC	
CHO DXB11	dhfr-	dhfr	DSMZ (ACC126)	Urlaub and Chasin (1980)
CHO DG44	dhfr-	dhfr	Chasin, LA	Urlaub et al. (1986)
NS0	no defect; very low GS	GS	with GS system: Lonza	Bebbington al. (1992)
SP2/0	hybrid BALB/c spleen + P3X63AG8	Antibiotic resistance markers	DSMZ (ACC146)	Shulman et al. (1978)
BHK	no defect	Antibiotic resistance markers	ATCC (CLL10)	
HEK 293	no defect	Antibiotic resistance markers	ATCC) (CRL1773)	Graham et al. (1977)
293T	with SV40T	Transient expression of SV40ori plasmid	ATCC (CRL11268)	Sena-Esteves et al. (1999)
PER.C6	no defect	Antibiotic resistance markers (neo)	Crucell	Fallaux et al. (1998)

References

Bebbington CR, Renner G, Thomson S, King D, Abrams D, Yarranton GT (1992) High-level expression of a recombinant antibody from myeloma cells using a glutamine synthetase gene as an amplifiable selectable marker. Biotechnology (NY) 10: 169–175

Berk AJ (1999) Activation of RNA polymerase II transcription. Curr Opin Cell Biol 11: 330–335

Boorsma M, Nieba L, Koller D, Bachmann MF, Bailey JE, Renner WA (2000) A temperature-regulated replicon-based DNA expression system. Nat Biotechnol 18: 429–432

Brecht R, Bushnaq-Josting H, Zietze S, Nuck R, Kloth C, Schwadtke A, Obermayer N, Marx U, Koch S (2003) Continuous GMP production in hollow fibre bioreactors – a viable alternative for the production of low volume biopharmaceuticals. 18th ESACT meeting, Spain

Calos MP (1998) Stability without a centromere. Proc Natl Acad Sci USA 95: 4084–4085

Castilho LR, Medronho RA (2002) Cell retention devices for suspended-cell perfusion cultures. Adv Biochem Eng Biotechnol 74: 129–169

Chu L, Robinson DK (2001) Industrial choices for protein production by large-scale cell culture. Curr Opin Biotechnol 12: 180–187

Coquelle A, Pipiras E, Toledo F, Buttin G, Debatisse M (1997) Expression of fragile sites triggers intrachromosomal mammalian gene amplification and sets boundaries to early amplicons. Cell 89: 215–225

Denome RM, Cole CN (1988) Patterns of polyadenylation site selection in gene constructs containing multiple polyadenylation signals. Mol Cell Biol 8: 4829–4839

Durocher Y, Perret S, Kamen A (2002) High-level and high-throughput recombinant protein production by transient transfection of suspension-growing human 293-EBNA1 cells. Nucleic Acids Res 30: E9

Fallaux FJ, Bout A, van der Velde I, van den Wollenberg DJ, Hehir KM, Keegan J, Auger C, Cramer SJ, van Ormondt H, van der Eb AJ, Valerio D, Hoeben RC (1998) New helper cells and matched early region 1-deleted adenovirus vectors prevent generation of replication-competent adenoviruses. Hum Gene Ther 9: 1909–1917

Gollogly L, Castronovo V (1996) A possible role for the alpha 1_3 galactosyl epitope and the natural anti-gal antibody in oncogenesis. Neoplasma 43: 285–289

Graham FL, Smiley J, Russell WC, Nairn R (1977) Characteristics of a human cell line transformed by DNA from human adenovirus type 5. J Gen Virol 36: 59–74

Haensler J and Szoka FC, Jr. (1993) Polyamidoamine cascade polymers mediate efficient transfection of cells in culture. Bioconjug Chem 4: 372–379

Hounsell EF, Davies MJ, Renouf DV (1996) O-linked protein glycosylation structure and function. Glycoconj J 13: 19–26

Jefferis R, Lund J (1997) Glycosylation of antibody molecules: structural and functional significance. Chem Immunol 65: 111–128

Jones D, Kroos N, Anema R, van Montfort B, Vooys A, van der Kraats S, van der Helm E, Smits S, Schouten J, Brouwer K, Lagerwerf F, van Berkel P, Opstelten DJ, Logtenberg T, Bout A (2003) High-level expression of recombinant IgG in the human cell line PerC6. Biotechnol Prog 19: 163–168

Kim SY, Lee JH, Shin HS, Kang HJ, Kim YS (2002) The human elongation factor 1 alpha (EF-1 alpha) first intron highly enhances expression of foreign genes from the murine cytomegalovirus promoter. J Biotechnol 93: 183–187

Kozak M (1984) Compilation and analysis of sequences upstream from the translational start site in eukaryotic mRNAs. Nucleic Acids Res 12: 857–872

Kozak M (2002) Pushing the limits of the scanning mechanism for initiation of translation. Gene 299: 1–34

Lindell J, Girard P, Muller N, Jordan M, Wurm F (2004) Calfection: a novel gene transfer method for suspension cells. Biochim Biophys Acta 1676: 155–161

Liu R, Baillie J, Sissons JG, Sinclair JH (1994) The transcription factor YY1 binds to negative regulatory elements in the human cytomegalovirus major immediate early enhancer/promoter and mediates repression in non-permissive cells. Nucleic Acids Res 22: 2453–2459

Lullau E, Kanttinen A, Hassel J, Berg M, Haag-Alvarsson A, Cederbrant K, Greenberg B, Fenge C, Schweikart F (2003) Comparison of batch and perfusion culture in combination with pilot-scale expanded bed purification for the production of soluble recombinant beta-secretase. Biotechnol Prog 19: 37–44

McBurney MW, Mai T, Yang X, Jardine K (2002) Evidence for repeat-induced gene silencing in cultured Mammalian cells: inactivation of tandem repeats of transfected genes. Exp Cell Res 274: 1–8

Miller DL, Dibbens JA, Damert A, Risau W, Vadas MA, Goodall GJ (1998) The vascular endothelial growth factor mRNA contains an internal ribosome entry site. FEBS Lett 434: 417–420

Molowa DT, Mazanet R (2003) The state of biopharmaceutical manufacturing. Biotechnol Annu Rev 9: 285–302

Monaco L, Tagliabue R, Soria MR, Uhlen M (1994) An *in vitro* amplification approach for the expression of recombinant proteins in mammalian cells. Biotechnol Appl Biochem 20: 157–171

Nanbru C, Lafon I, Audigier S, Gensac MC, Vagner S, Huez G, Prats AC (1997) Alternative translation of the proto-oncogene c-myc by an internal ribosome entry site. J Biol Chem 272: 32061–32066.

Nelson KL, Geyer S (1991) Bioreactor and process design for large-scale mammalian cell culture manufacturing. Bioprocess Technol 13: 112–143

Peri S, Pandey A (2001) A reassessment of the translation initiation codon in vertebrates. Trends Genet 17: 685–687

Piechaczek C, Fetzer C, Baiker A, Bode J, Lipps HJ (1999) A vector based on the SV40 origin of replication and chromosomal S/MARs replicates episomally in CHO cells. Nucleic Acids Res 27: 426–428

Prosch S, Stein J, Staak K, Liebenthal C, Volk HD, Kruger DH (1996) Inactivation of the very strong HCMV immediate early promoter by DNA CpG methylation *in vitro*. Biol Chem Hoppe Seyler 377: 195–201

Quinn CM, Wiles AP, El-Shanawany T, Catchpole I, Alnadaf T, Ford MJ, Gordon S, Greaves DR (1999) The human eukaryotic initiation factor 4AI gene (EIF4A1) contains multiple regulatory elements that direct high-level reporter gene expression in mammalian cell lines. Genomics 62: 468–476

Ray MK, Fagan SP, Brunicardi FC (2000) The Cre-loxP system: a versatile tool for targeting

genes in a cell- and stage-specific manner. Cell Transplant 9: 805–815

Rodriguez CI, Buchholz F, Galloway J, Sequerra R, Kasper J, Ayala R, Stewart AF, Dymecki SM (2000) High-efficiency deleter mice show that FLPe is an alternative to Cre-loxP. Nat Genet 25: 139–140

Rotondaro L, Mele A, Rovera G (1996) Efficiency of different viral promoters in directing gene expression in mammalian cells: effect of 3'-untranslated sequences. Gene 168: 195–198

Schlake T, Bode J (1994) Use of mutated FLP recognition target (FRT) sites for the exchange of expression cassettes at defined chromosomal loci. Biochemistry 33: 12746–12751

Sena-Esteves M, Saeki Y, Camp SM, Chiocca EA, Breakefield XO (1999) Single-step conversion of cells to retrovirus vector producers with herpes simplex virus-Epstein–Barr virus hybrid amplicons. J Virol 73: 10426–10439

Shulman M, Wilde CD, Kohler G (1978) A better cell line for making hybridomas secreting specific antibodies. Nature 276: 269–270

Urlaub G, Chasin LA (1980) Isolation of Chinese hamster cell mutants deficient in dihydrofolate reductase activity. Proc Natl Acad Sci USA 77: 4216–4220

Urlaub G, Mitchell PJ, Kas E, Chasin LA, Funanage VL, Myoda TT, Hamlin J (1986) Effect of gamma rays at the dihydrofolate reductase locus: deletions and inversions. Somat Cell Mol Genet 12: 555–566

Voisard D, Meuwly F, Ruffieux PA, Baer G, Kadouri A (2003) Potential of cell retention techniques for large-scale high-density perfusion culture of suspended mammalian cells. Biotechnol Bioeng 82: 751–765

White RE, Wade-Martins R, James MR (2001) Sequences adjacent to oriP improve the persistence of Epstein–Barr virus-based episomes in B cells. J Virol 75: 11249–11252

Wright A, Morrison SL (1997) Effect of glycosylation on antibody function: implications for genetic engineering. Trends Biotechnol 15: 26–32

Zahn-Zabal M, Kobr M, Girod PA, Imhof M, Chatellard P, de Jesus M, Wurm F, Mermod N (2001) Development of stable cell lines for production or regulated expression using matrix attachment regions. J Biotechnol 87: 29–42

12
Plant Cells

RAINER FISCHER, RICHARD M TWYMAN, JÜRGEN DROSSARD, STEPHAN HELLWIG and
STEFAN SCHILLBERG

12.1
General Biology of Plant Cells

12.1.1
Advantages of Plant Cells for the Production of Recombinant Proteins

Cultured plant cells possess a valuable combination of advantages for the production of recombinant proteins (Fischer et al. 1999; Sharp and Doran 2000). First, like microbial cells, plant cells are comparatively inexpensive to maintain. This is because their energy is derived by sugar dissimilation, and their relatively simple nutrient requirements can be satisfied with a fully synthetic medium. Plant cells can therefore be cultured under precisely defined conditions, allowing the production of industrial or pharmaceutical proteins according to current good manufacturing practice (cGMP). Second, unlike microbes, plant cells are highly evolved eukaryotic systems. Therefore, they have the ability to synthesize, fold, and assemble complex multi-subunit proteins and glycoproteins (Fischer and Emans 2000). The N-glycan structures synthesized in plants are not the same as those found in mammals (see Section 12.1.2), but recombinant glycoproteins produced in plants are generally functional and are much closer to their mammalian counterparts than those synthesized in filamentous fungi or yeast. They are also vastly superior to the aglycosylated derivatives synthesized in bacteria (Gomord 2004). Finally, plant cells do not contain the endotoxins present in many bacteria, nor do they harbor potential human pathogens such as viruses or prions (which can be present in mammalian cell lines and transgenic animals). Plant cells are therefore inherently safe, which presents another strong advantage for the production of therapeutic proteins.

While there has been considerable interest in the development of terrestrial plants for the production of recombinant proteins, the advantages of agricultural-scale production might be outweighed by the long development times, the inconsistent product yield and quality, the ever-tightening regulatory framework governing the use of field crops, and the difficulty in applying cGMP to the early stages of the production pipeline (Twyman et al. 2003; Ma et al. 2003; Fischer et al. 2004). In field plants, one must consider the possibility of contamination with agrochemicals and fertilizers, the impact of pests and diseases, and the variable cultivation conditions (e. g., weather, soil composition, and interaction with other organisms). Another important issue is the effect of the transgenic crop on the environment – for example, the possibility of transgene spread, environmental contamination with the recombinant product, and the risk of transgenic plant material entering the food and feed chains (Commandeur et al. 2003). These problems can be addressed without discarding the advantages of plant systems if plant cell cultures are used as the expression platform.

12.1.2
N-Glycan Synthesis in Plants

Since more than 50% of mammalian proteins are glycosylated, the majority of therapeutic proteins are glycoproteins. The glycans have a number of important structural and functional roles, such that they often affect the protein's stability, solubility, folding, biological activity, longevity, interactions with cellular receptors, and immunogenicity. Mammalian glycoproteins produced in plants and mammals are glycosylated on the same Asn residues, but the N-glycan structures are distinct. In all eukaryotic cells, N-glycosylation begins in the endoplasmic reticulum (ER) and involves the co-translational addition of an oligosaccharide precursor ($Glc_3Man_9GlcNAc_2$) to specific Asn residues within the consensus sequence (Asn-X-Ser/Thr). The N-glycans then undergo several maturation reactions, including the removal of certain residues in the ER, and the addition of further residues in the ER and Golgi apparatus. The steps occurring in the ER are conserved in mammals and plants, but they diverge in the late Golgi apparatus so that core $\alpha(1,6)$-linked fucose and terminal sialic acid residues are added in mammals, whereas bisecting $\beta(1,2)$-xylose and core $\alpha(1,3)$-fucose residues are added in plants. Such differences in glycan structure have been described when recombinant proteins produced in plants are compared to their native counterparts. Most of the evidence for differential glycosylation has been obtained through the analysis of antibodies produced in cultured mammalian cells and transgenic plants. Interestingly, not all plants produce the same glycosylation profiles. For example, antibodies produced in tobacco (*Nicotiana tabacum*) contain very heterogeneous mixtures of high-mannose-type and complex-type N-glycans, whereas those produced in alfalfa (*Medicago sativa*) contain mostly complex-type glycans (Bardor et al. 2003). Recently, it has been reported that sialylated glycoconjugates can be produced in *Arabidopsis thaliana* suspension cells (Shah et al. 2003).

12.2
Description of the Expression Platform

12.2.1
Culture Systems and Expression Hosts

A number of alternative culture systems can be used for the in-vitro cultivation of plant cells. These include hairy roots (Hilton and Rhodes 1990), shooty teratomas (Sharp and Doran 2001), immobilized cells (Archambault 1991), and suspension cell cultures (Kieran et al. 1997). While there are individual examples of all these systems being used for the production of recombinant proteins, attention has focused on suspension cells because these are the most amenable to cGMP procedures and they can be cultivated relatively easily in large-scale bioreactors (Schlatman et al. 1996; Wen et al. 1995). Suspension cell cultures have been derived from a number of different plant species including the model plant *Arabidopsis thaliana* (Desikan et al. 1996), plants used to produce small-molecule drugs such as *Catharanthus roseus* and *Taxus brevia* (Seki et al. 1997; Van Der Heijden et al. 1989), and important domestic crops such as tobacco, alfalfa, rice (*Oryza sativa*), tomato (*Lycopersicon esculentum*), and soybean (*Glycine max*) (Chen et al. 1994; Daniell and Edwards 1995; Hoehl et al. 1988; Kwon et al. 2003a; Nagata et al. 1992). It is the members of this latter group that have been used the most widely to produce recombinant proteins, and we focus on them below.

12.2.2
Derivation of Suspension Cells

Plant cell suspensions are typically derived from undifferentiated callus tissue which has been induced on a solid medium. The transfer of friable callus pieces into liquid medium is followed by agitation on rotary shakers or in fermentors to break the larger pieces into single cells and small aggregates of fewer than 20 cells. The de-differentiated state is maintained by growth regulators, which are added to the medium to promote rapid growth and maintain culture morphology. Transgenic cell suspensions can be seeded from callus derived from the explants of transgenic plants, or wild-type suspension cells can be transformed by co-cultivation with *Agrobacterium tumefaciens* (Horsch et al. 1985; Koncz and Schell 1986), particle bombardment (Christou 1993), protoplast transformation (Lindsey and Jones 1987), or indeed infection with recombinant plant viruses (Porta and Lomonossoff 2002). If a homogeneous culture can be generated and maintained in this manner, the cells can be grown in much the same way as lower eukaryote microbial cultures, although the cell densities are lower and the generation times are longer. Even so, it is quite possible to cultivate plant cell suspensions using conventional fermentor equipment with only minor adjustments, and many of the fermentor modes applied to microbial cultures (for example, batch, fed-batch, perfusion and continuous fermentation) can also be applied to plants (Hooker et al. 1990; ten Hoopen et al. 1992).

12.2.3
Optimizing Protein Accumulation and Recovery

The major limitations of plant cells include the comparatively poor cell growth rates compared to microbial cultures, wall growth and the formation of aggregates, somaclonal variation and other forms of genetic instability (e. g., gene silencing), and (at least for some cell types) shear sensitivity (Meijer et al. 1994; Offringa et al. 1990; Yu et al. 1996). Many of these issues have been addressed through improved fermentor design and agitation and aeration conditions (Bohme et al. 1997; Doran 1993), the optimization of the nutrient supply (Sakamoto et al. 1993; Sato et al. 1996), or by feeding the cultures with precursors or elicitors (Ishikawa et al. 1994; Ketchum et al. 1999; McCormack et al. 1997; Mirjalili and Linden 1996; Yukimune et al. 1996). Where these limitations are intrinsic to the host species and cultivar, they can only be overcome through the careful selection of callus cell lines with respect to product formation, growth characteristics and genetic stability (e. g., Huang et al. 2002). The yield of recombinant protein can also be enhanced by designing the expression construct to optimize all stages of gene expression and by targeting the recombinant protein appropriately (see Figure 12.1 for an example). Product recovery can be improved by adjusting the culture conditions to favor the recovery of particular products, and by choosing a suitable purification strategy for protein isolation. Some of the major factors that need to be considered are discussed in the following sections.

12.2.4
Expression Construct Design

Recombinant protein production in cultured plant cells depends initially on the high-level transcriptional activity of the expression construct. In most cases, the

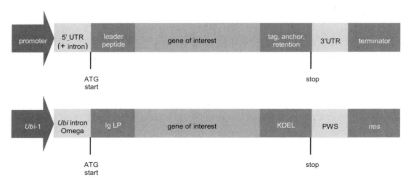

Fig. 12.1 Expression constructs used in plant cell cultures. A) A generic construct layout, showing the components that must be included. B) One of the constructs used by Torres et al. (1999) for the expression of scFv antibodies in rice cells. The ubiquitin-1 promoter is followed by the first intron of the ubiquitin gene, the tobacco mosaic virus omega sequence (which is thought to enhance translation), the murine IgG leader peptide, the antibody cDNA sequence, the KDEL C-terminal tag for retention of the protein within the ER, the pseudoknot 3'-UTR, which is thought to enhance mRNA stability, and finally the terminator from the *Agrobacterium tumefaciens nos* (nopaline synthase) gene.

transgene is driven by a strong, constitutive promoter, which causes the product to accumulate continuously. The cauliflower mosaic virus (CaMV) 35S promoter has been used extensively in tobacco cells (by far the most popular expression host), sometimes with a duplicated enhancer to increase its activity. A hepatitis B virus surface antigen has been produced in both tobacco and soybean cells, and in this case a hybrid *(ocs)₃mas* promoter from *A. tumefaciens* was used, which was also constitutive (Smith et al. 2002). Finally, the constitutive ubiquitin-1 (*Ubi1*) promoter from maize has been used in rice callus cultures to produce a single-chain antibody (Torres et al. 1999). As an alternative to constitutive promoters, inducible promoters can be used to restrict protein expression to the culture phases that are most favorable for high-level product accumulation. Among the various inducible promoters that have been developed for plants (Padidam 2003), only the rice α-amylase promoter has been used to produce recombinant proteins. The native protein expressed under the control of this promoter is secreted abundantly from cultured rice cells in response to sucrose starvation, so the promoter represents an ideal inducible system for use with rice suspension cells. Thus far, three proteins have been produced using this type of construct: (i) granulocyte-macrophage colony stimulating factor (GM-CSF) (Shin et al. 2003); (ii) α₁-antitrypsin (Huang et al. 2001; Terashima et al. 1999 a,b; Trexler et al. 2002); and (iii) lysozyme (Huang et al. 2003). A glucocorticoid-inducible promoter has been used in *Catharanthus roseus* suspension cells to control the expression of the gene for green fluorescent protein (Hughes et al. 2002).

Another aspect of construct design that is important for protein accumulation is the leader peptide, which is encoded at the 5′-end of the transgene open reading frame and is added to the N-terminus of the recombinant protein. The leader peptide targets the protein to the secretory pathway, ultimately resulting in secretion of the protein into the culture medium. Various different leader peptides have been used in plant cell culture systems, and those from plants and animals appear to work equally well. For mammalian proteins that are normally secreted from their natural source tissues, the native leader peptide is often retained. This has been the case for most recombinant antibodies, which generally contain murine heavy chain or light chain leader peptides (Sharp and Doran 2001; Torres et al. 1999), as well as specific examples of other secreted therapeutic proteins such as erythropoietin (Matsumoto et al. 1995) and GM-CSF (Lee et al. 2002). For other products, a heterologous leader peptide of plant origin has been used. Examples include the leader peptides from tobacco pathogenesis-related protein PR1a (Firek et al. 1993) and rice α-amylase (Shin et al. 2003; Terashima et al. 1999a,b; Huang et al. 2002).

In terrestrial plants, proteins directed to the secretory pathway can be retained in the ER using the C-terminal tetrapeptide tag KDEL, or directed to post-ER compartments such as protein storage vacuoles, protein bodies, or oil bodies (Fischer and Emans 2000; Ma et al. 2003; Twyman et al. 2003). However, these strategies are not frequently used in suspension cells because it is easier to recover proteins from the culture medium than the wet biomass. Targeting proteins for accumulation within the cell makes disruption necessary prior to protein purification. Cell disruption has several drawbacks, including the need for additional equipment and labor, and the release of phenolic substances and proteases that can reduce protein yield. In the absence of other sorting information, proteins directed to the secretory pathway in cultured cells will eventually reach the apoplast, from where they will either diffuse through the cell wall and into the culture medium or lodge in the cell wall matrix. The fate of the protein depends mostly on its size, since the pores in the plant cell wall allow the passage of globular proteins of 20 kDa or less. However, some larger pores are present, which allow the diffusion of larger proteins, albeit at a slower rate (Carpita et al. 1979). Recombinant proteins with molecular weights significantly in excess of 20 kDa have been recovered efficiently from the culture medium (Sharp and Doran 2001; Terashima et al. 1999b; Kwon et al. 2003b), while in other cases even fairly small proteins have remain trapped in the cell wall and have needed to be released by disruption or mild enzymatic digestion (Matsumoto et al. 1995; Tsoi and Doran 2002). The lack of strict size-dependence suggests that the secretion of proteins larger than 20 kDa may reflect physico-chemical properties such as charge and hydrophobicity, as well as size.

12.2.5
Foreign Protein Stability

Proteins undergo degradation both within the cell and in the extracellular environment, resulting in the appearance of specific fragments and nonspecific degradation 'smears' when either the medium or cell extracts are analyzed by western blot (Sharp and Doran 2001). Ultimately, the more degradation occurs, the lower the protein yields; hence, much research has been carried out to determine the causes of degradation and methods to prevent it. Intracellular degradation is a general phenomenon, not one that is restricted to particular host systems and expression strategies. Therefore, many different explanations have been offered. One general explanation is that when the protein is extracted from wet cell biomass, degradation results from the release of proteases during homogenization. In the case of multimeric proteins such as antibodies, Sharp and Doran (2001) have suggested that specific fragments could represent folding or assembly intermediates, perhaps reflecting the heterogeneity of glycosylation. Interestingly, in the case of an IgG_1 antibody expressed in cultured tobacco cells, these authors showed that most of the degradation occurred between the ER and Golgi apparatus, and not within the apoplast as might have been expected.

Several researchers have reported the poor stability of proteins in the culture medium, but again it is difficult to pin this on a single cause. It is likely that several factors contribute to protein loss, including the presence of proteases in the medium (Kwon et al. 2003b; Lee et al. 2002; Sharp and Doran 2001; Shin et al. 2003; Terashima et al. 1999a), the tendency for some proteins to precipitate or aggregate under non-physiological conditions (Sharp and Doran 2001), and the adsorption of certain charged proteins to the walls of the reactor vessel (Magnuson et al. 1996). The activity of proteases in the culture medium can be reduced using various strategies such as including protease inhibitors, or by adding gelatin to act as an alternative substrate (Kwon et al. 2003b). A very successful strategy is to use rice cell suspensions with the transgene under the control of the inducible α-amylase promoter. Rice cells are thought to secrete less protease into the culture medium than tobacco cells, and the promoter can be used to restrict protein expression to a defined production phase which is separate from the growth phase, when most proteases are produced (Shin et al. 2003).

12.2.6
Medium Additives that Enhance Protein Accumulation

The recovery of recombinant proteins from plant cell cultures can be enhanced by adding various agents to the growth medium. Substances that have been tested include simple inorganic compounds such as sodium chloride and potassium nitrate (Wongsamuth and Doran 1997), simple organic compounds such as amino acids (Fischer et al. 1999) and dimethylsulfoxide (DMSO) (Wahl et al. 1995), more complex organic polymers such as polyethylene glycol (Kwon et al. 2003 b; Lee et al. 2002), haemin (Tsoi and Doran 2002) and polyvinylpyrrolidone (PVP) (Kwon et al. 2003 b; LaCount et al. 1997; Lee et al. 2002; Magnuson et al. 1996; Sharp and Doran 2001; Wongsamuth and Doran 1997), plant hormones such as gibberellic acid (Tsoi and Doran 2002), and proteins such as gelatin (Kwon et al. 2003 b; Lee et al. 2002; Wongsamuth and Doran 1997) and bovine serum albumin (BSA) (James et al. 2000). These act variously as enhancers of protein synthesis or secretion, inhibitors of intracellular protein degradation, and extracellular stabilizing agents. There is no way to predict which, if any, of these agents will help in any particular case, and the best course of action is to test them empirically. Some of the additives may even be counterproductive in certain cases; for example, BSA and gelatin can interfere with downstream processing, and this must be balanced against their positive influence on protein yields.

12.2.6.1 Simple Inorganic Compounds
Supplementing the growth medium with simple inorganic compounds has, in a few cases, led to improved recombinant protein accumulation. For example, increasing the amount of salt (NaCl) in tobacco cell suspension cultures to a concentration between 50 mM and 100 mM increased the levels of GM-CSF in the culture medium by up to 50%. The intracellular level of the protein was not affected, which suggested that the salt functioned by increasing the rate or efficiency of protein secretion (James et al. 2002). Similarly, hairy root cultures maintained in a medium containing nitrate (KNO_3) produced almost twice the amount of a recombinant antibody compared to cultures lacking nitrate, and the culture medium contained nearly three times the level of antibody found in the nitrate-free medium (Wongsamuth and Doran 1997).

12.2.6.2 Amino Acids
Amino acids have been used as medium supplements with mixed results. In one report, the addition of several essential and nonessential amino acids to tobacco suspension cells at 10 h before cell harvest led to a threefold increase in antibody production (Fischer et al. 1999). $CaCl_2$ was also present in the medium to promote

amino acid uptake. In other studies, the addition of glutamine had a negligible impact on product yields (Liu and Lee 1999; Tsoi and Doran 2002).

12.2.6.3 Dimethylsulfoxide

Dimethylsulfoxide (DMSO) is a medium additive that increases the permeability of cells, thereby enhancing the release of recombinant proteins from the apoplast into the culture medium (Delmer 1979). However, the presence of DMSO in tobacco suspension cell cultures has been shown to increase both the intracellular and extracellular product levels, indicating a more general role for this organic compound, perhaps as a protein-stabilizing agent (Wahl et al. 1995). The presence of 2.8% DMSO in a tobacco cell culture expressing a recombinant antibody increased the total amount of antibody from 150 µg L^{-1} to 210 µg L^{-1}, while the presence of 4% DMSO resulted in a further yield increase to 430 µg L^{-1} (Wahl et al. 1995).

12.2.6.4 Organic Polymers

Several types of organic polymers have been added to plant cell cultures as potential stabilizing agents. These include haemin, an inhibitor of ubiquitin-dependent proteolysis (Tsoi and Doran 2001), polyethylene glycol (PEG) (Kwon et al. 2003b; Lee et al. 2002) and PVP. While haemin and PEG appeared to have no significant effect, PVP has enhanced the production of several proteins while having little or no impact on cell growth (LaCount et al. 1997; Magnuson et al. 1996). Indeed, LaCount and colleagues showed that the addition of PVP (M$_r$ 360 000) to a final concentration of 0.75 g L^{-1} resulted in a remarkable 35-fold increase in the level of extracellular foreign protein. Both the concentration of the polymer and its average molecular weight were important variables, with low-molecular-weight polymers having a less potent effect on product yields, and concentrations greater than 1 g L^{-1} showing no significant improvements.

12.2.6.5 Proteins

Proteins are added to cell suspension cultures as stabilizing agents. They are thought to function in a variety of ways, for example by acting as substrates for proteases, and by coating and adsorbing to exposed internal surfaces of the culture vessel, thereby keeping the target recombinant protein in solution. Both gelatin and BSA have been used in this manner, and have resulted in markedly increased product yields (James et al. 2000; Kwon et al. 2003b; Lee et al. 2002; Wongsamuth and Doran 1997). For example, the addition of 2% gelatin to a tobacco suspension cell culture producing interleukin-12 resulted in a fourfold increase in the yield (Kwon et al. 2003b), while the addition of 5% gelatin to cell suspensions expressing GM-CSF resulted in a 4.6-fold yield increase (Lee et al. 2002). Similarly, the addition of BSA to tobacco cell suspensions producing GM-CSF resulted in a twofold yield increase in the medium (James et al. 2000). However, the positive effects of these proteins must be balanced against other potentially negative factors. Gelatin has been shown to inhibit the growth of some cells, which could counteract the yield increases resulting from increased protein stability (Wongsamuth and Doran 1997). Both gelatin and BSA have the potential to interfere with downstream processing, which could reduce the quality and quantity of the final product. Finally, as BSA is derived from animal sources, there is a small risk that mammalian pathogens could be introduced into pharmaceutical products expressed in plant cells.

12.2.7
Other Properties of the Culture Medium

The use of additives is only one way in which product yields can be improved in plant cell cultures. Even simple modifications, such as changing the pH or osmolarity of the medium, can have positive effects. The pH of the medium influences the electrostatic properties of expressed proteins, and has the potential to affect protein folding and solubility, and ultimately stability. However, where the impact of pH changes has been assessed in plant cell and tissue cultures, it appears that moderate changes are well tolerated. The osmolarity of the medium has a much more significant effect on production, and can be controlled effectively through the addition of substances such as mannitol and sorbitol. It is thought that hyperosmolarity increases the rate at which proteins are secreted from the cells, but it may also influence protein synthesis, as reports have indicated that both intracellular and extracellular levels of recombinant protein can be enhanced when osmoticum is present. When tobacco cells expressing a recombinant antibody were shifted to a hyperosmolar medium, the antibody yields increased by about 2.5-fold compared to standard medium (Tsoi and Doran 2001). The inducible α-amylase promoter system used in rice suspension cells provides a further interesting example of the osmolarity effect. The activity of the α-amylase promoter is controlled by regulating the supply of sucrose in the medium, and promoter induction is achieved by sucrose starvation. However, while this results in elevated transcription, the absence of sucrose reduces the osmolarity of the medium and reduces the amount of protein recovered. This problem has been overcome in the case of α_1-antitrypsin production by supplementing the medium with mannitol, which increases osmolarity without repressing the promoter (Terashima et al. 1999 a).

Another property that has a significant effect on recombinant protein yields is the concentration of manganese ions in the culture medium. By omitting cobalt, copper, manganese, and zinc both individually and in combination, it was shown that media lacking manganese but containing the other three metals resulted in a significant increase in antibody levels. The reasons for this are unclear (Tsoi and Doran 2001).

12.2.8
Culture and Harvest Processes

As discussed above, improvements in product yields can be gained not only by modifying the media components and conditions, but also by modifying the culture process itself, including the use of continuous harvesting techniques that remove the product from the culture medium before degradation becomes a problem.

During the 1980s, large-scale plant cell cultures were used primarily for the production of secondary metabolites. Therefore, culture process development was directed towards dedicated bioreactor designs suitable for very large working volumes or providing adequate agitation at high cell densities with low shear stress (Singh and Curtis 1995; Tanaka et al. 1983). As the focus shifted towards the production of recombinant proteins, and especially those with therapeutic applications, regulatory considerations became important as well. It was realized that standard technologies might simplify the approval of each production process.

The use of plant cell culture for the production of recombinant proteins also had an impact on the fermentation strategies that could be applied. Generally, all major fermentation strategies (batch, fed-batch, repeated batch or draw/fill, continuous culture and perfusion culture) can be applied to plant cell culture (Chang and Sim 1995; ten Hoopen et al. 1992; van Gulik et al. 2001). However, since plant cells are difficult to clone and genetic stability is difficult to maintain, long-term cultivation strategies such as continuous and perfusion processes may not applicable. Therefore, batch and fed-batch processes will most likely dominate the industrial use of plant suspension cultures for the production of recombinant proteins. Another issue arising from this consideration is the absolute necessity to establish cryopreservation techniques and to generate master working cell banks for commercial processes. Thus, process development for plant cell cultures has focused on the optimization of media, medium supplements and physical parameters.

The most commonly used media for plant cell cultures are MS medium (Murashige and Skoog 1962), Gamborg's B5 medium (Gamborg et al. 1968) and White's medium (White 1963). The composition of these has been modified in order to optimize plant cell cultures with respect to growth, and the production of secondary metabolites (Gao et al. 2000; Raval et al. 2003) or recombinant proteins (James et al. 2000). Media additives, which can be used to optimize recombinant protein production, were discussed earlier in this section.

Another way to improve fermentation yields is continuous product removal by the addition of an inert solid absorber to the fermentation broth, or by using an absorber in a bypass. In the case of recombinant antibodies, this can be achieved simply by cycling the medium through an in-line Protein G column during the fermentation, and eluting the bound antibody at the end of the production cycle. Using this procedure, the yields of recombinant antibody can be increased up to eightfold, depending on variables such as the flow rate, culture period, and column pH (James et al. 2002). Similar principles can be applied to any protein expressed with a suitable affinity tag; for example, an epitope tag, a specific affinity partner such as glutathione-S-transferase, or a His_6 tag. The latter has been used to purify GM-CSF produced in tobacco suspension cells by continuous cycling of the culture medium through a column packed with metal affinity resin, resulting in a twofold increase in yields compared to cultures that were harvested by conventional methods (James et al. 2000, 2002).

One aspect of plant cell culture that plays a critical role in determining product yields is the rate of oxygen transfer. Inadequate aeration has a limiting effect on cell growth, and has been shown to limit product yields in, for example, an antibody-producing tobacco cell line grown in shake-flasks (Liu and Lee 1999). However, while increased aeration in a stirred 5-L bioreactor helped to increase the protein yields, excessive aeration caused foaming and had a counterproductive effect on both cell growth and product yields (Liu and Lee 1999).

12.3
Examples of Recombinant Proteins Produced in Plant Cell Suspension Cultures

The number of proteins that have been produced using plant cells and tissue cultures is moderate compared to the many proteins that have been expressed in transgenic plants (Twyman et al. 2003; Fischer et al. 2004). However, because of the advantages of plant cell cultures in terms of containment, batch-to-batch consistency and cGMP compliance, many of the proteins that have been produced in cell culture are potential diagnostics or therapeutics, with the focus on recombinant antibodies, enzymes and cytokines (Table 12.1). In most cases, the yields achieved in cell culture have been insufficient to make commercial production feasible at the current time. However, this situation may change with improvements in expression construct design and bioprocess development, and as the critical shortage in production capacity in other fermentor systems drives manufacturers to seek alternative processes.

Tobacco remains the most popular host species for cell culture-based protein production, reflecting the fact that this is a well-established and well-characterized model system with proven expression cassettes and facile transformation procedures. However, the rice cell system has been used on several occasions and benefits from the availability of the inducible α-amylase promoter. This has been used for the production of GM-CSF (Shin et al. 2003), α_1-antitrypsin (Huang et al. 2001; Terashima et al. 1999a,b; Trexler et al. 2002), and lysozyme (Huang et al. 2003). A direct comparison of rice and tobacco showed that GM-CSF accumulated to significantly higher levels in rice (Shin et al. 2003). Other proteins have been produced in soybean and tomato suspension cells (Smith et al. 2002; Kwon et al. 2003a). In many cases, recombinant proteins have been produced efficiently for extended periods in suspension cells (e.g., Magnuson et al. 1996), but in others the yields have decreased over time, probably as a result of somaclonal variation, gene silencing, or other forms of genetic instability (e.g., Smith et al. 2002).

12.4
Methodology Using Model Tobacco Cell Suspension Lines to Produce Recombinant Antibodies

We routinely use *Agrobacterium*-mediated transformation to generate transgenic tobacco suspension cells. For cultivar Petit Havana SR1, the cells are derived from transgenic callus, produced from leaf disks of transgenic plants. In contrast, the BY-2 cell line has suitable growth characteristics for direct transformation by co-cultivation of suspension cells with *A. tumefaciens* (An 1985). The latter procedure has the significant advantage that transient expression of the foreign gene can be detected 2 to 3 days after co-cultivation, the selection of transgenic callus on the basis of kanamycin resis-

Tab. 12.1 Production of recombinant proteins in plant suspension cells.

Host species	Protein	Promoter	Leader sequence	Maximum yield	Reference
Rice	Human α_1-antitrypsin	Rice α-amylase	Rice α-amylase	4.6–5.7 mg g^{-1} dry cell weight	Terashima et al. (1999 b)
	GM-CSF	Rice α-amylase	Rice α-amylase	150 µg L^{-1} supernatant	Shin et al. (2003)
	scFv recognizing carcinoembryonic antigen	Maize ubiquitin-1	Murine Ig heavy and light chain	3.8 µg g^{-1} callus fresh cell weight	Torres et al. (1999)
N. tabacum cv. NT-1	Bryodin 1	CaMV 35S	Extensin	30 mg L^{-1} supernatant	Francisco et al. (1997)
	Interleukin-2	CaMV 35S	Interleukin-2	100 µg L^{-1} supernatant, 800 µg L^{-1} fresh cell weight	Magnuson et al. (1998)
	Interleukin-4	CaMV 35S	Interleukin-4	180 µg L^{-1} supernatant, 280 µg L^{-1} fresh cell weight	Magnuson et al. (1998)
N. tabacum cv. Bright Yellow 2	Human erythropoietin	CaMV 35S	Human erythropoietin	26 ng g^{-1} TSP	Matsumoto et al. (1995)
	GM-CSF	Enhanced CaMV 35S	Tobacco etch virus	150 µg g^{-1} fresh cell weight, 250 µg L^{-1} supernatant	James et al. (2000)
N. tabacum cv. Petit Havana SR1	GM-CSF	Enhanced CaMV 35S	GM-CSF	180 µg L^{-1} supernatant, 738 µg L^{-1} supernatant (after addition of gelatin)	Lee et al. (2002)
	Interleukin-12	Enhanced CaMV 35S	Interleukin-12	175 µg L^{-1} supernatant	Kwon et al. (2003)
N. tabacum	Human serum albumin	Enhanced CaMV 35S	PR-S	250 µg g^{-1} TSP	Sijmons et al. (1990)
	G-CSF	Enhanced CaMV 35S		105 µg L^{-1} supernatant	Hong et al. (2002)
	Murine IgG$_1$, recognizing *Streptococcus mutans* surface antigen	CaMV 35S	Murine IgG$_1$ heavy and light chain	7.4 mg L^{-1} supernatant	Sharp and Doran (2001)

Abbreviations: CaMV 35S = Cauliflower mosaic virus 35S RNA promoter; U = enzyme units; G-CSF = granulocyte-colony stimulating factor; GM-CSF = granulocyte macrophage-colony stimulating factor; Ig = immunoglobulin; TSP = total soluble protein; scFv = single chain fragment variable.

tance takes approximately 3 weeks, and the initiation of suspension cell lines from this callus can be performed in less than 2 months. In comparison, leaf disc transformation, selection, regeneration and assay of transgenic Petit Havana SR1 plants, followed by the establishment of callus and suspension cultures requires a whole year. However, we generally obtain higher production levels in cell suspensions derived from transgenic Petit Havana SR1 plants.

As discussed above, proteins can be targeted to accumulate within a subcellular compartment, or they can be secreted into the culture medium. The preferred approach is secretion, as this makes cell disruption unnecessary, and routinely we capture secreted proteins from the culture supernatant following free diffusion from the apoplast, or release them from the cell by mild enzymatic digestion of the cell wall with pectinase.

We have produced a wide panel of recombinant antibodies in tobacco-suspension cells, including full-size immunoglobulins, Fab fragments, single chains Fv fragments, bispecific antibody fragments and fusion proteins. Using our standard plant-expression vector, we routinely obtain expression levels of 2 to 20 µg of recombinant antibody per gram of fresh cell weight. Using the rapidly dividing tobacco BY-2 cell line for fermentation, we have scaled-up the cultivation of transgenic suspension cells to a working volume of 30 L. With a 10% (v/v) inoculum, fermentation times of 150 h resulted in a yield of 7.5 kg fresh cell weight, corresponding to 0.4 kg dry weight. Some of the production parameters for the production of recombinant anti-

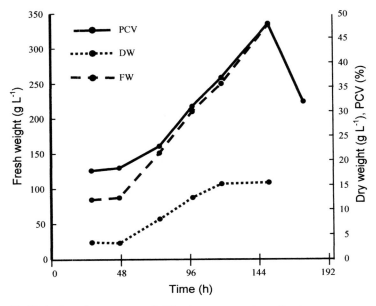

Fig. 12.2 Growth parameters of BY-2 tobacco suspension cells, showing variation in dry weight (DW), fresh weight (FW) and packed cell volume (PCV) over time. Note that, towards the end of the growth cycle, cells take up water and vacuolize without further increase in DW.

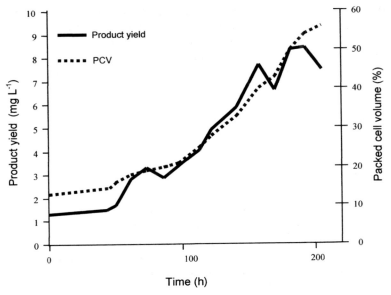

Fig. 12.3 The packed cell volume (PCV) of cultured BY-2 cells corre-lates well with the total product levels, allowing the harvest time to be determined using the PCV as an easily accessible parameter.

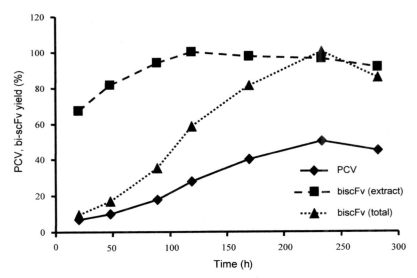

Fig. 12.4 Production of a bispecific scFv antibody in a 30-L reactor, showing the packed cell volume (PCV) and the percentage maximum product yields in the whole culture and the cell extract. In the extract, the maximum yield (100%) = 2.3 μg g^{-1} of cells. In the total culture, the maximum yield (100%) = 2.5 mg L^{-1}.

bodies in BY-2 cells are shown in Figures 12.2–12.4. Apart from some growth on the fermentor walls and impeller axis, no engineering problems were experienced. The equipment was a standard stirred-tank reactor with a three-bladed impeller running at 50 r.p.m., and an aeration rate of 0.1 volumes of air for every volume of culture medium per minute. Compared with other expression systems, the major differences in purifying recombinant proteins from plant suspension cells arise in the very first steps of the procedure. If the protein of interest is contained in the culture supernatant, removal of cell material can be achieved easily by vacuum filtration followed by clarification of the filtrate before the initial purification steps. However, if the target protein is located within the cell, a suitable method for gentle and efficient cell disruption is essential for large-scale protein production. Mechanical cell-disruption devices such as bead mills, although very efficient, give rise to problems due to heat generation, disruption of subcellular organelles accompanied by liberation of noxious chemicals (alkaloids, phenolics), and the generation of fine cell debris, which can be difficult to remove. In general, when disruption of the cells is necessary, large buffer volumes and various additives such as PVP, dithiothreitol (DTT), ascorbic acid, and EDTA are used to counteract oxidation and proteolytic activity, although these can limit the choice of initial purification methods. For example, using DTT or EDTA is incompatible with immobilized metal-ion affinity chromatography (IMAC). Our preferred methods for protein extraction are either cell sonication or cell-wall digestion using technical-grade pectinase (Fischer et al. 1999). Enzymatic digestion has the advantage that it quantitatively releases secreted full-size antibodies that cannot cross the cell wall. After extraction, bulk cell material can be removed by centrifugation or filtration through cheesecloth, and the filtrate clarified by cross-flow or hollow-fiber filtration. Alternatively, we use expanded-bed chromatography (Galliot et al. 1990) to process the unclarified filtrate. For the concentration of normally highly dilute recombinant antibody solutions and efficient initial purification, we use affinity chromatography as a first capturing step. We use Protein A or Protein G, antigen affinity or IMAC depending on the expressed recombinant proteins and, within the limitations mentioned above, have never encountered problems specific to the processing of plant suspension cell extracts. To minimize the time that the target proteins are exposed to the cell-extract components, the use of chromatography media allowing high flow rates is recommended. Using affinity chromatography for initial purification followed, if necessary, by an ion-exchange step and gel filtration for final purification, we generally obtain yields of 80% intact functional recombinant antibody from the starting material.

12.5
Conclusions

Plant suspension cell cultures carry most of the advantages of terrestrial plants, and can be used for the production of low or medium amounts of proteins. Although the scale of production is lower than for field plants, major benefits include the ability to produce proteins under cGMP conditions, the ability to isolate proteins continuously from the culture medium, and the ability to use sterile conditions that are ideal for pharmaceutical production. A growing number of recombinant proteins have been expressed in plant cell cultures, but improvements in yield, optimization of downstream processing and demonstration of biological equivalence are required before this platform becomes commercially feasible.

Appendix

Tab. A12.1   Selection of plant suspension cell lines used for the production of recombinant proteins.

Species	Strain	Comments	Source
Tobacco	BY-2	Rapid growth, direct *Agrobacterium*-mediated transformation of cells	Matsumoto et al. (1995)
	NT1	Derived from *Agrobacterium*-transformed leaf discs	Magnuson et al. (1998)
	Petit Havana SR1	Derived from *Agrobacterium*-transformed leaf discs	Lee et al. (2002)
Rice	Taipei 309	Derived from callus transformed by particle bombardment	Huang et al. (2001, 2002)
	Bengal	Derived from callus transformed by particle bombardment	Torres et al. (1999)
Soybean	Williams 82	Transformed by particle bombardment of cells	Smith et al. (2002)
Tomato	Seokwang	Derived from *Agrobacterium*-transformed leaf discs	Kwon et al. (2003)

Tab. A12.2   Selection of plant expression cassettes used in cell suspension cultures.

Promoter	Terminator	Leader peptide	Other elements	Reference
CaMV 35S	CaMV 35S	Native	–	Matsumoto et al. (1995)
CaMV 35S	T7	Native	–	Magnuson et al. (1998)
CaMV 35S	*nos*	Extensin	–	Francisco et al. (1997)
Ubi-1	*nos*	Native	*Ubi*-1 first intron TMV omega sequence KDEL	Torres et al. (1999)
Ramy3D	*Ramy3D*	*Ramy3D*	*Ramy3D* first intron	Terashima et al. (1999b)
Enhanced CaMV 35S	*nos*	Native	TMV omega sequence	James et al. (2000)
CaMV 35S	T7	–	TEV 5′-UTR	James et al. (2000)

References

An G (1985) High efficiency transformation of cultured tobacco cells. Plant Physiol 79: 568–570

Archambault J (1991) Large-scale (20-L) culture of surface-immobilized *Catharanthus roseus* cells. Enzyme Microbiol Technol 13: 882–892

Bardor M, Loutelier-Bourhis C, Paccalet T, Cosette P, Fitchette AC, Vezina LP, Trepanier S, Dargis M, Lemieux R, Lange C, Faye L, Lerouge P (2003) Monoclonal C5–1 antibody produced in transgenic alfalfa plants exhibits a N-glycosylation that is homogenous and suitable for glyco-engineering into human-compatible structures. Plant Biotechnol J 1: 451–462

Bohme C, Schroder MB, JungHeiliger H, Lehmann J (1997) Plant cell suspension culture in a bench-scale fermentor with a newly designed membrane stirrer for bubble-free aeration. Appl Microbiol Biotechnol 48: 149–154

Carpita N, Sabularse D, Montezinos D, Delmer DP (1979) Determination of the pore size of cell walls of living plant cells. Science 205: 1144–1147

Chang, HN, Sim SJ (1995) Extractive plant cell culture. Curr Opin Biotechnol 6: 209–212

Chen MH, Liu LF, Chen YR, Wu HK, Yu SM (1994) Expression of α-amylases, carbohydrate-metabolism, and autophagy in cultured rice cells is coordinately regulated by sugar nutrient. Plant J 6: 625–636

Christou P (1993) Particle gun-mediated transformation. Curr Opin Biotechnol 4: 135–141

Commandeur U, Twyman RM, Fischer R (2003) The biosafety of molecular farming in plants. AgBiotechNet 5: ABN 110

Daniell T, Edwards R (1995) Changes in protein methylation associated with the elicitation response in cell cultures of alfalfa (*Medicago sativa* L). FEBS Lett 360: 57–61

Delmer DP (1979) Plant Physiol 64: 623–629

Desikan R, Hancock JT, Neill SJ, Coffey MJ, Jones OT (1996) Elicitor-induced generation of active oxygen in suspension cultures of *Arabidopsis thaliana*. Biochem Soc Trans 24: 199S

Doran P (1993) Adv Biochem Eng 48: 115–168

Doran PM (2000) Foreign protein production in plant tissue cultures. Curr Opin Biotechnol 11: 199–204

Firek S, Draper J, Owen MRL, Gandecha A,

Cockburn B, Whitelam GC (1993) Secretion of a functional single-chain Fv protein in transgenic tobacco plants and cell suspension cultures. Plant Mol Biol 23: 861–870

Fischer R, Emans N (2000) Molecular farming of pharmaceutical proteins. Transgenic Res 9: 279–299

Fischer R, Emans N, Schuster F, Hellwig S, Drossard J (1999) Towards molecular farming in the future: using plant-cell-suspension cultures as bioreactors. Biotechnol Appl Biochem 30: 109–112

Fischer R, Liao YC, Drossard J (1999) Affinity-purification of a TMV-specific recombinant full-size antibody from a transgenic tobacco suspension culture. J Immunol Methods 226: 1–10

Fischer R, Stoger E, Schillberg S, Christou P, Twyman RM (2004) Plant-based production of biopharmaceuticals. Curr Opin Plant Biol 7: 152–158

Gamborg OL, Miller RA, Ojima K (1968) Nutrient requirements of suspension culture of soya bean root cells. Exp Cell Res 50: 151–158

Gailliot FP, Gleason C, Wilson JJ, Zwarick J (1990) Fluidized-bed adsorption for whole broth extraction. Biotechnol Prog 6: 370–375

Gao WY, Fan L, Paek KY (2000) Yellow and red pigment production by cell cultures of *Carthamus tinctorius* in a bioreactor. Plant Cell Tiss Org Cult 60: 95–100.

Gomord V, Faye L (2004) Posttranslational modification of therapeutic proteins in plants. Curr Opin Plant Biol 7: 171–181

Hilton MG, Rhodes MJC (1990) Growth and hyoscyamine production of hairy root cultures of *Datura stramonium* in a modified stirred tank reactor. Appl Microbiol Biotechnol 33: 132–138

Hoehl U, Upmeier B, Barz W (1988) Appl Microbiol Biotechnol 28: 319–323

Hooker BS, Lee JM, An GH (1990) Cultivation of plant-cells in a stirred vessel – effect of impeller design. Biotechnol Bioeng 35: 296–304

Horsch RB, Fry JE, Hoffmann NL, Eichholtz D, Rogers SG, Fraley RT (1985) A simple and general method for transferring genes into plants. Science 227: 1229–1231

Huang JM, Sutliff TD, Wu LY, Nandi S, Benge K, Terashima M, Ralston AH, Drohan W, Huang N, Rodriguez RL (2001)

Expression and purification of functional human alpha-1-antitrypsin from cultured plant cells. Biotechnol Prog 17: 126–133

Huang JM, Wu LY, Yalda D, Adkins Y, Kelleher SL, Crane M, Lonnerdal B, Rodriguez RL, Huang N (2002) Expression of functional recombinant human lysozyme in transgenic rice cell culture. Transgenic Res 11: 229–239

Hughes EH, Hong SB, Shanks JV, San KY, Gibson SI (2002) Characterization of an inducible promoter system in *Catharanthus roseus* hairy roots. Biotechnol Prog 18: 1183–1186.

Ishikawa A, Yoshihara T, Nakamura K (1994) Jasmonate-inducible expression of a potato cathepsin-D inhibitor-GUS gene fusion in tobacco cells. Plant Mol Biol 26: 403–414

James E, Mills DR, Lee JM (2002) Increased production and recovery of secreted foreign proteins from plant cell cultures using an affinity chromatography bioreactor. Biochem Eng J 12: 205–213

James EA, Wang CL, Wang ZP, Reeves R, Shin JH, Magnuson NS, Lee JM (2000) Production and characterization of biologically active human GM-CSF secreted by genetically modified plant cells. Protein Express Purif 19: 131–138

Ketchum REB, Gibson DM, Croteau RB, Shuler L (1999) The kinetics of taxoid accumulation in cell suspension cultures of *Taxus* following elicitation with methyl jasmonate. Biotechnol Bioeng 62: 91–105

Kieran PM, MacLoughlin PF, Malone DM (1997) Plant cell suspension cultures: some engineering considerations. J Biotechnol 59: 39–52

Koncz C, Schell J (1986) The promoter of T_L-DNA gene 5 controls the tissue-specific expression of chimeric genes carried by a novel type of *Agrobacterium* binary vector. Mol Gen Genet 204: 383–396

Kwon TH, Kim YS, Lee JH, Yang MS (2003a) Production and secretion of biologically active human granulocyte-macrophage colony stimulating factor in transgenic tomato suspension cultures. Biotechnol Lett 25: 1571–1574

Kwon TH, Seo JE, Kim J, Lee JH, Jang YS, Yang MS (2003b) Expression and secretion of the heterodimeric protein interleukin-12 in plant cell suspension culture. Biotechnol Bioeng 81: 870–875

LaCount W, An GH, Lee JM (1997) The effect of polyvinylpyrrolidone (PVP) on the heavy chain monoclonal antibody production from plant suspension cultures. Biotechnol Lett 19: 93–96

Lee JH, Kim NS, Kwon TH, Jang YS, Yang MS (2002) Increased production of human granulocyte-macrophage colony stimulating factor (hGM-CSF) by the addition of stabilizing polymer in plant suspension cultures. J Biotechnol 96: 205–211

Lindsey K, Jones MGK (1987) Transient gene-expression in electroporated protoplasts and intact cells of sugar-beet. Plant Mol Biol 10: 43–52

Liu F, Lee JM (1999) Effect of culture conditions on monoclonal antibody production from genetically modified tobacco suspension cultures Biotechnol Bioprocess Eng 4: 259–263

Ma JKC, Drake PMW, Christou P (2003) The production of recombinant pharmaceutical proteins in plants. Nature Rev Genet 4: 794–805

Magnuson NS, Linzmaier PM, Gao JW, Reeves R, An GH, Lee JM (1996) Enhanced recovery of a secreted mammalian protein from suspension culture of genetically modified tobacco cells. Protein Express Purif 7: 220–228

Matsumoto S, Ikura K, Ueda M, Sasaki R (1995) Characterization of a human glycoprotein (erythropoietin) produced in cultured tobacco cells. Plant Mol Biol 27: 1163–1172

McCormack BA, Gregory ACE, Kerry ME, Smith C, Bolwell GP (1997) Purification of an elicitor-induced glucan synthase (callose synthase) from suspension cultures of French bean (*Phaseolus vulgaris* L.): purification and immunolocation of a probable Mr-65 000 subunit of the enzyme. Planta 203: 196–203

Meijer JJ, ten Hoopen HJG, Vangameren YM, Luyben KCAM, Libbenga KR (1994) Effects of hydrodynamic stress on the growth of plant-cells in batch and continuous-culture. Enzyme Microbiol Technol 16: 467–477

Mirjalili N, Linden JC (1996) Methyl jasmonate induced production of taxol in suspension cultures of *Taxus cuspidata*: ethylene interaction and induction models. Biotechnol Prog 12: 110–118

Murashige T, Skoog F (1962) A revised medium for rapid growth and bio assay for tobacco tissue cultures. Plant Physiol 15: 473–497

Nagata T, Nemoto Y, Hasezawa S (1992) Tobacco BY-2 cell line as the HeLa cell in the cell biology of higher plants. Int Rev Cytol 132: 1–30

Offringa R, Degroot MJA, Haagsman HJ, Does MP, Vandenelzen PJM, Hooykaas PJJ (1990) Extrachromosomal homologous recombination and gene targeting in plant cells after *Agrobacterium*-mediated transformation. EMBO J 9: 3077–3084

Padidam M (2003) Chemically regulated gene expression in plants. Curr Opin Plant Biol 6: 169–177

Porta C, Lomonossoff GP (2002) Viruses as vectors for the expression of foreign sequences in plants. Biotechnol Genet Eng 19: 245–291

Raval KN, Hellwig S, Prakash G, Ramos-Plasencia A, Srivastava A, Buchs J (2003) Necessity of a two-stage process for the production of azadirachtin-related limonoids in suspension cultures of *Azadirachta indica*. J Biosci Bioeng 96: 16–22

Sakamoto K, Iida K, Sawamura K, Hajiro K, Asada Y, Yoshikawa T, Furuya T (1993) Effects of nutrients on anthocyanin production in cultured cells of *Aralia cordata*. Phytochemistry 33: 357–360

Sato K, Nakayama M, Shigeta J (1996) Culturing conditions affecting the production of anthocyanin in suspended cell cultures of strawberry. Plant Sci 113: 91–98

Schlatmann JE, ten Hoopen HJG, Heijnen JJ (1996) Large-scale production of secondary metabolites by plant cell cultures. In: Plant cell culture secondary metabolism: Toward industrial application (DiCosmo F, Misawa M, Eds), CRC Press, Boca Raton, FL, pp 11–52

Seki M, Ohzora C, Takeda M, Furusaki S (1997) Taxol (Paclitaxel) production using free and immobilized cells of *Taxus cuspidata* Biotechnol Bioeng 53: 214–219

Sharp JM, Doran PM (2001) Strategies for enhancing monoclonal antibody accumulation in plant cell and organ cultures. Biotechnol Progr 17: 979–992

Shin YJ, Hong SY, Kwon TH, Jang YS, Yang MS (2003) High level of expression of recombinant human granulocyte-macrophage colony stimulating factor in transgenic rice cell suspension culture. Biotechnol Bioeng 82: 778–783

Singh G, Curtis WR (1995) Reactor design for plant cell suspension cultures. In: Biotechnological applications of plant cultures (Shargool PD, Ngo TT, Eds), CRC Press, Boca Raton, FL, pp 151–183

Smith ML, Mason HS, Shuler ML (2002) Hepatitis B surface antigen (HBsAg) expression in plant cell culture: Kinetics of antigen accumulation in batch culture and its intracellular form. Biotechnol Bioeng 80: 812–822

Tanaka H, Nishijima F, Suwa M, Iwamoto T (1983) Rotating drum fermentor for plant-cell suspension-cultures. Biotechnol Bioeng 25: 2359–2370

Ten Hoopen HJG, van Gulik WM, Heijnen JJ (1992) Continuous culture of suspended plant cells. In Vitro Cell Dev Biol Plant 28: 115–120

Terashima M, Ejiri Y, Hashikawa N, Yoshida H (1999a) Effect of osmotic pressure on human alpha(1)-antitrypsin production by plant cell culture. Biochem Eng J 4: 31–36

Terashima M, Murai Y, Kawamura M, Nakanishi S, Stoltz T, Chen L, Drohan W, Rodriguez RL, Katoh S (1999b) Production of functional human alpha(1)-antitrypsin by plant cell culture. Appl Microbiol Biotechnol 52: 516–523

Torres E, Vaquero C, Nicholson L, Sack M, Stoger E, Drossard J, Christou P, Fischer R, Perrin Y (1999) Rice cell culture as an alternative production system for functional diagnostic and therapeutic antibodies. Transgenic Res 8: 441–449

Trexler MM, McDonald KA, Jackman AP (2002) Bioreactor production of human alpha(1)-antitrypsin using metabolically regulated plant cell cultures. Biotechnol Progr 18: 501–508

Tsoi BMY, Doran PM (2002) Effect of medium properties and additives on antibody stability and accumulation in suspended plant cell cultures. Biotechnol Appl Biochem 35: 171–180

Twyman RM, Stoger E, Schillberg S, Christou P, Fischer R (2003) Molecular farming in plants: host systems and expression technology. Trends Biotechnol 21: 570–578

Van Der Heijden R, Verpoorte R, ten Hoopen HJG (1989) Cell and tissue cultures of *Catharanthus roseus* (L) Don G. – A literature survey. Plant Cell Tiss Org Cult 18: 231–280

van Gulik WM, ten Hoopen HJG, Heijnen JJ (2001) The application of continuous culture for plant cell suspensions. Enzyme Microbiol Tech 28: 796–805

Wahl MF, An GH, Lee JM (1995) Effects of dimethylsulfoxide on heavy-chain monoclonal antibody production from plant cell culture. Biotechnol Lett 17: 463–468

WEN WS (1995) Appl Biochem Biotechnol 50: 189–216

WHITE PR (1963) The cultivation of animal and plant cells. Ronald Press, NY

WONGSAMUTH R, DORAN PM (1997) Production of monoclonal antibodies by tobacco hairy roots. Biotechnol Bioeng 54: 401–415

YU SX, KWOK KH, DORAN PM (1996) Effect of sucrose, exogenous product concentration, and other culture conditions on growth and steroidal alkaloid production by *Solanum aviculare* hairy roots. Enzyme Microbiol Technol 18: 238–243

YUKIMUNE Y, TABATA H, HIGASHI Y, HARA Y (1996) Methyl jasmonate-induced overproduction of paclitaxel and baccatin III in *Taxus* cell suspension cultures. Nature Biotechnol 14: 1129–1132

13

Wide-Range Integrative Expression Vectors for Fungi, based on Ribosomal DNA Elements

Jens Klabunde, Gotthard Kunze, Gerd Gellissen, and Cornelis P. Hollenberg

List of Genes

Gene	Encoded gene product
ALEU2	Arxula β-isopropyl malate dehydrogenase
Conphys	phytase (Aspergillus niger)
FMD	formate dehydrogenase (H. polymorpha)
GFP	green fluorescent protein (Aequorea victoria)
HARS	autonomously replicating sequence (H. polymorpha)
HSA	human serum albumin
hph	hygromycin B resistance (E. coli)
Insulin	insulin (human)
lacZ	β-galactosidase (E. coli)
MFα1	mating factor MFα1 (S. cerevisiae)
MOX	methanol oxidase (H. polymorpha)
PHO5	acid phosphatase (S. cerevisiae)
TEF1	elongation factor 1α
TPS	trehalose-6-phosphate synthase (H. polymorpha)
URA3 (ODC1)	ornithidine decarboxylase (S. cerevisiae)
18S rDNA	18S ribosomal RNA (H. polymorpha)
25S rDNA	25S ribosomal RNA (A. adeninivorans)

Abbreviations

ETS	external transcribed spacer
IFN-β	human interferon beta
IFN-γ	human interferon gamma
ITS	internal transcribed spacer
NTS	non-transcribed spacer
rDNA	ribosomal DNA

Production of Recombinant Proteins. Novel Microbial and Eucaryotic Expression Systems. Edited by Gerd Gellissen
Copyright © 2005 WILEY-VCH Verlag GmbH & Co. KGaA, Weinheim
ISBN: 3-527-31036-3

13.1
Why is a Wide-range Expression Vector Needed?

In the previous chapters a plethora of expression platforms has been presented, ranging from bacteria, yeasts and filamentous fungi to cells of higher eukaryotes. All of these platforms have particular favorable characteristics. Specific product examples have been presented that attest to the advantages of the individual system. Some of the systems presented are distinguished by a growing track record as established producers of valuable proteins that have already reached the market. Other newly defined systems have yet to establish themselves, but demonstrate great potential for industrial applications. However, all systems have drawbacks and limitations: sometimes, attempts to produce a heterologous protein fail completely; in other cases, productivity, secretion or modification and processing are severely impaired, thereby preventing the development of a competitive production process or a marketable product. Criteria such as ease of genetic manipulation, safety, economic fermentation and process design, and the need for certain post-translational modifications must be taken into account in selecting a particular system (see Chapter 1). However, it is evident that no single system is optimal for all proteins. Hence, predictions for a successful strain and process development can only be made to a certain extent, and misjudgments cannot be excluded. This in turn means that the initial selection may result in costly time- and resource-consuming failures. It would therefore be useful if one could assess several selected organisms in parallel for criteria such as appropriate protein processing or secretion in a given case. The availability of a vector that could be targeted to the various platform candidates would greatly facilitate such a comparison. While it seems very unlikely that it will be possible to design a vector that is suited for the whole range of organisms and cells selected for this book, this goal might be feasible for related organisms. The yeasts and filamentous fungi include a great diversity of organisms, as is already evident among the six examples selected for this book. In general, fungi are excellent hosts for the production of recombinant proteins, as detailed in the previous chapters. They offer the desired ease of genetic manipulation and rapid growth to high cell densities (Romanos et al. 1992; Heinisch and Hollenberg 1993; Sudbery 1996; Gellissen 2000, 2002). As eukaryotes, they are able to perform multiple post-translational modifications, thus producing even complex foreign proteins that are often identical or very similar to native products from plant or mammalian sources (Ruetz and Gros 1994; Gilbert et al. 1994; Wittekindt et al. 1995; Vozza et al. 1996; Gellissen 2000, 2002; Valenzuela et al. 1982; Sudbery 1996). Nevertheless, these organisms exhibit differences in productivity, processing or glycosylation, although few examples are available of the production of the same protein in a range of fungal species. A few such differences are briefly listed in the following; for a more detailed description, the reader is referred to the previous chapters in which the individual fungal systems are described. Some important disadvantages of the traditional *Saccharomyces cerevisiae* system quickly became apparent, which limit its general use in biotechnology. Glycoproteins are often over-glycosylated, and terminal mannose residues in N-linked glycans are added by an α-1,3 bond which is suspected to be allergenic (Jigami and Odani 1999; Guengerich et al. 2004; see Chapters 6 and 7 on *Hansenula polymorpha* and *Pichia pastoris*, respectively). The narrow substrate specificity of *S. cerevisiae* hampers fermentation design (Bruinenberg 1986 ; Romanos et al. 1992). In *Arxula adeninivorans*, patterns of *O*-glycosylation vary, depending on morphological status (Wartmann et al. 2002 a). One particular hydrophobic protein has been found to be secreted by this species, but not by *H. polymorpha* (unpublished results). The two methylotrophic species, *H. polymorpha* and *P. pastoris*, differ in their methanol requirement for the activation of promoters derived from genes of the methanol pathway.

These few arbitrary examples already illustrate the necessity of carefully considering a range of fungal organisms before deciding on an expression platform.

13.2
Which Elements are Essential for a Wide-range Expression Vector?

The design of a vector suited for a wide range of fungal organisms must meet several prerequisites. Such a plasmid must contain a targeting element suitable for all test species. The promoter that drives heterologous gene expression must be functional in all these organisms. The vector/host system must employ a dominant selection marker or a sequence that can complement the auxotrophy in all selected organisms.

Some *A. adeninivorans* and *H. polymorpha*-derived sequences fulfill all of these criteria. For selection, the *A. adeninivorans*-derived *LEU2* gene or an *E. coli*-derived resistance marker (*hph*) (Wartmann et al. 2003; Rösel and Kunze 1998) that confers resistance against hygromycin were chosen. rDNA is an obvious universal target for integration, and rDNA targeting has been described for a range of yeast species, including *S. cerevisiae* (Lopes et al. 1989, 1991), *Kluyveromyces lactis* (Bergkamp et al. 1992), and *Yarrowia lipolytica* (Le Dall et al. 1994; see also Chapter 8). However, only recently have conserved rDNA sequences of both organisms been defined as targeting elements with appropriate characteristics. For the control of expression, an *A. adeninivorans*-derived *TEF1* promoter was chosen (Rösel and Kunze 1995). Before describing the development and the application of these vectors, we first focus in the following sections on the structure of the rDNA and its suitability as a target for the integration of foreign DNA.

13.3
Structure of the Ribosomal DNA and its Utility as an Integration Target

All living cells follow the same basic principles for protein biosynthesis, and the ribosomes constitute the central cellular component of the biosynthetic machinery. Thus, ribosomal components are highly conserved in nature. In growing cells, ribosomes can contribute up to 15 % of the total cellular mass (Long and Dawid 1980). This requires that the corresponding genes are present in high copy numbers, are readily accessible for efficient transcription, and share a high degree of homology. All of these features make this genetic system attractive for DNA targeting.

Besides the ribosomal proteins, ribosomes contain conserved RNA species, the ribosomal RNAs (rRNAs). The genes that encode these rRNAs are present in the desired high copy number and are typically clustered as head-to-tail tandem arrays of identical units (rDNA). Transcription occurs in a special compartment in the nucleus, called the nucleolus (Warner et al. 1972). The copy number of rDNA repeats ranges from seven copies in *E. coli*, to 30–200 in yeasts, and up to 600 copies in the clawed toad *Xenopus* (Alberts et al. 2004).

In yeast, heterologous DNA sequences can be stably integrated in high copy number, as described later.

13.3.1
Organization of the rDNA in Yeast

rDNA clusters can vary considerably in size and copy number of the repeats among different yeast species; in addition, the copy number can vary by about twofold in a given strain (Warner 1989). For example, in the baker's yeast *S. cerevisiae*, about 150 copies of a 9.1-kb unit are localized at a single locus on chromosome XII (Petes 1979; Maleszka and Clark-Walker 1993), while *H. polymorpha* contains about 50 copies of an 8.1-kb repeat on chromosome II (Waschk et al. 2002; Klabunde et al. 2003a).

The rDNA repeats are in most instances organized as arrays of rRNA genes and noncoding intergenic spacer regions, as detailed in Figure 13.1A. Each rRNA gene is transcribed into a single precursor molecule by RNA polymerase I. Subsequently, this precursor is processed to form the 18S, 5.8S, and 25S (28S) rRNAs. During this process the external transcribed spacer (ETS) and the internal transcribed spacers (ITS1/2) are excised. Precursor transcription starts at the leader sequence of the 5' ETS and stops at the 3'-end of the 25S (28S) rRNA gene. The intergenic non-tran-

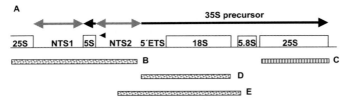

Fig. 13.1 Yeast rDNA unit and the position of derived rDNA-targeting elements. A) The rDNA unit contains genetic elements in the following order: NTS1 (non-transcribed spacer), 5S rRNA, NTS2, the sequence of 35S precursor with the ETS1 (external transcribed spacer), followed by the sequence of the 18S, 5.8S, and 25S rDNA. This gene order has been confirmed for *S. cerevisiae*, *H. polymorpha* and other yeast species; for *A. adeninivorans* the presence of an identical order still has to be shown. B,D,E) represents the position of the targeting segments of *H. polymorpha* and (C) *A. adeninivorans* assessed for transformation. Inclusion of this element in targeting vectors resulted in low transformation efficiency and unstable transformants in case of (D), in high transformation efficiency and stable transformants in case of (B, C, and E).

scribed spacers (NTS1/2) include promoter, enhancer and suppressor elements to control RNA polymerase I (Pol. I)-directed transcription (Udem and Warner 1972). In yeasts, the coding region for the 5S subunit – which is transcribed by RNA polymerase III (Pol. III) – is in most instances located in the NTS located between successive rRNA precursors, as has been shown for *S. cerevisiae* (Johnston et al. 1997), *Ashbya gossypii* (Wendland et al. 1999), and *H. polymorpha* (Klabunde et al. 2002).

13.3.2
Sequence Characteristics of rDNA

The rDNA sequences are highly conserved during evolution. However, this conservation is restricted to the sequences encoding the various rRNA species; the sequences of the noncoding segments can be quite divergent. Therefore, elements derived from coding sequences must be employed in the construction of a vector for wide-range application. Noncoding elements are likely to function in a species-specific manner. For instance, non-conserved sequences that are important for mitotic stability have been described for *S. cerevisiae* (Lopes et al. 1996), and sequences that modulate expressibility have been postulated in *H. polymorpha* (Klabunde et al. 2003a). The vector described in the following sections uses a *H. polymorpha*-derived targeting element that comprises both coding and noncoding sequences. Most of the regulatory elements in the noncoding sequences have been analyzed in *S. cerevisiae*. In light of the low extent of homology and the lack of conclusive experimental data, it can only be assumed that such functional sequences are also present at equivalent locations in the *H. polymorpha* unit.

In *S. cerevisiae*, several *cis*-acting elements and *trans*-acting factors are known co-ordinately to facilitate transcription of rRNA genes by RNA polymerase I (Vogelauer et al. 1998). Special recombinogenic elements, termed host spot elements (*HOT1*), promote homologous recombination (Lin and Keil 1991). The inclusion of such elements in the targeting segment was found to have a major impact on the maintenance and the mitotic stability of heterologous DNA inserts in *S. cerevisiae* (Lopes et al. 1996).

13.4
Transformation Based on rDNA Integration

In the late 1980s, a plasmid type was described that could be stably integrated in high copy number into the rDNA locus of *S. cerevisiae* (Lopes et al. 1989), forming clusters of tandemly repeated plasmid copies (Lopes et al. 1991). The novel feature of these plasmids was the presence of an rDNA fragment instead of the commonly used 2 μm DNA or *CEN/ARS* sequences. It has since been shown that this principle can also be applied to other yeast species, including *Kl. lactis*, *Y. lipolytica*, *C. utilis*, *Phaffia rhodozyma*, *A. adeninivorans*, and *H. polymorpha* (Table 13.1).

In all of these early examples, the rDNA fragments used for targeting were recovered from the species to be transformed. More recent examples of homologous

Tab. 13.1 rDNA targeting in yeast.

Species	Gene product	Marker	rDNA	Copies	Reference(s)
S. cerevisiae	α-amylase	TRP1-d	NTS2	50–100	Nieto et al. (1999)
S. cerevisiae	–	LEU2-d	versch.	>300	Lopes et al. (1989); Lopes et al. (1991, 1996)
Kl. lactis	human lysozyme	HIS3	25S	4–40	Rossolini et al. (1992)
Kl. lactis	α-galactosidase	TRP1	25S	60	Bergkamp et al. (1992)
C. utilis	–	cycloheximide	n.d	6–10	Kondo et al. (1995)
Y. lipolytica	β-galactosidase, extracellular protease	URA3-d	ETS-18S	3–40	Le Dall et al. (1994); Juretzek et al. (2001)
P. ciferrii	endogenous tetra-acetyl phytosphingosine	cycloheximide	NTS2	5–7	Bae et al. (2003)
Ph. rhodozyma	–	geneticin[R]	ETS-18S	60	Wery et al. (1997)
A. adeninivorans	β-galactosidase, HSA, GFP	hph	25S	2–10	Rösel et al. (1998); Wartmann and Kunze (2000); Wartmann et al. (2002 b); Klabunde et al. (2003 a)
	β-galactosidase, GFP, phytase	hph	ETS-18S	n.d.	Wartmann et al. (2003)
	HSA, GFP	ALEU2	25S	1–5	
H. polymorpha	GFP, β-galactosidase, phytase, human insulin	URA3	25S-NTS	5–50	Klabunde et al. (2002)
	phytase	hph	ETS-18S	n.d.	Klabunde et al. (2003 a)
P. stipitis	phytase	hph	ETS-18S	n.d.	Klabunde et al. (2003 a)
H. polymorpha	human interferon β	hph	ETS-18S	1–22	Klabunde (2003 b)
A. adeninivorans	human interferon γ				
P. stipitis	human lactoferrin				
S. cerevisiae					

rDNA-mediated transformation from *H. polymorpha* and *A. adeninivorans* are presented for illustration. In *H. polymorpha*, a 2.4-kb segment of rDNA encompassing parts of the 25S, the complete 5S and the non-transcribed spacer region between the 25S and 18S sequences (Figure 13.1 B) was inserted into conventional integrative *H. polymorpha* plasmids harboring the *S. cerevisiae*-derived *URA3* gene, which permits selection of recombinant strains derived from a *H. polymorpha* (*odc1*) host (Klabunde et al. 2002). Most interestingly, these rDNA plasmids were found to be stably integrated into the rDNA as several independent clusters. This eventually allowed co-integration and co-expression of several (up to three) different genes upon transformation with independent rDNA vectors that contained the same selection marker (Klabunde et al 2002; see Section 13.5).

In *A. adeninivorans*, the 25S rDNA was used as the targeting element (Rösel et al. 1998; Wartmann et al. 2002 b, 2003; Figure 13.1 C). For selection, the plasmids were equipped either with the *E. coli*-derived *hph* gene conferring hygromycin B resistance (pAL-HPH1) or with the *ALEU2* gene (pAL-ALEU2m) as selection marker. All transformants harbored several copies (pAL-HPH1) or a single copy (pAL-ALEU2m) of the vector integrated into the genomic rDNA cluster.

Two recent studies describe the use of elements derived from the two yeasts for wide-range application. An rDNA-based *A. adeninivorans* integration vector was successfully used with a range of alternative yeast species, including *S. cerevisiae*, *Debaryomyces hansenii*, *D. polymorphus*, *H. polymorpha*, and *P. pastoris*. This vector harbors the conserved 25S rDNA sequence for targeting, the *A. adeninivorans*-derived *TEF1* promoter for expression control of the reporter sequence, and the *E. coli*-derived *hph* gene mentioned above for selection of transformants. Due to the presence of the dominant selection marker this vector type was successfully employed for transformation of wild-type strains of all species tested. In all cases the vector was found to be integrated either as a single copy or in a few copies (Terentiev et al. 2004).

This plasmid type was also assessed for heterologous gene expression. For this purpose, a *GFP* reporter gene was employed, which was inserted between the constitutive *A. adeninivorans*-derived *TEF1* promoter and the *S. cerevisiae*-derived *PHO5* terminator for expression control. Again, the resulting plasmid (pAL-HPH-TEF-GFP) was successfully used to transform *A. adeninivorans*, *S. cerevisiae*, *D. hansenii*, *D. polymorphus*, *H. polymorpha*, and *P. pastoris* strains. It was found to be integrated in low copy numbers in all transformants. Transformants were tested for the recombinant product either by Western blot analysis or by fluorescence microscopy. The amounts varied only slightly among various transformants. *A. adeninivorans* transformants contained the highest levels of GFP, followed by the *S. cerevisiae* and *D. polymorphus* strains. The lowest GFP concentrations were detected in the *D. hansenii*, *H. polymorpha*, and *P. pastoris* transformants (Terentiev et al. 2004) (Figure 13.2).

In *H. polymorpha*, a vector of similar design was employed with a newly defined targeting element and the same *hph* selection marker. In the first series of vectors a *H. polymorpha*-derived targeting fragment was incorporated encompassing part of the 18S rDNA, the entire ETS region and 200 bp of the NTS2 region including promoter elements (Klabunde et al. 2003 a; Figure 13.1D). This series further harbored expression cassettes for GFP, a synthetic phytase or β-galactosidase, retaining the *TEF1* promoter and the *PHO5* terminator of the *A. adeninivorans* vector. However, a low transformation frequency and a high level of plasmid loss were observed for both *H. polymorpha* and *A. adeninivorans*. Production of the reporter proteins was stable for 2 to 5 days only.

Therefore, the integration segment was modified by adding an additional 200 bp of the NTS2 sequence (now including the whole putative promoter of the 35S rRNA precursor), the ETS and the full-length 18S coding sequence – resulting in the modified vector pTHpH18lHp (Figure 13.1 E and Figure 13.3). This was successfully em-

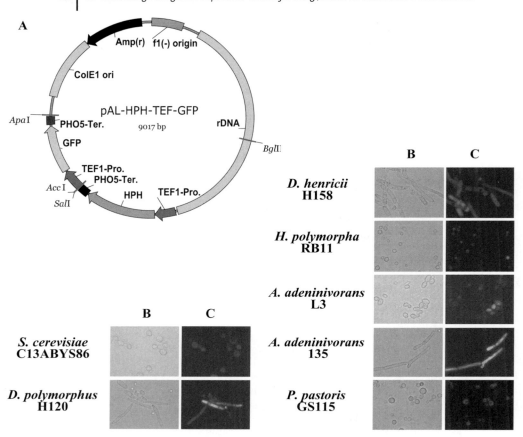

Fig. 13.2 A) Physical map of the expression/integration vector pAL-HPH-TEF-GFP. The vector contains the 25S rDNA sequence of *A. adeninivorans* (rDNA, gray segment) for targeting. For selection, it harbors the *E. coli*-derived *hph* gene inserted between the *A. adeninivorans*-derived *TEF1* promoter (TEF1-Pro., black arrow) and the *S. cerevisiae*-derived *PHO5* terminator (PHO5-Ter., black bar). The vector further contains an expression cassette for the reporter gene in the order *TEF1* promoter – *GFP* – *PHO5* terminator. B, C) Detection of recombinant GFP-producing yeast cells by fluorescence microscopy. Recombinant yeast strains transformed with plasmid pAL-HPH-TEF-GFP were cultured for 48 h in YEPD medium at 30 °C and subsequently analyzed by staining [(B) transmission electron microscopy; (C) GFP-fluorescence].

ployed with another set of yeast species, comprising *H. polymorpha* and *A. adeninivorans*, *S. cerevisiae*, and *P. stipitis*. Again, the transformants of all species contained the plasmid stably integrated into the genomic DNA (Klabunde et al. 2003a). *H. polymorpha*, *A. adeninivorans*, and *P. stipitis* transformants produced high levels of the reporter protein phytase over a period of 30 days (up to 25 U mL^{-1} OD^{-1}).

A

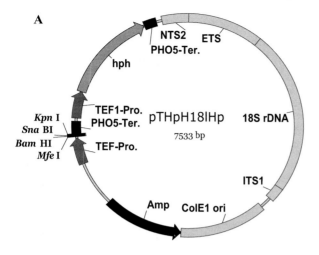

B [U mL^{-1}OD^{-1}]

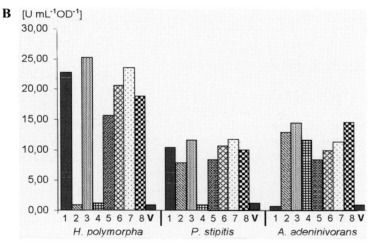

Fig. 13.3 A) Physical map of the expression/integration vector pTHpH18lHp. The vector contains as targeting element the NTS2-ETS-18SrDNA-ITS1 region of the *H. polymorpha* rDNA (NTS2-ETS-18SrDNA-ITS1, gray segment). The selection marker and the components of the expression cassette for the reporter gene are identical to that in pAL-HPH-TEF-GFP as described in Figure 13.3. B) Detection of recombinant phytase-producing *H. polymorpha*, *P. stipitis*, and *A. adeninivorans* transformants. *H. polymorpha* RB11/pTHpH18lHP-Phytase, *P. stipitis* CBS5774/pTHpH18lHP-Phytase and *A. adenini-* *vorans* LS3/pTHpH18lHP-Phytase were cultured for 48 h in YEPD medium at 30 °C and subsequently inspected for phytase secretion. *H. polymorpha* RB11/pTHpH18lHP-Phytase transformants exhibited the highest phytase activity (up to 25 U mL^{-1} OD^{-1}). For unknown reasons, in two transformants (2, 4) only background activity was detectable comparable to the vector control (V). In the *P. stipitis* strains of the CBS5774/pTHpH18lHP type and *A. adeninivorans* strains of the LS3/pTHpH18lHP-Phytase series, lower phytase activities of about 10 U mL^{-1} OD^{-1} were observed.

13.5
rDNA Integration as a Tool for Targeting Multiple Expression Cassettes

Following the observation that the integrated heterologous DNA was present as multiple clusters inserted in rDNA, we tested whether rDNA plasmids – each equipped with the same targeting element and the same selection marker, but bearing different reporter genes – could be integrated simultaneously into the rDNA. The *H. polymorpha* strain RB11 (*odc1*) was co-transformed with a mixture of up to four different plasmids all containing the same (*URA3*) selection marker but different expression cassettes for GFP (*Aequorea victoria*), β-galactosidase (*E. coli*), insulin (*Homo sapiens*), or phytase (*Asp. niger*). Transformation indeed resulted in mitotically stable strains harboring one, two, or three different types of plasmids integrated in the rDNA (Klabunde et al. 2002). The overall copy number of the heterologous DNA was found not to exceed the number of rDNA repeats, irrespective of the number of different plasmids involved. Strains harboring more than a single type of plasmid co-expressed the introduced genes, resulting in functional reporter proteins. Thus, this approach provides an attractive tool for the rapid generation of recombinant strains that simultaneously co-produce several proteins in desired stoichiometric ratios.

13.5.1
Co-integration of Reporter Plasmids in *A. adeninivorans*

Co-integration of various plasmids was also assessed in *A. adeninivorans* following the example of *H. polymorpha*. To do so, the heterologous reporter genes *GFP*, *lacZ*, and *Conphys* were introduced into the wide-range plasmid pTHpH18lHP harboring the *hph* gene as selection marker (Klabunde et al. 2003a). The resulting transformants were mitotically stable. Upon culture for up to 30 days in nonselective medium, no loss or rearrangement of the integrated plasmid sequences was observed. Co-integration was observed after co-transformation in the following combinations: pTHpH18lHP-lacZ–pTHpH18lHP-GFP or pTHpH18lHP–Phytase–pTHpH18lHP-GFP. In the respective recombinant strains both encoded proteins were co-produced over the period analyzed, additionally demonstrating the successful expression of both types of genes (Figure 13.4).

13.5.2
Approaches to the Production of Pharmaceuticals by Co-integration of Different Genes

The *H. polymorpha* expression system excels by a growing track record as a producer of recombinant pharmaceuticals (see Chapter 6). In all cases, traditional vectors were used for the generation of the recombinant strains. Integration of a second gene was achieved by supertransformation of an existing recombinant strain using a different selection marker. In this way, a recombinant whole-cell biocatalyst was generated that co-expresses a gene for spinach glycolate oxidase (*GO*) and a *S. cerevisiae*-derived *CTT1* gene in a given stoichiometric ratio (Gellissen et al. 1996b). Pharmaceutical examples include the production of mixed hepatitis B particles containing

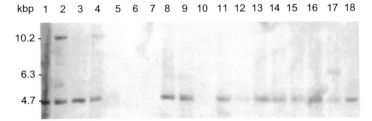

Fig. 13.4 Southern blot analysis of *A. adeninivorans* transformants generated with a mixture of three rDNA-targeting vectors. *A. adeninivorans* LS3 was transformed with a mixture of the plasmids pTHpH18lHp-Phytase (11.3 kb), pTHpH18lHp-GFP (9.3 kb), and pTHpH18lHp-lacZ (10.2 kb) linearized with ApaI. Genomic DNA was isolated from 18 representative transformants, restricted with PvuII, separated on 0.8% agarose, transferred to a nylon membrane and hybridized to a *hph*-fragment-derived probe. The probe hybridizes to a fragment of about 10.2 kb in pTHpH18lHp-lacZ, 6.3 kb in pTHpH18lHp-Phytase, and 4.7 kb in pTHpH18lHp-GFP. Transformants 1, 3, 8–9, 11–16, 18 (lanes 1, 3, 8–9, 11–16, 18) harbored a single integrated pTHpH18lHp-GFP sequence, transformants 2 and 4 (lanes 2, 4) the plasmids pTHpH18lHp-GFP and pTHpH18lHp-lacZ, transformant 17 the plasmids pTHpH18lHp-GFP and pTHpH18lHp-Phytase. Strains 5–7 and 10 (lanes 5–7, 10) showed no signals due to an insufficient DNA load.

the small and the large surface antigen inserted into host-derived membranes (Janowicz et al 1991; Gellissen and Melber 1996a; see also Chapter 15). In another instance, this approach was used to design a strain for the production of authentically processed IFNa-2a (Müller et al. 2002). In all of these cases the approach was both tedious and time-consuming. The co-integration of two or more genes via rDNA targeting offers a possible alternative to the tedious and time-consuming supertransformation approach. To test this, IFN-γ was selected as example where production is impaired by poor secretion and over-glycosylation of the secreted product (Gellissen et al. 2002). The rDNA integration approach was applied to generate strains in which the gene for the cytokine was co-integrated and co-expressed together with candidate genes that could potentially influence and improve secretion and glycosylation. Of several candidate genes investigated, the *H. polymorpha*-derived *CNE1* gene encoding calnexin was found to improve production of the cytokine considerably, when co-expressed with the interferon gene. Overproduction of the calnexin homologue also enabled an improved secretion of a core-glycosylated species of human IFN-γ.

13.6
Conclusions and Perspectives

Integration of heterologous DNA using a single rDNA targeting element has been successfully applied to a range of yeast species, including *S. cerevisiae*, *K. lactis*, *Y. lipo-*

lytica, C. utilis, Ph. rhodozyma, A. adeninivorans, and *H. polymorpha.* Two targeting elements of different design have now been defined, derived from two different yeast species.

For expression control, the *A. adeninivorans*-derived *TEF1* promoter has proved most useful. This promoter element elicits high expression levels in all yeast species tested so far.

Initially the *E. coli*-derived *hph* gene was chosen as a dominant selection marker, thus removing any dependency on the availability of suitable auxotrophic strains as transformation hosts. This gene is now being replaced by a *LEU2* segment, as leucine-auxotrophic strains are now available for most of the species of interest.

The design of the vector thus meets all requirements for wide-range targeting. The vector can be applied to all tested yeast species. Furthermore, it provides for easy and fast integration of several different genes into an individual species. This option could be used to generate strains for the co-production of several enzymes, thereby creating new metabolic pathways, or for the production of complex proteins that consist of several subunits. Most interestingly, it provides a useful tool for the rapid co-expression of genes that can then be assessed for their effects on the synthesis and secretion of a specific product.

The range of organisms that can be employed with the plasmids described here is currently being extended. Preliminary results indicate that filamentous fungi can also be transformed with these vectors.

Acknowledgments

These experimental studies were supported by grants from the Ministry of Economy, Nordrhein-Westfalen (TPW-9910v08), Deutsche Bundesstiftung Umwelt (AZ 13048), and by Funds of the Chemical Industry (GK).

References

ALBERTS B, JOHNSON A, LEWIS J, RAFF M, RO-
BERTS K, WALTER P (2004) Synthese und Bearbeitung nicht-kodierender RNAs. In: Molekularbiologie der Zelle (Alberts B, Ed). Wiley-VCH, Weinheim, Germany, pp 380–389

BAE JH, SOHN JH, PARK CS, RHEE JS, CHOI ES (2003) Integrative transformation system for the metabolic engineering of the sphingoid base-producing yeast *Pichia ciferrii*. Appl Environ Microbiol 69: 812–819

BERGKAMP RJ, KOOL IM, GEERSE RH, PLANTA RJ (1992) Multiple-copy integration of the alpha-galactosidase gene from *Cyamopsis tetragonoloba* into the ribosomal DNA of *Kluyveromyces lactis*. Curr Genet 21: 365–370

BRUINENBERG PM (1986) The NADP(H) redox couple in yeast metabolism. Antonie van Leeuwenhoek 52: 411–429

GELLISSEN G (2000) Heterologous protein production in methylotrophic yeasts. Appl Microbiol Biotechnol 54: 741–750

GELLISSEN G (2002) *Hansenula polymorpha* – Biology and Applications. Weinheim, Wiley-VCH

GELLISSEN G, MÜLLER F, SIENBER H, TIEKE A, JENZELEWSKI V, DEGELMANN A, STRASSER A (2002) Production of cytokines in *Hansenula polymorpha*. In: *Hansenula polymorpha* – Biology and Applications (Gellissen G, Ed). Weinheim, Wiley-VCH, pp 229–254

GELLISSEN G, MELBER K (1996a) Methylotrophic yeast *Hansenula polymorpha* as production organism for recombinant pharmaceuticals. Drug Res 46: 943–948

GELLISSEN G, PIONTEK M, DAHLEMS U, JENZELEWSKI V, GAVAGAN JE, DICOSIMO R, ANTON DL, JANOWICZ ZA (1996b) Recombinant *Hansenula polymorpha* as a biocatalyst: coexpression of spinach glycolate oxidase (*GO*) and the *S. cerevisiae* catalase T (*CTT1*) gene. Appl Microbiol Biotechnol 46: 46–54

GILBERT SC, VAN URK H, GREENFIELD AJ, MCAVOY MJ, DENTON KA, COGHLAN D, JONES GD, MEAD DJ (1994) Increase in copy number of an integrated vector during continuous culture of *Hansenula polymorpha* expressing functional human haemoglobin. Yeast 10: 1569–1580

GUENGERICH L, KANG HA, GELLISSEN G, SUCKOW M (2004) A platform of heterologous gene expression based on the methylotrophic yeast *Hansenula polymorpha*. In: The Mycota II – Genetics and Biotechnology, 2nd edition (Kück U, Ed). Springer-Verlag, Berlin Hamburg, pp 273–287

HEINISCH J, HOLLENBERG CP (1993) Yeasts. In: Biotechnology, Vol 1 – Biological Fundamentals, 2nd edition (Rehm HJ, Reed G, Pühler A, Stadler P, Eds) VCH Verlagsgesellschaft, Weinheim, pp 470–514

JANOWICZ ZA, MELBER K, MERCKELBACH A, JACOBS E, HARFORD N, COMBERBACH M, HOLLENBERG CP (1991) Simultaneous expression of the S and L surface antigens of hepatitis B, and formation of mixed particles in the methylotrophic yeast, *Hansenula polymorpha*. Yeast 7: 431–443

JIGAMI Y, ODANI T (1999) Mannosylphosphate transfer to yeast mannan. Biochim Biophys Acta 1426: 335–345

JOHNSTON M, HILLIER L, RILES L, ALBERMANN K, ANDRE B, ANSORGE W, BENES V, et al. (1997) The nucleotide sequence of *Saccharomyces cerevisiae* chromosome XII. Nature 387: 87–90

JURETZEK T, LE DALL M, MAUERSBERGER S, GAILLARDIN C, BARTH G, NICAUD J (2001) Vectors for gene expression and amplification in the yeast *Yarrowia lipolytica*. Yeast 18: 97–113

KLABUNDE J, DIESEL A, WASCHK D, GELLISSEN G, HOLLENBERG CP, SUCKOW M (2002) Single-step co-integration of multiple expressible heterologous genes into the ribosomal DNA of the methylotrophic yeast *Hansenula poly-*

morpha. Appl Microbiol Biotechnol 58: 797–805

KLABUNDE J, KUNZE G, GELLISSEN G, HOLLENBERG CP (2003a) Integration of heterologous genes in several yeast species using vectors containing a *Hansenula polymorpha*-derived rDNA-targeting element. FEMS Yeast Res 4: 185–193

KLABUNDE J (2003b). Koproduktion pharmazeutischer Proteine und Hilfsfaktoren zur Optimierung mikrobieller Expressionssysteme bei Beschränkung auf ein einziges integratives Vektorsystem. Department of microbiology. Düsseldorf, Heinrich-Heine University, pp 55–94

KONDO K, SAITO T, KAJIWARA S, TAKAGI M, MISAWA N (1995) A transformation system for the yeast *Candida utilis*: use of a modified endogenous ribosomal protein gene as a drug-resistant marker and ribosomal DNA as an integration target for vector DNA. J Bacteriol 177: 7171–7177

LE DALL MT, NICAUD JM, GAILLARDIN C (1994) Multiple-copy integration in the yeast *Yarrowia lipolytica*. Curr Genet 26: 38–44

LIN YH, KEIL RL (1991) Mutations affecting RNA polymerase I-stimulated exchange and rDNA recombination in yeast. Genetics 127: 31–38

LONG EO, DAWID IB (1980) Repeated genes in eukaryotes. Annu Rev Biochem 49: 727–764.

LOPES TS, KLOOTWIJK J, VEENSTRA AE, VAN DER AAR PC, VAN HEERIKHUIZEN H, RAUE HA, PLANTA RJ (1989) High-copy-number integration into the ribosomal DNA of *Saccharomyces cerevisiae*: a new vector for high-level expression. Gene 79: 199–206

LOPES TS, HAKKAART GJ, KOERTS BL, RAUE HA, PLANTA RJ (1991) Mechanism of high-copy-number integration of pMIRY-type vectors into the ribosomal DNA of *Saccharomyces cerevisiae*. Gene 105: 83–90

LOPES TS, DE WIJS IJ, STEENHAUER SI, VERBAKEL J, PLANTA RJ (1996) Factors affecting the mitotic stability of high-copy-number integration into the ribosomal DNA of *Saccharomyces cerevisiae*. Yeast 12: 467–477

MALESZKA R, CLARK-WALKER GD (1993) Yeasts have a four-fold variation in ribosomal DNA copy number. Yeast 9: 53–58

MÜLLER F II, TIEKE A, WASCHK D, MÜHLE C, MÜLLER F I, SEIGELCHIFER M, PESCE A, JENZELEWSKI V, GELLISSEN G (2002) Production of IFNα-2a in *Hansenula polymorpha*. Process Biochem 38: 15–25

NIETO A, PRIETO JA, SANZ P (1999) Stable high-copy-number integration of *Aspergillus oryzae* alpha-amylase cDNA in an industrial baker's yeast strain. Biotechnol Prog 15: 459–466

PETES TD (1979) Yeast ribosomal DNA genes are located on chromosome XII. Proc Natl Acad Sci USA 76: 410–414

ROMANOS MA, SCORER CA, CLARE JJ (1992) Foreign gene expression in yeast: a review. Yeast 8: 423–488

RÖSEL H, KUNZE G (1995) Cloning and characterization of a *TEF1* gene coding for elongation factor 1α from the yeast *Arxula adeninivorans*. Curr Genet 28: 360–366

RÖSEL H, KUNZE G (1998) Integrative transformation of the dimorphic yeast *Arxula adeninivorans* LS3 based on hygromycin B resistance. Curr Genet 33: 157–163

ROSSOLINI GM, RICCIO ML, GALLO E, GALEOTTI CL (1992) *Kluyveromyces lactis* rDNA as a target for multiple integration by homologous recombination. Gene 119: 75–81

RUETZ S, GROS P (1994) Functional expression of P-glycoproteins in secretory vesicles. J Biol Chem 269: 12277–12284

SUDBERY PE (1996) The expression of recombinant proteins in yeasts. Curr Opin Biotechnol 7: 517–524

TERENTIEV Y, PICO AH, BÖER E, WARTMANN T, KLABUNDE J, BREUER U, BABEL W, SUCKOW M, GELLISSEN G, KUNZE G (2004) A wide-range integrative yeast expression vector system based on *Arxula adeninivorans*-derived elements. J Ind Microbiol Biotechnol 31: 223–228

UDEM SA, WARNER JR (1972) Ribosomal RNA synthesis in *Saccharomyces cerevisiae*. J Mol Biol 65: 227–242

VALENZUELA P, MEDINA A, RUTTER WJ, AMMERER G, HALL BD (1982) Synthesis and assembly of hepatitis B virus surface antigen particles in yeast. Nature 298: 347–350

VOGELAUER M, CIOCI F, CAMILLONI G (1998) DNA protein-interactions at the *Saccharomyces cerevisiae* 35 S rRNA promoter and in its surrounding region. J Mol Biol 275: 197–209

VOZZA LA, WITTWER L, HIGGINS DR, PURCELL TJ, BERGSEID M, COLLINS-RACIE LA, LAVALLIE ER, HOEFFLER JP (1996) Production of a recombinant bovine enterokinase catalytic subunit in the methylotrophic yeast *Pichia pastoris*. Biotechnology (NY) 14: 77–81

WARNER JR, KUMAR A, UDEM SA, WU RS (1972) Ribosomal proteins and the assembly of ribosomes in eukaryotes. Biochem J 129: 29P–30P

WARNER JR (1989) Synthesis of ribosomes in *Saccharomyces cerevisiae*. Microbiol Rev 53: 256–271

WARTMANN T, KUNZE G (2000) Genetic transformation and biotechnological application of the yeast *Arxula adeninivorans*. Appl Microbiol Biotechnol 54: 619–624

WARTMANN T, STEPHAN UW, BUBE I, BÖER E, MELZER M, MANTEUFFEL R, STOLTENBURG R, GUENGERICH L, GELLISSEN G, KUNZE G (2002a) Post-translational modifications of the *AFET3* gene product – a component of the iron transport system in budding cells and mycelia of the yeast *Arxula adeninivorans*. Yeast 19: 849–862

WARTMANN T, BÖER E, HUARTO PICO A, SIEBER H, BARTELSEN O, GELLISSEN G, KUNZE G (2002b) High-level production and secretion of recombinant proteins by the dimorphic yeast *Arxula adeninivorans*. FEMS Yeast Res 2: 363–369

WARTMANN T, STOLTENBURG R, BOER E, SIEBER H, BARTELSEN O, GELLISSEN G, KUNZE G (2003) The *ALEU2* gene – a new component for an *Arxula adeninivorans*-based expression platform. FEMS Yeast Res 3: 223–232

WASCHK D, KLABUNDE J, SUCKOW M, HOLLENBERG CP (2002). Characteristics of the *Hansenula polymorpha* genome. In: *Hansenula polymorpha* – biology and applications (Gellissen G, Ed). Wiley-VCH, Weinheim, Germany, pp 95–104

WENDLAND J, POHLMANN R, DIETRICH F, STEINER S, MOHR C, PHILIPPSEN P (1999) Compact organization of rRNA genes in the filamentous fungus *Ashbya gossypii*. Curr Genet 35: 618–625

WERY J, GUTKER D, RENNIERS AC, VERDOES JC, VAN OOYEN AJ (1997) High copy number integration into the ribosomal DNA of the yeast *Phaffia rhodozyma*. Gene 184: 89–97

WITTEKINDT NE, WURGLER FE, SENGSTAG C (1995) Functional expression of fused enzymes between human cytochrome P4501A1 and human NADPH-cytochrome P450 oxidoreductase in *Saccharomyces cerevisiae*. DNA Cell Biol 14: 273–283

14
Comparative Fermentation

Stephan Hellwig, Christoph Stöckmann, Gerd Gellissen, and Jochen Büchs

List of abbreviations

135	Mutant strain of *Arxula adeninivorans* with altered behaviour with respect to dimorphism
AOX1	(promoter of the) alcohol oxidase gene
BMM	Complex medium for small-scale cultures of *Sordaria macrospora*
CM, CCM	Semisynthetic media for small-scale cultures of *Sordaria macrospora*
dO_2	dissolved oxygen concentration (in % of saturation)
DW	dry weight [g L^{-1}]
FTU	Phytase unit, enzyme amount releasing 1 µmol of inorganic phosphate per minute from sodium phytate at pH 5.5 at 37 °C.
GFP	Green fluorescent protein
HCDF	high-cell-density fermentation
HSA	human serum albumin
IDA	imino-diacetic acid
IMAC	immobilized metal affinity chromatography
IPTG	Isopropyl-β-D-thiogalactopyranoside
LB	Luria-Bertani broth
LS3	Wild-type strain of *Arxula adeninivorans*
MES	2-[*N*-Morpholino]ethanesulfonic acid
MM, GM	Synthetic media for small-scale cultures of *Sordaria macrospora*
MOX	(promoter of the) methanol oxidase gene
OD_{600}	optical density at 600 nm
OTR_{max}	Maximum oxygen transfer capacity [mol L^{-1} h^{-1}]
PMA1	(promoter of the) gene encoding the plasma-membrane-residing ATPase
SCP	Single cell protein
STR	stirred tank reactor
SYN6	Synthetic medium for *Hansenula polymorpha*
SYN6-MES	Synthetic SYN6 medium, modified for *Arxula adeninivorans* shake-flask cultures

Production of Recombinant Proteins. Novel Microbial and Eucaryotic Expression Systems. Edited by Gerd Gellissen
Copyright © 2005 WILEY-VCH Verlag GmbH & Co. KGaA, Weinheim
ISBN: 3-527-31036-3

List of abbreviations (continued)

TB	Terrific Broth (an *E. coli* medium)
TEF1	(Promoter of the) gene for elongation factor 1α from *Arxula adeninivorans*
TY	Tryptone Yeast extract (an *E. coli* medium)
vvm	volume per volume per minute (volumetric aeration rate as L L^{-1} min^{-1})
YMM	Synthetic Yeast Minimal Medium for *Arxula adeninivorans*
YMM*	Synthetic Yeast Minimal Medium, modified for *Arxula adeninivorans*

14.1
Introduction

A plethora of expression platforms has been presented in the previous chapters. Some of the systems are distinguished by a growing track record as established production systems for recombinant compounds that have already reached the market. Other newly defined platforms have yet to establish themselves, but demonstrate great potential for industrial applications. All systems exhibit not only advantages, but also drawbacks and limitations. However, the commercial success of a recombinant product does not depend exclusively on the characteristics of the microbial or cellular host and the underlying genetic and biochemical properties – it also depends on the definition of efficient fermentation and downstream procedures. More than twenty years of experience in the production of recombinant proteins has taught us that every combination of a particular gene, control elements, mRNA, target protein and expression host presents unique challenges to establishing effective, robust – and thus industrially applicable – expression strategies (Balbas 2001). Therefore, the definition of an expression system, the feasibility and implications of fermentation and purification procedures, and its upscaling to an anticipated industrial scale should be assessed – at least in theory – at an early stage. Such an assessment requires an integrated approach that combines aspects on methods with issues such as costs, compliance with regulatory affairs, and a careful consideration of the particular (often very complex) IP situation.

This chapter aims at providing an overview of the large-scale cultivation of the recombinant strains and cells described in this book in stirred tank reactors (STR) (hereafter referred to as "fermentation"), and focuses for the most part on the systems developed within the TPW program (see Editorial).

14.2
Escherichia coli

For detailed information on *Escherichia coli*, the reader is referred to Chapter 2. The current typical host strain bears a number of modifications, for example, to avoid gene transfer, to reduce proteolytic activities, to allow certain post-translational modifications and protein folding, and to support the translation of rare codons by overexpressing certain tRNA genes. Clearly, the definition of the promoter element is of great impact in fermentation design. The expansive range of promoter elements – and thus the large number of options for the resulting fermentation procedures – can be detailed only to a limited degree in the following sections.

14.2.1
Media and Fermentation Strategies

The standard medium for *E. coli* propagation in test tubes or shake-flasks is the Luria-Bertani broth (LB), a complex medium without specific carbon sources (see Appendix A14.1). All known strains used for the propagation of plasmids and employed as hosts for heterologous gene expression can grow on LB (Doig et al. 2001; Job et al. 2002). In non-supplemented LB, growth is restricted due to nutrient or energy limitations to OD_{600} 2–4. A strong increase of pH values up to 8.8 is generally observed when *E. coli* is grown without any additional carbon sources in LB due to the accumulation of ammonia in the medium. Supplying a carbon source (e.g., glycerol) in TB avoids a rise in pH and leads to higher biomass formation (Losen et al. 2004) and acidification of the medium. Similar effects have been observed for *Staphylococcus carnosus* (see Section 14.3.1). For growth to higher densities – and thus to a higher volumetric productivity in STR cultivations – the medium must be supplemented with carbon sources (e.g., glycerol or glucose), and a pH control should be employed. Supplementation of complex media with minerals allows further increases in biomass formation.

LB medium supplemented with carbon source and minerals has been used among others by Qiu et al. (1998), Simmons et al. (2002), Tong et al. (2001), and Xu et al. (2002). Minerals as well as complex compounds can be contained in the medium from the point of inoculation, or they can be fed at later stages of fermentation.

Completely avoiding the use of complex media constituents reduces the process cost and may aid in simplifying the fermentation process and downstream processing, especially when a high purity of the recombinant protein is required. For certain therapeutic applications the absence of complex compounds (especially animal-derived peptones) in the entire process is desirable. One simple synthetic medium for the growth of *E. coli* strains is M9, which can be used for *E. coli* BL21 cultures (Chao et al. 2002). Mineral media have been described that enable growth to very high cell densities (more than 145 g L^{-1} DW) without any complex additions (Horn et al. 1996; Yee and Blanch 1993). In this case, additional mineral N and P sources and trace element solutions are fed to the fermentor in a relatively complex feed strategy during the later phases of the fermentation. The respective medium described by Horn et al. (1996) is detailed in Appendix A14.1.

The highest growth rate obtained in fermentations in LB-based medium is 0.77 h^{-1} (doubling time ~54 min) when using glucose as carbon and casamino acids as a complex N-source. However, the higher costs and the presence of undesired animal-derived components prevent its general use in the pharmaceutical industry. Maximal growth rates in MM medium are 0.57 h^{-1} with glucose and 0.32 h^{-1} with glycerol as carbon sources (Paalme et al. 1997).

A wide range of fermentation strategies has been assessed for *E. coli* employing batch, fed-batch, and continuous fermentation modes. Most commonly, a fed-batch fermentation mode is executed using a medium composed of basal salts. Fed-batch fermentation is the strategy of choice for a number of reasons. First, the total amount of minerals needed for growth to very high cell densities cannot be added in the required concentrations at the beginning of a fermentation due to inhibitory effects, solubility limitations, and precipitation problems. Second, under anaerobic conditions *E. coli* tends to produce large amounts of acetate, formate, and other metabolic by-products. These metabolites are similarly seen when supplementing glucose in higher concentrations, derived from so-called "glucose overflow pathways". Both anaerobic or microxic conditions, as well as high local carbon source concentrations, can occur in large and poorly mixed fermentors. In fed-batch fermentations glucose concentrations and oxygen availability can be easily controlled (Becker et al. 1997; Xu et al. 1999). Third, fed-batch fermentation allows rate-controlled growth to a desired cell density prior to and after imposing inducing conditions for a chosen promoter. The cell density at the time of induction should be established in such a way that it ensures provision of adequate mixing and aeration until the anticipated end of the fermentation, thereby supporting continuous further growth. A carbon source feed can be fixed as growth-limiting during induction such that, for example, glucose fed at a limiting rate does not inhibit expression of a foreign gene driven by a lactose (or isopropyl-thiogalactopyranoside; IPTG)-dependent promoter.

Cell densities as high as 145 g L^{-1} DW can be obtained in fed-batch fermentations if several different feed solutions are added at certain intervals (Horn et al. 1996; Yee and Blanch 1993). Such a sophisticated fermentation design might not be generally applicable to other expression strains, but simpler processes exist.

Aeration and agitation should be defined in such a way that the pO$_2$ is maintained above a threshold level at which oxygen becomes limiting. However, in special cases such as the production of an *Acinetobacter calcoaceticus*-derived cyclohexanone monooxygenase in a continuous fermentation, higher productivity can be achieved under conditions of oxygen limitation (Doig et al. 2001).

One specific problem during the fermentation of *E. coli* is its propensity for acetate formation, and the resultant growth inhibition as discussed above (Lin 2000; van de Walle and Shiloach 1998; Xu et al. 1999). This can be avoided by using adequate fermentation strategies. In addition, some strains such as RV308 (Horn et al. 1996) are reported to be less prone than others to the production of acetate.

One particularly important aspect of fermentation development is the optimization of induction conditions. Despite a stringent dependency on the selected promoter elements, variations are possible even in identical genetic set-ups. Initially, the question of induction temperature should be addressed. Although recombinant *E. coli* may exhibit its fastest growth at 37 °C, the optimal temperature for heterologous gene expression is often lower, typically between 25 and 30 °C; however, temperatures as low as 21 °C (Wlad et al. 2001) or even 15 °C (K. Uhde, Institute of Molecular Biotechnology, RWTH Aachen, personal communication) may be necessary. The idea behind lowering the induction temperature is to slow down the overall metabolic rate and allow more time for the correct folding and transportation of recombinant proteins; this avoids deposition of the product as inclusion bodies, or insoluble protein particles. In some fermentation strategies a lower temperature is also applied during the growth phase (e. g., Wlad et al. 2001). This strategy follows a rationale to minimize metabolic stress prior to the production phase, especially when using strong promoters for expression control.

Another parameter to consider is the duration of induction. Induction times ranging from 4 to 40 hours have been reported. This parameter should be evaluated for each individual protein, because the productivity or accumulation of a given recombinant protein is determined by the specific speed of its formation and degeneration.

In the case of some very strong promoters – for example the viral T7 promoter – it has been shown that higher productivities can be obtained when applying a down-modulated induction. Using the production of human superoxide dismutase as a model system, the metabolic stress imposed by the overexpression of the recombinant gene has been analyzed in more detail. It was found that a lowered induction rate may help to divert sufficient energy and monomeric molecules (e. g., nucleotides and tRNAs) towards growth and maintenance, and that this may lead to a more sustained induction period (Cserjan-Puschmann et al. 1999).

14.2.2
Downstream Processing

Although inclusion bodies may present an attractive start for purification (Falconer et al. 1998; Park et al. 1998; Sinacola and Robinson 2002), subsequent procedures can be time-consuming and inefficient, and also create a wastewater problem – at least at an industrial scale. Simple and efficient methods of protein secretion to the fermentation medium are not available for the Gram-negative bacterium *E. coli*. Several signal peptides can be employed that direct the recombinant proteins to the bacterial periplasm, for example those derived from pectate lyase, β-lactamase, or from several proteases. Further export into the medium depends largely on protein size (smaller proteins seem to pass the peptidoglycan layer more easily), and often requires the autolysis of cells. Thus, even in the case of fermentation supernatants as a starting material, many more contaminating proteins are encountered in *E. coli* compared to the situation in eukaryotic microbes (e. g., yeasts) which have efficient secretion mechanisms (Figure 14.1).

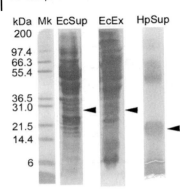

Fig. 14.1 SDS-PAGE of typical downstream processing starting points from various expression systems. Mk = Molecular weight marker; EcSup = *E. coli* supernatant from a medium cell density fermentation after induction; EcEX = *E. coli* cell extract from a medium cell density fermentation after induction; HpSup = *Hansenula polymorpha* culture supernatant from a HCDF. Arrowheads indicate the position of the protein of interest.

14.2.3
Case Study: Production of GFP in a Medium Cell Density Fermentation of *E. coli*

In a particular case study, recombinant GFP was produced intracellularly in *E. coli* BL21(DE3) using a pRSet vector for transformation (Figure 14.2). A 7-L stainless steel bioreactor (Applikon, The Netherlands) was operated at 4-L working volume. Glucose (1.5%, w/v) was used as a carbon source in a defined mineral medium. The culture conditions were 37 °C, 0.5 vvm, 750 r.p.m.; the pH was maintained constant at 6.9 by addition of 25% NH_4OH. After 11.5 h or OD_{600} of 10, a feed containing 50% (w/v) glucose was initiated. The feed was set to a glucose-limiting rate until a desired OD_{600} of 40 was reached. At this point, the feed was stopped and the temperature was lowered from 37 to 25 °C. At 30 min after the pO_2 rose sharply, indicating carbon source depletion, expression of the T7-driven GFP gene was initiated by adding 0.5 mM IPTG. At 30 min after IPTG addition, the glucose feed was continued at the same rate as before. OD_{600} did not increase after an induction phase of 3 h. Approximately equal amounts (based on culture volume) of GFP were found in the cell homogenate and in the supernatant. The total amount of his$_6$-tagged GFP purified via Ni-IDA IMAC from both fractions was 28 mg L^{-1} of fermentation broth.

14.3
Staphylococcus carnosus

The Gram-positive, nonpathogenic bacterium *Staphylococcus carnosus* TM300 was initially described as a recombinant protein expression host by Goetz (1990) and Falk et al. (1991). The protease-protected secretion of recombinant proteins to the medium can be accomplished by using the pre-pro secretion signal of *Staphylococcus hyicus* (Demleitner and Goetz 1994, Meens et al. 1997). Secretion to the medium is highly desirable as the purification of proteins from clarified fermentation supernatant can be significantly easier than purification from cell extracts; moreover, this presents an advantage over the production in Gram-negative bacteria such as *E. coli*.

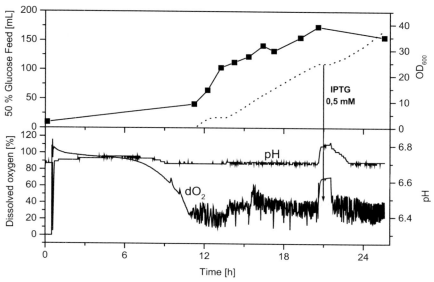

Fig. 14.2 Typical *E. coli* medium cell density fermentation using a mineral medium (4-L scale). Upper panel: ■ = OD_{600} determined after appropriate dilution of the culture broth; $\cdots$ = cumulative feed of 50% (w/v) glucose. Lower panel: the pH set-point was 6.8, pH was maintained above this value by the addition of 25% (w/v) NH_4OH. dO_2 was measured online. A limiting glucose feed rate maintained dO_2 above 10% of saturation. $\cdots$ = temperature set to 37 °C during the growth phase and lowered to 25 °C prior to induction.

Secretion of recombinant proteins has been shown for the variable domain of an immunoglobulin light chain (Pschorr et al. 1994), the *E. coli* outer membrane protein OmpA (Meens et al. 1997), and a human calcitonin fusion protein (Dilsen et al. 2000, 2001).

14.3.1
Media and Fermentation Strategies

Dilsen et al. (2000, 2001) have investigated cultivation strategies for *Staphylococcus carnosus* in shake-flask and parallel fed-batch bubble column reactors, and subsequently scaled-up the process of human calcitonin production to the 150-L scale in a STR, with excellent results. Using a xylose-inducible promotor, they obtained up to 2000 mg L^{-1} of the recombinant fusion protein functionally secreted to the fermentation supernatant.

Although these authors were able to define a mineral medium without complex constituents in which *S. carnosus* was able to grow, they concluded that for the effective expression of a recombinant protein the addition of complex compounds may be inevitable. Consequently, they developed a process strategy that takes these findings into account.

S. carnosus TM300 bearing the expression vector pXPhCT2 used by Dilsen et al. (2000) showed auxotrophies for L-alanine, L-arginine, L-aspartate, and L-cysteine. Additionally, they could show that the strain lost the expression plasmid when cultured under non-repressing conditions, thus necessitating the use of glucose as a carbon source during preculture. They established a pH optimum of 7, and a temperature optimum of 37 °C. Additionally, cultivation at high dO_2 concentrations (>30% of saturation) proved beneficial for high product yields.

As the use of glucose or lactose as carbon source led to the increased formation of organic acids, the final strategy relied on glycerol as the carbon and energy source and xylose for induction (Dilsen et al. 2001).

The fermentation strategy established by Dilsen et al. (2001) incorporated a pH-controlled feed of 100 g L^{-1} yeast extract prefixed to pH 10 to maintain a steady supply of complex compounds at a low level and to avoid acidification of the culture, which occurs when a vital compound of the yeast extract becomes limited. Using this strategy, these authors obtained a cell dry weight of up to 45 g L^{-1}. Growth rates of up to 0.3 h^{-1} were observed. The process was successfully transferred to pilot scale, and the volumetric productivity compared favorably to the production of human calcitonin precursor peptide in *E. coli* (30 mg $L^{-1} h^{-1}$ versus 6.7 mg $L^{-1} h^{-1}$).

To summarize, *S. carnosus* presents an attractive expression system for the production of recombinant proteins for therapeutic processes. It is nonpathogenic, it features a functional system for secretion of the protein of interest to the culture supernatant, and it has been shown to be capable of attractive volumetric productivities.

14.4
Arxula adeninivorans

In Chapter 5, a complete set of components is presented for heterologous gene expression in *Arxula adeninivorans*. *A. adeninivorans* exhibits several beneficial properties for the (industrial) fermentation of heterologous proteins. *A. adeninivorans* is a robust system. Its osmotolerance (Yang et al. 2000) permits the use of high buffer concentrations for pH stabilization in non-titrated shake-flask cultures. Highly concentrated fed-batch media for high-cell-density fermentations can be developed. The broad range of carbon and nitrogen sources utilized by *A. adeninivorans* (Middelhoven et al. 1984, 1991) provides many options for substrate-based fermentation control. *A. adeninivorans* can grow at temperatures up to 48 °C (Wartmann et al. 1995), thereby enabling fermentation (phases) to be conducted at higher temperatures (Böttcher et al. 1988). As a yeast, rapid metabolism and high growth rates reduce the fermentation times and can also increase productivity. As the secretion of proteins is provided (see Chapter 5, Table 5.1), secreted protein products can easily be separated from the fermentation broth. The mitotically stable genomic integration of heterologous genes (Rösel and Kunze 1998) supports the stable reproducible fermentation and production processes.

Due to the high potential of recombinant *A. adeninivorans* as a producer of industrial relevant proteins, a cultivation system for screening in shaken bioreactors and for high-cell-density fermentations in stirred tank bioreactors was developed.

14.4.1
Current Status of Media and Fermentation Strategies

Soon after its identification, *A. adeninivorans* strain LS3 (PAR-4) was employed in the production of single cell protein (SCP) (Böttcher et al. 1988). This early process took advantage of the thermotolerance of *A. adeninivorans*, which allowed higher cultivation temperatures than those applied to established SCP organisms. Recently, recombinant *A. adeninivorans* strains were cultivated in stirred tanks on a laboratory scale for the production of polyhydroxyalkanoates (Terentiev et al. 2004).

Until now, most observations on the culturing properties of *A. adeninivorans* have been based on shake-flask experiments (for examples, see Chapter 5; see also Yang et al. 2000; Wartmann et al. 2002, 2003). In these studies, either the selective and synthetic Yeast Minimal Medium (YMM) supplemented with vitamins and 20 g L^{-1} glucose (derived from Tanaka et al. 1967), or the nonselective complex YEPD medium (5 g L^{-1} yeast extract, 5 g L^{-1} peptone, 10 g L^{-1} glucose) was used.

14.4.2
Development of Media and Fermentation Strategies

A. adeninivorans wild-type strain LS3 (Kunze and Kunze 1994) and mutant strain 135 (Wartmann et al. 2000) were each investigated. Both strains and derivatives can serve as hosts for heterologous gene expression (see Chapter 5; see also Wartmann et al. 2002, 2003. The investigation focused on the establishment of synthetic media and operating conditions for non-limited growth in shake-flask cultures and STRs. Defined synthetic media ease precise culture evaluation, thereby enabling material balancing as well as an adjustment of culture conditions. Furthermore, the use of complex media may impair recombinant protein purification, which is much less pronounced when applying synthetic media.

The shake-flask cultures were analyzed using a device for online measurement of the oxygen transfer rate in shake-flasks (RAMOS) as a measure for the respiration rates of the cultures (Anderlei and Büchs 2001; Anderlei et al. 2004). This device has already been applied successfully to several investigations of alternative platforms (Silberbach et al. 2003; Stöckmann et al. 2003 a,b; Losen et al. 2004).

The assessment of synthetic YMM medium was started with an evaluation of nitrogen supplementation. In theoretical considerations, the nitrogen content of a well-balanced synthetic medium was calculated as 3.3 mmol N g glucose^{-1}, based on biomass composition and a biomass yield of about 0.55 g DW g glucose^{-1} (see below). However, unmodified YMM medium contains only 2.2 mmol N g glucose^{-1}. Therefore, the nitrogen content of YMM medium was raised to 4.6 mmol N g glucose^{-1}. However, shake-flask cultures of strain LS3 and strain 135 remained limited, despite this nitrogen addition. In Figure 14.3, the open circles show the representative respiration rate for strain LS3 in YMM medium with ammonium as the nitrogen source. Initially, the respiration rate increased exponentially, as would be expected for a non-limited culture (Anderlei et al. 2001; Stöckmann et al. 2003 b). However, after a 13-h period of cultivation the respiration rate was limited to 9 mmol L^{-1} h^{-1}, and continued to decline over the fermentation time. A final cell mass of only ca. 6.6 g DW L^{-1} was obtained. Re-assessment of mass balancing revealed severe calcium and iron limitations in the YMM medium. Increased calcium and iron concentrations improved biomass formation and respiration. Nevertheless, a sigmoidal shape of the respiration rate over fermentation time indicated an inhibitory factor in the cultures (data not shown; Anderlei et al. 2001). As the pH value also decreased over fermentation time to 2.0, it was assumed that growth was inhibited by these low pH values. Therefore, the pH of YMM medium was stabilized with MES (2-[N-morpholino]ethanesulfonic acid) buffer. A buffer concentration of 0.14 mol L^{-1} and an initial pH value of 6.4 were found to be suitable for pH stabilization, as calculated by model predictions according to Stöckmann et al. (2003 b). The resulting YMM* medium with increased nitrogen, calcium and iron content, as well as MES buffer, is described in Appendix A14.3. In this medium only thiamine is added, since it was found to be the only essential vitamin for *A. adeninivorans* growth. This corroborates early findings by Middelhoven et al. (1991). In Figure 14.3, the closed circles depict the respiration rate of strain LS3 in YMM* medium. Strain 135 showed very similar fermentation characteristics. The pH was successfully stabilized to a final value of ~5.3 (data not shown), and initially the respiration rate increased exponentially. Maximal specific growth rates of 0.32 ± 0.01 h^{-1} (strain LS3) and 0.31 ± 0.01 h^{-1} (strain 135) were calculated from the respiration rates (Stoeckmann et al. 2003 b). After 15 h, the specific growth rates decreased to ≤ 0.15 h^{-1}. The stationary phase started after 24 h, when the glucose was depleted and the respiration rate dropped.

SYN6 medium as described for *H. polymorpha* (see Appendix A14.3) was also tested for shake-flask cultures of *A. adeninivorans*. As glycerol was found not to be a suitable carbon source for growth (data not shown), the glycerol normally contained in SYN6 medium was substituted by glucose. Nutrient concentrations of the standard SYN6 medium remained unchanged for use in shake-flask cultures, although the relatively high amounts are optimized for stirred, high-cell-density, fed-batch fer-

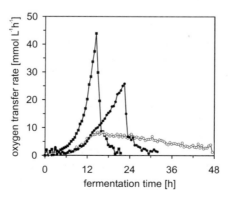

Fig. 14.3 Oxygen transfer rates of *Arxula adeninivorans* shake-flask cultures in original YMM medium with 0.338 mg L^{-1} Ca and 0.041 mg L^{-1} Fe ($\bigcirc$), in YMM* medium with 272.8 mg L^{-1} Ca and 2.1 mg L^{-1} Fe, buffered with 0.14 M MES (pH 6.4) ($\bullet$) and in SYN6-MES medium with 0.14 M MES (pH 6.4) ($\blacksquare$); cultivation temperature: 30 °C; operating conditions: 250-mL shake-flasks with cotton plug and culture volumes of 10 mL for SYN6-MES medium and 20 mL for YMM medium and YMM* medium, 350 r.p.m. shaking frequency and 50 mm shaking diameter of the rotary shaker.

mentations. As *A. adeninivorans* can tolerate such high concentrations, it is possible to assess to some extent the stirred tank bioreactors conditions on a shake-flask scale, thus facilitating any subsequent adjustments to an industrial scale. Shake-flask cultures of strains LS3 and 135 exhibited exponentially increasing respiration rates in SYN6 medium. However, at respiration rates of about 35 mmol $L^{-1} h^{-1}$ the shape of the respiration rate curve became sigmoidal. This was accompanied by a low final pH of 2.1, as in the case of YMM medium, and again indicated the presence of pH inhibition. The conventional pH buffering of this medium with 0.12 mol L^{-1} phosphate is not feasible, as the medium's initial pH of 4.1 is far from the effective buffer ranges of phosphate buffers (Blanchard 1984). As with the YMM* medium, the pH of SYN6 medium was successfully stabilized by buffering with 0.14 mol L^{-1} MES to an initial pH of 6.4. The final pH of the cultures with buffered SYN6-MES medium (for a detailed description, see Appendix A14.3) was about 5.3 (data not shown). In Figure 14.3, the closed squares represent the respiration rates of strain LS3 in a shake-flask culture with SYN6-MES medium. The respiration rates increased exponentially from the beginning to the end of the culture, reaching a maximum of 44 mmol $L^{-1} h^{-1}$. The stationary phase was reached after only 15 h, as indicated by the fall in respiration rate.

Optimal growth of *A. adeninivorans* is provided at pH values between 2.8 and 6.5. Within this pH range, maximal specific growth rates of 0.32 ± 0.01 (strain LS3) and 0.31 ± 0.01 (strain 135) are obtained (Stöckmann et al. (1)). In cultures with SYN6-MES medium and YMM* medium, high biomass yields of about 0.55 g DW g glucose^{-1} for both strains LS3 and 135 are reached. The high mineral salt concentrations in SYN6-MES medium ensured a non-limiting nutrient supply of the cultures. In contrast, the YMM* medium, with its lower component concentrations, was presumably still affected by an insufficient supply with microelements, resulting in partially lower growth rates (Figure 14.3, closed circles).

The maximal oxygen transfer capacity (OTR$_{max}$) determines the maximal possible respiration rate of the shake-flask cultures. OTR$_{max}$ is determined by the operating conditions and the physico-chemical properties of the medium (Maier and Büchs 2001). Respiration of *A. adeninivorans* cultures, which were limited to an OTR$_{max}$ of about 0.02 mol $L^{-1} h^{-1}$, resulted in greatly extended culturing times and in a partially anaerobic metabolism of the cultured cells. Up to 3.5 g L^{-1} ethanol was generated, which was subsequently metabolized again. As oxygen-limitations lead to non-defined conditions, they should be avoided. The operating conditions chosen for YMM* medium and SYN6-MES medium cultures (see Figure 14.3) allow OTR$_{max}$ values of about 0.036 mol $L^{-1} h^{-1}$ and 0.058 mol $L^{-1} h^{-1}$, respectively. These values are well above the maximal respiration rate of the cultures, and ensure a non-limiting oxygen supply.

Strain LS3 was also cultivated in a fed-batch STR with a 2-L working volume (VSF, Bioengineering AG, Wald, Switzerland). Conventional SYN6 medium as described for *H. polymorpha* was employed (see Appendix A14.3), and glycerol was substituted by glucose. The cultures were aerated between 0.4 and 1.5 vvm and stirred at between 400 and 2000 r.p.m. The cultivation temperature was 30 °C, and the pH was maintained at 6.0 with ammonia (12.5% w/w), which simultaneously served as a nitrogen supply. The fermentations were started as a batch. After the depletion of glucose, a glucose-limiting feed (620 g L^{-1} glucose) was started. The stirrer speed was controlled in accordance with the increasing turbidity of the culture broth. Glucose-limiting feeding maintained the oxygen at a concentration of 40% of air saturation, which ensured that the conditions were not limited by oxygen. The course of respiration, specific growth rate and biomass yield of the batch phase were found to be very similar to those seen in shake-flasks experiments. In the feeding phase, a maximal respiration rate of 0.15 mol L^{-1} h^{-1} was reached, while the final biomass concentration was 112 g L^{-1}, corresponding to a biomass yield of 0.57 g DW g glucose^{-1}.

14.4.3
Case Study: Production of Heterologous Phytase in Shake-flask Cultures and a High-cell- density, Fed-batch Fermentation of *A. adeninivorans*

A. adeninivorans G1211 (*aleu2*) was transformed as reported by Wartmann et al. (2003). The resulting recombinant strain secretes a heterologous phytase mutein similar to that described by Mayer et al. (1999) under the control of the promoter of the *TEF1* gene (Rösel and Kunze 1995). The recombinant strain was cultivated in shake-flasks with SYN6-MES medium, and in a fed-batch STR as described for wild-type strain LS3. Batch shake-flask cultures and fed-batch fermentations in stirred tanks of recombinant *A. adeninivorans* with SYN6(-MES) medium exhibited the same or similar growth characteristics as in the case of wild-type strain LS3. Clearly, transformation did not burden the metabolism and growth of recombinant *A. adeninivorans*. Phytase secretion by the recombinant strain was growth-coupled, as would be expected for the constitutive *TEF1* promoter. Similar productivities with phytase secretion of about 13 FTU mL^{-1} (FTU = phytase unit, see abbreviations) were observed for both, the batch phase of the fed-batch fermentation as well as the batch cultures in shake-flasks. During the glucose-limited feeding phase, phytase was secreted superproportionally to the formation of biomass, reaching a maximum of ca. 900 FTU mL^{-1}. These results demonstrate the high potential of recombinant *A. adeninivorans* when cultivated under correctly selected and optimized conditions.

14.5
Hansenula polymorpha

The methylotrophic yeast *Hansenula polymorpha* is an established host for heterologous gene expression (Gellissen 2000, 2002; see also Chapter 6). *H. polymorpha* and the closely related species *Pichia pastoris* share many key features: both yeasts grow to extremely high cell densities in simple, defined mineral media, and both can use a variety of carbon sources (Parshina et al. 1983). Both organisms can tolerate low pH conditions and fairly high concentrations of glucose, without exhibiting crabtree-like effects. *H. polymorpha* tolerates higher temperatures (up to 50 °C) than *Pichia pastoris* (Gellissen et al. 1995; Ryabova et al. 2003). Fundamental differences exist in the regulation of the strong promoters derived from the methanol utilization pathway that are commonly employed in expression cassettes. These differences have an impact on the development of fermentation strategies.

14.5.1
Media and Fermentation Strategies

The "gold standard" in terms of defined mineral medium for *H. polymorpha* is SYN6 medium, which has been developed at RheinBiotech GmbH (Jenzelewski 2002). Details of the medium are specified in Appendix A14.3. This medium contains all basal salts to support growth to high cell densities. The original fermentation strategy is to grow the cells using glycerol as a carbon source in a batch phase, to boost the cell mass to a desired density in a fed-batch phase, and finally to induce expression of the recombinant gene driven by the *MOX* or the *FMD* promoter, the most prominent elements derived from genes of the methanol utilization pathway (Janowicz et al. 1991; Hodgkins et al. 1993; Kang et al. 2001; Jenzelewski 2002). Cell densities up to 100 g L^{-1} DW have been reported (Gellissen and Melber 1996; Hodgkins et al. 1993). The growth rates observed during growth on glycerol were 0.2–0.25 h^{-1} (Parshina et al. 1983) and 0.18 h^{-1} during growth on methanol (Kim et al. 2001).

An efficient and robust strategy to control the glycerol feed during the growth phase of *H. polymorpha* is pH-controlled feeding of a mixture of glycerol and ammonia at a fixed ratio of 0.048 g ammonia per g glycerol. If provided as a balanced mixture, a convenient carbon source feed rate that adapts automatically to the glycerol consumption rate can be realized. However, it is clear that standard configurations with a manually adjusted glycerol feed rate and "normal" pH-control using ammonia as a base can also be employed.

For induction, several strategies have been described. Basically, the feeding of methanol as the sole carbon source, feeding glycerol or even glucose at a limiting rate (sometimes controlled by the dO_2 signal) and mixed feeds of glycerol/methanol have been used successfully. The latter "two-carbon source" fermentation has been utilized in the production of *H. polymorpha*-derived hepatitis B vaccines as the final phase of the induction. This mode was executed in this example as methanol is an inducer of membrane proliferation, a component of the recombinant HBsAg particles (see Chapter 15).

In an early study of hepatitis B pre-S2 antigen production in *H. polymorpha*, a medium was employed that contained yeast extract as a complex additive. Use of this medium resulted in a dramatic increase in productivity (de Roubin et al. 1991). Further optimization of the process led to fermentation at elevated pH with reduced amounts of yeast extract (de Roubin et al. 1992).

The presence of methanol is not required for full strength of methanol pathway-derived promoters in *H. polymorpha*. Both *MOX* and *FMD* promoters are rendered active when maintaining glycerol at certain low levels, or even by controlled low supplementation with the repressive glucose ("glucose starvation") (Jenzelewski 2002; Guengerich et al. 2004; see also Chapter 6). In contrast, promoter activity is strictly dependent on the presence of methanol in *AOX1*-driven expression in *P. pastoris* (Hellwig et al. 2001). This is not an inherent property of the *H. polymorpha* promoter as the *P. pas-*

toris-derived *AOX1*-promoter is rendered active under glycerol-limiting conditions when transferred to the *H. polymorpha* environment (Raschke et al. 1996).

In consequence of this greater versatility, fermentation strategies have been developed that evade the presence of undesired toxic methanol (indeed, the production process for the hepatitis B vaccine is the only industrial process that uses methanol for the above-mentioned reasons). A "one-carbon fermentation" mode, in which the methanol fed-batch phase is replaced by a limiting glycerol fed-batch phase, has been applied to several processes for secreted proteins, among others for the production of recombinant hirudin (Weydemann et al. 1995; Avgerinos et al. 2001; Bartelsen et al. 2002). Glycerol can easily be fed at a limiting rate by using the pO_2 signal in a feed pump control. Depletion of the carbon source in high-cell-density fermentations on defined media leads to a rapid increase in pO_2, which plunges rapidly upon the addition of carbon source. In this way, an oxygen transfer-controlled cycle is created by adjusting a limiting feed rate automatically to consumption of the carbon source (Gellissen and Melber 1996; Jenzelewski 2002; Guengerich et al. 2004).

In one particularly efficient process, *H. polymorpha* has been used in the production of the feed additive phytase using the *FMD* promoter for expression control. Significantly, the use of glycerol in the initial batch phase was found not to be required, and in fact could be substituted with low-cost glucose, without drastically affecting the product yield. Derepression – and thus activation – of the promoter was brought about by glucose starvation (a continuous feed of minimal levels of glucose). At a 2000-L scale, fermentation with glucose as the sole carbon source led to high product yields and an 80% reduction in raw material costs compared to glycerol-based fermentation. The strain was found to secrete the recombinant phytase at levels up to 13.5 g L^{-1}. This extremely high product yield, in combination with the successful development of a methanol and glycerol-free fermentation process, illustrates how this platform can be applied as economically competitive production system for technical enzymes (Mayer et al. 1999; Papendieck et al. 2002).

Aside from promoters derived from genes of the methanol utilization pathway, several strong constitutive control elements are available that imply the use of glucose-containing media in fermentation processes (Suckow and Gellissen 2002; see also Chapter 6). As such, the constitutive *PMA1* promoter was used for the production of glucose oxidase and human serum albumin (HSA). The productivities compared favorably with those obtained with strains harboring the *MOX* promoter for expression control (Cox et al. 2000).

Another constitutive promoter, the *GAP* promoter, was also assessed using a *HSA* reporter gene. Again, it was found that higher specific and volumetric productivities could be achieved when comparing it with the expression of a *MOX* promoter-controlled sequence (Heo et al. 2003).

The *TPS1* promoter (Reinders et al. 1999; Suckow and Gellissen 2002) is a strong constitutive promoter derived from the key gene of the trehalose synthetic pathway. Its strength can further be boosted by cultivation at elevated temperatures (Reinders et al. 1999; Amuel et al. 2000).

The generation of a pre-culture is of great impact for the subsequent production runs. It has been observed that *H. polymorpha* strains exhibit an extended lag phase and significantly decreased maximal growth rates in a synthetic fermentation medium when pre-culturing has been performed in a complex glucose-containing medium (Stöckmann et al. 2003 b; S. Hellwig, unpublished results).

14.5.2
Downstream Processing

Secretion is a preferred option to produce a foreign protein, as it avoids the need for cell breakage and the use of crude cell extracts as a starting material (see Figure 14.1). Several signal peptides have been successfully employed to target foreign proteins to the secretory apparatus in *H. polymorpha*, such as the *S. cerevisiae*-derived MFα1 pre-pro sequence, signal sequences from *Carcinus maenas* and *Schwanniomyces occidentalis*, and the native pre-pro secretion signal of HSA (Gellissen and Melber 1996; Kang et al. 2001).

14.5.3
Case Study: Production of a Secreted Protein in *H. polymorpha*

A production strain expressing the gene for a secreted small, glycosylated protein under control of the *MOX* promoter was assessed in fermentation runs at a 4-L working scale (Figure 14.4). Culturing during the growth phase was carried out using a balanced ammonia/glycerol feed mixture supplemented by the pH control cycle as described before. Thus, the N-source and C-source were fed in a coupled fashion, and the feed rate was automatically adjusted to the glycerol consumption rate. The growth phase was fairly long, due to the fact that the pre-culture had been grown on a complex glucose medium. After reaching a desired cell density, the pH set point was lowered to 4.0. In previous experiments this had been shown to increase product accumulation. For induction, the N- and C-supplies were then uncoupled. Glycerol was fed at a limiting rate in a pO_2-controlled manner, and ammonia was fed controlled by the pH control cycle. Induction was carried out for more than 50 h.

The recombinant product accumulated in the fermentation supernatant to a level of 1200 mg L^{-1}, as determined by HPLC.

14.6
Sordaria macrospora

Sordaria macrospora exhibits several beneficial properties for the production of heterologous proteins:

1. As no conidiospores are formed (Fields 1970; Esser 1982), and even sexual spore formation can be prevented by the utilization of sterile mutant strains (see Chapter 10), spore contaminations can be avoided.
2. Auxotrophic host strains are available, thus avoiding the need for undesired antibiotic resistance genes as selection markers.

Typcial *Hansenula polymorpha* HCD fermentation

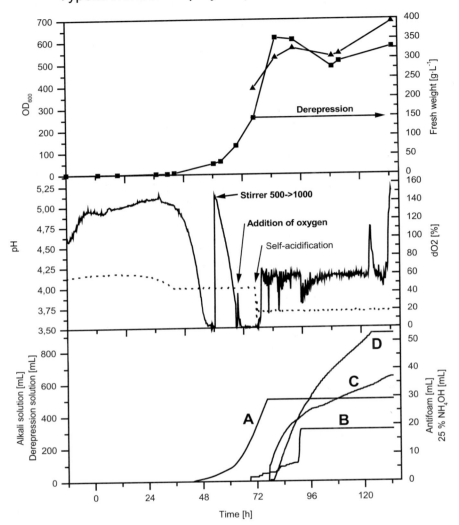

Fig. 14.4 Typical *Hansenula polymorpha* high-cell-density fermentation on the synthetic medium SYN6. Upper panel: OD_{600} (■), FW
(▲). Center panel: dO_2 readout (solid line), pH (broken line). Lower
panel: A = mixed NH_4/glycerol feed; B = antifoam dosage; C = 25 %
(w/v) ammonia; D = 90 % (w/v) glycerol. Cells were grown on mixed
NH_4/glycerol feed controlled by the pH control cycle until the desired cell density was reached. At this point, the pH set-point was
lowered and glycerol was fed at limiting rate controlled by dO_2, while
pH was controlled by automatic addition of NH_4OH.

3. Due to the presence of a secretory pathway, heterologous protein products can be secreted into the culture broth and can be modified by components associated with the secretory pathway.

The filamentous or pelleted growth of filamentous fungi and *S. macrospora* can hamper submerged fermentation in several ways:

1. Increased viscosities of the fermentation broth may impair mixing and mass transfer, resulting in heterogeneous and sub-optimal conditions (Gbewonyo and Wang 1983; Olsvik and Kristiansen 1994; Peter et al. 2004).
2. Since the natural habitat of *S. macrospora* is the dung of herbivores (Fields 1970), *S. macrospora* is evolutionarily adapted to superficial growth on stagnant solid substrates. Hence, the development of conditions for optimal growth in submerged and agitated cultures appears to be challenging. Often, morphology, growth and product formation is influenced by the type of liquid medium and type and intensity of agitation, i.e. hydromechanical stress (Bushell et al. 1997; Büchs and Zoels 2001; Katzer et al. 2001).
3. Compared to yeast and bacteria, the metabolism of filamentous fungi is slower, and growth rates are lower. Thus, the duration of a typical fermentation can be quite long and the productivity can be relatively low. This is another reason to aim for optimal conditions for growth and production.
4. Cultures of filamentous fungi are frequently inoculated with a defined titer of conidiospores (Davis and de Serres 1970; Metz and Kossen 1977). As *S. macrospora* lacks these spores, alternative strategies must be developed.

14.6.1
Current Status of Media and Fermentation Strategies

Hitherto, the submerged cultivation of *S. macrospora* has rarely been investigated. Most experiences have been accrued in biomass generation for genetic studies (Walz and Kück 1995), and for that purpose *S. macrospora* is prevalently cultivated on the surface of stationary liquid media or on solid media, inoculated with pieces of vegetative mycelium taken from agar plates (U. Kück, personal communication; Pöggeler et al. 1997). Currently, several complex and synthetic media are used for the cultivation of *S. macrospora*. BMM medium, containing corn meal extract and malt extract, is used for fructification and spore germination (Fields 1970; Esser 1982). Semisynthetic CCM medium (Walz and Kück 1995) and CM (Nowrousian et al. 1999) medium contain relevant amounts of non-defined complex compounds, and are used in genetic experiments. Synthetic GM medium and MM medium (Le Chevanton and Leblon 1989) solely contain mineral salts supplemented with defined amounts of vitamins. With the exception of CCM medium, which contains sucrose, all (semi)synthetic media contain glucose as carbon source.

14.6.2
Development of Media and Strategies for Submerged Cultivation

The development of shake-flask culture conditions is of great importance. Conditions for the submersion of homogeneously dispersed cells, which are not limited by mass transfer, hydromechanical stress, or nutrients have to be found. In additional, well-defined synthetic growth media lacking complex compounds except for vitamins must be formulated. Complex compounds such as cell extracts or peptones are not preferred, as these compounds can stimulate protease formation, which subsequently can degrade the targeted protein product (Van den Hombergh et al. 1997). Protein recovery may also be affected by large amounts of complex compounds. Finally, the quantitative evaluation of cultures is facilitated by well-defined synthetic media, as the balancing and determination of process parameters can be based on known substrates and stoichiometries.

Again, shake-flask cultures were analyzed using devices for the online measurement of oxygen transfer rates in shake-flasks to determine the respiration rates of the cultures, as described previously (RAMOS; Anderlei and Büchs 2001; Anderlei et al. 2004). *Sordaria macrospora* k-hell was used for the culture experiments (kindly provided by Prof. Ulrich Kück); this strain has a wild-type phenotype (Esser and Straub 1958). Cultivations were carried out at a standard temperature of 27 °C (Walz and Kück 1995).

Cultivation in synthetic GM and MM medium inoculated with pieces of BMM plate cultures resulted in very low respiration rates (maximum 0.3 mmol L^{-1} h^{-1}) (Figure 14.5; open and filled circles). After one week of cultivation, cell dry weights of typically 0.5 g L^{-1} (GM) and 1.7 g L^{-1} (MM) were measured, accompanied by high concentrations of residual glucose. As the pH was severely decreased over the cultivation time, several buffers and initial pH values were applied. A stabilization of the pH was achieved with MES buffer at an initial pH of about 6.3. This was found to promote growth in MM and especially GM medium; hence, the latter was selected for further investigations. In buffered GM medium, the glucose was completely consumed and a DW of typically 9 g L^{-1} could be achieved. Maximal respiration rates increased to almost 1 mmol L^{-1} h^{-1}. Nevertheless, the cultivations still took one week. The mycelium became blackened over the cultivation time, and grew as aggregated clumps. This blackening may have been due to melanin formation, as melanin can be formed in response to environmental stress (Juzlova et al. 1996, Eagen et al. 1997, Henson et al. 1999), nutritional limitations were assumed as a possible stress factor. In fact, an increased yeast nitrogen base concentration of the GM medium could shift the mycelial blackening to the end of the cultivations. The composition of this medium, called GM*, is specified in Appendix A14.4. This medium served as provisional working medium for further investigations.

Growth in the form of aggregated clumps (Figure 14.5, open squares) caused mass transfer-limiting and heterogeneous conditions in the inner regions of the clumps. A sufficient nutrient supply, as well as the removal of potentially inhibiting products, were only ensured at the surface of the clumps. For this reason, the respiration rates increased only linearly and remained low, as shown in Figure 14.5 (open squares). Clumpy growth could successfully be prevented by inoculation with

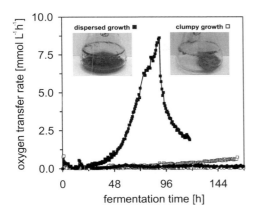

Fig. 14.5 Oxygen transfer rates of *Sordaria macrospora* shake-flask cultures in original MM medium (●), original GM medium (○) and GM* medium with clumpy growing mycelium (□) and homogeneously dispersed mycelium (■). Cultivation temperature 27 °C; operating conditions: 250-mL shake-flasks with cotton plug and 30 mL culture volume, 150 r.p.m. shaking frequency and 50 mm shaking diameter of the rotary shaker. Cultures in original GM medium and MM medium were inoculated with pieces of agar plate cultures (solid BMM medium; Esser 1982). In the case of GM* medium, dispersed pieces of crushed mycelium were used for inoculation.

dispersed pieces of crushed mycelium. A reproducible standard procedure for inoculum preparation was established. Vegetative mycelium was taken from stationary cultures and resuspended in 0.9% (w/v) sodium chloride. This suspension was "whirl-mixed" together with glass spheres, which resulted in dispersed mycelium units of millimeter range (Stöckmann et al. (2)). Each viable unit served as an origin for growth, similar to the inoculation with conidiospores (Metz and Kossen 1977). The hyphae extended radially and were ramified during cultivation, resulting in loose "fluffs" with an increased area exposed to the medium. In Figure 14.5, the closed squares depict the respiration rate of a homogeneously dispersed shake-flask culture inoculated with dispersed vegetative mycelium. This culture exhibited a distinctive exponential growth phase with a maximum specific growth rate of about $0.07\ h^{-1}$, derived from the oxygen transfer rate (Stöckmann et al. 2003 b). The growing fluffs (Figure 14.5, closed squares) clearly remained small enough to avoid mass transfer limitation and to grow at a constant rate. The maximal respiration rate of $8.5\ mmol\ L^{-1}\ h^{-1}$ was one order of magnitude higher compared to clumpy growth (Figure 14.5, open squares). After depletion of the glucose, the oxygen transfer rate fell, introducing the stationary phase of the culture after only 90 h.

Although this inoculation with dispersed vegetative mycelium was quite successful, cultures occasionally reaggregated and formed clumps, resulting in all the adverse characteristics described above. Several additional methods were investigated to obtain dispersed growth at high probability.

Dispersed growth can be stabilized by surfactants and polymers (Metz and Kossen 1977). Rajan and Vikar (1987) reported clumpy growth of the related Sordariaceae, *Neurospora crassa*. By adding $100\ mg\ L^{-1}$ Triton to the cultures, pelleted growth could be stimulated. In contrast, the addition of Triton to cultures of *S. macrospora* proved to be unfeasible, as growth and respiration were reduced even at a Triton concentration of only $10\ mg\ L^{-1}$. Carboxypolymethylene can also be used to stabilize dispersed growth; this anionic polymer is able to coat the hyphae and thereby provide inter-mycelial ionic repulsion (Elmayergi et al. 1973; Morrin and Ward 1989). However, when applied to *S. macrospora* cultures, reaggregation of mycelium still occasionally recurred.

As *S. macrospora* grows naturally on solid materials, support particles can potentially serve as a basis for dispersed growth in submerged cultures (Gbewonyo and Wang 1983). Hydrophobic (polypropylene, polyacryl), hydrophilic (silica gel, sintered glass), and porous structures (Poraver, Perlit) were tested in *S. macrospora* cultures. No specific affinity of the mycelium to any one of the tested materials could be observed, and clumpy growth was still partially observed.

A ring of mycelium regularly adhered to the inner glass wall of the shake-flasks at the upper level of the rotating culture broth (Figure 14.5, closed squares). Wall growth was found to support clumpy growth, while aggregates of adhering biomass were observed to slide down into the culture broth, continuing to grow as a clump. An investigation was made to determine whether the mycelial adhesion could be prevented by changing the physico-chemical surface properties of the flask walls. Hence, flasks with hydrophobic walls (glass flasks treated with alkylsilane, or polycarbonate or siliconized flasks) were utilized. For comparison, normal flasks treated with nonionic hydroxylsilanes or with sulfuric acid provided hydrophilic walls. Growth experiments revealed that *S. macrospora* mycelium could adhere to, and grow on, all tested flask walls. With respect to growth on surfaces, *S. macrospora* is clearly an extraordinarily versatile organism.

The morphology and growth of filamentous fungi can be affected by the level of conidiospores, when used for inoculation (Davis and de Serres 1970; Metz and Kossen 1977; Tucker and Thomas 1994; Cui et al. 1998). As a rule of thumb, increasing spore titers can produce filamentous growth, while low spore titers can produce pelleted growth. Thus, the effect of different concentrations of dispersed mycelium used for the inoculation of *S. macrospora* cultures was investigated. Ultimately, it was determined that high concentrations of inoculum are a necessary pre-requisite against the formation of clumps, and this almost always leads to a dispersed filamentous growth.

Appendix

A14.1 *Escherichia coli* Media

(i) Luria-Bertani broth contains:
1% (w/v) peptone
0.5% (w/v) yeast extract
0.5% (w/v) NaCl

Luria-Bertani broth is a simple complex medium suitable for the small-scale expression of recombinant proteins in *E. coli*. It does not contain a dedicated carbon source, and does not support growth to cell densities higher than OD_{600} 2–4. Nevertheless, it has been used for the cultivation of *E. coli* in STR (Job et al. 2002). LB is frequently enriched by the addition of glucose (i.e., to prevent leaky expression from lac-operon-based promoters) or glycerol or minerals (Qiu et al. 1998; Xu et al. 2002). Other complex media frequently used for the cultivation of *E. coli* are TB, $2 \times$ TB, TY, $2 \times$ TY, and YT.

(ii) Defined shake-flask medium (Yee and Blanch 1993) contains:

1.85 g L^{-1}	NH_4Cl
0.1 g L^{-1}	KH_2PO_4
0.5 g L^{-1}	K_2HPO_4
0.2 g L^{-1}	$MgSO_4 \cdot 7H_2O$
0.02 g L^{-1}	$CaCl_2 \cdot 2H_2O$
0.04 g L^{-1}	$FeSO_4 \cdot 7H_2O$
0.2 g L^{-1}	L-arginine $\cdot$ HCl
1 mL L^{-1}	Trace metal solution

(iii) Defined shake-flask medium (Horn et al. 1996) contains:

8.6 g L^{-1}	$Na_2HPO_4 \cdot 2H_2O$
3.0 g L^{-1}	KH_2PO_4
1.0 g L^{-1}	NH_4Cl
0.5 g L^{-1}	NaCl
60 mg L^{-1}	FeIII citrate hydrate
3.0 mg L^{-1}	H_3BO_3
15 mg L^{-1}	$MnCl_2 \cdot 4H_2O$
8.4 mg L^{-1}	EDTA $\cdot$ $2H_2O$
1.5 mg L^{-1}	$CuCl_2 \cdot 2H_2O$
2.5 mg L^{-1}	$Na_2Mo_4 \cdot 2H_2O$
2.5 mg L^{-1}	$CoCl_2 \cdot 6H_2O$
8.0 mg L^{-1}	$Zn(CH3COO)_2 \cdot 2H2O$
10 g L^{-1}	Glucose
0.6 g L^{-1}	$MgSO_4 \cdot 7H_2O$

(iv) M9 minimal medium (DSMZ, strain collection) contains:

6 g L^{-1}	Na_2HPO_4
3 g L^{-1}	KH_2PO_4
1 g L^{-1}	NH_4Cl
0.5 g L^{-1}	NaCl
0.246 g L^{-1}	$MgSO_4 \cdot 7H_2O$
0.015 g L^{-1}	$CaCl_2 \cdot 2H_2O$
2 g L^{-1}	Glucose
20 mg L^{-1}	Proline
2 mM	Thiamine-HCl

M9 is a defined medium frequently used for the cultivation of *Escherichia coli* JM strains.

(v) Minimal medium (Horn 1996) contains:

16.6 g L^{-1}	KH2PO4
4.0 g L^{-1}	$(NH_4)_2HPO_4$
2.1 g L^{-1}	Citric acid
75 mg L^{-1}	FeIII citrate $\cdot$ $2H_2O$

3.8 mg L^{-1}	H$_3$BO$_3$
18.8 mg L^{-1}	MnCl$_2$ · 4H$_2$O
10.5 mg L^{-1}	EDTA · 2H$_2$O
1.9 mg L^{-1}	CuCl$_2$ · 2H$_2$O
3.1 mg L^{-1}	Na$_2$Mo$_4$ · 2H$_2$O
3.1 mg L^{-1}	CoCl$_2$ · 6H$_2$O
10 mg L^{-1}	Zn(CH$_3$COO)$_2$ · 2H$_2$O
25 g L^{-1}	Glucose
1.5 g L^{-1}	MgSO$_4$ · 7H$_2$O

(vi) Glucose feed solution (Horn 1996) contains:

669.6 g L^{-1}	Glucose
19.8 g L^{-1}	MgSO$_4$ · 7H$_2$O

(vii) N/P feed solution (Horn et al. 1996) contains:

227.0 g L^{-1}	(NH$_4$)$_2$HPO$_4$
169.5 g L^{-1}	(NH$_4$)H$_2$PO$_4$

(viii) Trace element feed solution (Horn et al. 1996) contains:

5 g L^{-1}	FeIII citrate hydrate
250 mg L^{-1}	H$_3$BO$_3$
125 mg L^{-1}	MnCl$_2$ · 4H$_2$O
700 mg L^{-1}	EDTA · 2H$_2$O
125 mg L^{-1}	CuCl$_2$ · 2H$_2$O
213 mg L^{-1}	Na$_2$Mo$_4$ · 2H$_2$O
213 mg L^{-1}	CoCl$_2$ · 6H$_2$O
668 mg L^{-1}	Zn(CH$_3$COO)$_2$ · 2H$_2$O

A14.2 *Staphylococcus carnosus* Media

(i) Fed-batch cultivation initial medium (Dilsen et al. 2000) contains:

750 mM	Glycerol
5 g L^{-1}	Yeast extract
200 mL L^{-1}	5 × Salt solution (see below)

(ii) 5 × Salt solution contains:

13.21 g L^{-1}	NH$_4$(SO$_4$)
4.066 g L^{-1}	MgCl$_2$ · 6H$_2$O
1.029 g L^{-1}	CaCl$_2$ · 2H$_2$O
0.1 g L^{-1}	MnCl$_2$ · 4H$_2$O
13.5 mg L^{-1}	FeCl$_3$ · 6H$_2$O
1.4 mg L^{-1}	ZnCl$_2$
10 mL L^{-1}	Trace salts solution (see below)

(iii) Trace salts solution contains:

0.3 g L^{-1}	H$_3$BO$_3$
0.1 g L^{-1}	ZnSO$_4$ · 7H$_2$O
0.2 g L^{-1}	CoCl$_{2\_}$ · 6H$_2$O
0.01 g L^{-1}	CuCl$_2$ · 2H$_2$O
0.02 g L^{-1}	NiCl$_2$ · 6H$_2$O
0.03 g L^{-1}	Na$_2$MoO$_4$ · 2H$_2$O

(iv) pH-controlled yeast extract feed solution (adjust to pH 10) contains:

100 g L^{-1}	Yeast extract

The fed-batch media used by Dilsen et al. (2000, 2001) rely on glycerol as a carbon and energy source and the use of a pH-controlled yeast-extract feed to obtain high cell densities. *Staphylococcus carnosus* also grows well on complex media such as LB. For pre-cultures, and possibly also during prolonged growth phases, the medium should be supplemented with glucose (1%, w/v) to prevent plasmid-loss. Dilsen et al. used a xylose-inducible promotor. Xylose (0.5%, w/v) was added for induction.

A14.3 Yeast Media

(i) YEB Medium contains:

5 g L^{-1}	Nutrient Broth/Beef extract
1 g L^{-1}	Yeast extract
5 g L^{-1}	Peptone 140 (Tryptic digest of casein)
5 g L^{-1}	Sucrose

YEB is a rich medium that is frequently used for the cultivation of yeast in shake-flasks.

(ii) BMGY (as recommended by Invitrogen) contains:

1% (w/v)	Yeast extract
2% (w/v)	Peptone
1.34% (w/v)	YNB w/o amino acids
0.1 M	Phosphate buffer (autoclaved)
0.02% (w/v)	Biotin (filter sterilized)
1% (w/v)	Glycerol

BMGY is a buffered, rich medium containing glycerol as carbon source. It supports growth of *Pichia pastoris* to relatively high cell densities, and is used in small-scale expression screening experiments. For induction of *AOX1*-controlled genes, cells are grown on BMGY and transferred to BMMY, in which glycerol is replaced by methanol as the sole carbon source.

(iii) *Pichia pastoris* minimal medium (Invitrogen) contains:

2.67 % (v/v)	Phosphoric acid (85 %)
14.9 g L^{-1}	MgSO$_4$ · 7H$_2$O
0.93 g L^{-1}	CaSO$_4$
4.13 g L^{-1}	KOH
18.2 g L^{-1}	K$_2$SO$_4$
0.4 % (v/v)	Ptm1 trace salts
4.0 % (w/v)	Glycerol (ultrapure, 1.26 g mL^{-1})

This medium is suggested by Invitrogen as a *Pichia pastoris* HCDF medium. It supports the growth of *P. pastoris* to cell densities >100 g L^{-1} DW in fed-batch fermentations.

(iv) *Pichia pastoris* Ptm1 trace element solution (Invitrogen) contains:

6 g L^{-1}	CuSO$_4$ · 5H$_2$O
0.08 g L^{-1}	NaI
3 g L^{-1}	MnSO$_4$ · H$_2$O
0.2 g L^{-1}	Na$_2$MoO$_4$ · 2H$_2$O
0.02 g L^{-1}	Boric acid
0.5 g L^{-1}	CoCl$_2$ · 6H$_2$O
20 g L^{-1}	ZnCl$_2$
65 g L^{-1}	FeSO$_4$ · 7H$_2$O
0.2 g L^{-1}	Biotin
5 mL	H$_2$SO$_4$ conc.

(v) SYN6 Medium (*Hansenula polymorpha*) (Rhein Biotech) contains:
(a) SYN6 basal salts:

13.3 g L^{-1}	(NH$_4$)H$_2$PO$_4$
3 g L^{-1}	MgSO$_4$ · 7 H$_2$O
3.3 g L^{-1}	KCl
0.33 g L^{-1}	NaCl
20 g L^{-1}	Glycerol

(b) SYN6 supplementary solutions:
Calcium chloride 100 × stock solution
Micro Elements 100 × stock solution
Vitamin 100 × stock solution
Trace Elements 100 × stock solution

(c) SYN6 100 × Micro Elements stock solution contains:

6.65 g L^{-1}	EDTA
6.65 g L^{-1}	(NH$_4$)FeSO$_4$ · 6H$_2$O
0.55 g L^{-1}	CuSO$_4$ · 5H$_2$O
2.0 g L^{-1}	ZnSO$_4$ · 7H$_2$O
2.65 g L^{-1}	MnSO$_4$ · H$_2$O

(d) SYN6 100 x Calcium chloride stock solution:
100 g L^{-1} CaCl$_2$ 2H$_2$O

(e) SYN6 100 × Vitamin stock solution contains:
6.65 g L^{-1} 2-Propanol/H$_2$O (1:1)
0.04 g L^{-1} Biotin 0.04 g
13.35 g L^{-1} Thiamine

(f) SYN6 100 × Trace Elements stock solution contains:
0.066 g L^{-1} NiSO$_4$ 6H$_2$O
0.066 g L^{-1} CoCl$_2$ 6H$_2$O
0.066 g L^{-1} HBO$_3$
0.066 g L^{-1} KI
0.066 g L^{-1} NaMO$_4$ · 2H$_2$O

(vi) YMM* (derived from Tanaka et al. 1967)
(a) YMM* basal salt solution contains:
6 g L^{-1} (NH$_4$)$_2$SO$_4$
1 g L^{-1} KH$_2$PO$_4$
27.3 g L^{-1} MES (2-[N-Morpholino]ethanesulfonic acid)
2.04 g L^{-1} MgSO$_4$ · 7H$_2$O
1 g L^{-1} CaCl$_2$ · 2H$_2$O
0.01 g L^{-1} FeCl$_3$ · 6H$_2$O
20 g L^{-1} Glucose

The solution is titrated with sodium hydroxide to pH 6.4.

(b) YMM* 1000 × Micro Elements stock solution contains:
0.5 g L^{-1} H$_3$BO$_3$
0.107 g L^{-1} CuSO$_4$ · 5H$_2$O
0.1 g L^{-1} KI
0.4 g L^{-1} ZnSO$_4$ · 7H$_2$O
0.303 g L^{-1} MnSO$_4$ · H$_2$O
0.234 g L^{-1} Na$_2$MoO$_4$ · 2H$_2$O
0.183 g L^{-1} CoCl$_2$ · 6H$_2$O

(c) YMM* 200 × thiamine stock solution
0.4 g L^{-1} Thiaminedichloride

YMM* is a shake-flask medium for *Arxula adeninivorans*.

(vii) SYN6-MES (for *Arxula adeninivorans*, modified from Jenzelewski 2002)
(a) SYN6 basal salts solution contains:
27.3 g L^{-1} MES
7.66 g L^{-1} (NH$_4$)$_2$SO$_4$

1 g L^{-1}	KH$_2$PO$_4$
3 g L^{-1}	MgSO$_4$ · 7 H$_2$O
3.3 g L^{-1}	KCl
0.33 g L^{-1}	NaCl
20 g L^{-1}	Glucose

The solution is titrated with sodium hydroxide to pH 6.4.

(b) SYN6 supplementary solutions (for preparation, see SYN6 Medium of *Hansenula polymorpha*)
Calcium chloride 100 × stock solution
Micro Elements 100 × stock solution
Vitamin 100 × stock solution
Trace Elements 100 × stock solution

SYN6-MES is a buffered synthetic shake-flask medium for *Arxula adeninivorans*.

A14.4 *Sordaria macrospora* Media

(i) GM* medium
(a) GM* 2 × MES stock solution
39.04 g L^{-1} MES (2-[*N*-Morpholino]ethanesulfonic acid)

The solution is titrated to pH 6.5 with sodium hydroxide.

(b) GM* 2 × YNB stock solution contains:

42 g L^{-1}	yeast nitrogen base w/o amino acids
40 g L^{-1}	glucose

The solution is titrated to pH 5.5 with sodium hydroxide.

(c) GM* 40 000 × mineral salt stock solution (Zickler et al. 1984) contains:

50 g L^{-1}	Ascorbinic acid
50 g L^{-1}	ZnSO$_4$ · 7H$_2$O
10 g L^{-1}	Fe(NH$_4$)$_2$(SO$_4$)2 · 6H$_2$O
2.5 g L^{-1}	CuSO$_4$ · 5H$_2$O
0.5 g L^{-1}	MnSO$_4$ · H$_2$O
0.5 g L^{-1}	H$_3$BO$_3$
0.5 g L^{-1}	Na$_2$MoO$_4$ · 2H$_2$O

(d) GM* 1000 × biotin stock solution
0.098 g L^{-1} Biotin

GM* medium was developed as working medium for growth studies on *Sordaria macrospora* in shake-flasks.

References

AMUEL C, GELLISSEN G, HOLLENBERG CP, SUCKOW M (2000) Analysis of heat shock promoters in *Hansenula polymorpha*: TPS1, a novel element for heterologous gene expression. Biotechnol Bioprocess Eng 5: 247–252

ANDERLEI T, BÜCHS J (2001) Device for sterile online measurement of the oxygen transfer rate in shaking flasks. Biochem Eng J 7: 157–162

ANDERLEI T, ZANG W, PAPASPYROU M, BÜCHS J (2004) Online respiration activity measurement (OTR, CTR, RQ) in shake flasks. Biochem Eng J 17(3): 187–194

AVGERINOS GC, TURNER BG, GORELICK KJ, PAPENDIECK A, WEYDEMANN U, GELLISSEN G (2001) Production and clinical development of a *Hansenula polymorpha*-derived PEGylated hirudin. Semin Thromb Hemost 27: 357–372

BALBAS P (2001) Understanding the art of producing protein and nonprotein molecules in *Escherichia coli*. Mol Biotechnol 19: 251–267

BARTELSEN O, BARNES CS, GELLISSEN G (2002) Production of coagulants in *Hansenula polymorpha*. In: *Hansenula polymorpha* – biology and applications (Gellissen G, Ed), Wiley-VCH, Weinheim, pp 211–228

BECKER S, VLAD D, SCHUSTER S, PFEIFFER P, UNDEN G (1997) Regulatory O2 tensions for the synthesis of fermentation products in *Escherichia coli* and relation to aerobic respiration. Arch Microbiol 168: 290–296

BLANCHARD JS (1984) Buffers for enzymes. Methods Enzymol 104: 404–414

BÖTTCHER F, KLINNER U, KÖHLER M, SAMSONOVA IA, KAPULTSEVICH J, BLIZNIK X (1988) Verfahren zur Futterhefeproduktion in zuckerhaltigen Medien. DD 278 354 A1

BÜCHS J, ZOELS B (2001) Evaluation of maximum to specific power consumption ratio in shaking bioreactors. J Chem Eng Jpn 34(5): 647–653

BUSHELL ME, DUNSTAN GL, WILSON GC (1997) Effect of small scale culture vessel type on hyphal fragment size and erythromycin production in *Saccharopolyspora erythraea*. Biotechnol Lett 19(9): 849–852

CHAO YP, CHIANG CJ, HUNG WB (2002) Stringent regulation and high-level expression of heterologous genes in *Escherichia coli* using T7 system controllable by the araBAD promoter. Biotechnol Prog 18: 394–400

COX H, MEAD D, SUDBERY P, ELAND RM,

MANNAZZU I, EVANS L (2000) Constitutive expression of recombinant proteins in the methylotrophic yeast *Hansenula polymorpha* using the *PMA1* promoter. Yeast 16: 1191–1203

CSERJAN-PUSCHMANN M, KRAMER W, DUERRSCHMID E, STRIEDNER G, BAYER K (1999) Metabolic approaches for the optimisation of recombinant fermentation processes. Appl Microbiol Biotechnol 53: 43–50

CUI YQ, OUWEHAND JNW, VAN DER LANS RGJM, GIUSEPPIN MLF, LUYBEN KCAM (1998) Aspects of the use of complex media for submerged fermentation of *Aspergillus awamori*. Enzyme Microb Tech 23: 168–177

DAVIS RH, DE SERRES FJ (1970) Genetic and microbial research techniques for *Neurospora crassa*. Method Enzymol 17A: 79–143

DE ROUBIN MR, BASTIEN L, SHEN SH, GROLEAU D (1991) Fermentation study for the production of hepatitis B virus pre-S2 antigen by the methylotrophic yeast *Hansenula polymorpha*. J Ind Microbiol 8: 147–156

DE ROUBIN MR, CAILAS MD, SHEN SH, GROLEAU D (1992) Influence of pH, nitrogen and phosphorus sources on the production of hepatitis B virus pre-S2 antigen by *Hansenula polymorpha*. Appl Microbiol Biotechnol 38: 158–164

DEMLEITNER G, GOTZ F (1994) Evidence for importance of the *Staphylococcus hyicus* lipase pro-peptide in lipase secretion, stability and activity. FEMS Microbiol Lett 121: 189–197

DILSEN S, PAUL W, HERFORTH D, SANDGATHE A, ALTENBACH-REHM J, FREUDL R, WANDREY C, WEUSTER-BOTZ D (2001) Evaluation of parallel operated small-scale bubble columns for microbial process development using *Staphylococcus carnosus*. J Biotechnol 88: 77–84

DILSEN S, PAUL W, SANDGATHE A, TIPPE D, FREUDL R, THOMMES J, KULA MR, TAKORS R, WANDREY C, WEUSTER-BOTZ D (2000) Fed-batch production of recombinant human calcitonin precursor fusion protein using *Staphylococcus carnosus* as an expression-secretion system. Appl Microbiol Biotechnol 54: 361–369

DOIG SD, O'SULLIVAN LM, PATEL S, WARD JM, WOODLEY JM (2001) Large scale production of cyclohexanone monooxygenase from *Escherichia coli* TOP10 pQR239. Enzyme Microbiol Technol 28: 265–274

EAGEN R, BRISSON A, BREUIL C (1997) The sap-staining fungus Ophiostoma piceae synthesizes different types of melanin in different growth media. Can J Microbiol 43(6): 592–595

ELMAYERGI H, SCHARER JM, MOO-YOUNG M (1973) Effect of polymer additives on fermentation parameters in a culture of *A. niger*. Biotechnol Bioeng 15: 845–859

ESSER K, STRAUB J (1958) Genetische Untersuchungen an *Sordaria macrospora* Auersw., Kompensation und Induktion bei genbedingten Entwicklungsdefekten. Z Vererbungslehre 89: 729–746

ESSER K (1982) Cryptogams – cyanobacteria, algae, fungi, lichens. Cambridge University Press, London, United Kingdom

FALCONER RJ, O'NEILL BK, MIDDELBERG APJ (1998) Chemical treatment of *Escherichia coli*. II. Direct extraction of recombinant protein from cytoplasmic inclusion bodies in intact cells. Biotechnol Bioeng 57: 381–386

FALK MP, SANDERS EA, DECKWER WD (1991) Studies on the production of lipase from recombinant *Staphylococcus carnosus*. Appl Microbiol Biotechnol 35: 10–13

FIELDS WG (1970) An introduction to the genus *Sordaria*. Neurospora Newsletter 16: 14–17

GBEWONYO K, WANG DIC (1983) Enhancing gas-liquid mass transfer rate in non-newtonian fermentations by confining mycelial growth to microbeads in a bubble column. Biotechnol Bioeng 25: 2873–2887

GELLISSEN G (2000) Heterologous protein production in methylotrophic yeasts. Appl Microbiol Biotechnol 54: 741–750

GELLISSEN G (ED) *Hansenula polymorpha* – biology and applications. Wiley-VCH, Weinheim, 2002

GELLISSEN G, HOLLENBERG CP, JANOWICZ ZA (1994) Gene expression in methylotrophic yeasts. In: Smith A (ed): Gene expression in recombinant microorganisms, Marcell Dekker Inc., New York, 195–239

GELLISSEN G, MELBER K (1996) Methylotrophic yeast *Hansenula polymorpha* as production organism for recombinant pharmaceuticals. Drug Res 46: 943–948

GOETZ F (1990) Applied genetics in the gram-positive bacterium *Staphylococcus carnosus*. Food Biotechnol 4: 505–513

GUENGERICH L, KANG HA, BEHLE B, GELLISSEN G, SUCKOW M (2004) A platform for heterologous gene expression based on the methylotrophic yeast *Hansenula polymorpha*. In: The mycota II: Genetics and biotechnology, 2nd edition (Kueck U, Ed), Springer-Verlag, Berlin, pp 273–287

HELLWIG S, EMDE F, RAVEN NPG, HENKE M, VAN DER LOGT P, FISCHER R (2001) Analysis of single-chain antibody production in *Pichia pastoris* using on-line methanol control in fed-batch and mixed-feed fermentations. Biotechnol Bioeng 74: 344–352

HENSON JM, BUTLER MJ, DAY AW (1999) The dark side of the mycelium: melanins of phytopathogenic fungi. Annu Rev Phytopathol 37: 447–471

HEO JH, HONG WK, CHO EY, KIM MW, KIM JY, KIM CH, RHEE SK, KANG HA (2003) Properties of the *Hansenula polymorpha*-derived constitutive *GAP* promoter, assessed using an HSA reporter gene. FEMS Yeast Res 4: 175–184

HODGKINS M, MEAD D, BALLANCE D J, GOODEY A, SUDBERY P (1993) Expression of the glucose oxidase gene from *Aspergillus niger* in *Hansenula polymorpha* and its use as a reporter gene to isolate regulatory mutations. Yeast 9: 625–356

HORN U, STRITTMATTER W, KREBBER A, KNUPFER U, KUJAU M, WENDEROTH R, MULLER K, MATZKU S, PLÜCKTHUN A, RIESENBERG D (1996) High volumetric yields of functional dimeric miniantibodies in *Escherichia coli*, using an optimized expression vector and high-cell-density fermentation under non-limited growth conditions. Appl Microbiol Biotechnol 46: 524–532

JANOWICZ ZA, MELBER K, MERCKELBACH A, JACOBS E, HARFORD N, COMBERBACH M, HOLLENBERG C P (1991) Simultaneous expression of the S and L surface antigens of hepatitis B, and formation of mixed particles in the methylotrophic yeast, *Hansenula polymorpha*. Yeast 7: 431–443

JENZELEWSKI V (2002) Fermentation and primary product recovery. In: *Hansenula polymorpha* – biology and applications (Gellissen G, Ed), Wiley-VCH, Weinheim, pp 156–174

JOB V, MOLLA G, PILONE MS, POLLEGIONI L (2002) Overexpression of a recombinant wild-type and His-tagged *Bacillus subtilis* glycine oxidase in *Escherichia coli*. Eur J Biochem 269: 1456–1463

JUZLOVA P, MARTINKOVA L, KREN V (1996) Secondary metabolites of the fungus Monascus: a review. J Ind Microbiol 16(3): 163–170

KANG HA, KANG W, HONG WK, KIM MW, KIM

JY, SOHN JH, CHOI ES, CHOE KB, RHEE SK (2001) Development of expression systems for the production of recombinant human serum albumin using the MOX promoter in *Hansenula polymorpha* DL-1. Biotechnol Bioeng 76: 175–185

KATZER W, BLACKBURN M, CHARMAN K, MARTIN S, PENN J, WRIGLEY S (2001) Scale-up of filamentous microorganisms from tubes and shake-flasks into stirred vessels. Biochem Eng J 7: 127–134

KIM MD, KANG HA, RHEE SK, SEO JH (2001) Effects of methanol on expression of an anticoagulant hirudin in recombinant *Hansenula polymorpha*. J Ind Microbiol Biotechnol 27: 58–61

KUNZE G, KUNZE I (1994) Characterization of *Arxula adeninivorans* from different habitats. Antonie van Leeuwenhoek 65: 29–34

LE CHEVANTON L, LEBLON G (1989) The ura5 gene of the ascomycete *Sordaria macrospora*: molecular cloning, characterization and expression in *Escherichia coli*. Gene 77(1): 39–49

LIN HY (2000) Influence of controlled glucose oscillations on a fed-batch process of recombinant *Escherichia coli*. J Biotechnol 79: 27–37

LOSEN M, LINGEN B, POHL M, BÜCHS J (2004) Effect of oxygen-limitation and medium composition on *Escherichia coli* in small-scale cultures. Biotechnol. Progress. (in press)

MAIER U, BÜCHS J (2001) Characterisation of the gas-liquid mass transfer in shaking bioreactors. Biochem Eng J 7: 99–106

MAYER AF, HELLMUTH K, SCHLIEKER H, LOPEZ-ULIBARRI R, OERTEL S, DAHLEMS U, STRASSER AWM, VAN LOON APGM (1999) An expression system matures: a highly efficient and cost-effective process for phytase production by recombinant strains of *Hansenula polymorpha*. Biotechnol Bioeng 63: 373–381

MEENS J, HERBORT M, KLEIN M, FREUDL R (1997) Use of the pre-pro part of *Staphylococcus hyicus* lipase as a carrier for secretion of *Escherichia coli* outer membrane protein A (OmpA) prevents proteolytic degradation of OmpA by cell-associated protease(s) in two different gram-positive bacteria. Appl Environ Microbiol 63: 2814–2820

METZ B, KOSSEN NWF (1977) Biotechnology Review: The growth of molds in the form of pellets – a literature review. Biotechnol Bioeng 29: 781–799

MIDDELHOVEN WJ, DE JONG IM, DE WINTER M (1991) *Arxula adeninivorans*, a yeast assimilating many nitrogenous and aromatic compounds. Antonie van Leeuwenhoek 60: 129–137

MIDDELHOVEN WJ, HOOGKAMER-TE NIET MC, KREGER VAN RIJ NJW (1984) *Trichosporon adeninovorans* sp. nov., a yeast species utilizing adenine, xanthine, uric acid, putrescine and primary n-alkylamines as sole source of carbon, nitrogen and energy. Antonie van Leeuwenhoek 50: 369–378

MORRIN M, WARD OP (1989) Studies on interaction of Carbopol-934 with hyphae of *Rhizopus arrhizus*. Mycol Res 92(3): 265–272

NOWROUSIAN M, MASLOFF S, PÖGGELER S, KÜCK U (1999) Cell differentiation during sexual development of the fungus *Sordaria macrospora* requires ATP citrate lyase activity. Mol Cell Biol 19(1): 450–460

OLSVIK E, KRISTIANSEN B (1994) Rheology of filamentous fermentations. Biotech Adv 12: 1–39

PAALME T, ELKEN R, KAHRU A, VANATALU K, VILU R (1997) The growth rate control in *Escherichia coli* at near to maximum growth rates: the A-stat approach. Antonie Van Leeuwenhoek 71: 217–230

PAPENDIECK A, DAHLEMS U, GELLISSEN G (2002) Technical enzyme production and whole-cell biocatalysis: application of *Hansenula polymorpha*. In: *Hansenula polymorpha* – biology and applications (Gellissen G, Ed). Wiley-VCH, Weinheim, pp 255–271

PARK CJ, LEE JH, HONG SS, LEE HS, KIM SC (1998) High-level expression of the angiotensin-converting-enzyme-inhibiting peptide, YG-1, as tandem multimers in *Escherichia coli*. Appl Microbiol Biotechnol 50: 71–76

PARSHINA SN, BLIZNIK KM, KAPULTSEVICH IG, STERKIN VE (1983) Growth characteristics of haploid and diploid *Hansenula polymorpha* yeasts on methanol in continuous culture. Mikrobiologia 52: 812–815

PETER CP, LOTTER S, MAIER U, BÜCHS J (2004) Impact of out-of-phase conditions on screening results in shaking flask experiments. Biochem Eng J 17: 205–215

PÖGGELER S, NOWROUSIAN M, JACOBSEN S, KÜCK U (1997) An efficient procedure to isolate fungal genes from an indexed cosmid library. J Microbiol Methods 29: 49–61

PSCHORR J, BIESELER B, FRITZ HJ (1994) Production of the immunoglobulin variable domain REIv via a fusion protein synthesized and secreted by *Staphylococcus carnosus*. Biol Chem Hoppe Seyler 375: 271–280

Qiu J, Swartz JR, Georgiou G (1998) Expression of active human tissue-type plasminogen activator in *Escherichia coli*. Appl Environ Microbiol 64: 4891–4896

Rajan JS, Virkar PD (1987) Induced pelletized growth of *Neurospora crassa* for tyrosinase biosynthesis in airlift fermentors. Biotechnol Bioeng 24: 770–772

Raschke WC, Neiditch BR, Hendricks M, Cregg J M (1996) Inducible expression of a heterologous protein in *Hansenula polymorpha* using the alcohol oxidase 1 promoter of *Pichia pastoris*. Gene 177: 163–167

Reinders A, Romano I, Wiemken A, de Virgilio C (1999) The thermophilic yeast *Hansenula polymorpha* does not require trehalose synthesis for growth at high temperatures but does for normal acquisition of thermo-tolerance. J Bacteriol 181: 4665–4668

Rösel H, Kunze G (1995) Cloning and characterization of a *TEF1* gene for elongation factor 1α from the yeast *Arxula adeninivorans*. Curr Genet 28: 360–366

Rösel H, Kunze G (1998) Integrative transformation of the dimorphic yeast *Arxula adeninivorans* based on hygromycin B resistance. Curr Genet 33: 157–163

Ryabova OB, Chmil OM, Sibirny AA (2003) Xylose and cellobiose fermentation to ethanol by the thermotolerant methylotrophic yeast *Hansenula polymorpha*. FEMS Yeast Res 4: 157–164

Simmons LC, Reilly D, Klimowski L, Shantha Raju T, Meng G, Sims P, Hong K, Shields RL, Damico LA, Rancatore P, Yansura DG (2002) Expression of full-length immunoglobulins in *Escherichia coli*: rapid and efficient production of aglycosylated antibodies. J Immunol Methods 263: 133–147

Silberbach M, Maier B, Zimmermann M, Büchs J (2003) Glucose oxidation by *Gluconobacter oxydans*: characterization in shaking flasks, scale-up and optimization of the pH profile, Appl Microbiol Biotechnol 62(1): 92–98

Sinacola JR, Robinson AS (2002) Rapid refolding and polishing of single-chain antibodies from *Escherichia coli* inclusion bodies. Protein Expr Purif 26: 301–308

Stöckmann C, Kück U, Gellissen G, Büchs J, (2) Submerged cultivation of the filamentous

fungus *Sordaria macrospora* in shake flasks. In preparation for FEMS Microbiol Lett

Stöckmann C, Losen M, Dahlems U, Knocke C, Gellissen G, Büchs J (2003a) Effect of oxygen supply on passaging, stabilising and screening of recombinant *Hansenula polymorpha* production strains in test tube cultures. FEMS Yeast Res 4: 195–205

Stöckmann C, Maier U, Anderlei T, Knocke C, Gellissen G, Büchs J (2003b) The oxygen transfer rate as key parameter for the characterization of *Hansenula polymorpha* screening cultures. J Ind Microbiol Biotechnol 30: 613–622

Stöckmann C, Midderhoff G, Knoll A, Kunze G, Gellissen G, Büchs J, (1) Definition of screening and fermentation conditions for recombinant *Arxula adeninivorans* strains. In preparation for J Appl Microbiol

Suckow M, Gellissen G (2002) The expression platform based on *H. polymorpha* strain RB11 and its derivatives – history, status and perspectives. In: *Hansenula polymorpha* – biology and applications (Gellissen G, Ed). Wiley-VCH, Weinheim, pp 105–123

Tanaka A, Ohnishi N, Fukui S (1967) Studies on the formation of vitamins and their function in hydrocarbon fermentation. Production of vitamins and their function in hydrocarbon medium. J Ferment Technol 45: 617–623

Terentiev Y, Breuer U, Babel W, Kunze G (2004) Non-conventional yeasts as producers of polyhydroxyalkanoates – Genetic engineering of *Arxula adeninivorans*. Appl Microbiol Biotechnol 64: 376–381

Tong WY, Yao SJ, Zhu ZQ, Yu J (2001) An improved procedure for production of human epidermal growth factor from recombinant *E. coli*. Appl Microbiol Biotechnol 57: 674–679

Tucker KG, Thomas CR (1994) Inoculum effects on fungal morphology: shake flasks versus agitated bioreactors. Biotechnol Tech 8(3): 153–156

van de Walle M, Shiloach J (1998) Proposed mechanism of acetate accumulation in two recombinant *Escherichia coli* strains during high density fermentation. Biotechnol Bioeng 57: 71–78

Van den Hombergh JPTW, van de Vondervoort JI, Faissinet-Tachet L, Visser J (1997) *Aspergillus* as a host for heterologous protein production: the problem of proteases. Trends Biotechnol 15: 256–263

WALZ M, KÜCK U (1995) Transformation of *Sordaria macrospora* to hygromycin resistance: characterisation of transformation by electrophoretic karyotyping and tetrad analysis. Curr Genet 29: 88–95

WARTMANN T, BÖER E, HUARTO PICO A, SIEBER H, BARTELSEN O, GELLISSEN G, KUNZE G (2002) High-level production and secretion of recombinant proteins by the dimorphic yeast *Arxula adeninivorans*. FEMS Yeast Res 2: 363–369

WARTMANN T, ERDMANN J, KUNZE I, KUNZE G (2000) Morphology-related effects on gene expression and protein accumulation of the yeast *Arxula adeninivorans* LS3. Arch Microbiol 173: 253–261

WARTMANN T, KRÜGER A, ADLER K, BUI MD, KUNZE I, KUNZE G (1995) Temperature dependent dimorphism of the yeast *Arxula adeninivorans* LS3. Antonie van Leeuwenhoek 68: 215–223

WARTMANN T, STOLTENBURG R, BÖER E, SIEBER H, BARTELSEN O, GELLISSEN G, KUNZE G (2003) The *ALEU2* gene – a new component for an *Arxula adeninivorans*-based expression platform. FEMS Yeast Res 3: 223–232

WEYDEMANN U, KEUP P, GELLISSEN G, JANOWICZ ZA (1995) Ein industrielles Herstellungsverfahren von rekombinantem Hirudin in der methylotrophen Hefe *Hansenula polymorpha*. Bioscope 7–14

WLAD H, BALLAGI A, BOUAKAZ L, GU Z, JANSON JC (2001) Rapid two-step purification of a recombinant mouse Fab fragment expressed in *Escherichia coli*. Protein Expr Purif 22: 325–329

XU B, JAHIC M, BLOMSTEN G, ENFORS SO (1999) Glucose overflow metabolism and mixed-acid fermentation in aerobic large-scale fed-batch processes with *Escherichia coli*. Appl Microbiol Biotechnol 51: 564–571

XU R, DU P, FAN JJ, ZHANG Q, LI TP, GAN RB (2002) High-level expression and secretion of recombinant mouse endostatin by *Escherichia coli*. Protein Expr Purif 24: 453–459

YANG XX, WARTMANN T., STOLTENBERG R, KUNZE G (2000) Halotolerance of the yeast *Arxula adeninivorans*, Antonie van Leeuwenhoek 77: 303–311

YEE L, BLANCH HW (1993) Recombinant trypsin production in high cell density fed-batch cultures in *Escherichia coli*. Biotechnol Bioeng 41: 781–790

ZICKLER D, LEBLON G, HAEDENS V, COLLARD A, THURIAUX P (1984) Linkage group-chromosome correlations in *Sordaria macrospora*: chromosome identification by three-dimensional reconstruction of their synaptonemal complex. Curr Genet 8: 57–67

15
Recombinant Hepatitis B Vaccines: Disease Characterization and Vaccine Production

Pascale Brocke, Stephan Schaefer, Karl Melber, Volker Jenzelewski, Frank Müller, Ulrike Dahlems, Oliver Bartelsen, Kyung-Nam Park, Zbigniew A. Janowicz, and Gerd Gellissen

15.1
Introduction

Hepatitis B vaccines belong to the most prominent examples for recombinant pharmaceuticals. Therefore, the description of a particular product development based on heterologous genes expression in the yeast *Hansenula polymorpha* is provided in the present chapter. This chapter is an updated and modified version of an article previously published in a monograph on *H. polymorpha* (Schäfer et al. 2002).

The advent of gene technology has provided new and powerful methods for the safe, efficient production of pharmaceuticals. Early examples include human growth hormone (Goeddel et al. 1979a) and insulin (Goeddel et al. 1979b) produced in recombinant strains of *E. coli*. Among the most important available recombinant pharmaceuticals are yeast-derived hepatitis B vaccines based on particles containing the hepatitis B surface antigen (HBsAg) inserted into the host-derived membrane (Emmini et al. 1986; Harford et al. 1987). Indeed, the success of current vaccination programs against hepatitis B is a result of the development of effective, yeast-derived recombinant hepatitis B surface proteins. Initially, the production of such vaccines was restricted to baker's yeast, *Saccharomyces cerevisiae*, but with improvements in biotechnological methods many new expression systems have been identified and developed. In particular, the methylotrophic yeast *Hansenula polymorpha* (Gleeson et al. 1986; Roggenkamp et al. 1986; Gellissen et al. 1990; Hollenberg and Gellissen 1997; Gellissen and Hollenberg 1997, 1999; Gellissen 2002) has been found to exhibit many superior expression characteristics (see Chapter 6), and is currently being used in the production of several vaccines against different subtypes of hepatitis B virus (Gellissen and Melber 1996).

This chapter will briefly describe the hepatitis B virus, its subtypes, and the disease it causes. Following this, recombinant protein production will be discussed, focusing in particular on the application of the *H. polymorpha* expression system. We describe how a heterologous *H. polymorpha* strain expressing HBsAg is constructed, and how efficient vaccine production is developed based on such recombinant strains. Finally, we will examine the current status of available vaccines and provide an outlook to alternative vaccine strategies.

15.2
Virus and Disease Characteristics

15.2.1
The Hepadnaviruses

Hepatitis B virus (HBV) was identified as the causative agent of serum hepatitis in the 1970s (Dane et al. 1970) after B. Blumberg discovered the Australia antigen (Blumberg et al. 1968). Blumberg first recognized this antigen as a serum protein specific for aborigines in Australia – it was only later that the infectious nature of Australia antigen was identified. It turned out to be the surface protein of HBV that is secreted into the bloodstream of infected patients in large excess over viral particles (Mahoney 1999). HBV was found to be endemic in many parts of the world, with more than 2 billion people having had contact with the virus, and more than 350 million chronic carriers of the virus (Zuckerman and Zuckerman 2000). Several viruses closely related to HBV were subsequently discovered in various primates, in members of the Sciuridae in Northern America as well as in the more distantly related members of the Aves (Table 15.1).

Tab. 15.1 Known hepadnaviruses and their hosts. Not all hepadnaviruses listed have been acknowledged officially by the International Committee for the Taxonomy of Viruses. For viruses infecting nonhuman primates, the classification and acronym is under debate.

Orthohepadnavirus	*Host*	*Reference*
Hepatitis B Virus (HBV)	Man *Homo sapiens sapiens*	Dane et al. (1970)
Chimpanzee Hepatitis B Virus (ChHBV)	Chimpanzee *Pan troglodytes*	Vaudin et al. (1988)
Gibbon Hepatitis B Virus (GiHBV)	White-Handed Gibbon *Hylobates lar*	Norder et al. (1996)
Orangutan Hepatitis B Virus (OuHBV)	Orangutan *Pongo pygmaeus pygmaeus*	Warren et al. (1999)
Gorilla Hepatitis B Virus (GoHBV)	Gorilla *Gorilla gorilla*	Grethe et al. (2000)
Woolly Monkey Hepatitis B Virus (WMHBV)	Woolly Monkey *Lagothrix lagotricha*	Lanford et al. (1998)
Woodchuck Hepatitis Virus (WHV)	Woodchuck *Marmota monax*	Summers et al. (1978)
Ground Squirrel Hepatitis Virus (GSHV)	Ground Squirrel *Spermophilus beecheyi*	Marion et al. (1980)

Tab. 15.1   (continued)

Orthohepadnavirus	Host	Reference
Arctic Squirrel Hepatitis Virus (ASHV)	Arctic Squirrel *Spermophilus parryi kennicotti*	Testut et al. (1996)
Avihepadnavirus		
Duck Hepatitis B Virus (DHBV)	Pekin duck *Anas domesticus*	Mason et al. (1980)
Heron Hepatitis B Virus (HHBV)	Grey Heron *Adrea cinerea*	Sprengel et al. (1988)
Snow Goose Hepatitis B Virus (SGHBV)	Snow Goose *Anser caerulescens*	Chang et al. (1999)
Stork Hepatitis B Virus (STHBV)	White Stork *Ciconia ciconia*	Pult et al. (1998)
Ross Goose Hepatitis Virus (RGHV)	Ross Goose *Anser rossi*	Shi et al. (1993)
Grey Teal Hepatitis B Virus (GTHBV)	Grey Teal *Anas gibberifrons gracilis*	Li et al. (1998)
Maned Duck Hepatitis B Virus (MDHBV)	Maned Duck *Chenonetta jubata*	Li et al. (1998)

All viruses are now united in the family of hepadnaviridae (Burrell et al. 2000), which is divided into the genus *Orthohepadnavirus* in mammals and the genus *Avihepadnavirus* in birds (Table 15.1). The relatedness of the viruses based on comparison of entire viral genomes is shown in Figure 15.1.

Woodchuck, ground squirrel, and arctic squirrel belong to the Sciuridae, and their respective hepadnaviruses are phylogenetically related. Clearly different, but closer to human HBV, is the recently discovered hepadnavirus of the woolly monkey (WMHBV). Human HBV can be grouped into eight genotypes, A to H, which differ by at least 8% (Norder et al. 1994; Stuyver et al. 2000; Arauz-Ruiz et al. 2002). Genotype F, which is found in Brazil, Colombia, and Polynesia is the most divergent genotype (Norder et al. 1994; Naumann et al. 1993), and is grouped into genotypes F1 and F2 (Norder et al. 2003). Because the genomes of the hepadnaviruses found in nonhuman primates are so similar to HBV they have been considered genotypes of the primate hepadnaviruses (Takahashi et al. 2000).

The genome of hepadnaviruses codes for four groups of proteins; all are encoded on the minus strand (Figure 15.2):

- The core protein (HBcAg) forming the nucleocapsid and a shorter form of it (HBeAg) which is secreted by the host cell. HBeAg is produced by usage of the in-frame start codon situated upstream of the core start, but being proteolytically truncated afterwards at its C-terminus. The function of the latter is unknown.
- The three hepatitis B surface antigen (HBsAg) proteins of different size having different initiation sites of transcription but being identical at their carboxy-ter-

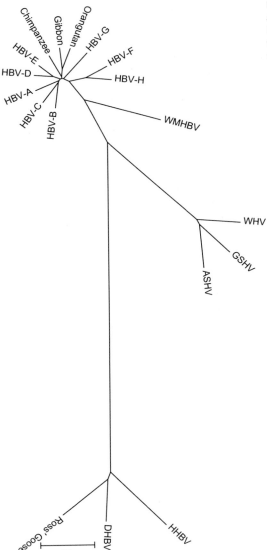

Fig. 15.1 Phylogenetic relatedness of all available completely sequenced hepadnaviruses. The sequences of the viral genomes were analyzed using the program Dnastar.

minus. These are the small surface proteins (SHBs) of 24 kDa, which are the most abundant; the middle HBs of 32 kDa generated by start of translation from a start codon in frame upstream of the SHBs start (MHBs); and the large surface proteins (LHBs) of 39 kDa coding for the largest surface protein of HBV which contains all domains of S- and MHBsAg plus additional amino acids derived from usage of the third in-frame start codon of the nested set of surface ORFs.

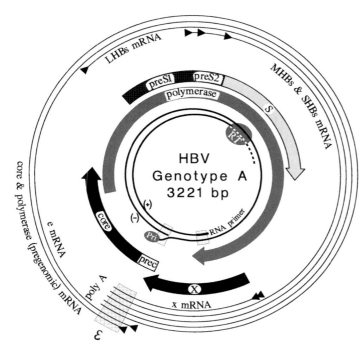

Fig. 15.2 Schematic diagram of the HBV genome and genetic organization. The inner circles represent the viral DNA as found in virions. The arrows represent the four different open reading frames.
Outer circles represent the co-terminal viral mRNAs as found in infected cells. The 5'-end of (–) DNA strand is linked with the priming
domain (Pri), and the 3'-end of the (+) strand DNA is associated
with the reverse transcriptase domain (RT) of the viral polymerase.
(Modified after Kann and Gerlich 1998.)

- The DNA polymerase, which is also a reverse transcriptase with a primer function
  and an RNAseH domain.
- Protein X (HBx) is a protein with unknown function for the virus and a plethora
  of reported properties *in vitro*.

While the DNA-minus strand encapsidated in the virions has full length, the plus
strand is incomplete and varies in length. The viral DNA is held in circular form by
an overlap between plus and minus strands of about 240 bp for orthohepadnaviruses
and about 60 bp for avihepadnaviruses.

The hepadnaviridae are round, enveloped viruses (Figure 15.3). HBV particles
have a hydrated diameter of 52 nm (Jursch 2000) which appears as 45 nm in negative staining. Diameters of 40–47 nm have been reported for the other hepadnaviridae (Schaefer et al. 1998). The viral genome is packed, together with the viral polymerase and a cellular kinase, into a capsid with a diameter of 34 nm (as determined

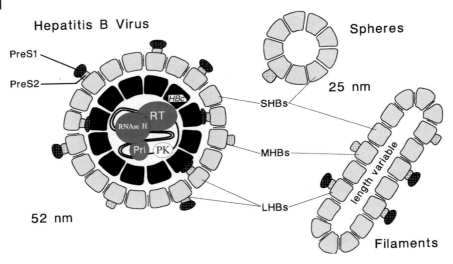

Fig. 15.3 Schematic diagram of hepadnavirus particles. The virus particles contain an internal nucleocapsid (HBc), the viral genome, the polymerase consisting of domains with reverse transcriptase activity (RT), RNaseH and a domain serving as primer for the synthesis of minus strand DNA (Pri). The subviral particles shown on the right, are made up only of surface proteins in different composition. (Modified after Kann and Gerlich 1998.)

by cryo-electron microscopy; Crowther et al. 1994). In the serum of chronic carriers the viral surface protein is found as DNA-free spherical or filamentous particles in large excess over infectious virions (Heermann and Gerlich 1991).

The hepatitis B virus is taken up by an as-yet unknown mechanism/receptor by the hepatocyte (Figure 15.4). Somewhere within the cytoplasm the viral envelope is removed and the free nucleocapsid moves to the nuclear pores (Kann et al. 1997), where the HBV DNA-genome leaves the capsid and translocates into the nucleus. In the nucleus, the short plus strand is completed (by the attached viral DNA polymerase), leading to a covalently closed circular double-stranded DNA (cccDNA). This cccDNA serves as template for the transcription of viral RNAs in the nucleus by cellular RNA-polymerase II. The different mRNA species are exported from the nucleus into the cytoplasm, where translation occurs. Synthesis of the HBsAg takes place at the endoplasmic reticulum (ER), and the HBsAg is anchored in the ER-membrane. The core protein and the viral polymerase are translated by free ribosomes in the cytoplasm from the largest mRNA (3.5 kb). These two proteins form a complex with their mRNA whereby the core protein encapsidates the viral pre-genome in the cytoplasm. The encapsidated viral RNA is reverse-transcribed into the complete DNA minus strand by the viral polymerase, and synthesis of the incomplete plus strand occurs. Thereafter, the viral capsid is enveloped at the ER with the HBs-containing membrane, and finally the infectious virus is secreted. De-

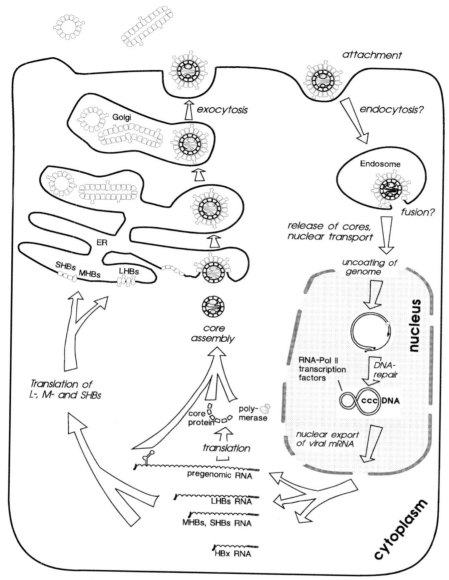

Fig. 15.4 Simplified model of the hepadnaviral life cycle.
For details, see the text.

tailed reviews on HBV replication can be found in Kann and Gerlich (1998) and Nassal and Schaller (1996).

15.2.2
Subtypes of HBV

Four major antigenic determinants of HBs can be distinguished with antibodies that recognize different epitopes on particles formed by SHBs. All known subtypes contain the *a*-determinant (Le Bouvier 1972), which is encoded between amino acid residues 124 and 147 (Ashton-Rickhardt and Murray 1989). The difference between the mutually exclusive subtype-specific determinant d/y (Le Bouvier 1972) and w/r (Bancroft et al. 1972) is generated by amino acid exchanges from K to R at residues 122 (Peterson et al. 1984) and 160 (Okamoto et al. 1987), respectively (Figure 15.5).

Additional sub-determinants allowed the differentiation of four serotypes of ayw and two of adw (Couroucé et al. 1976). Thus, according to the Paris workshop on

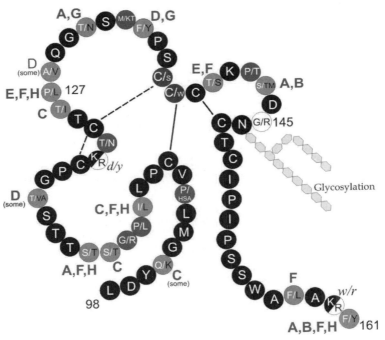

Fig. 15.5 Hypothetical model of the a-determinant and its adjacent sequences formed by the major surface protein of the hepatitis B virus. The a-determinant comprises residues 124 to 147. Conserved amino acid residues are shown in black. Non-conserved residues are shown in gray. If residues vary subtype-specifically, the minor genotype is indicated in bold beside the amino acid. Residues 122 and 160 which confer subtype changes from d/y and w/r, respectively, are indicated in black and white. The position of the frequently described escape variant G145→R is shown in white. The figure is based on unpublished data and on Weinberger et al. (2000) and Günther et al. (1999).

Tab. 15.2 Correlation of subtype and genotype of HBV. Subtype adw2 of genotype B is found mainly in the Far East.

Genotype	Subtype
A	adw2
B	adw2
C	adr and ayr
D	ayw1, 2, 3
E	ayw4
F	adw4
G	adw2 [a]

a) The probable subtype of genotype G was deduced from the DNA sequence.

HBV surface antigen subtypes, eight serotypes exist (adr, ayr, ayw_1, ayw_2, ayw_3, ayw_4, adw_2, and adw_4). By use of the determinant q^+/q^- found in subtype adr, nine subtypes can be distinguished (Courouce et al. 1976). Later, several other determinants on SHBs (t/i) (Ohnuma et al. 1993) or the preS2-epitope of MHBs (Usuda et al. 1999) have been identified using monoclonal antibodies.

Because several subtypes of SHBs are distinguished only by the exchange of a single amino acid residue, the typing of HBV has increasingly been carried out by analyzing DNA sequences from complete genomes or parts of HBV (Okamoto et al. 1988). With the recent discovery of genotype H of HBV (Arauz-Ruiz et al. 2002), eight genotypes, A to H, can be distinguished. Each genotype differs by more than 8% at the nucleotide level from the others (see Figure 15.1). The data in Table 15.2 show that, except for adw_2 and adw_4 (which can be divided into genotypes A, B and G, or F and H, respectively), all other subtypes can be grouped to specific genotypes. In single cases, the exchange of one amino acid altered the subtype such that HBV strains with subtype adw_2 turned out to be of genotype D and not of genotype A, as would be expected from serotyping (Blitz et al. 1998).

The distribution of HBV subtypes in the world has been determined using sera of 5337 silent HBsAg carriers (Courouce-Pauty et al. 1983), and a distinct geographic prevalence was found. The worldwide distribution of HBV genotypes A to H is shown in Figure 15.6.

In 2000, Kao et al. reported that HBV genotype C was associated with more severe liver disease, and further evidences are now accumulating that HBV genotypes may influence the course of disease (Chu and Lok 2002; Bartholomeusz and Schaefer 2003). In addition, a better response rate to interferon treatment in patients chronically infected with HBV genotype B compared to genotype C has been reported (Wai et al. 2002), pointing towards a clinical relevance of genotypes.

How did the existence of different HBV genotypes/serotypes influence the design of a prophylactic vaccine? The cross-protection of chimpanzees was achieved when animals were immunized with SHBsAg of a subtype (ad) differing from the HBV strain used for challenging (ay) (Purcell and Gerin 1975). However, in these experiments it was also found that the majority of antibodies generated immediately after vaccination were subtype-specific. The immunization of human vaccinees or chim-

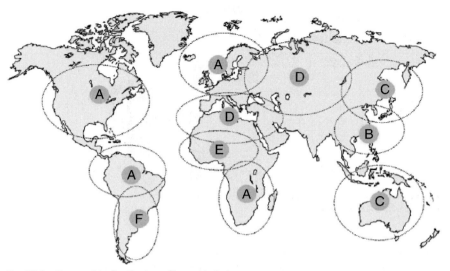

Fig. 15.6 Geographic distribution of hepatitis B virus genotypes.

panzees with SHBs of adw specificity first gave rise to *d*-specific antibodies. This response broadened to include antibodies against the *a*-determinant with time (Purcell and Gerin 1975; Legler et al. 1983). Interestingly, the first immune escape variant of HBV after vaccination with genotype A was found in a region with prevalence of genotoype D (Carman et al. 1990). In studies where sub-typing was carried out with serologic analysis, it was found that vaccine breakthrough more commonly occurred in cases with the *y* than with *d* allele (Lelie et al. 1991). In addition, Ngui et al. reported that immune escape in properly vaccinated children born to HBV DNA-positive mothers were less likely to occur with HBV genotype C (Ngui et al. 1998).

However, statistical analysis has not yet revealed any differences in immune response or frequency of vaccine breakthrough dependent on the SHBs subtype used for vaccine production (Günther et al. 1999; Assad and Francis 2000).

15.2.3
Pathogenesis

The hepatitis B virus is transmitted parenterally via infected blood and blood products, via mucosal routes, organ transplantation, or perinatally during birth. The virus mainly infects liver tissue. Hepatitis B virus infections can be either transient or chronic. About 90–95 % of infected adults recover completely after apparent or inapparent hepatitis, and are regarded as cured (Figure 15.7). Damage to hepatocytes infected by HBV is not caused by the virus itself – HBV *per se* is not cytopathic – but by the host immune response.

Chronic hepatitis B (SHBs antigenemia for more than 6 months) develops in up to 5–10 % of infected healthy adults, and in up to 90 % of newborns (Mahoney 1999;

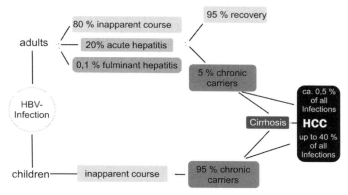

Fig. 15.7 Schematic diagram of the course of hepatitis B virus infection. HCC = hepatocellular.

Kann and Gerlich 1998). After persisting hepatitis B infection, cirrhosis of the liver eventually develops. Even without preceding cirrhosis the development of hepatocellular carcinoma (HCC) is possible (Bréchot et al. 1998). It has been estimated that about 0.5% of all people infected with HBV develop HBV-associated HCC between 20 and 40 years after infection (Buendia et al. 1998). Thus, it is estimated that about 25% of chronic HBV carriers – that is, up to 1 million people per year, die from cirrhosis or HCC as sequelae of hepatitis B (Zuckerman and Zuckerman 2000).

15.2.4
Immune Response

The immune response in self-limited acute hepatitis is characterized by a strong co-ordinated activation of innate as well as cellular (CD4+ and CD8+ T cells), and humoral adaptive responses that finally allow the control of HBV, although at the expense of possible clinical disease. In the early innate response, interferon-γ (IFNγ) and tumor necrosis factor-α (TNFα) produced by NK and NKT cells, contribute to the initial control of HBV replication non-cytolytically (Guidotti et al. 1999; Kakimi et al. 2000; Kakimi et al. 2001; Webster et al. 2000). Shortly afterwards, a vigorous adaptive immune response develops. The production of antibodies reduces the quantity of circulating virus outside the cells, lowering the risk of viral spread and protecting from reinfection later on. The appearance of detectable antibodies against SHBs shows the seroconversion that reflects the clearance of HBV infection by the infected host, and this is used in diagnosis. Anti-HBc-antibodies (IgG) are a diagnostic marker for contact with the virus as they are present in acute, chronic and resolved infection – that is, they are retained for life. The cellular arm of the response is characterized by a CD4+ T-cell response with a type 1 profile of cytokine production and a multi-specific polyclonal CD8+ T-cell response which contributes to lysis of the infected liver cells, as well as to the suppression of viral replication by non-cytolytic mechanisms (Guidotti 2002). In particular, the CD8+ T cells have an essential role, as shown by depletion experiments conducted in HBV-infected chimpanzees (Thimme et al. 2003); notably, these cells are specific for several epitopes within the viral surface and core antigens and the polymerase. Effective viral control is associated with these multi-specific cytotoxic T lymphocytes (Penna et al. 1991; Nayersina et al. 1993; Rehermann et al. 1995, 1996; Bertoletti and Naoumov 2003).

The immune response observed in chronic hepatitis is weaker and of different quality compared to that in acute self-limiting hepatitis. The cytotoxic T lymphocyte (CTL) response in chronic patients is only oligoclonal, relatively weak, and has a different hierarchy of epitope specificity (Bertoletti and Maini 2000; Maini and Bertoletti 2000).

Antibodies against the surface antigen (HbsAg) are absent in patients who progress to a chronic state (Budkowska et al. 1986; Theilmann et al. 1987; Coursaget et al. 1988; Alberti et al. 1990) (Figure 15.8). The liver damage occurring in chronic hepatitis B is not proportional to the number of HBV-specific T cells (Maini et al. 2000), but is rather caused by recruited non-antigen-specific cells (Sitia et al. 2002). The extent of cell death in an infected liver of a seemingly healthy chronic carrier can be enormous, but without any signs of liver disease apparent in the serum. It has been estimated that in chronic infection about 10^9 hepatocytes (0.3–3 % of all hepatocytes) are killed each day, and must be replenished (Nowak et al. 1996). This constant cell replacement, as well as the possible integration of gene segments of viral DNA, may be reasons for the development of hepatocellular carcinoma (Hillemann 2003).

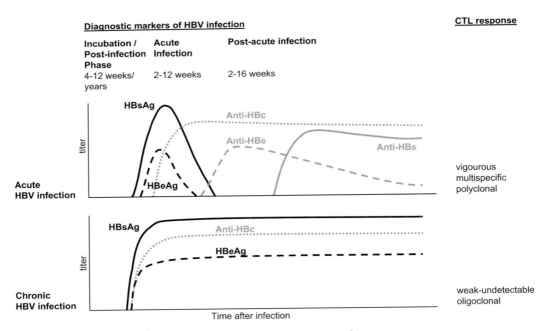

Fig. 15.8 Diagnostic markers of hepatitis B virus (HBV) infection (antigens and antibodies) and type of cytotoxic T lymphocyte (CTL) response in acute versus chronic HBV infection.

15.2.5
Prophylactic HBV Vaccines

In many countries of the world, the first licensed vaccines against HBV became available during the early 1980s. As there is no possibility for propagation of the virus *in vitro*, this vaccine was produced by harvesting and purifying HBsAg from the serum of chronic carriers. However, while they are effective and safe, serum-derived vaccines are expensive and in relatively short supply due to a shortage of human carrier plasma that meets the requirements for vaccine production. Since 1986, recombinant hepatitis B vaccines have been used as a more practical alternative, with the HBsAg being produced in yeast. Immunization with HBsAg results in antibody production against the antigen that protects against an infection with HBV. The newborn of chronically HBV-infected mothers receive a combined active and passive immunization.

15.3
Recombinant Vaccine Production

Heterologous HBV antigen production has been developed in several different host expression systems, including yeasts, bacteria, insect cells, plant cells, mammalian cells, and transgenic animals (Figure 15.9). Bacterial systems such as *Escherichia coli* (Billman-Jacobe 1996; Makrides 1996), whilst cost-effective and simple to work with, are often unable to produce pharmaceutical proteins in a correctly folded form, and this can sometimes compromise vaccine efficacy. The suitability of plant systems such as potato, tobacco, lupin, or lettuce (Mason and Arntzen 1995) for the production of pharmaceutical-grade proteins has yet to be shown. Today, all recombinant hepatitis B vaccines approved for the market are produced in either yeast or mammalian cells (Assad and Francis 2000) (Table 15.3).

15.3.1
Yeasts as Production Organisms

Currently, there are two recombinant hepatitis B vaccines which are approved by the FDA and available for use (Table 15.3). Both are S-antigen vaccines produced in the yeast *S. cerevisiae*. This system has limitations however (as have been pointed out in the previous chapters), and recombinant antigen production is therefore currently under investigation in several alternative yeast host systems (Reiser et al. 1990; Buckholz and Gleeson 1991; Romanos et al. 1992; Gellissen et al. 1992a), namely the methylotrophs *H. polymorpha* (Roggenkamp et al. 1986) and *Pichia pastoris* (Cregg et al. 1985; Cregg and Madden 1987). To date, two *H. polymorpha*-based systems for the production of the adw$_2$ and adr subtypes of HBsAg have been developed, one of which has been approved for use by the World Health Organization (WHO) (Table 15.3).

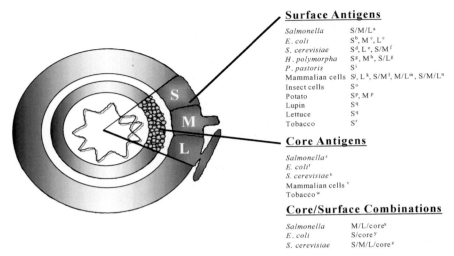

Surface Antigens

Salmonella	S/M/L[a]
E. coli	S[b], M[c], L[c]
S. cerevisiae	S[d], L[e], S/M[f]
H. polymorpha	S[g], M[h], S/L[g]
P. pastoris	S[i]
Mammalian cells	S[j], L[k], S/M[l], M/L[m], S/M/L[n]
Insect cells	S[o]
Potato	S[p], M[p]
Lupin	S[q]
Lettuce	S[q]
Tobacco	S[r]

Core Antigens

Salmonella[s]
E. coli[t]
S. cerevisiae[u]
Mammalian cells[v]
Tobacco[w]

Core/Surface Combinations

Salmonella	M/L/core[x]
E. coli	S/core[y]
S. cerevisiae	S/M/L/core[z]

Fig. 15.9 Heterologous expression of hepatitis B genes. The various recombinant antigens produced to date are shown in a schematic drawing of the virus. They have been produced in the expression system indicated. References are as follows: [a]Wu et al. (1989); [b]Lee et al. (1986); [c]Kim et al. (1996); [d]Harford et al. (1987); [e]Korec et al. (1989); [f]Yoshida et al. (1991); [g]Janowicz et al. (1991); [h]Shen et al. (1989); [i]Cregg et al. (1987); [j]Laub et al. (1983); [k]Yu et al. (1992); [l]Shouval et al. (1994); [m]Youn and Samanta (1989); [n]Diminsky et al. (1997); [o]Deml et al. (1999); [p]Ehsani et al. (1997); [q]Kapusta et al. (1999); [r]Liu et al. (1994); [s]Schödel et al. (1996); [t]Wizemann and von Brunn (1999); [u]Miya-Nohara et al. (1986); [v]Roossinck et al. (1986); [w]Tsuda et al. (1998); [x]Schödel et al. (1994); [y]Shiau and Murray (1997); [z]Shiosaki et al. (1991). Commercially available *S. cerevisae*- and *H. polymorpha*-derived hepatitis B vaccines are listed in Table 15.3.

Tab. 15.3 Commercially available *S. cerevisiae* and *H. polymorpha*-derived hepatitis B vaccines.

Product	Trade name	Company	Approval, date	Recombinant host organism
HBsAg vaccine	Recombivax	Merck and Co., Inc.	FDA, July 1986	*S. cerevisiae*
HBsAg vaccine	Engerix B	SmithKline Beecham Biologicals	FDA, September 1989	*S. cerevisiae*
HBsAg vaccine	AgB	Laboratorio Pablo Cassará; (LPC)	Argentina, September 1995	*H. polymorpha*
HBsAg vaccine	Hepavax-Gene	Green Cross Vaccine	WHO, 1997	*H. polymorpha*
HBsAg vaccine	Biovac-B	Wockhardt	India, 2001	*H. polymorpha*
HBsAg vaccine	Gene Vac-B	Serum Institute of India	India, 2001	*H. polymorpha*

15.3.2
Construction of a *H. polymorpha* Strain Expressing the Hepatitis B S-antigen

The construction of recombinant *H. polymorpha* strains generally follows a standard protocol, which was described previously in Chapter 6:

- construction of the expression cassette and plasmid vector;
- transformation of *H. polymorpha*; and
- isolation and characterization of recombinant strains.

15.3.2.1 Expression Cassette and Vector Construction

The construction of the S-antigen-expressing H415 strain by Janowicz et al. (1991) is a typical example of this process. A 683-bp S-antigen coding sequence was derived from plasmid pRIT10616 (Harford et al. 1987), and a *MOX* promoter fragment as well as signals for transcription termination were derived from the *H. polymorpha* *MOX* gene (Ledeboer et al. 1985a; Eckart 1988). These three elements were combined to form a *MOX* promoter-*HBsAg* gene-*MOX* terminator expression cassette. The functional cassette was then inserted into a plasmid vector with the following features: a chloramphenicol resistance gene for propagation in *E. coli*, a *H. polymorpha* autonomously replicating sequence (*HARS*1) (Roggenkamp et al. 1986), and the *URA3* gene from *S. cerevisiae* (Stinchcomb et al. 1980) as a selection marker for transformation of *H. polymorpha* (plasmid pRBS-269). Plasmids containing *HARS* sequences have been shown to exhibit a high frequency of integration into the genome. It has been found that use of the *HARS* vector in *H. polymorpha* characteristically yields multiple integrants. Strains have been identified containing up to 60 copies of foreign expression cassettes. The plasmid used for generation of the *H. polymorpha* strain producing the *Hansenula*-derived hepatitis B vaccine, Hepavax-Gene®, is similar to pRBS-269. Production strains were also constructed incorporating the *FMD* promoter instead of *MOX*. A physical map of pFPMT-sadw$_2$, the plasmid vector harboring the *FMD* promoter used to generate strains for the production of AgB (another *H. polymorpha*-derived HBV vaccine), is shown in Figure 15.10.

15.3.2.2 Transformation of *H. polymorpha*

Using the polyethylene glycol method (Cregg et al. 1985), *H. polymorpha* strain RB10 (*odc1*) was transformed with pRBS-269 and screened for plasmid integration. Several transformant strains with stably integrated expression cassettes were generated, and strain H415 was one of several which were isolated and tested for S-antigen expression under nonselective conditions (Janowicz et al. 1991).

15.3.2.3 Strain Characterization

Expression characteristics were tested by growing the transformed strains on semi-rich media containing glucose, glycerol, or methanol. The amount of HBsAg produced relative to a standard amount of purified HBsAg was measured by quantitative immunoblot assay. The level of production in strain H415 grown in methanol was several milligrams HBsAg per 100 mg of soluble protein. Synthesis was de-

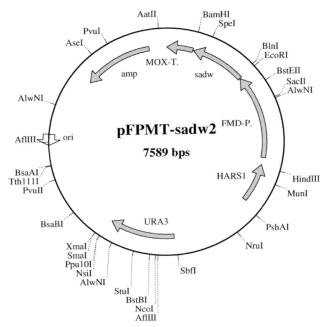

Fig. 15.10 Map of plasmid vector pFPMT-sadw2 containing a *FMD*-promoter/HBsA-g(adw2)/*MOX*-terminator expression cassette. pFPMT-sadw2 is composed of the following DNA fragments, starting from the unique *Hind*III site in a counter-clockwise direction: the *FMD* promoter; a fragment coding for HBsAg (sub-type adw2); a *MOX* sequence for transcriptional termination; a sequence containing a gene for ampicillin resistance and an origin of replication for propagation in *E. coli*; the *URA3* gene as a transformation marker in ura3 mutants of *H. polymorpha*; and a *Hansenula* autonomously re-plicating sequence (HARS1).

creased by 70% when the cells were cultured on glycerol, and no S-antigen synthesis was detected in glucose-cultured cells, indicating that antigen production was as effectively controlled as the natural *MOX* gene.

Antigenicity of the crude protein extract was also tested using commercially available monoclonal AUSZYME and polyclonal AUSIA antibody tests. The extract was highly reactive with antibodies specific for conformational epitopes of the 22-nm HBsAg particle. Density determination by centrifugation through cesium chloride or sucrose gradients and subsequent AUSZYME and Western blot assays, along with size analysis utilizing electron microscopy, showed the presence of the expected 22-nm, 1.17–1.20 g cm^{-3} particle (Figure 15.11).

In *H. polymorpha*, the extracted viral surface antigen is found to be assembled into yeast-derived lipid membranes, similar to the situation in other yeasts. As mentioned earlier, peroxisomal proliferation – and membrane proliferation in general – is associated with methanol induction, and previous studies have indicated that this lipoprotein particle structure is essential for the antigenicity of the HBsAg (Rutgers et al. 1988). Furthermore, a clear gene dosage effect has been observed (Janowicz

A.

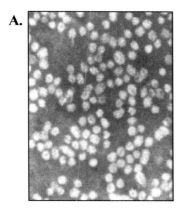

B.

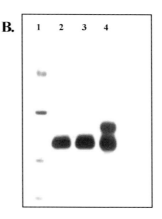

Fig. 15.11 Characterization of recombinant HBsAg particles produced in *H. polymorpha*. HBsAg particles were purified and analyzed as described in the text (see Section 15.3.1). A) Electron microscopy analysis (original magnification, ×142 000); B) SDS-PAGE analysis of purified HBsAg. Two batches of HBsAg were separated on 12% SDS gels and visualized by silver staining. Lane 1 = MW marker; lanes 2 and 3 = two batches of purified r-HBsAg; lane 4 = commercial serum-derived HBsAg.

et al. 1991; Gellissen et al. 1992a), and high-copy number strains are found to be mitotically stable under nonselective conditions (Janowicz et al. 1991). The previously mentioned *H. polymorpha*-derived HBsAg vaccines are produced in recombinant strains similar to H415, but with higher numbers of integrated cassettes. For example, the strain used for the production of AgB contains 32 functional copies of the integrated HBsAg expression cassette.

15.3.3
H. polymorpha-derived HBsAg Production Process

The HBsAg production process in *H. polymorpha* consists of 11 steps, and the schedule is illustrated in Figure 15.12. Currently, cultivation on a 50-L fermentor scale yields HBsAg in a multigram range per batch (Piontek 1998).

15.3.3.1 Fermentation (Upstream Process)

The starting material for each individual batch consists of one vial of a working cell bank. A working cell bank is a 2-mL aliquot of a production strain culture mixed with glycerol to a final concentration of 17% and stored at –70 °C. A statistically significant number of vials are tested for viable cell content, presence of the HBsAg gene, copy number of the integrated expression cassette, as well as for mitotic stability of the vector copies at the start and the end of a typical fermentation process. The high homogeneity of the cell seed vials, together with the reproducible fermentation conditions, ensures batch-to-batch uniformity. Furthermore, the genetic stability of the host/vector system facilitates approval by the relevant regulatory authorities.

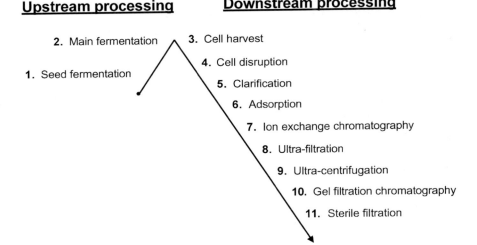

Fig. 15.12 Production process for HBsAg particles in recombinant *H. polymorpha*. Recombinant strains of *H. polymorpha* expressing HBsAg are fermented, and the antigen is purified as described in the text. The process yields purified HBsAg integrated onto yeast-derived membrane particles which may then be adsorbed to aluminum hydroxide for administration as a vaccine.

Product-containing cells are generated via a two-fermentor cascade, consisting of a 5-L seed fermentor used to inoculate the 50-L main fermentor. Seed cultivation is performed in a batch fermentation mode, without oxygen limitation. The whole fermentation process, starting from the single vial of working cell bank, yields a biomass of more than 10 g dry cell weight per liter, within 55 h. The production fermentation is carried out in a two-carbon-source mode (Figure 15.13). The initial cultivation is performed with glycerol feeding in fed-batch mode, with subsequent semi-continuous glycerol feeding controlled by dissolved oxygen level. While the first phase of this cultivation is performed to obtain high cell densities, the purpose of the second phase is to derepress the promoter controlling expression of the HBsAg gene. Thus, a low glycerol concentration is maintained by oxygen-controlled feeding. The derepression results in high levels of intracellular HBsAg, and batch-wise addition of methanol in the final fermentation phase increases the product amount into the multi-gram range. As pointed out previously, the addition of methanol also serves to induce the massive membrane proliferation essential for the formation of the antigenic lipoprotein particles (see also Chapters 6 and 14).

The complete process can be controlled by a timer to determine the batch additions, and by an oxygen probe controlling the feeding pump during the derepression phase. After less than 70 h of total fermentation time this procedure yields 100 g dry cell weight per liter, and a total product amount in multi-gram ranges. Both seed and production fermentations are performed with fully synthetic medium, free of

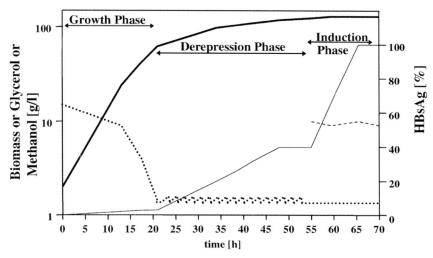

Fig. 15.13 Fermentation of a HBsAg-producing *H. polymorpha* strain (schematic). The fermentation procedure follows the description provided in the text. Thick line = biomass; thin line = HBsAg; dashed line = methanol; dotted line = glycerol.

complex and undefined additives. Microscopic examination, turbidity measurement and dry weight determination are performed off-line, as in-process controls.

15.3.3.2 Purification (Downstream Processing)

Tangential flow filtration is applied to harvest the culture. In this step, media components are removed from the cells and exchanged against the buffer appropriate for the cell disruption step. The filtration can either be performed in a continuous diafiltration mode, or by repeated cycles of concentration and dilution. In any case, the tangential flow filtration step yields a concentrated cell suspension with a much higher cell density than the final culture, allowing for a substantial decrease in the process time required for cell disruption.

Two different mechanical cell disruption principles have been applied for release of product from the cells: 1) grinding in a glass bead mill; and 2) disruption in a high-pressure homogenizer. In several independent test batches, the amount of product released per gram of dry cells has been found to be substantially higher using pressure homogenization compared to the grinding processes. Analysis of the final purified antigen did not show any difference with respect to the disruption method used. Furthermore, the grinding process depends strongly on the quality and consistency of the glass beads in order to obtain reproducible results, while the high-pressure homogenization does not require any additional raw material for disruption of the cells. For both processes, the buffer system must be supplemented with inhibitors to block the action of the intracellular proteases which are released together with the product. Detergent must also be added to the buffer to solubilize the antigenic particles. Regardless of the cell disruption procedure used, the subsequent purification steps yield Final Aqueous Bulk conforming to the WHO technical guidelines.

A particular disadvantage of the high degree of cell homogenization required for maximal product release is the heterogeneous composition of the crude cell extract. In addition to a low percentage of more or less intact cells, a mass of cell debris of varying shape and size is created. The high amount of protein and DNA released results in an increased viscosity of the extract. High-speed centrifugation of such an extract yields a turbid supernatant, a soft pellet and an interphase, which easily contaminates the supernatant during harvest. This problem is minimized by precipitation of the total crude extract with polyethylene glycol. The resulting supernatant has a drastically reduced level of contaminating cell debris and, in parallel, a slightly reduced amount of contaminating host proteins. As in the later purification steps, continuous high-speed centrifugation with automated solid harvest is introduced for solid/liquid separation, allowing for closed operation throughout the complete process.

The most efficient way to remove host proteins is the specific binding of HBsAg particles to a suitable matrix. Whereas HBsAg binds almost quantitatively, only about 10% of the total host proteins co-purify with the antigen particles. This purification method involves an initial adsorption to the matrix in a batch operation, the subsequent removal of unbound proteins, washing of the adsorbed HBsAg, and finally desorption. Adsorption and desorption are controlled via pH and temperature shifts. Step yields of 80% are obtained with 10-fold purification. Separation of the solid matrix from the liquid supernatant is the most critical operation in this process. The matrix material is very sensitive to mechanical stress, for example shear forces applied during suspension of the cells, and especially during entrance into the centrifuge bowl.

Based on the initial volume of the crude cell extract, it is calculated that an approximately threefold reduction in volume is achieved. The subsequent steps – ion-exchange chromatography and ultrafiltration – are designed to further reduce the volume and remove a major part of the host-derived lipids. Since the active principle, the HBsAg particle, is a mixture of small surface antigen integrated into the host's membrane structure, the lipid content is an important characteristic to be monitored during purification.

Volume reduction by ultrafiltration is crucial, as the final step for removal of host contaminants is a cesium chloride ultracentrifugation. Filtering the product in a diluted form would require increased investment in equipment and raw materials. The gradient required for separation of particles from contaminants is self-forming within 48 h of operation. The particles accumulate as a distinct band, which can be identified visually as a brownish product fraction. Harvesting can be performed in different modes depending on the use of either tube-type centrifugation or zonal operation:

- via a canula injected into the side wall of the tube and visually monitoring the removal;

- via a canula injected into the bottom of the tube and fractionation controlled by UV monitoring; or
- via displacement with more dense cesium chloride solution in the zonal operation.

The last step of purification – gel filtration chromatography – is designed to remove the cesium salt from the product. Separation of the HBsAg particles from the cesium is monitored using ultraviolet absorbance and conductivity. The specifications characterizing the HBsAg, in the form of Final Aqueous Bulk, derived from the above summarized purification process, are listed in Table 15.4.

The purified HBsAg is formulated by adsorption to an aluminum hydroxide adjuvant and addition of a preservative. A single adult dose containing 20 µg rHBsAg may be administered in three single injections at 0, 1, and 6 months.

Tab. 15.4 Specifications characterizing purified *H. polymorpha*-derived HBsAg, in the form of Final Aqueous Bulk.

Purity	>95 % according to SDS-PAGE monomeric, dimeric and trimeric bands
Identity	Reactive with anti/HBsAg antibodies according to Western blot analysis
Reactivity	>1 mg according to AUSZYME per mg protein according to Lowry
Lipid content	0.5–1.8 mg mg^{-1} protein according to Merckotest
Nucleic acid content	<100 pg per 20 µg protein according to Threshold or Dot-blot analysis
Cesium content	<10 µg per 20 µg protein

15.4
HepavaxGene®

HepavaxGene®, a vaccine based on HBsAg of adr subtype, has been tested extensively in preclinical and clinical studies to assess its safety and immunogenicity. In addition to controlled clinical data, there is post-marketing experience in 36 countries where HepavaxGene® is registered. Since its introduction into the market in 1996, more than 350 million doses of HepavaxGene® have been administered to humans in more than 90 countries (Jan 2004).

15.4.1
Preclinical Studies

In order to obtain toxicity data, a single intramuscular injection was given to 16 male and 16 female Crl:CD(SD)BR rats. The dosage of 30 µg kg^{-1} was based upon a 100-fold factor over the likely human dose. There was no evidence of any toxic effect on all of the parameters measured, indicating that the test substance was well tolerated. Antigen (i.e., the test substance as administered) was measurable in eight of ten animals at 6 h after injection, and was seen in another animal at 8 days after injection; five animals (one male, four females) produced measurable levels of antibody to rHBsAg by day 15 (GCVC Korea; report on file).

In a study to obtain information on the acute toxicity of the test substance, four cynomolgus monkeys were injected intramuscularly with a single dose of HepavaxGene® at 10 µg kg^{-1} – that is, a 35-fold factor over the human dose. The data confirmed the vaccine to be nontoxic to the test animals. There were no clinical signs associated with treatment, and – more particularly – there was no sign of any reaction to treatment at the injection site in all animals (GCVC Korea; report on file).

The immunogenicity, safety, and efficacy of HepavaxGene® were assessed in four chimpanzees. Initially, two chimpanzees received three intramuscular injections at monthly intervals, after which the sera were tested for antibodies to HBV. No adverse effects were noted in two chimpanzees during a three-injection course of immunization with HepavaxGene®. Both vaccinated chimpanzees were protected from infection by a challenge dose of HBsAg ayw that infected both positive control animals (GCVC Korea; report on file).

15.4.2
Clinical Studies

The safety and immunogenicity of HepavaxGene® have been investigated in a variety of clinical studies in neonates, children, and adults. The most relevant data obtained from these studies are listed in Table 15.5.

Tab. 15.5 Clinical studies with HepavaxGene®.

Study no.	Country	No. of subjects	Subject age group	Dosage (µg)	Schedule	Protection rate (%)
1	Korea[a]	47	Neonates	10	0-1-6	100
2	Korea	104	Neonates and children 0–10 years	10	0-1-6	94–100
3	Korea[b]	113	Adults 21–69 years	20	0-1-2	85
4	Korea	93	Adults 21–75 years	20	0-1-6	97
5	UK	20	Adults 18–45 years	20	0-1-6	95
6	Vietnam[c]	124	Neonates	10	0-1-2	97
7	Vietnam[d]	112	Children 0–5 years	NA	Follow up	98
8	Vietnam[e]	105	Neonates with sero + mother	10	HBIG+	94
				ExB 10	0-1-6	94
9	Turkey	230	Neonates	10–20	Various	100
				ExB 10		100
				GHe 20		100
10	Turkey[f]	76	Adults 17–22 years	20	0-1-2	100
				ExB	0-1-2	100
				GHe 20	0-1-2	91

ExB = Engerix B (Glaxo SmithKline, GSK); GHe = Gen Hevac B (Pasteur Merieux).
HBIG = hepatitis B immune globulin.
a) Cho et al. (1996); **b)** Lee et al. (2000); **c)** Hieu (1996); **d)** Hieu et al. (2001); **e)** Hieu et al. (2002);
f) Eyigün et al. (1998).

Tab. 15.6 Recommended dosage of HepavaxGene®.

Group	Formulation	Initial dose (mL)	Dose at 1 month (mL)	Dose at 6 months (mL)
Neonates	10 μg per 0.5 mL	0.5	0.5	0.5
0–10 years	10 μg per 0.5 mL	0.5	0.5	0.5
>10 years	20 μg per 1.0 mL	1.0	1.0	1.0

In terms of efficacy criteria, the usual parameter to assess immunogenicity was the level of anti-HBs. The seroprotection rate – defined as anti-HBs levels over 10 mIU mL^{-1} – was generally high and consistent throughout the clinical studies (Cho et al. 1996; Hieu 1996; Lee et al. 2000) (Table 15.5).

The majority of clinical studies were conducted using the 5, 10 or 20 μg dosage intramuscularly with a 0-1-6 months vaccination schedule. The alternative 0-1-2 months vaccination schedule was tested in some studies, and proved equally efficient.

Recommendations for HepavaxGene® dosage, based on the results of the clinical development program, are detailed in Table 15.6.

In clinical studies, all vaccinated subjects were followed-up and evaluated for the occurrence of adverse events. Overall, there were few adverse reactions, and these were mild and transient. As could be expected from a vaccination, the most common adverse events were soreness, erythema, and swelling at the injection site, but these symptoms usually subsided within 2 days after vaccination. Uncommon systemic complaints such as fever, headache, nausea, dizziness, and fatigue have been observed in some vaccinated subjects, but a causal relationship with the vaccine has never been established.

As HepavaxGene® is a DNA recombinant product, concomitant administration with other killed vaccines is not likely to cause interference with the immune response to these vaccines. In the clinical development program, HepavaxGene® was given concurrently with diphtheria, tetanus toxoids and pertussis (DTP), or with inactivated poliovirus vaccine (IPV), or live attenuated measles, mumps and rubella virus vaccine (MMR). There was no evidence of any interference with other simultaneously administered vaccines. Conversely, passively acquired antibody to hepatitis B surface antigen (anti-HBs), which is present in hepatitis B immune globulin (HBIG), does not appear to interfere with the active immune response produced by HepavaxGene® (Hieu 1996).

In three separate studies, the safety and immunogenicity of HepavaxGene® has been compared to that of other hepatitis B vaccines, such as the S. cerevisiae-derived EngerixB and GenHevac B. In terms of adverse effects, there was no significant difference between the mentioned vaccines. The level of immunogenicity measured by anti-HBs levels showed that HepavaxGene® appeared at least as effective as the other two vaccines.

The long-term effect of HepavaxGene® has been investigated by a 5-year follow-up of a cohort of 124 neonates following a primary 0–1-2 months vaccination schedule. Among a total of 124 children, 112 were assessed at the age of 5 years. Anti-HBs levels were positive in 110 of 112 children (seroconversion rate 98%), with a stabilizing geometric mean titer of 152 mIU mL^{-1} – well above the protection level of 10 mIU mL^{-1} (Hieu 2001). These results suggest that the protective effect of HepavaxGene® may last longer than the 5-year period, and a follow-up study is currently in progress to investigate this proposal.

15.4.3
Second-generation Prophylactic Vaccine@ SUPERVAX

A second-generation vaccine has been developed by Berna Biotech Ltd based on the hepatitis B antigen produced through the *H. polymorpha* expression system, as described above. It is combined with the synthetic adjuvant RC-529, a nontoxic lipid A mimetic. Clinical trials have demonstrated a solid protection against hepatitis B in more than 95% of vaccinees after only two vaccinations, given one month apart. The vaccine is already registered for use in Argentina.

15.5
Hepatitis B Vaccines: Past, Present, and Future

15.5.1
Use and Success of Prophylactic Hepatitis B Vaccination

Over a decade of experience with HBV vaccines with several 100 million recipients has shown that the vaccination is safe, and very effective. In the expanded program on immunization, the WHO aims at eradicating the virus. In 2001, 126 countries had integrated HBV universal vaccination of children and newborns into their national immunization program (CDC 2003).

A striking success of HBV vaccination is the reduction of HBV-associated hepatocellular carcinoma (HCC). The first epidemiological data show a sharp decrease in the incidence of HCC in vaccinees (Blumberg 1997; Chang et al. 1997; Zuckerman 1997; Lee et al. 1998). While executing universal vaccination in Taiwan, the carrier rate in children aged 6 years declined from 10% to 0.9% (Chang et al. 1997). Concomitantly, the incidence of HCC in 6- to 14-year-old children fell significantly, from 0.7 per 100 000 between 1981 and 1986 to 0.36 per 100 000 between 1990 and 1994. Later studies showed that children vaccinated as newborns were protected for at least 15 years against HBV infection (Lin et al. 2003 a,b).

15.5.2
Current Shortcomings of Hepatitis B Vaccines

Despite the efficacy of current recombinant hepatitis B vaccines, several problems persist and deserve to be addressed.

15.5.2.1 Non-responders

Between 5 and 10% of all adults are low- or non-responders after a full course of vaccination. The highest risk factors for low response are age, smoking, male gender, and chronic diseases (for reviews, see Jilg 1998; Mahoney 1999; Assad and Francis 2000). In addition, genetic differences appear to influence response since some ethnic groups (Hsu et al. 1996) and carriers of some HLA antigens (Craven et al. 1986; Alper et al. 1989; Kruskall et al. 1992; McDermott et al. 1997; Desombere et al. 1998) appear to be predisposed to become non- or low-responders (for a review, see Wang et al. 2003).

The problem of low response can (at least partially) be overcome by repeated vaccination with higher vaccine doses (Kim et al. 2003). The inclusion of additional epitopes such as preS1 and preS2 of HBsAg is of advantage, but these antigens must be produced in mammalian cells (e.g., Chinese hamster ovary cells). Another approach is to include novel more potent adjuvants, such as the afore-mentioned lipid A mimetics. In addition, alternative routes of administration such as intradermal injection have been investigated (Rahman et al. 2000; for reviews, see Jilg 1998; Mahoney 1999; Assad and Francis 2000).

15.5.2.2 Incomplete Vaccination

The standard HBV vaccination regimen requires three injections at 0, 1, and 6 months. Under many settings this course remains incomplete, with the risk of low response and, most likely, an insufficient protection against infection. Recently, reduced vaccination schedules with two injections at 0 and 6 months using the conventional vaccine were reported to result in acceptable serum conversion rates and protective antibody titers (Cassidy et al. 2001). Several other approaches using a higher vaccine dose, more potent adjuvants, different routes of administration, or slow-antigen release formulations are currently being tested to overcome the disadvantages of a three-injection regimen.

15.5.2.3 Escape Variants

Soon after use of the HBV vaccines became widespread, an escape variant was identified, which carried an exchange of glycine at position 145 in the a-determinant of SHBs to arginine (Carman et al. 2000). This G145→R variant was later identified in many countries at a rate of 3–10% of all vaccinees (for reviews, see Günther et al. 1999; Carman 1997). Most studies found that this variant arises *de novo* in vaccinees who receive their first dose of HBV vaccine post exposition. This variant is fit for survival in the host, does not revert to wild-type (Carman 1997; Günther et al. 1999), and can be transmitted horizontally (Oon et al. 2000). The G145→R variant also arises under therapy with hyper-immune globulin after liver transplantation (Protzer-Knolle et al. 1998; Shouval and Samuel 2000). Thus, it is assumed that in newborns with a high level of HBV replication after perinatal infection, the high level of anti-HBs administered by passive immunization delivers the selective pressure for the generation of the G145→R variant (Shouval and Samuel 2000). Among the vaccine breakthroughs observed, the variant G145→R comprises more than 50%. However, chimpanzees have been shown to be protected by standard commercial HBV vaccine against experimental infection with the variant G145 →R virus (Ogata et al. 1999; Purcell et al. 2000).

In addition to the G145R variant, several other variants have been described after therapy with hyper-immune globulin, or after vaccine escape (Carman 1997; Günther et al. 1999). These variants are recognized by commercial assays.

The widespread use of the nucleoside analogue lamivudine in the treatment of chronic hepatitis B has led to the emergence of several variants with resistance against this drug (Stuyver et al. 2001; Feld and Locarnini 2002). These resistant variants harbor mutations in the polymerase gene which affect the overlapping reading frame of the HBsAg ORF (Torresi 2002). The experience with lamivudine shows that the use of antivirals in general may lead to the induction of mutations in the HBsAg. As HBsAg is the basis of current hepatitis B vaccines, variants could arise where the standard vaccine would no longer prevent infection. In addition, such HBsAg variants might not be recognized by current diagnostic assays. However, recombinant vaccine technology would allow for swift inclusion of variant virus-derived HBsAg in a prophylactic vaccine if necessary.

Even when current vaccines are being further improved and their use extended worldwide, vaccination programs are maintained with major efforts. Clearly, hepatitis B infection will not disappear in the foreseeable future, mainly due to the chronic state of the virus and the huge number of chronic carriers (Medley et al. 2001)

15.5.3
Alternative Vaccine Strategies

Due to the success of recombinant hepatitis B vaccines – both in improving public health and in economic terms – the field of hepatitis B has in the past few years provided a prolific "playground" for almost every technical platform arising in biotechnology. The vaccine is based on a single defined antigen (HBsAg) which is easy to employ, and the protective efficacy of a novel vaccine formulation correlates with the presence of antibodies against HBsAg. Thus, a plethora of novel vaccine approaches are being investigated in the laboratory, and several of these have advanced to late preclinical state, or have even entered clinical trials. As a comprehensive overview would by far exceed the scope of this article, only a few of these novel vaccine approaches are described briefly.

15.5.3.1 Oral Administration of Plant-derived, Edible Vaccines
A number of transgenic plants ranging, from tobacco to potato and lettuce, have been generated to produce HBsAg. It was shown that the antigen from plants was immunogenic in animals after parenteral administration. The optimal use of a plant-derived vaccine would, of course, be oral uptake of a consumable part of the transgenic plant; this would provide low cost, simple mass vaccination. In line with this idea, raw transgenic potato and lettuce have been tested in animals and humans (Kapusta et al. 1999; Richter et al. 2000). It was shown that, via an oral uptake, specific serum IgG responses were induced. Despite these promising results – and the undoubted attraction of this concept – there is a reluctance towards the uncritical

use of edible vaccines. Issues such as standardization and control of the antigen dose, antigen stability, adjuvant requirements and potential deleterious effects of high-dose oral antigen (risk of inducing tolerance) have so far prevented edible plant vaccines from becoming reality.

15.5.3.2 Oral Administration of Live Bacterial Vectors

Another design for an oral vaccine is based on expression of HBV antigens in avirulent strains of *Salmonella typhimurium/typhi* as a bacterial vaccine carrier. The latter is used as a registered oral vaccine against typhoid fever. A hepatitis B virus nucleocapsid (HBcAg)-preS2-HBsAg fusion protein was expressed in *Salmonella* strains. These strains were found to elicit potentially virus-neutralizing anti-preS-serum IgG as well as anti HBc-antibodies after a single oral dose in mice (Schödel et al. 1990, 1994). However, when tested in volunteers no immune response to the preS antigen was detectable after oral immunization (Nardelli-Haefliger et al. 1996).

15.5.3.3 Live Viral Vectors

In an early attempt to substitute the plasma-derived HBsAg vaccine recombinant, *vaccinia* virus strains were constructed harboring the gene for HBsAg. Chimpanzees vaccinated with a live recombinant HBsAg/*vaccinia* virus were shown to be protected against hepatitis following challenge with HBV (Moss et al. 1984). However, apart from a limited trial carried out in the Soviet Union, which demonstrated antigen priming by the recombinant *vaccinia* vector but induced no significant anti-HBs titers (Chernos et al. 1990), no efforts were made to advance this technology for routine vaccination. The major reason was the unsolved problems of severe adverse side effects caused by immunization with *vaccinia* virus, and the unpredictable risks inherent to live recombinant viruses in general.

15.5.3.4 DNA Vaccines

An elegant means of immunization materialized from the discovery that DNA encoding a protein antigen could act as a vaccine when injected into mice. Injection of an expression plasmid coding HBsAg was shown to protect chimpanzees from a challenge with an infectious dose of HBV (Davis et al. 1996). Despite the encouraging results, vaccine approaches using DNA in humans or larger mammals proved to be rather disappointing. Further efforts were made to improve the efficacy of DNA vaccines by the inclusion of CpG oligonucleotides or cytokine co-expression (Chow et al. 1997) as adjuvants, or by employing sophisticated particle-mediated epidermal delivery. The first demonstration of a DNA vaccine inducing protective anti-HBs antibody titers was achieved via three epidermal administrations of gold particles coated with 1–4 μg of plasmid DNA encoding HBsAg (Roy et al. 2000). However, the observed antibody titers were low compared to those of standard protein vaccines, probably reflecting the preferential induction of a Type 1 helper cell response.

15.5.3.5 Single-dose Vaccines

A single-dose vaccine would greatly facilitate extended vaccination programs. For an assessment of suitable vaccine formulations, the HBsAg was encapsulated into biodegradable polymer matrices such as PLGA (poly-lactide-*co*-glycolic acid). Immunization experiments in mice revealed that a single injection with HBsAg-PLGA microspheres combined with aluminum-adsorbed HBsAg yielded serum antibody titers comparable to two doses of aluminum-absorbed vaccine (Shi et al. 2002). The formulation of microspheres retaining HBsAg in an immunogenic form is a complicated process, and problems of stability are still to be addressed. In another approach, controlled-release microspheres were formulated with a HBsAg-derived 48-mer peptide containing the main immunogenic domain and a trehalose derivative (Moynihan et al. 2002). This formulation produced high anti-HBs antibody titers in mice after a single injection with 100 µg of peptide.

15.5.3.6 Mammalian Cell-derived, PreS1/preS2-containing Vaccines

Several mammalian cell-derived hepatitis B vaccines containing the large (LHBsAg) and the middle (MHBsAg) surface proteins of HBV in addition to the small (SHBsAg) have been tested during the past few years. Studies evaluating the efficacy of preS1/S2-containing vaccines showed that the protection rate at 7 months was essentially equivalent for the conventional yeast-derived (SHBsAg only) vaccine (three doses at 0, 1, and 6 months) and the preS1/S2 vaccine (two doses at 0 and 1 month). When administered three times (0, 1, and 6 months), the protection levels were superior compared to those elicited by the conventional vaccine (Zuckerman et al. 2002). Studies in previous non-responders to conventional SHBsAg seroconversion in up to 70% of recipients of the preS1/S2-containing vaccine was observed (Zuckerman et al. 1997). In addition, protective antibody levels in recipients of preS1/S2-containing vaccines were reached much faster than with the conventional vaccines (Heineman et al. 1999; Shapira et al. 2001). This performance profile predestines the preS1/S2 vaccines for specific risk groups and non-responders in developed countries where the higher price can be afforded. Alternatively, yeast-based systems could also be developed for production of preS- containing vaccines, as demonstrated with a *Hansenula polymorpha* strain efficiently co-expressing SHBsAg and LHBsAg which subsequently formed mixed virus-like particles (Janowicz et al. 1991).

15.5.3.7 Novel Adjuvants

A range of novel adjuvants has been evaluated for inclusion in hepatitis B vaccine formulations, and several of these have been tested in humans. These include the lipid A mimetics MPL (Thoelen et al. 2001) and its synthetic derivative RC-529 (see Section 15.4.3), CpG motif oligonucleotides (Halperin et al. 2003), and MF59 (Heineman et al. 1999) were tested alone or as supplements to current aluminum-based hepatitis B vaccines. This approach has been described for a *H. polymorpha*-derived HBsAg vaccine and RC-529 in Section 15.4.3.

15.5.4
Treatment of Chronic Hepatitis

Despite the success of vaccination against HBV, there remain about 350 million chronic carriers who are at risk of dying from HBV-associated liver cirrhosis or cancer. For these people, therapy with interferon and antiviral nucleoside analogues is available (Hoofnagle 1998; Dusheiko 1999). Both of these therapies have limitations, however. Although treatment with nucleoside analogues is well tolerated and highly efficient, only 27% of patients treated for 2 years remained HBV DNA-negative after withdrawal (Dusheiko 1999). Resistance to treatment develops in 14% of patients treated for 1 year with lamivudine, and in up to 38% of those treated for 2 years (Dusheiko 1999). Adefovir dipivoxil, as a newly licensed nucleotide analogue, shows a greater efficacy, with only rare drug resistance (Aloman and Wands 2003; Dusheiko 2003; Humphries and Dixon 2003; Papatheodoridis and Hadziyannis 2004) in comparison to lamivudine. Treatment with interferon has considerable side effects, is expensive, and has a less than 50% response rate (Hoofnagle 1998). Spontaneous and interferon-related clearance of chronic HBV infection (although rare), and successful clearance after transplantation of the bone marrow of donors with natural immunity to HBV, indicate that the immune system in principle could resolve chronic infection. This clearance is associated with a CTL class I-restricted response (Ilan 1993; Chisari 1997; Lau 2002). Therapeutic vaccination aims at inducing such HBV-specific Th1-type response/CTL.

Vaccine-based immunotherapy to broaden the weak virus-specific T-cell response has been reviewed recently (Michel 2002; Hilleman 2003). Several studies were conducted to elicit a CTL response in chronic HBV carriers (Livingston et al. 1997, 1999; Alexander et al. 1998; Couillin et al. 1999; Heathcote et al. 1999). For example, chronic HBV patients have been vaccinated with standard HBs-containing vaccines with six injections of the standard dose of vaccine over 1 year (Couillin et al. 1999; Pol et al. 2000), or by injection of pre S2/s or preS1/2/s antigenic components (Jung 2002; Yalcin et al. 2003). Vaccinations with peptides that stimulate CTL activity (Alexander et al. 1998; Livingston et al. 1999) have also been tried. In addition, prime-boost strategies or the inclusion of cytokines (interleukin-2, granulocyte-macrophage colony stimulating factor) are currently pursued (Pancholi 2001; Dahmen 2002; Wang 2002,). To date, there has been no breakthrough in the treatment of chronic hepatitis B by therapeutic vaccination. Nevertheless, T-cell activity was stimulated in a large proportion of the patients monitored in the studies mentioned, and continuing efforts to further improve the potency of specific immunotherapy holds promise for a causal cure of chronic hepatitis B.

For example, combination therapies with antivirals and vaccination with Woodchuck hepatitis B surface antigen (WHBs) (Menne et al. 2002) or DNA vaccination with expression vectors for the large-surface protein of duck hepatitis B virus (DHBV) (Thermet et al. 2003) have shown pronounced antiviral effects in chronic carrier woodchucks and ducks, respectively.

15.5.5
Combination Vaccines

Several combined vaccines including DTP + HB, DT + HB, HA + HB, and HB + Hib are already available for use in some countries, and DTP + HB + IPV and DTP + HB + IPV + Hib combined vaccines are currently under development (Papaevangelou 1998).

In a comparative study of the immunogenicity of DTPw-HBV combination vaccines versus separately administered DTPw and HB, a DTPw-HB combination vaccine containing 10 μg of HBsAg elicited significantly higher anti-HBs titers than the separately administered HB vaccine after the primary and booster vaccination course (Poovorawan et al. 1999). Although immunological interference between components of a vaccine combination may possibly affect immunogenicity testing (Sesardic et al. 1999), it is clear that HBV administered in combination with other childhood vaccines can, at the very least, lower the cost and complexity of pediatric vaccination programs.

15.6
Conclusions

To date, advances in gene technology have led to the production of a wide range of recombinant pharmaceuticals. Commercially available *H. polymorpha* vaccines are restricted so far to a single example, namely the hepatitis B vaccines presented in this chapter. The described product development has shown that yeasts in general, and *H. polymorpha* in particular, meet the characteristics and prerequisites necessary for the production of sophisticated vaccine products, namely a particle consisting of host-derived membranes with the recombinant antigens inserted into it. The emerging genomics and ongoing research will provide a variety of new antigens. This will not only lead to improvements in the currently available vaccines, but will also enable the production of a variety of new recombinant vaccines in the near future.

References

ALBERTI A, CAVALLETTO D, CHEMELLO L, BELUSSI F, FATTOVICH G, PONTISSO P, MILANESI G, RUOL A (1990) Fine specificity of human antibody response to the PreS1 domain of hepatitis B virus. Hepatology 12: 199–203

ALEXANDER J, FIKES J, HOFFMAN S, FRANKE E, SACCI J, ELLA A, CHISARI F,V, GUIDOTTI LG, CHESNUT RW, LIVINGSTON B, SETTE A (1998) The optimization of helper T lymphocyte (HTL) function in vaccine development. Immunol Res 18: 79–92

ALOMAN C, WANDS JR (2003) Resistance of HBV to adefovir dipivoxil: a case for combination antiviral therapy? Hepatology 38: 1584–1587

ANTHONY C (1982) The Biochemistry of Methylotrophs. Academic Press, London

ALPER CA, KRUSKALL MS, MARCUS-BAGLEY D, CRAVEN DE, KATZ AJ, BRINK SJ, DIENSTAG JL, AWDEH Z, YUNIS EJ (1989) Genetic prediction of nonresponse to hepatitis B vaccine. N Engl J Med 321: 708–712

ARAUZ-RUIZ P, NORDER H, ROBERTSON BH, MAGNIUS LO (2002) Genotype H: a new Amerindian genotype of hepatitis B virus revealed in Central America. J Gen Virol 83: 2059–2073

ASHTON-RICKHARDT PG, MURRAY K (1989) Mutations that change the immunological subtype of hepatitis B virus surface antigen and distinguish between antigenic and immunogenic determination. J Med Virol 29: 204–214

ASSAD S, FRANCIS A (2000) Over a decade of experience with a yeast recombinant hepatitis B vaccine. Vaccine 18: 57–67

AVANZINI MA, BELLONI C, DE SILVESTRI A, CASTELLAZZI AM, MARCONI M., MORETTA A, MONTAGNA D, MARTINETTI M, CUCCIA M, RONDINI G, CIARDELLI L, MACCARIO R (2003) Antigen-specific T cell response in infants after recombinant hepatitis B virus vaccination at birth: evaluation of T helper lympho-cyte diversity. Clin Immunol 107: 122–128

BANCROFT WH, MUNDON FK, RUSSELL K (1972) Detection of additional antigenic determinants of hepatitis B antigen. J Immunol 109: 842–888

BARTHOLOMEUSZ A, SCHAEFER S (2003) Hepatitis B Virus – Significance of Genotypes. Rev Med Virol 13: 1–14

BERTOLETTI A, MAINI MK (2000) Protection or damage: a dual role for the virus-specific cytotoxic T lymphocyte response in hepatitis B and C infection? Curr Opin Immunol 12: 403–408

BERTOLETTI A, NAOUMOV NV (2003) Translation of immunological knowledge into better treatments of chronic hepatitis B. J Hepatol 39: 115–124

BILLMAN-JACOBE H (1996) Expression in bacteria other than *E. coli*. Curr Opin Biotechnol 7: 500–504

BLITZ L, PUJOL FH, SWENSON PD, ATENCIO R, ARAUJO M, COSTA L, MONSALVE DC, TORRES JR, FIELDS HA, LAMBERT S, VAN GEYT C, NORDER H, MAGNIUS LO, ECHEVERRIA JM, STUYVER L (1998) Antigenic diversity of hepatitis B virus strains of genotype F in Amerindians and other population groups from Venezuela. J Clin Microbiol 36: 648–651

BLUMBERG BS (1997) Hepatitis B virus, the vaccine, and the control of primary cancer of the liver. Proc Natl Acad Sci USA 94: 7121–7125

BRÉCHOT C, JAFFREDO F, LAGORCE D, GERKEN G, MEYER ZUM B, PAPAKONSTONTINOU A, HADZIYANNIS S, ROMEO R, COLOMBO M, RODES J, BRUIX J, NAOUMOV N (1998) Impact of HBV, HCV, and GBV C/HGV on hepatocellular carcinomas in Europe: results of a European concerted action. J Hepatol 29: 173–183

BUCKHOLZ RG, GLEESON MA (1991) Yeast systems for the commercial production of heterologous proteins. Bio/Technology 9: 1067–1072

BUDKOWSKA A, DUBREUIL P, CAPEL F, PILLOT J (1986) Hepatitis B virus pre-S gene-encoded antigenic specificity and anti-pre-S antibody: relationship between anti-pre-S response and recovery. Hepatology 6: 360–368

BUENDIA MA, PATERLINI P, TIOLLAIS P, BRECHOT C (1998) Hepatocellular carcinoma: molecular aspects. In: Viral Hepatitis (Zuckerman AJ, Thomas CT, Eds). Churchill Livingstone, London, pp 179–200

BURRELL CJ (1995) Hepadnaviridae. In: Classification and Nomenclature of Viruses. Sixth Report on Taxonomy of the International Committee of Viruses (Murphy FA, Fauquet CM, Bishop DHL, Ghabrial SA, Jarvis AW, Martelli GP, Mayo MA, Summers MD, Eds). Springer-Verlag, Wien, pp 179–184

BURRELL CJ, CHISARI FV, GERLICH WH, GOWANS EJ, HOWARD CR, KANN M, MARION PL (2000) Hepadnaviridae. In *Virus Taxonomy*. Seventh Report of the International Committee on Taxonomy of Viruses (van Regenmortel MHV, Fauquet CM, Bishop DHL, Eds). Academic Press, San Diego, pp 325–334

CDC (2003) Global progress toward universal childhood hepatitis B vaccination. MMWR 52: 868–870

CARMAN WF (1997) The clinical significance of surface antigen variants of hepatitis B virus. J Viral Hepatol 1: 11–20

CASSIDY WM, WATSON B, IOLI VA, WILLIAMS K, BIRD S, WEST DJ (2001) A randomized trial of alternative two- and three- dose hepatitis B vaccination regimens in adolescents: antibody responses, safety, and immunogenic memory. Pediatrics 107: 626–631

CARMAN WF, ZANETTI AR, KARAYIANNIS P, WATERS J, MANZILLO G, TANZI E, ZUCKERMAN AJ, THOMAS HC (1990) Vaccine induced escape mutant of hepatitis B virus. Lancet 336: 325–329

CHANG M-H, CHEN C-J, LAI M-S, HSU H-M, WU T-C, KONG M-S, LIANG D-C, SHAU W-Y, CHEN D-S (1997) Universal hepatitis B vaccination in Taiwan and the incidence of hepatocellular carcinoma in children. N Engl J Med 336: 1855–1859

CHANG SF, NETTER HJ, BRUNS M, SCHNEIDER R, FRÖHLICH K, WILL H (1999) A new avian hepadnavirus infecting snow geese (*Anser caerulescens*) produces a significant fraction of virions containing single-stranded DNA. Virology 262: 39–54

CHERNOS VI, CHELIAPOV NV, ANTONOVA TP, et al. (1990) Verification of the safety, inoculability, reactogenicity and antigenic properties of a live recombinant smallpox-hepatitis B vaccine in an experiment in volunteers. Vopr Virusol. 35: 132–135

CHISARI FV (1997) Cytotoxic T cells and viral hepatitis. J Clin Invest 99: 1472–1477

CHISARI FV, FERRARI C (1995) Hepatitis B virus immunopathology. Springer Semin Immunopathol 17: 261–281

CHO YK, JUNG YS, YI GW, KIM YO, CHOI MB, SHIN SK, PARK CH, WOO HO, YOUN HS,

MAENG KY (1996) Phase III clinical trial for immunogenicity and safety of a new recombinant hepatitis B virus vaccine (HG-II) in the newborn. Kor J Clin Pharmacol Ther 4: 197–205

CHOW Y-H, HUANG W-L, CHI W-K, CHU Y-D, TAO M-H (1997) Improvement of hepatitis B virus DNA vaccines by plasmids coexpressing hepatitis B surface antigen and interleukin 2. J Virol 71: 169–178

CHU CJ, LOK AS (2002) Clinical significance of hepatitis B virus genotypes. Hepatology 35: 1274–1276

COUDERC R, BARATTI J (1980) Oxidation of methanol by the yeast *Pichia pastoris*: purification and properties of alcohol oxidase. Agric Biol Chem 44: 2279–2289

COUILLIN I, POL S, MANCINI M, DRISS F, BRÉCHOT C, TIOLLAIS P, MICHEL ML (1999) Specific vaccine therapy in chronic hepatitis B: induction of T cell proliferative responses specific for envelope antigens. J Infect Dis 180: 15–26

COUROUCÉ AM, HOLLAND P, V, MULLER JY, SOULIER JP (1976) HBs antigen subtypes: Proceedings of the International Workshop on HBs Antigen Subtypes. Bibl Haematol 42: 1–158

COUROUCÉ-PAUTY AM, PLANCON A, SOULIER JP (1983) Distribution of HBsAg subtypes in the World. Vox Sanguinis 44: 197–211

COURSAGET P, ADAMOWICZ P, BOURDIL C, YVONNET B, BUISSON Y, BARRES JL, SALIOU P, CHIRON JP, MAR ID (1988) Anti-pre-S2 antibodies in natural hepatitis B virus infection and after immunization. Vaccine 6: 357–361

CRAVEN DE, AWDEH ZL, KUNCHES LM, YUNIS EJ, DIENSTAG JL, WERNER BG, POLK BF, SYNDMAN DR, PLATT R, CRUMPACKER CS (1986) Nonresponsiveness to hepatitis B vaccine in health care workers. Results of revaccination and genetic typings. Ann Intern Med 105: 356–360

CREGG J, MADDEN K (1987) Development of yeast transformation systems and construction of methanol utilizing defective mutants of *Pichia pastoris* by gene disruption. In: Biological Research on Industrial Yeasts (Stewart G, Russel I, Klein R, Hiebsch R, Eds). CRC Press, Boca Raton, FL, USA, pp 1–18

CREGG JM, BARRINGER KJ, HESSLER AY, MADDEN KR (1985) *Pichia pastoris* as a host system for transformations. Mol Cell Biol 5: 3376–3385

CREGG JM, TSCHOPP JF, STILLMAN C, SIEGEL R, AKONG M, CRAIG WS, BUCKHOLZ RG, MADDEN KR, KELLARIS PA, DAVIS GR, SMILEY BL, CRUZE J, TORREGOSSA R, VELICELEBI G, THILL GP (1987) High-level expression and efficient assembly of hepatitis B surface antigen in the methylotrophic yeast, *Pichia pastoris*. Bio/Technology 5: 479–485

CROWTHER RA, KISELEV NA, BOTTCHER B, BERRIMAN JA, BORISOVA GP, OSE V, PUMPENS P (1994) Three-dimensional structure of hepatitis B virus core particles determined by electron cryomicroscopy. Cell 77: 943–950

DAHMEN A, HERZOG-HAUFF S, BOCHER WO, GALLE PR, LOHR HF (2002) Clinical and immunological efficacy of intradermal vaccine plus lamivudine with or without interleukin-2 in patients with chronic hepatitis B. J Med Virol 66: 452–460

DANE DS, CAMERON CH, BRIGGS M (1970) Virus like particles in serum of patients with Australia-antigen-associated hepatitis. Lancet 1: 695–698

DAVIS HL, MCCLUSKIE MJ, GERIN JL, PURCELL RH (1996) DNA vaccine for hepatitis B: evidence for immunogenicity in chimpanzees and comparison with other vaccines. Proc Natl Acad Sci USA 93: 7213–7218

DAVIS HL, SUPARTO II, WEERATNA RR, JUMINTARTO DD, ISKANDRIATI SS, CHAMZAH AA, MARUF AA, NENTE CC, PAWITRI DD, KRIEG AM, HERIYANTO, SMITS W, SAJUTHI DD (2000) CpG DNA overcomes hyporesponsiveness to hepatitis B vaccine in orangutans. Vaccine 18: 1920–1924

DEML L, SCHIRMBECK R, REIMANN J, WOLF H, WAGNER R (1999) Purification and characterization of hepatitis B virus surface antigen particles produced in *Drosophila* Schneider-2 cells. J Virol Methods 79: 205–217

DESOMBERE I, WILLEMS A, LEROUX-ROELS G (1998). Response to hepatitis B vaccine: multiple HLA genes are involved. Tissue Antigens 51: 593–604

DIMINSKY D, SCHIRMBECK R, REIMANN J, BARENHOLZ Y (1997) Comparison between hepatitis B surface antigen (HBsAg) particles derived from mammalian cells (CHO) and yeast cells (*Hansenula polymorpha*): composition, structure and immunogenicity. Vaccine 15: 637–647

DUSHEIKO G (1999) A pill a day or two for hepatitis B. Lancet 353: 1032.

DUSHEIKO G (2003) Adefovir dipivoxil for the treatment of HBeAg-positive chronic hepatitis B: a review of the major clinical studies. J Hepatol 39 (Suppl 1): S116–S123

ECKART M (1988) Klonierung und Charakterisierung der Gene für Dihydroxyacetonesynthetase und Methanoloxidase aus der methylotrophen Hefe *Hansenula polymorpha*. Thesis, Universität Duesseldorf, Germany.

EDELMAN R, WASSERMAN SS, BODISON SA, PERRY JG, O'DONNOGHUE M, DeTOLLA LJ (2003) Phase II safety and immunogenicity study of type F botulinum toxoid in adult volunteers. Vaccine 21: 4335–4347

EYIGÜN CP, YILMAZ S, GÜL C, SENGÜL A, HACIBEKTASOGLU A, VAN THIEL DH (1998) A comparative trial of two surface subunit recombinant hepatitis B vaccines vs. a surface and PreS subunit vaccine for immunization of healthy adults. J Viral Hepatitis 5: 265–269

EHSANI P, KHABIRI A, DOMANSKY NN (1997) Polypeptides of hepatitis B surface antigen produced in transgenic potato. Gene 190: 107–111

ELLIS SB, BRUST PF, KOUTZ PJ, WATERS AF, HARPOLD MM, GINGERAS TR (1985) Isolation of alcohol oxidase and two other methanol regulatable genes from the yeast *Pichia pastoris*. Mol Cell Biol 5: 1111–1121

EMMINI E, ELLIS R, MILLER W, McALEER W, SCOLNICK E, GERETY R (1986) Production and analysis of recombinant hepatitis B vaccine. J Infect 13 (Suppl. A): 3–9

FELD J, LOCARNINI S (2002) Antiviral therapy for hepatitis B virus infections: new targets and technical challenges. J Clin Virol 25: 267–283

GALLE PR (1997) Apoptosis in liver disease. J Hepatol 27: 405–412

GALLE PR, HOFMANN WJ, WALCZAK H, SCHALLER H, OTTO G, STREMMEL W, KRAMMER PH, RUNKEL L (1995) Involvement of the CD95 Apo 1/Fas receptor and ligand in liver damage. J Exp Med 182: 1223–1230

GELLISSEN G, HOLLENBERG CP (1997) Application of yeasts in gene expression studies: a comparison *of Saccharomyces cerevisiae, Hansenula polymorpha* and *Kluyveromyces lactis* – a review. Gene 190: 87–97

GELLISSEN G, HOLLENBERG CP (1999) *Hansenula*. In: Encyclopedia of Food Microbiology Vol 2 (Robinson RK, Batt CA, Patel PD, Eds). Academic Press, San Diego, CA, USA, pp 976–982

GELLISSEN G, MELBER K (1996) Methylotrophic yeast *Hansenula polymorpha* as production organism for recombinant pharmaceuticals. Arzneim-Forsch/Drug Res 46: 943–948

GELLISSEN G, STRASSER A, MELBER K, MERCKELBACH A, WEYDEMANN U, KEUP P, DAHLEMS U, PIONTEK M, HOLLENBERG C, JANOWICZ Z (1990) Die methylotrophe Hefe als Expressionssystem für heterologe Proteine. BioEngineering 5: 20–26

GELLISSEN G, JANOWICZ Z, WEYDEMANN U, MELBER K, STRASSER A, HOLLENBERG C (1992 a) High level expression of foreign genes in *Hansenula polymorpha*. Biotechnol Adv 10: 179–189

GELLISSEN G, MELBER K, JANOWICZ ZA, DAHLEMS UM, WEYDEMANN U, PIONTEK M, STRASSER AW, HOLLENBERG CP (1992 b) Heterologous protein production in yeast. Antonie van Leeuwenhoek 62: 79–93

GELLISSEN G, HOLLENBERG CP, JANOWICZ Z (1995) Gene expression in methylotrophic yeasts. In: Gene Expression in Recombinant Microorganisms (Smith A, Ed). Marcel Dekker, New York, pp 195–239

GELLISSEN G (ED) *Hansenula polymorpha* – biology and applications. Wiley-VCH, Weinheim, 2002

GLEESON M, SUDBERY P (1988) The methylotrophic yeasts. Yeast 4: 1–15.

GLEESON M, ORTORI S, SUDBERY P (1986) Transformation of the methylotrophic yeast *Hansenula polymorpha*. J Gen Microbiol 132: 3459–3465

GOEDDEL D, HEYNECKER H, HOZUMI T, ARENTZEN R, ITAKURA K, YANSURA D, ROSS M, MIOZZARI G, CREA R, SEEBURG P (1979 a) Direct expression in *Escherichia coli* of a DNA sequence coding for human growth hormone. Nature 281: 544–548

GOEDDEL D, KLEID DG, BOLIVAR F (1979 b) Expression in *E. coli* of chemically synthesized genes for human insulin. Proc Natl Acad Sci USA 76: 106–110

GRETHE S, HECKEL JO, RIETSCHEL W, THOMSSEN R, HUFERT FT (2000) Molecular epidemiology of hepatitis B virus variants in non-human primates. J Virol 74: 5377–5381

GUIDOTTI LG (2002) The role of cytotoxic T cells and cytokines in the control of hepatitis B virus infection. Vaccine 20 Suppl 4: A80–A82

GÜNTHER S, FISCHER L, PULT I, STERNECK M, WILL H (1999) Naturally occurring variants of hepatitis B virus. Adv Virus Res 52: 25–137

GUIDOTTI LG, ROCHFORD R, CHUNG J, SHAPIRO M, PURCELL R, CHISARI FV (1999) Viral clearance without destruction of infected cells during acute HBV infection. Science 284: 825–829

GUO JT, ZHOU H, LIU C, ALDRICH C, SAPUTELLI J, WHITAKER T, BARRASA MI, MASON WS, SEEGER C (2000) Apoptosis and regeneration of hepatocytes during recovery from transient hepadnavirus infections. J Virol. 74: 1495–1505

HALPERIN SA, VAN NEST G, SMITH B, ABTAHI S, WHILEY H, EIDEN JJ (2003) A phase I study of the safety and immunogenicity of recombinant hepatitis B surface antigen co-administered with an immunostimulatory phosphorothioate oligonucleotide adjuvant. Vaccine 21: 2461–2467

HARDER W, VEENHUIS M (1989) Metabolism of one-carbon compounds. In: Metabolism and Physiology of Yeasts (Rose A, Harrison J, Eds). Academic Press, San Diego, CA, USA, pp 289–313

HARFORD N, CABEZON T, COLAU B, DELISSE A-M, RUTGERS T, DE WILDE M (1987) Construction and characterization of a *Saccharomyces cerevisiae* strain (RIT4376) expressing hepatitis B surface antigen. Postgrad Med J 63: 65–70

HEATHCOTE J, MCHUTCHISON J, LEE S, TONG M, BENNER K, MINUK G, WRIGHT T, FIKES J, LIVINGSTON B, SETTE A, CHESTNUT R (1999) A pilot study of the CY-1899 T-cell vaccine in subjects chronically infected with hepatitis B virus. Hepatology 30: 531–536

HEERMANN KH, GERLICH WH (1991) Surface Proteins of Hepatitis B Viruses. In: Molecular Biology of the Hepatitis B Virus (MacLachlan A, Ed). CRC Press, Boca Raton, FL, USA, pp 109–143

HEINEMAN TC, CLEMENTS-MANN ML, POLAND GA, JACOBSON RM, IZU AE, SAKAMOTO D, EIDEN J, VAN NEST GA, HSU HH (1999) A randomized, controlled study in adults of the immunogenicity of a novel hepatitis B vaccine containing MF59 adjuvant. Vaccine 17: 2769–2778

HIEU NT (1996) News Med Pharm (Vietnam) 10: 27–32

HIEU NT, KIM K-H, TIMMERMANS I (2001) J Infect Disease Pharmacother 5: 53–65

HIEU NT, KIM KH, JANOWICZ Z, TIMMERMANS I (2002) Comparative efficacy, safety and immunogenicity of Hepavax-Gene and Engerix-B, recombinant hepatitis B vaccines, in infants born to HBsAg and HBeAg positive mothers in Vietnam: an assessment at 2 years. Vaccine 20: 1803–1808

HILLEMAN MR. (2003) Critical overview and outlook: pathogenesis, prevention, and treatment of hepatitis and hepatocarcinoma caused by hepatitis B virus. Vaccine 21: 4626–4649

HINNEN A, BUXTON F, CHAUDHURI B, HEIM J, HOTTIGER T, MEYHACK B, POHLIG G (1995) Gene expression in recombinant yeast. In: Gene Expression in Recombinant Microorganisms (Smith A, Ed). Marcel Dekker, New York, pp 121–193

HO M-S, MAU Y-C, HUANG S-F, HSU L-C, LIN S-R, HSU H-M (1998) Patterns of circulating hepatitis B surface antigen variants among vaccinated children born to hepatitis B surface antigen carrier and non-carrier mother, J Biomed Sci 5: 355–362

HOLLENBERG CP, GELLISSEN G (1997) Gene expression in methylotrophic yeasts. Curr Opin Biotechnol 8: 554–560

HUMPHRIES JC, DIXON JS (2003) Antivirals for the treatment of chronic hepatitis B: current and future options. Intervirology 46: 413–420

HOOFNAGLE JH (1998) Therapy of viral hepatitis. Digestion 59: 563–578

HSU LC, LIN SR, HSU HM, CHAO WH, HSIEH JT, WANG MC, LU CF, CHANG YH, HO MS (1996) Ethnic differences in immune responses to hepatitis B vaccine. Am J Epidemiol 143: 718–724

ILAN Y, NAGLER A, ADLER R, TUR-KASPA R, SLAVIN S, SHOUVAL D (1993) Ablation of persistent hepatitis B by bone marrow transplantation from a hepatitis B-immune donor. Gastroenterology 104: 1818–1821

JACQUES P, MOENS G, DESOMBERE I, DEWIJNGAERT J, LEROUX-ROELS G, WETTENDORFF M, THOELEN S (2002) The immunogenicity and reactogenicity profile of a candidate hepatitis B vaccine in an adult vaccine non-responder population. Vaccine 20: 3644–3649

JANOWICZ ZA, ECKART MR, DREWKE C, ROGGENKAMP RO, HOLLENBERG CP, MAAT J, LEDEBOER AM, VISSER C, VERRIPS CT (1985) Cloning and characterization of the *DAS* gene encoding the major methanol assimilatory enzyme from the methylotrophic yeast *Hansenula polymorpha*. Nucleic Acids Res 13: 3043–3062

JANOWICZ ZA, MELBER K, MERCKELBACH A, JACOBS E, HARFORD N, COMBERBACH M, HOLLENBERG CP (1991) Simultaneous expression of the S and L surface antigens of hepatitis B, and formation of mixed particles in the methylotrophic yeast, *Hansenula polymorpha*. Yeast 7: 431–443

JILG W (1998) Novel hepatitis B vaccines. Vaccine 16: S65–S68.

JURSCH CA (2000) Größenbestimmung von Viren in menschlichem Blutplasma durch Ausschlußchromatographie. Universität of Gießen, Germany

KAKIMI K, GUIDOTTI LG, KOEZUKA Y, CHISARI FV (2000) Natural killer T cell activation inhibits hepatitis B virus replication in vivo. J Exp Med 192: 921–930

KAKIMI K, LANE TE, CHISARI FV, GUIDOTTI LG (2001) Cutting edge: inhibition of hepatitis B virus replication by activated NK T cells does not require inflammatory cell recruitment to the liver. J Immunol 167: 6701–6705

KANE MA (1996) Global status of hepatitis B immunisation. Lancet 348–696

KANN M, GERLICH WH (1998) Hepatitis B. In: Virology (Mahy BWJ, Collier L, Eds). Arnold, London, pp 745–773

KANN M, BISCHOF A, GERLICH W (1997) In vitro model for the nuclear transport of the hepadnavirus genome. J Virol 71: 1310–1316

KAO JH, CHEN PJ, LAI MY, CHEN DS (2000a) Hepatitis B genotypes correlate with clinical outcomes in patients with chronic hepatitis B. Gastroenterology 118: 554–559

KAO JH, WU NH, CHEN PJ, LAI MY, CHEN DS (2000b) Hepatitis B genotypes and the response to interferon therapy. J Hepatol 33: 998–1002

KAPUSTA J, MODELSKA A, FIGLEROWICZ M, PNIEWSKI T, LETELLIER M, LISOWA O, YUSIBOV V, KOPROWSKI H, PLUCIENNICZAK A, LEGOCKI AB (1999) A plant-derived edible vaccine against hepatitis B virus. FASEB J 13: 1796–1799

KATO N, TAMAOKI H, TANI Y, OGATA K (1972) Purification and characterization of formaldehyde dehydrogenase in the methanol utilizing yeast; *Kloeckera* sp. no. 2201. Agric Biol Chem 36: 2411–2419

KATO N, OMORI Y, TANI Y, OGATA K (1976) Alcohol oxidases of *Kloeckera* sp. no 2201 and *Hansenula polymorpha*. Catalytic properties and subunit structures. Eur J Biochem 64: 341–350

KIM HS, KIM YK, RYU SE, HONG HJ (1996) Production of hepatitis B virus preS polypeptide in *Escherichia coli* by mutation of the 5′-end

coding sequence and its purification and characterization. Gene 177: 173–177

KIM MJ, NAFZIGER AN, HARRO CD, KEYSERLING HL, RAMSEY KM, DRUSANO GL, BERTINO JS JR (2003) Revaccination of healthy nonresponders with hepatitis B vaccine and prediction of seroprotection response. Vaccine 21: 1174–1179

KOREC E, KORCOVA J, PALKOVA Z, VONDREJS V, KORINEK V, REINIS M, BICHKO VV, HLOZANEK I (1989) Expression of hepatitis B virus large envelope protein in *Escherichia coli* and *Saccharomyces cerevisiae*. Folia Biol (Praha) 35: 315–327

KRUSKALL MS, ALPER CA, AWDEH Z, YUNIS EJ, MARCUS-BAGLEY D (1992) The immune response to hepatitis B vaccine in humans: inheritance patterns in families. J Exp Med 175: 495–502

LANFORD RE, CHAVEZ D, BRASKY KM, BURNS III RB, RICO HR (1998) Isolation of a hepadnavirus from the woolly monkey a New World primate. Proc Natl Acad Sci USA 95: 5757–5761

LAU GK, SURI D, LIANG R, RIGOPOULOU EI, THOMAS MG, MULLEROVA I, NANJI A, YUEN ST, WILLIAMS R, NAOUMOV NV (2002) Resolution of chronic hepatitis B and anti-HBs seroconversion in humans by adoptive transfer of immunity to hepatitis B core antigen. Gastroenterology 122: 614–624

LAUB O, RALL LB, TRUETT M, SHAUL Y, STANDRING DN, VALENZUELA P, RUTTER WJ (1983) Synthesis of hepatitis B surface antigen in mammalian cells: expression of the entire gene and the coding region. J Virol 48: 271–280

LE BOUVIER GL (1972) Seroanalysis by immune diffusion: the subtypes of type B hepatitis virus. In: Hepatitis and Blood Transfusion (Vyas GN, Perkins A, Schmid R, Eds). Grune & Stratton, New York, pp 97–110

LEDEBOER AM, EDENS L, MAAT J, VISSER C, BOS JW, VERRIPS CT, JANOWICZ Z, ECKART M, ROGGENKAMP R, HOLLENBERG CP (1985a) Molecular cloning and characterization of a gene coding for methanol oxidase in *Hansenula polymorpha*. Nucleic Acids Res 13: 3063–3082

LEE J, KOMAGATA K (1980) Taxonomic study of methanol assimilating yeasts. J Appl Microbiol 26: 133–158

LEE YH, TUNG YT, LO SJ (1986) Expression and secretion of hepatitis B viral surface antigen in

E. coli. Biochem Biophys Res Commun 135: 1042–1049

LEE MS, KIM DH, KIM H, LEE HS, KIM CY, PARK TS, YOO KY, PARK BJ, AHN YO (1998) Hepatitis B vaccination and reduced risk of primary liver cancer among male adults: a cohort study in Korea. Int J Epidemiol 27: 316–319

LEE OJ, PARK SM, LEE JI, DONG SH, CHO SW, KIM YS (2000) Immunogenicity and safety of recombinant hepatitis B vaccine (HG-II) in healthy adults according to 0–1-2 months vaccination schedule. Kor J Med 58: 12–18

LEGLER K, STROHMEYER H, RITTER S, GERLICH WH, THOMSSEN R (1983) Kinetics, subtype specificity and immunoglobulin class of anti-HBs induced by hepatitis B vaccine. Dev Biol Stand 54: 179–189

LELIE P, IP H, REESNINK H, WONG V, KUHNS M (1991) Prevention of the hepatitis B virus carrier state in infants of mothers with and low serum levels of HBV-DNA. In: Viral hepatitis and liver disease (Hollinger F, Lemon S, Margolis H, Eds) Williams and Wilkins, Baltimore, pp 753–756

LI L, DIXON RJ, GU X, NEWBOLD JE (1998) Comparison of the sequences of the grey teal, maned duck and duck hepatitis B viruses. The molecular biology of hepatitis B virus. University of California, San Diego, CA, USA, pp 13

LIDSTROM M (Ed) (1990) Hydrocarbons and methylotrophy, in: Methods in Enzymology. Academic Press, San Diego, CA, USA

LIN HH, WANG LY, HU CT, HUANG SC, HUANG LC, LIN SS, CHIANG YM, LIU TT, CHEN CL (2003 a) Decline of hepatitis B carrier rate in vaccinated and unvaccinated subjects: sixteen years after newborn vaccination program in Taiwan. J Med Virol 69: 471–474

LIN YC, CHANG MH, NI YH, HSU HY, CHEN DS (2003 b) Long-term immunogenicity of universal hepatitis B virus vaccination in Taiwan. J Infect Dis 187: 134–138

LIU YL, WANG JF, QIU BS, ZHAO SZ, TIAN B (1994) Expression of human hepatitis B virus surface antigen gene in transgenic tobacco. SCI CHINA B 37: 37–41

LIVINGSTON BD, CRIMI C, GREY H, ISHIOKA G, CHISARI FV, FIKES J, CHESNUT RW, SETTE A (1997) The hepatitis B virus-specific CTL responses induced in humans by lipopeptide vaccination are comparable to those elicited by acute viral infection. J Immunol 159: 1383–1392

LIVINGSTON BD, ALEXANDER J, CRIMI C, OSEROFF C, CELIS E, DALY K, GUIDOTTI LG, CHISARI FV, FIKES J, CHESNUT RW, SETTE A (1999) Altered helper T lymphocyte function associated with chronic hepatitis B virus infection and its role in response to therapeutic vaccination in humans. J Immunol 162: 3088–3095

LOBAINA Y, GARCIA D, ABREU N, MUZIO V, AGUILAR JC (2003) Mucosal immunogenicity of the hepatitis B core antigen. Biochem Biophys Res Commun 300: 745–750

McDERMOTT AB, ZUCKERMAN NJ, SABIN CA, MARSH SG, MADRIGAL JA (1997) Contribution of human leukocyte antigens to the antibody response to hepatitis B vaccination. Tissue Antigens 50: 8–14

MAHONEY FJ (1999) Update on Diagnosis, Management, and Prevention of Hepatitis B Virus Infection. Clin Microbiol Rev 12: 351–366

MAKRIDES S (1996) Strategies for achieving high-level expression of genes in *Escherichia coli.* Microbiol Rev 60: 512–538

MAINI MK, BERTOLETTI A (2000) How can the cellular immune response control hepatitis B virus replication? J Viral Hepatol 7: 321–326

MAINI MK, BONI C, LEE CK, LARRUBIA JR, REIGNAT S, OGG GS, KING AS, HERBERG J, GILSON R, ALISA A, WILLIAMS R, VERGANI D, NAOUMOV NV, FERRARI C, BERTOLETTI A (2000) The role of virus-specific CD8(+) cells in liver damage and viral control during persistent hepatitis B virus infection. J Exp Med 191: 1269–1280

MARION PL, OSHIRO LS, REGNERY DC, SCULLARD GH, ROBINSON WS (1980) A virus in Beechey ground squirrels that is related to hepatitis B virus of humans. Proc Natl Acad Sci USA 77: 2941–2945

MASON WS, SEAL G, SUMMERS J (1980) Virus of Pekin ducks with structural and biological relatedness to human hepatitis B virus. J Virol 36: 829–836

MASON HS, LAM DMK, ARNTZEN CJ (1992) Expression of hepatitis B surface antigen in transgenic plants. Proc Natl Acad Sci USA 89: 11745

MEDLEY GF, LINDOP NA, EDMUNDS WJ, NOKES DJ (2001) Hepatitis B virus endemicity. heterogeneity, catastrophic dynamics and control. Nature Medicine 7: 619–624

MENNE S, RONEKER CA, TENNANT BC, KORBA BE, GERIN JL, COTE PJ (2002) Immunogenic effects of woodchuck hepatitis virus surface antigen vaccine in combination with antiviral therapy: breaking of humoral and cellular immune tolerance in chronic woodchuck hepatitis virus infection. Intervirology 45: 237–250

MICHEL ML (2002) Towards immunotherapy for chronic hepatitis B virus infections. Vaccine 20 Suppl 4: A83–A88

MILICH DR (1997) Immune response to the hepatitis B virus: Infection, animal models, vaccination. Viral Hepatitis Rev 3: 63–103

MIYANOHARA A, IMAMURA T, ARAKI M, SUGAWARA K, OHTOMO N, MATSUBARA K (1986) Expression of hepatitis B virus core antigen gene in *Saccharomyces cerevisiae*: synthesis of two polypeptides translated from different initiation codons. J Virol 59: 176–180

MOSS B, SMITH GL, GERIN JL, PURCELL RH (1984) Live recombinant vaccinia virus protects chimpanzees against hepatitis B virus. Nature 311: 67–69

MOYNIHAN JS, BLAIR J, COOMBES A, D'MELLO F, HOWARD CR (2002) Enhanced immunogenicity of a hepatitis B virus peptide vaccine using oligosaccharide ester derivative microparticles. Vaccine 20: 1870–1876

NARDELLI-HAEFLIGER D, KRAEHENBUHL JP, CURTISS R, SCHÖDEL F, POTTS A, KELLY S DE GRANDI P (1996) Oral and rectal immunization of adult female volunteers with a recombinant attenuated *Salmonella typhii* vaccine strain. Infect Immun 64: 5219–5224

NASSAL M, SCHALLER H (1996) Hepatitis B virus replication – an update. J Viral Hepatitis 3: 217–226

NAUMANN H, SCHAEFER S, YOSHIDA CF, GASPAR AM, RE R, GERLICH WH (1993) Identification of a new hepatitis B virus HBV genotype from Brazil that expresses HBV surface antigen subtype adw4. J Gen Virol 74: 1627–1632

NAYERSINA R, FOWLER P, GUILHOT S, MISSALE G, CERNY A, SCHLICHT HJ, VITIELLO A, CHESNUT R, PERSON JL, REDEKER AG (1993) HLA A2 restricted cytotoxic T lymphocyte responses to multiple hepatitis B surface antigen epitopes during hepatitis B virus infection. J Immunol 150: 4659–4671

NGUI SL, ANDREWS NJ, UNDERHILL GS, HEPTONSTALL J, TEO CG (1998) Failed postnatal immunoprophylaxis for hepatitis B: characteristics of maternal hepatitis B virus as risk factors. Clin Infect Dis 27: 100–106

NORDER H, ARAUZ-RUIZ P, BLITZ L, PUJOL FH, ECHEVARRIA JM, MAGNIUS L (2003) The T1858 variant predisposing for the pre-core stop mutation correlates with one of two major genotype F hepatitis B virus clades. J Gen Virol 84: 2083–2087

NORDER H, COUROUCÉ AM, MAGNIUS LO (1994) Complete genomes, phylogenetic relatedness, and structural proteins of six strains of the hepatitis B virus, four of which represent two new genotypes. Virology 198: 489–503

NORDER H, EBERT JW, FILDS H, MUSHAWAR IK, MAGNIUS LO (1996a) Complete sequencing of a gibbon hepatitis B virus genome reveals a unique genotype distantly related to the chimpanzee hepatitis B virus. Virology 218: 214–223

NOWAK MA, BONHOEFFER S, HILL AM, BOEHME R, THOMAS HC, McDADE H (1996) Viral dynamics in hepatitis B virus infection. Proc Natl Acad Sci USA 93: 4398–4402

OGATA N, COTE PJ, ZANETTI AR, MILLER RH, SHAPIRO M, GERIN J, PURCELL RH (1999) Licensed recombinant hepatitis B vaccines protect chimpanzees against infection with the prototype surface gene mutant of hepatitis B virus. Hepatology 30: 779–786

OHNUMA H, MACHIDA A, OKAMOTO H, TSUDA F, SAKAMOTO M, TANAKA T, MIYAKAWA Y, MAYUMI M (1993) Allelic subtypic determinants of hepatitis B surface antigen (i and t) that are distinct from d/y or w/r. J Virol 67: 927–932

OKAMOTO H, IMAI M, TSUDA F, TANAKA T, MIYAKAWA Y, MAYUMI M (1987) Point mutation in the s gene of hepatitis B virus for a d/y or w/r subtypic change in two blood donors carrying a surface antigen of compound subtype adry or adwr. J Virol 61: 3030–3034

OKAMOTO H, TSUDA F, SAKUGAWA H, SASTROSEWINJO R, IMAI M, MIYAKAWA Y, MAYUMI M (1988) Typing hepatitis B virus by homology in nucleotide sequence: comparison of surface antigen subtypes. J Gen Virol 69: 2575–2583

OON CJ, CHEN WN, GOO KS, GOH KT (2000) Intra-familial evidence of horizontal transmission of hepatitis B virus surface antigen mutant G145R. J Infect 41: 260–264

PANCHOLI P, LEE DH, LIU Q, TACKNEY C, TAYLOR P, PERKUS M, ANDRUS L, BROTMAN B, PRINCE AM. (2001) DNA prime/canarypox boost-based immunotherapy of chronic hepatitis B virus infection in a chimpanzee. Hepatology 33: 448–454

PAPATHEODORIDIS GV, HADZIYANNIS SJ (2004) Review article: current management of chronic hepatitis B. Aliment Pharmacol Ther 19: 25–37

PAPAEVANGELOU G (1998) Current combined vaccines with hepatitis B. Vaccine 16: S69–S72

PENNA A, ARTINI M, CAVALLI A, LEVRERO M, BERTOLETTI A, PILLI M, CHISARI F,V, REHERMANN B, DEL PRETE G, FIACCADORI G, FERRARI C (1996) Long-lasting memory T cell responses following acute self-limited acute hepatitis B. J Clin Invest 98: 1185–1194

PENNA A, CHISARI FV, BERTOLETTI A, MISSALE G, FOWLER P, GIUBERTI T, FIACCADORI F, FERRARI C (1991) Cytotoxic T lymphocytes recognize an HLA-A2-restricted epitope within the hepatitis B virus nucleocapsid antigen. J Exp Med 174: 1565–1570

PETERSON DL, PAUL DA, LAM L, I, ACHORD DT (1984) Antigenic structure of hepatitis B surface antigen: identification of the 'D' subtype determinant by chemical modification and use of monoclonal antibodies. J Immunol 132: 920–927

PHUMIAMORN S, SATO H, KAMIYAMA T, KUROKAWA M, SHIRAKI K (2003).Induction of humoral and cell-mediated immunity to hepatitis B surface antigen by a novel adjuvant activity of Oka varicella vaccine. J Gen Virol 84: 287–291

PIONTEK M (1998) Von der Zelle zur Anlage – vom Laborprozess in die industrielle Fertigung. Process 11: 60–61

POL S, MICHEL ML, BRÉCHOT C (2000) Immune therapy of hepatitis B virus chronic infection. Hepatology 31: 548–549

POOVORAWAN Y,THEAMBOONLERS A, SANPAVAT S, CHONGSRISAWAT V,WILLEMS P, SAFARY A (1999) Comparison study of combined DTPw-HB vaccines and separate administration of DTPw and HB vaccines in Thai children. Asian Pac J Allergy Immunol 17: 113–120

PROTZER-KNOLLE U, NAUMANN U, BARTENSCHLAGER R, BERG T, HOPF U, MEYER ZUM BUSCHENFELDE KH, NEUHAUS P, GERKEN G (1998) Hepatitis B virus with antigenically altered hepatitis B surface antigen is selected by high-dose hepatitis B immune globulin after liver transplantation. Hepatology 27: 254–263

PULT I, NETTER HJ, FRÖHLICH K, KALETA EF, WILL H (1998) Identification structural and functional analysis of a new avian Hepadnavirus from storks (STHBV). The molecular biology of Hepatitis B Virus. University of California, San Diego, CA, USA, pp 2–12

PURCELL RH, GERIN JL (1975) Hepatitis B subunit vaccine: a preliminary report of safety and efficacy tests in chimpanzees. Am J Med Sci 270: 395–399

PURCELL RH (2000) Hepatitis B virus mutants and efficacy of vaccination. Lancet 356: 769–770

RAHMAN F, DAHMEN A, HERZOG-HAUFF S., BOCHER WO, GALLE PR, LOHR HF (2000) Cellular and humoral immune responses induced by intradermal or intramuscular vaccination with the major hepatitis B surface antigen. Hepatology 31: 521–527

REHERMANN B, FERRARI C, PASQUINELLI C, CHISARI FV (1996) The hepatitis B virus persists for decades after patients' recovery from acute viral hepatitis despite active maintenance of a cytotoxic T-lymphocyte response. Nat Med 2: 1104–1108

REHERMANN B, FOWLER P, SIDNEY J, PERSON J, REDEKER A, BROWN M, MOSS B, SETTE A, CHISARI FV (1995) The cytotoxic T lymphocyte response to multiple hepatitis B virus polymerase epitopes during and after acute viral hepatitis. J Exp Med 181: 1047–1058

REISER J, GLUMOFF V, KALIN M, OCHSNER U (1990) Transfer and expression of heterologous genes in yeasts other than *Saccharomyces cerevisiae*. Adv Biochem Eng Biotechnol 43: 75–102

REN F, HINO K,YAMAGUCHI Y, FUNATSUKI K, HAYASHI A, ISHIKO H, FURUTANI M, YAMASAKI T, KORENAGA K,YAMASHITA S, KONISHI T, OKITA K (2003) Cytokine-dependent anti-viral role of CD4-positive T cells in therapeutic vaccination against chronic hepatitis B viral infection. J Med Virol 71: 376–384

RICHTER LJ,THANAVALA Y, ARNTZEN CJ, MASON HS (2000) Production of hepatitis B surface antigen in transgenic plants for oral immunization. Nature Biotechnol 18: 1167–1171

ROGGENKAMP R, SAHM H, HINKELMANN W, WAGNER F (1975) Alcohol oxidase and catalase in peroxisomes of methanol-grown *Candida boidinii*. Eur J Biochem 59: 231–236

ROGGENKAMP R, HANSEN H, ECKART M, JANOWICZ Z, HOLLENBERG C (1986) Transformation of the methylotrophic yeast *Hansenula polymorpha* by autonomous replication and integration vectors. Mol Gen Genet 202: 302–308

Romanos MA, Scorer CA, Clare JJ (1992) Foreign gene expression in yeast: a review. Yeast 8: 423–488

Roos S, Fuchs K, Roggendorf M (1989) Protection of woodchucks from infection with woodchuck hepatitis virus by immunization with recombinant core protein. J Gen Virol 70: 2087–2095

Roossinck MJ, Jameel S, Loukin SH, Siddiqui A (1986) Expression of hepatitis B viral core region in mammalian cells. Mol Cell Biol 6: 1393–1400

Roy MJ, Wu MS, Barr LJ, Fuller JT, Tussey LG, Speller S, Culp J, Burkholder JK, Swain WF, Dixon RM, Widera G, Vessey R, King A, Ogg G, Gallimore A, Haynes JR, Heydenburg Fuller D (2000) induction of antigen specific CD8+T cells, T helper cells, and protective levels of antibody in humans by particle-mediated administration of a hepatitis B virus DNA vaccine. Vaccine 22: 764 – 778

Rutgers T, Cabezon T, Harford N, Vanderbrugge D, Descurieux M, Van Opstal O, Van Wijnendaele F, Hauser P, Voet P, De Wilde M (1988) Expression of different forms of hepatitis B virus envelope proteins in yeast. In: Viral Hepatitis and Liver Disease (Zuckerman A, Ed). A.R. Liss, New York, pp 304–308

Safadi R, Israeli E, Papo O, Shibolet O, Melhem A, Bloch A, Rowe M, Alper R, Klein A, Hemed N, Segol O, Thalenfeld B, Engelhardt D, Rabbani E, Ilan Y (2003) Treatment of chronic hepatitis B virus infection via oral immune regulation toward hepatitis B virus proteins. Am J Gastroenterol 98: 2505–2515

Sahm H (1977) Metabolism of methanol by yeasts. In: Advances in Microbiological Engineering (Ghose T, Frechter A, Blankenbrough H, Eds). Springer-Verlag, Berlin, pp 77–103

Sahm H, Wagner F (1973) Microbial assimilation of methanol. Properties of formaldehyde dehydrogenase and formate dehydrogenase from *Candida boidinii*. Arch Mikrobiol 90: 263–268

Schaefer S, Piontek M,, Ahn S-J, Papendieck A, Janowicz ZA, Timmermans I, Gellissen G (2002) Recombinant hepatitis B vaccines – disease characterization and vaccine production. In: *Hansenula polymorpha* – biology and applications (Gellissen G, Ed). Wiley-VCH, Weinheim, pp 175–210

Schaefer S, Tolle T, Lottmann S, Gerlich W (1998) Animal models and experimental systems in hepatitis B virus research. In: Molecular Mechanisms in Disease and Novel Strategies for Therapy (Koshy R, Caselmann W, Eds). Imperial College Press, London, pp 51–74

Schödel F, Milich DR, Will H (1990) Hepatitis B virus nucleocapsid/preS2 fusion proteins expressed in attenuated *Salmonella* for oral vaccination. J Immunol 145: 4317–4321

Schödel F, Kelly SM, Peterson DL, Milich DR, Curtiss R, III (1994) Hybrid hepatitis B virus core-pre-S proteins synthesized in avirulent *Salmonella typhimurium* and *Salmonella typhi* for oral vaccination. Infect Immun 62: 1669–1676

Schödel F, Kelly S, Tinge S, Hopkins S, Peterson D, Milich D, Curtiss R, III (1996) Hybrid hepatitis B virus core antigen as a vaccine carrier moiety. II. Expression in avirulent *Salmonella* spp. for mucosal immunization. Adv Exp Med Biol 397: 15–21

Schodel F, Peterson D, Hughes J, Milich DR (1993) A virulent *Salmonella* expressing hybrid hepatitis B virus core/pre-S genes for oral vaccination. Vaccine 11: 143–148

Schütte H, Flossdorf J, Sahm H, Kula M (1976) Purification and properties of formaldehyde dehydrogenase and formate dehydrogenase from *Candida boidinii*. Eur J Biochem 62: 151–160

Sesardic D, Dawes CS, Mclellan K, Durrani Z, Yost SE, Corbel MJ (1999) Nonpertussis components of combination vaccines: problems with potency testing. Biologicals 27: 177–181

Shapira MY, Zeira E, Adler R, Shouval D (2001) Rapid seroprotection against hepatitis B following the first dose of a Pre-S1/Pre-S2/S vaccine. J Hepatol 34: 123–127

Shen SH, Bastien L, Nguyen T, Fung M, Slilaty SN (1989) Synthesis and secretion of hepatitis B middle surface antigen by the methylotrophic yeast *Hansenula polymorpha*. Gene 84: 303–309

Shi H, Cullen JM, Newbold JE (1993) A novel isolate of duck hepatitis B virus: GenBank accession no. M95589

Shi L, Caulfield MJ, Chern RT, Wilson RA, Sanyal G, Volkin DB (2002) Pharmaceutical and immunological evaluation of a single-shot hepatitis B vaccine formulated with PLGA microspheres. J Pharm Sci 91: 1019–1035

SHIAU AL, MURRAY K (1997) Mutated epitopes of hepatitis B surface antigen fused to the core antigen of the virus induce antibodies that react with the native surface antigen. J Med Virol 51: 159–166

SHIOSAKI K, TAKATA K, NISHIMURA S, MIZOKAMI H, MATSUBARA K (1991) Production of hepatitis B virion-like particles in yeast. Gene 106: 143–149

SHOUVAL D, SAMUEL D (2000) Hepatitis B immune globulin to prevent hepatitis B virus graft reinfection following liver transplantation: a concise review. Hepatology 32: 1189–1195

SHOUVAL D, ILAN Y, ADLER R, DEEPEN R, PANET A, EVEN-CHEN Z, GORECKI M, GERLICH WH (1994) Improved immunogenicity in mice of a mammalian cell-derived recombinant hepatitis B vaccine containing pre-S1 and pre-S2 antigens as compared with conventional yeast-derived vaccines. Vaccine 12: 1453–1459

SITIA G, ISOGAWA M, KAKIMI K, WIELAND SF, CHISARI FV, GUIDOTTI LG. (2002) Depletion of neutrophils blocks the recruitment of antigen-nonspecific cells into the liver without affecting the antiviral activity of hepatitis B virus-specific cytotoxic T lymphocytes. Proc Natl Acad Sci USA 99: 13717–13722

SPRENGEL R, KALETA EF, WILL H (1988) Isolation and characterization of a hepatitis B virus endemic in herons. J Virol 62: 3832–3839

STINCHCOMB D, THOMAS M, KELLY I, SELKER E, DAVIES R (1980) Eucaryotic DNA segments capable of autonomous replication in yeasts. Proc Natl Acad Sci USA 77: 4559–4563

STUYVER L, DE GENDT S, VAN GEYT C, ZOULIM F, FRIED M, SCHINAZI RF, ROSSAU R (2000) A new genotype of hepatitis B virus: complete genome and phylogenetic relatedness. J Gen Virol 81: 67–74

STUYVER L, LOCARNINI S, LOK A, RICHMAN D, CARMAN W, DIENSTAG J, SCHINAZI R (2001) Nomenclature for antiviral-resistant human hepatitis B virus mutations in the polymerase region. Hepatology 33: 751–757

SUMMERS J, SMOLEC JM, SNYDER R (1978) A virus similar to human hepatitis B virus associated with hepatitis and hepatoma in woodchucks. Proc Natl Acad Sci USA 75: 4533–4537

SUTNICK AI, LONDON WT (1968) Hepatitis and leukemia: their relation to Australia antigen. Bull NY Acad Med 44: 1566–1586

TAKAHASHI K, BROTMAN B, USUDA S, MISHIRO S, PRINCE AM (2000) Full-genome sequence analyses of hepatitis B virus HBV strains recovered from chimpanzees infected in the wild: implications for an origin of HBV. Virology 267: 58–64

TANI Y (1984) Microbiology and biochemistry of methylotrophic yeasts. In: Methylotrophs: Microbiology, Biochemistry and Genetics (Hou C, Ed). CRC Press, Boca Raton, FL, USA, pp 33–85

TESTUT P, RENARD CA, TERRADILLOS O, VITVITSKI TL, TEKAIA F, DEGOTT C, BLAKE J, BOYER B, BUENDIA MA (1996) A new hepadnavirus endemic in arctic ground squirrels in Alaska. J Virol 70: 4210–4219

THEILMANN L, KLINKERT MQ, GMELIN K, KOMMERELL B, PFAFF E (1987) Detection of antibodies against pre-S1 proteins in sera of patients with hepatitis B virus (HBV) infection. J Hepatol 4: 22–28

THERMET A, ROLLIER C, ZOULIM F, TREPO C, COVA L (2003) Progress in DNA vaccine for prophylaxis and therapy of hepatitis B. Vaccine 21: 659–662

THIMME R, WIELAND S, STEIGER C, GHRAYEB J, REIMANN KA, PURCELL RH, CHISARI FV (2003) CD8(+) T cells mediate viral clearance and disease pathogenesis during acute hepatitis B virus infection. J Virol 77: 68–76

THOELEN S, DE CLERQ N, TORNIEPORTH N (2001) A prophylactic hepatitis B vaccine with a novel adjuvant system. Vaccine 19: 2400–2403

TORRESI J (2002) The virological and clinical significance of mutations in the overlapping envelope and polymerase genes of hepatitis B virus. J Clin Virol 25: 97–106

TSUDA S, YOSHIOKA K, TANAKA T, IWATA A, YOSHIKAWA A, WATANABE Y, OKADA Y (1998) Application of the human hepatitis B virus core antigen from transgenic tobacco plants for serological diagnosis. Vox Sanguinis 74: 148–155

USUDA S, OKAMOTO H, IWANARI H, BABA K, TSUDA F, MIYAKAWA Y, MAYUMI M (1999) Serological detection of hepatitis B virus genotypes by ELISA with monoclonal antibodies to type specific epitopes in the preS2-region product. J Virol Methods 80: 97–112

VAN DER KLEI I, HARDER W, VEENHUIS M (1991) Biosynthesis and assembly of alcohol oxidase, a peroxisomal matrix protein in methylotrophic yeasts: a review. Yeast 7: 195–209

van Dijken J, Otto R, Harder W (1976)
Growth of *Hansenula polymorpha* in a methanol-limited chemostat: physiological responses due to the involvement of methanol
oxidase as a key enzyme in methanol metabolism. Arch Microbiol 111: 137–144

Vaudin M, Wolstenholme AJ, Tsiquaye KN,
Zuckerman AJ, Harrison TJ (1988) The
complete nucleotide sequence of the genome
of a hepatitis B virus isolated from a naturally
infected chimpanzee. J Gen Virol 69: 1383–
1389

Veenhuis M, Harder W (1988) Microbodies in
yeasts: structure, function and biogenesis.
Microbiol Sci 5: 347–351

Veenhuis M, Van Dijken JP, Harder W (1983)
The significance of peroxisomes in the meta-
bolism of one-carbon compounds in yeasts.
Adv Microb Physiol 24: 1–82

Wai CT, Chu CJ, Hussain M, Lok AS (2002)
HBV genotype B is associated with better response to interferon therapy in HBeAg(+)
chronic hepatitis than genotype C. Hepatology
36: 1425–1430

Wang FS (2003) Current status and prospects
of studies on human genetic alleles associated
with hepatitis B virus infection. World J Gastroenterol 9: 641–644

Wang J, Zhu Q, Zhang T, Yu H (2002) A pilot
study on the combined therapy of granulocyte-macrophage colony-stimulating factor
and hepatitis B vaccine on chronic hepatitis B
virus carrier children. Chin Med J (Engl) 115:
1824–1828

Warren KS, Heeney JL, Swan RA, Heriyanto,
Verschoor EJ (1999) A new group of hepad-
naviruses naturally infecting orangutans
(*Pongo pygmaeus*). J Virol 73: 7860–7865

Webster GJ, Reignat S, Maini MK,
Whalley SA, Ogg GS, King A, Brown D,
Amlot PL, Williams R, Vergani D,
Dusheiko GM, Bertoletti A (2000) Incuba-
tion phase of acute hepatitis B in man: dy-
namic of cellular immune mechanisms.
Hepatology 32: 1117–1124

Weinberger KM, Bauer T, Bohm S, Jilg W
(2000) High genetic variability of the group-
specific a-determinant of hepatitis B virus surface antigen (HBsAg) and the corresponding
fragment of the viral polymerase in chronic
virus carriers lacking detectable HBsAg in
serum. J Gen Virol 81: 1165–1174

Wilson JN, Nokes DJ, Carman WF (1998) Current status of HBV vaccine escape variants –

a mathematical model of their epidemiology.
J Viral Hepat 5: 25–30

Wilson JN, Nokes DJ, Carman WF (1999) The
predicted pattern of emergence of vaccine-re-
sistant hepatitis B: a cause for concern? Vaccine 17: 973–978

Wizemann H, von Brunn A (1999) Purification of *E. coli*-expressed HIS-tagged hepatitis
B core antigen by Ni^{2+} chelate affinity chromatography. J Virol Methods 77: 189–197

Wu JY, Newton S, Judd A, Stocker B, Robinson WS (1989) Expression of immunogenic
epitopes of hepatitis B surface antigen with
hybrid flagellin proteins by a vaccine strain of
Salmonella. Proc Natl Acad Sci USA 86: 4726–
4730

Woo PC, Wong LP, Zheng BJ, Yuen KY (2001)
Unique immunogenicity of hepatitis B virus
DNA vaccine presented by live-attenuated *Sal-
monella typhimurium*. Vaccine 19: 2945–2954

Yalcin K, Danis R, Degertekin H, Alp MN,
Tekes S, Budak T (2003) The lack of effect of
therapeutic vaccination with a pre-S2/S HBV
vaccine in the immune tolerant phase of
chronic HBV infection. J Clin Gastroenterol
37: 330–335

Yoshida I, Takamizawa A, Fujita H, Manabe S,
Okabe A (1991) Expression of the hepatitis B
surface antigen gene containing the preS2 region in *Saccharomyces cerevisiae*. Acta Med
Okayama 45: 1–10

Youn BW, Samanta H (1989) Purification and
characterization of pre-S-containing hepatitis
B surface antigens produced in recombinant
mammalian cell culture. Vaccine 7: 60–68

Yu XM, Wang Y, Li ZP (1992) An HBV large
surface antigen protein which can be secreted
from mammalian cells. SCI CHINA B 35:
455–462

Zuckerman AJ (1997) Prevention of primary
liver cancer by immunization. N Engl J Med
336: 1906–1907

Zuckerman JN, Sabin C, Craig FM,
Williams A, Zuckerman AJ (1997) Immune
response to a new hepatitis B vaccine in
healthcare workers who had not responded to
standard vaccine: randomised double blind
dose–response study. Br Med J 314: 329–333

Zuckerman JN, Zuckerman AJ (2000) Current
topics in hepatitis B. J Infect 41: 130–136

Zuckerman JN, Zuckerman AJ (2002) Recombinant hepatitis B triple antigen vaccine:
Hepacare. Expert Rev Vaccines 1: 141–144

16
Biopharmaceuticals and the Industrial Environment
GEORG MELMER

16.1
Introduction

Just 50 years after the discovery of the double helix structure of desoxyribonucleic acid (DNA) by Watson and Crick (1953), new developments have revolutionized the pharmaceutical industry. A range of newly developed drugs has been produced, based on the application of biotechnological and gene technological methods. Consequently, numerous emerging biopharmaceutical companies have successfully taken advantage of the new options, and continue to do so. As companies such as Amgen or Biogen began to integrate recombinant techniques into the methods and requirements of the pharmaceutical industry, the pharmaceutical industry became one of the major economical forces worldwide. In the United States alone, prescription drug sales of $216.4 billion were achieved in 2003 – a healthy rise of 11.5% compared to the previous year (IMS Health 2003). Due not only to demographic changes – notably the increasing share of elderly people in the populations of industrialized nations – but also to major demands for improved healthcare in the emerging economies and the developing countries, the pharmaceutical industry will remain the subject of further growth. Indeed, several economics studies have suggested that worldwide, in the year 2010, the sales of drugs will amount to over $600 billion (Burrill 2003).

Although currently the sales of cardiovascular drugs heads the list of therapeutics, with far in excess of $50 billion in annual sales, the fastest growing segment is represented by antipsychotic medicines, with sales increased overall by 22.1%, and seizure treatments by 24.4% (see Table 16.1). Overall, cholesterol-reducing agents – though not biopharmaceuticals – remained the top-selling medications, generating $13.9 billion in the United States in 2003. Indeed, Lipitor® – the anti-cholesterol produced by Pfizer Inc. – maintained its lead as the top-selling drug and secured the title for Pfizer of the world's largest pharmaceutical company (Table 16.2).

Being largely research-based, the pharmaceutical industry has not only successfully provided valuable new drugs for treatment of diseases, but its research and development departments have also made significant contributions to the basic understanding of diseases and their causes. This combination of drug provision and the generation of knowledge had a major impact on the global improvement of health

Production of Recombinant Proteins. Novel Microbial and Eucaryotic Expression Systems. Edited by Gerd Gellissen
Copyright © 2005 WILEY-VCH Verlag GmbH & Co. KGaA, Weinheim
ISBN: 3-527-31036-3

Tab. 16.1 World pharmaceutical market by therapeutic areas.

Therapeutic category	Sales in 2003 ($ million)	Sales in 2002 ($ million)	Growth rate (%)
Cardiovascular	55 385	50 135	10
Central nervous system	50 635	44 431	14
Alimentary/metabolism	42 501	39 117	9
Respiratory	26 722	24 959	7
Anti-infectives	24 737	23 026	7
Musculoskeletal	17 392	15 847	10
Genitourinary	16 305	15 192	7
Cytostatics	13 127	10 951	20
Dermatologicals	8827	8390	5
Blood agents	9476	8152	16
Sensory agents	5795	5301	9
Diagnostic agents	5091	4611	10
Hormones	4446	4127	8

Tab. 16.2 Worldwide sales of top ten pharmaceutical companies in 2002.

Company	Revenues ($, 000)	R&D Expenditures ($, 000)
Pfizer + Pharmacia (merged in 2003)	28 228 + 12 037	5176 + 2359
GlaxoSmithKline	27 060	4108
Merck	20 130	2677
AstraZeneca	17 841	3069
Johnson & Johnson	17 151	3957
Aventis	16 639	3235
Bristol-Myers Squibb	14 705	2218
Novartis	13 547	2799
Wyeth	10,899	2080
Lilly	10 385	2149

and life expectancy. For a comprehensive overview of the pharmaceutical development before the establishment of gene technology methods, the reader is referred to Bindra and Lednicer (1983) and Thromber (1986).

Historically, the pharmaceutical industry emerged from a number of different approaches to identify new therapeutic agents, amongst others from the discovery and analysis of compounds from natural sources. A striking early example of this was the discovery of salicylates in extracts of the willow tree bark, and this approach has led to the discovery of medicinal agents of high value. In fact, this approach is still being pursued today, with an increased emphasis during the past decade on the exploration of areas of wide biodiversity such as rain forests or hot springs. The major breakthrough, however, in innovative drug discovery was provoked by systematic investigations of synthesized and modified organic molecules, and for almost a century these materials determined the fate of the pharmaceutical industry. An early exception to this was the finding by Banting and Best that insulin could reverse the ravages of diabetes, thus demonstrating that proteins could also be used as drugs.

Among the wide range of medicines used today, very few were known before the Second World War. At that time, morphine, digitalis, aspirin, and barbiturates were commonly prescribed, and the first sulfonamide antibacterial drugs were just beginning to appear. During the war, however, several new drug types were developed for the treatment of wounded soldiers, or to treat venereal diseases. Before the war, new drug development was dominated by the German chemical industry, but this predominance has been replaced during the past decades by a more global distribution of the research-based pharmaceutical industry, which has evolved into one of the most productive industrial segments worldwide. Indeed, today the pharmaceutical industry is dominated by Anglo-American companies (Table 16.2).

The focus on the production of new heterocycle-based compounds was supplemented by a range of other pharmaceuticals. Vaccine production became the strength of some pharmaceutical companies such as Aventis, GlaxoSmithKline, Chiron, or Berna Biotech. Due to a desperate need for the aforementioned insulin and the increasing demand for antibiotics, a new branch of the pharmaceutical industry emerged as companies began to use biotechnological methods rather than chemical syntheses.

The dependency on innovation and research has led to certain structural features that distinguish the pharmaceutical industry from others. Innovations require close interaction with universities and other cutting-edge research facilities. The research and development (R&D) of new drugs is a long-term task, and is therefore very expensive. Recent sources have estimated that the cost to develop a particular drug can be in excess of $800 million, and that the time required can sometimes exceed 12 years (Bos et al. 2001). Because of these enormous costs, pharmaceutical companies strive to develop and sell their products on a worldwide basis. Consequently, these companies tend to be supranationally or even globally structured, and are usually not restricted to an individual nation, though there are some exceptions. One very important point is imposed by the fact that the development and production of pharmaceuticals must meet stringent requirements for regulatory approval and, in many countries, for price approval. Price regulation is important not only because it affects profits and the ability to re-invest in new research, but also because it relates to the healthcare costs in general. Although drugs account for less than 10 % of total healthcare costs, the pricing of pharmaceuticals has frequently become a political issue to which governments around the world relate in their attempts to reduce their healthcare budgets.

16.2
Early Success Stories

Although proteins are widely used in research, medicine, and industry, their recovery from their natural sources can be both difficult and expensive. Occasionally, these natural sources (e.g., human or animal blood) are in insufficient supply to meet demands. During the early 1980s, this situation was predicted to become a major problem of the future, an example being the treatment of chronic diseases such as diabetes. In 2002, the biotechnology industry celebrated the 20th anniversary of the approval of the first biopharmaceutical drug – not surprisingly, recombinant insulin. The gene for human insulin had been cloned and the protein was expressed from the cloned gene by using a suitable expression system: *Escherichia coli*. The demand for insulin had steadily increased since Banting and Best first described its use in the treatment of diabetes. As it is a small protein of only 51 amino acids in length, it was possible to clone the gene (insulin is encoded as a precursor of less than 100 amino acids), even with the somewhat primitive methods available in the 1970s. Moreover, an established purification scheme for the protein could be used, thus easing the competitive production of a recombinant variant. This recombinant insulin provided the basis for one of the most successful biotech companies, namely Genentech, which currently ranks as number two in the list of the top ten biopharmaceutical companies (see Table 16.3). Genentech licensed the product to Eli Lilly, which sells it under the tradename, Humulin®. Humulin® received the status of a "blockbuster" product, with annual sales in excess of $1000 million (2002: $1004 million). Later, a second-generation product, Humalog® was approved and also reached blockbuster status, with sales of $834 million in 2002. Meanwhile, additional *Saccharomyces cerevisiae*-derived and *Hansenula polymorpha*-derived recombinant insulins have become available.

The use of pharmaceutical proteins from natural sources can impose additional health risks. For example, many people have contracted diseases from contaminated blood products such as hormones. In the case of growth hormone, rigid testing of the natural products caused a rise in price and led to a further shortage of supply, thereby creating a desperate need for a safe substitute. Once again, Genentech were the pioneers in cloning the gene for human growth hormone and producing the hormone heterologously in *E. coli*. As previously, the product (Humatrope®) was licensed to Eli Lilly, who sell it to treat growth deficiency in children. Humatrope® is a very successful product, with annual sales exceeding $300 million, despite the lower abundance of affected persons compared to those with diabetes.

Tab. 16.3 Worldwide sales of top ten biopharmaceutical companies in 2002.

Company	Revenues ($, 000)	R&D expenditure ($, 000)	Royalty revenues ($, 000)
Amgen	4991	1117	332
Genentech	2164	624	366
Serono	1423	358	123
Biogen	1034	368	114
Genzyme	858	305	not available
MedImmune	786	144	62
Chiron	766	326	47
Gilead	424	135	43
Millennium	159	338	193
Intermune	112	130	not available

A third early success also emerged from pioneering studies conducted by Genentech when, in 1984, the gene for Factor VIII was cloned (Gitschier et al. 1984). The gene spans 186 000 base pairs (bp) of genomic DNA, and is transcribed into a messenger RNA (mRNA) of about 9000 bp in length. This mRNA codes for a single-chain protein of 2351 amino acids that is secreted after N-terminal processing and glycosylation. Protein size – as well as the high degree of glycosylation – necessitated the use of mammalian rather than bacterial cells to ensure its authentic production. Due to the inherent complexity of the product and process, it was not before 1992 that the first recombinant Factor VIII products (Recombinate®, Baxter; Bioclate®, Aventis Behring) were approved by the Food and Drug Administration (FDA) and subsequently launched

These three examples all hint to the important prerequisites for the success of biopharmaceutical drugs: (i) the recombinant drug usually replaces an established product that is recovered from natural sources; and (ii) the presence or biological activity of the natural product is impaired due to deficiencies in an affected patient. In addition, the design and development of purification and production procedures required to meet increasingly stringent safety criteria are more straightforward when applying newly developed biotechnological methods.

16.3
The Bumpy Road Appeared

In many cases, previously unknown technologies had to be developed for particular product and process developments. However, it was not only the production methods that became a severe hurdle for the new biopharmaceuticals, as many very promising new products failed to pass clinical trials, due either to eliciting unexpected and detrimental side effects or to simply lacking the required efficacy.

The development of monoclonal antibodies in 1975 (Köhler and Milstein 1975) provided tremendous opportunities for the development of recombinant antibody-based therapeutics. The latter agents were, among other properties, designed to block the early stages of cancer or to inhibit receptors that were important for cell function.

With new efficient technologies in place for the large-scale production of recombinant proteins, it became possible to produce very rare cytokines, such as interferons and interleukins. A range of cytokine-based products was evaluated and, after some failures, several potent drugs emerged for the treatment of cancer, viral infections, or arthritis.

These early promising biotherapeutics caused great excitement and attracted investors to the new "money-printing" biotechnology. Accordingly, a wave of investments flooded into the new industrial branch, and these were used to build up biotechnology and contract manufacturing companies, thereby dramatically increasing the manufacturing capacities for biopharmaceuticals. However, when the first generation of murine monoclonals failed, and some early recombinant cytokines did not pass advanced clinical trials due to extreme side effects, redundant capacities were generated.

16.4
The Breakthrough in Many Areas

During the subsequent years, as knowledge of the biological processes was increased, and methods of expression and purification improved, new ground was laid for biopharmaceuticals that indeed successfully passed clinical examination. Between 1982 – when recombinant insulin was launched – and 2000, a total of 84 innovative biopharmaceuticals was approved in the United States and/or the European Union for administration to humans. For a selection of these biopharmaceutical products, see Table 16.4.

Tab. 16.4 Biopharmaceuticals approved in the United States and/or Europe.

Product	Organism utilized	Company	Therapeutic indication	Date approved
Recombinant interferons and interleukins				
Pegasys (Peginterferon α-2a	*E. coli*	Roche	Hepatitis C	2002 (EU, US)
PegIntron A (PEGylated rIFN-α-2b	*E. coli*	Schering-Plough	Chronic hepatitis C	2000 (EU), 2001 (US)
Viraferon (rIFN-α-2b)	*E. coli*	Schering-Plough	Chronic hepatitis B and C	2000 (EU)
ViraferonPEG (PEGylated rIFN-α-2b)	*E. coli*	Schering-Plough	Chronic hepatitis C	2000 (EU)
Alfatronol (rh IFN-α-2b	*E. coli*	Schering-Plough	Hepatitis B, C and various cancers	2000 (EU)
Viraferon (rh IFN-α-2b)	*E. coli*	Schering-Plough	Hepatitis B, C	2000 (EU)
Intron A (rIFN-α-eb	*E. coli*	Schering-Plough	Cancer, genital warts, hepatitis	1986 (US), 2000 (EU)
Alfatronol (rh IFN-α-2b	*E. coli*	Schering-Plough	Hepatitis B, C and various cancers	
Rebetron (combination of ribavirin and rh IFN-α-2b)	*E. coli*	Schering-Plough	Chronic hepatitis C	1999 (EU)
Infergen (r IFN-α, synthetic type I IFN)	*E. coli*	Schering-Plough	Chronic hepatitis C	1997 (US), 1999 (EU)
Roferon A (rh IFN-α-2b	*E. coli*	Schering-Plough	Hairy cell leukemia	1986 (US)
Rebif (rh IFN-β-1a)	CHO cells	Ares-Serono	Relapsing/remitting multiple sclerosis	1998 (EU), 2002 (US)
Avonex (rh IFN-β-1a)	CHO cells	Biogen	Relapsing multiple sclerosis	1997 (EU), 1996 (US)

Tab. 16.4 (continued)

Product	Organism utilized	Company	Therapeutic indication	Date approved
Betaseron (rh IFN-β-1 b, differs from human protein by C17 → S)	*E. coli*	Berlex Labs/Chiron	Relapsing/remitting multiple sclerosis	1993 (US)
Betaferon/rh IFN-β-1 b, differs from human protein by C17 → S)	*E. coli*	Schering AG	Multiple sclerosis	1995 (EU)
Kineret (anakinra; rh IL-1 receptor antagonist)	*E. coli*	Amgen	Rheumatoid arthritis	2001 (US)
Neumega (r IL-11, lacks N-terminal proline of native molecule)	*E. coli*	Genetics Institute	Prevention of chemo-therapy-induced thrombocytopenia	1997 (US)
Proleukin (r IL-2, differs from human molecule in that it is devoid of an N-terminal alanine and contains C125 → S substitution	*E. coli*	Chiron	Renal cell carcinoma	1992 (US)
Actimmune (rh IFN-γ-1 b)	*E. coli*	Genentech	Chronic granulo-matous disease	1990 (US)
Recombinant vaccines				
Ambirix	*S. cerevisae*	GlaxoSmithKline	Immunization against Hepatitis A and B	2002 (EU)
Pediarix	*S. cerevisae*	SmithKline Beecham	Immunization against various conditions inducing Hepatitis B (children)	2002 (US)
HBVAXPRO	*S. cerevisae*	Aventis Pharma	Immunization against Hepatitis B	2001 (EU)
Twinrix	*S. cerevisae*	SmithKline Beecham (EU), GlaxoSmithKline (US)	Immunization against Hepatitis A and B	1996 (EU) (adult), 1997 (EU) (pediatric), 2001 (US)
Infanrix-Hexa	*S. cerevisae*	SmithKline Beecham	Immunization against diphtheria, tetanus, pertussis, *Haemophilus influenzae* type B, Hepatitis B and polio	2000 (EU)

Tab. 16.4   (continued)

Product	Organism utilized	Company	Therapeutic indication	Date approved
Infanrix-Penta	*S. cerevisae*	SmithKline Beecham	Immunization against diphtheria, tetanus, pertussis, Hepatitis B and polio	2000 (EU)
Hepcare	Mammalian (murine) cell line	Medeva Pharma	Immunization against hepatitis B	2000 (EU)
Hexavac	*S. cerevisae*	Aventis Pasteur	Immunization against diphtheria, tetanus, pertussis, *H. influenzae* type B, hepatitis B and polio	2000 (EU)
Procomvax	*S. cerevisae*	Aventis Pasteur	Immunization against *H. influenzae* type B and hepatitis B	1999 (EU)
Primavax	*S. cerevisae*	Aventis Pasteur	Immunization against diphtheria, tetanus and hepatitis B	1998 (EU)
Infanrix Hep B	*S. cerevisae*	SmithKline Beecham	Immunization against diphtheria, tetanus, pertussis and hepatitis B	1997 (EU)
Twinrix	*S. cerevisae*	SmithKline Beecham	Immunization against hepatitis A and B	1996 (EU) (adult), 1997 (EU)
Comvax	*S. cerevisae*	Merck	Vaccination of infants against *H. influenzae* type B and hepatitis B	1996 (US)
Tritanrix-HB	*S. cerevisae*	SmithKline Beecham	Vaccination against hepatitis B, diphtheria, tetanus and pertussis	1996 (US)
Recombivax	*S. cerevisae*	Merck	Hepatitis B prevention	1986 (US)
Lymerix	*E. coli*	SmithKline Beecham	Lyme disease vaccine	1998 (US)
Tricelluvax		Chiron SpA	Immunization against diphtheria, tetanus and pertussis	1999 (EU)

Tab. 16.4 (continued)

Product	Organism utilized	Company	Therapeutic indication	Date approved
Recombinant blood factors				
Helixate NexGen	BHK cells	Bayer	Hemophilia A	2000 (EU)
ReFacto	CHO cells	Genetics Institute/ Wyeth Europa	Hemophilia A	1999 (EU), 2000 (US)
Kogenate	BHK cells	Bayer	Hemophilia A	1993 (US), 2000 (EU)
Bioclate	CHO cells	Aventis Behring	Hemophilia A	1993 (US)
Recombinate	Animal cell line	Baxter Healthcare/ Genetics Institute	Hemophilia A	1992 (US)
NovoSeven	BHK cells	Novo Nordisk	Some forms of hemophilia	1996 (EU), 1999 (US)
Benefix	CHO cells	Genetics Institute	Hemophilia B	1997 (US, EU)
Recombinant anticoagulants				
Tenecteplase	CHO cells	Boehringer Ingelheim	Myocardial infarction	2001 (EU)
TNKase	CHO cells	Genentech	Myocardial infarction	2000 (US)
Ecokinase	*E. coli*	Galenus Mannheim	Acute myocardial infarction	1996 (EU)
Rapilysin	*E. coli*	Roche	Acute myocardial infarction	1996 (EU)
Retavase	*E. coli*	Boehringer Mannheim/Centocor	Acute myocardial infarction	1996 (US)
Activase	CHO cells	Genentech	Acute myocardial infarction	1987 (US)
Refludan	*S. cerevisae*	Hoechst Marion Roussel/Behringwerke AG	Anticoagulation therapy for heparin-associated thrombocytopenia	1997 (EU), 1998 (US)
Revasc	*S. cerevisae*	Aventis	Prevention of venous thrombosis	1997 (EU)
Recombinant hormones				
Insulin				
Actrapid/Velosulin/ Monotard/Insulatard/ Protaphane/Mixtrad/ Actraphane/Ultratard	*S. cerevisae*	Novo Nordisk	Diabetes mellitus	2002 (EU)

Tab. 16.4 (continued)

Product	Organism utilized	Company	Therapeutic indication	Date approved
Novolog	S. cerevisae	Novo Nordisk	Diabetes mellitus	2001 (US)
Novolog Mix 70/30	S. cerevisae	Novo Nordisk	Diabetes mellitus	2001 (US)
Novomix 30	S. cerevisae	Novo Nordisk	Diabetes mellitus	2000 (EU)
Lantus	E. coli	Aventis	Diabetes mellitus	2000 (EU, US)
Optisulin	E. coli	Aventis	Diabetes mellitus	2000 (EU)
NovoRapid	E. coli	Novo Nordisk	Diabetes mellitus	1999 (EU)
Liprolog	E. coli	Eli Lilly	Diabetes mellitus	1997 (EU)
Insuman	E. coli		Diabetes mellitus	1997 (EU)
Humalog	E. coli	Eli Lilly	Diabetes mellitus	1996 (EU, US)
Novolin	E. coli	Novo Nordisk	Diabetes mellitus	1991 (US)
Humulin	E. coli	Eli Lilly	Diabetes mellitus	1982 (US)
Human growth hormone (hGH)				
Somavert	E. coli	Pfizer	Treatment of acromegaly	2003 (US), 2002 (E()U)
Nutropin AQ	E. coli	Schwarz Pharma	Growth failure/ Turner's syndrome	1994 (US), 2001 (EU)
Serostim		Serono Laboratories	Treatment of AIDS- associated catabolism/ wasting	1996 (US)
Saizen		Serono Laboratories	hGH deficiency in children	1996 (US)
Genotropin	E. coli	Pharmacia & Upjohn	hGH deficiency in children	1995 (US)
Norditropin		Novo Nordisk	Treatment of growth failure in children due to inadequate growth hormone secretion	1995 (US)
BioTropin		Savient Pharmaceuticals	hGH deficiency in children	1995 (US)
Nutropin	E. coli	Genentech	hGH deficiency in children	1994 (US)
Humatrope	E. coli	Eli Lilly	hGH deficiency in children	1987 (US)
Protropin	E. coli	Genentech	hGH deficiency in children	1985 (US)

Tab. 16.4 (continued)

Product	Organism utilized	Company	Therapeutic indication	Date approved
Follicle-stimulating hormone				
Follistim	CHO cells	NV Organon	Infertility	1997 (US)
Puregon	CHO cells	NV Organon	Anovulation and superovulation	1996 (EU)
Gonal F	CHO cells	Ares-Serono	Anovulation and superovulation	1995 (EU), 1997 (US)
Other hormones				
Forsteo (human parathyroid hormone)	*E. coli*	Eli Lilly	Treatment of established osteoporosis in post-menopausal women	2003 (EU)
Forteo (human parathyroid hormone)	*E. coli*	Eli Lilly	Treatment of osteoporosis in some post-menopausal women	2002 (US)
Ovitrelle/Ovidrelle (choriogonadotropin)	CHO cells	Serono	Used in selected assisted reproductive techniques	2000 (US), 2001 (EU)
Tyrogen (human TSH)	CHO cells	Genzyme	Detection/treatment of thyroid cancer	1998 (US), 2000 (EU)
Luveris (human luteinizing hormone	CHO cells	Ares-Serono	Some forms of infertility	2000 (EU)
Forcaltonin (salmon calcitonin)	*E. coli*	Unigene	Paget's disease	1999 (EU)
Glucagen (human glucagon)	*S. cerevisae*	Novo Nordisk	Hypoglycemia	1998 (US)
Recombinant hematopoietic growth factors				
Erythropoietin				
Aranesp	CHO cells	Amgen	Treatment of anemia	2001 (US, EU)
Nespo	CHO cells	Dompe Biotech	Treatment of anemia	2001 (EU)
Neorecormon	CHO cells	Roche	Treatment of anemia	1997 (EU)
Procrit	Mammalian cell line	Ortho Biotech	Treatment of anemia	1990 (US)
Epogen	Mammalian cell line	Amgen	Treatment of anemia	1989 (US)

Tab. 16.4 (continued)

Product	Organism utilized	Company	Therapeutic indication	Date approved
Granulocyte-macrophage colony stimulating factor				
Neulasta		Amgen/Dompe Biotech	Chemotherapy-induced neutropenia	2002 (US, EU)
Leukine	*E. coli*	Immunex (now Amgen)	Autologous bone marrow transplantation	1991 (US)
Neupogen	*E. coli*	Amgen	Chemotherapy-induced neutropenia	1991 (US)
Monoclonal antibody-based products				
Bexxar (against CD20)	Mammalian cell line	Corixa/GlaxoSmithKline	Treatment of CD20 positive follicular non-Hodgkin's lymphoma	2003 (US)
Xolair (binds IgE)	CHO cells	Genentech	Asthma	2003 (US)
Humira (against TNF)	Mammalian cell line	Abbott	Rheumatoid arthritis	2002 (US)
Zevalin (against CD20)	CHO cells	Idec Pharmaceuticals	Non-Hodgkin's lymphoma	2002 (US)
Mabcampth (EU), Campath (US) (against CD52)		Millenium, ILEX, Berlex	Chronic lymphocytic leukemia	2001 (EU, US)
Mylotarg (against CD33)		Wyeth	Acute myeloid leukemia	2000 (US)
Herceptin (against human epidermal growth factor receptor 2, HER2)		Genentech, Roche	Treatment of metastatic breast cancer, in case tumor over-expresses HER2 protein	1998 (US), 2000 (EU)
Remicade (against TNF-α)		Centocor	Treatment of Crohn's disease	1998 (US), 1999 (EU)
Synagis (against epitope on the surface of respiratory syntactical virus)		MedImmune, Abbott	Prophylaxis of lower-tract respiratory disease caused by syncytial virus in pediatric patients	1998 (US), 1999 (EU)
Mabthera (against CD20 surface antigen)		Hoffmann-La Roche	Non-Hodgkin's lymphoma	1998 (EU)
Rituxan (against CD20 surface antigen)		Genentech/IDEC Pharmaceuticals	Non-Hodgkin's lymphoma	1997 (US)
ReoPro (against the platelet surface receptor GPIIb/IIIa)		Centocor	Prevention of blood clots	1994 (US)
Orthoclone OKT3 (against CD3)		Ortho Biotech	Reversal of acute kidney transplant rejection	1986 (US)

Tab. 16.5 Number of approved biopharmaceuticals in six major markets.

Pharmaceutical market	Approved up to 2000	Approved since 2000
Cancer	11	5
Diabetes	6	6
Hemophilia	4	3
Hepatitis	10	16
Human growth hormone deficiency	8	2
Myocardial infarction	5	2

During the following three years, from 2001 to 2003, an additional 64 biopharmaceuticals (see Tables 16.4 and 16.5) were approved in North America and/or Europe. These drugs included hormones, blood factors, thrombolytics, vaccines, interferons, monoclonal antibodies, and therapeutic enzymes. They often constituted major breakthroughs for the treatment of severe diseases, with Avonex®, a recombinant interferon beta-1a for the treatment of multiple sclerosis, as a particularly striking example. By meeting the demands of a needful drug with appropriate indications, this cytokine became a stunning commercial success, similar to the examples during the early days of recombinant protein technology.

The above figure of 64 products excludes new indications for agents that initially were approved for another indication before the year 2000. Seven of the products that were approved for the first time in a particular region had in fact previously been launched elsewhere. Furthermore, several new products contained the same active ingredient. For example, Actrapid®, Velosulin®, Monotard®, Insulatard®, Protaphane®, Mixtard®, Actraphane®, and Ultratard® are different formulations of the identical recombinant insulin produced in *S. cerevisae* by Novo Nordisk. Similarly, several newly approved vaccines contain an identical *S. cerevisae*-derived hepatitis B surface antigen (rHbsAg). Taking this into account, thirty new genuine biopharmaceuticals have been launched during the past three years. During the same period, an estimated 80 new chemical entities (NCEs) were approved in the European Union and in the United States. Thus, biopharmaceuticals have accounted for more than one-quarter of all new drug approvals during the past three years, and this share is expected to further increase in future.

All 64 newly approved biopharmaceuticals represent protein-based drugs, and do not include compounds based on nucleic acids. Several of the biopharmaceuticals are unmodified recombinant proteins, and an increasing number of polypeptides have been modified by genetic engineering for improved or altered biological activity, immune tolerance, or clearance. Due to technical, manufacturing and regulatory difficulties, no gene therapy drug has yet gained marketing approval.

During the past three years, the number of new biotechnological drugs almost equals that of the previous 18 years. In total, an estimated 1 billion people are already benefiting from the modern biopharmaceutical drugs!

With over 500 products in various stages of clinical evaluation (http://www.phrma.org), there will be a constant increase in the entry of new biopharmaceuticals

Tab. 16.6 Selected biopharmaceutical drugs.

Class	Indication/ Therapeutic area	Market size 2000 ($ million)	Market size 2001 ($ million)
Erythropoetin	Anemia	5787	6803
Insulin	Diabetes	3490	4017
Blood clotting factor	Hemophilia	2400	2585
Colony stimulating factor	Neutropenia	2083	2181
Interferon beta	Multiple sclerosis, hepatitis	1735	2087
Interferon alpha	Cancer, hepatitis	1769	1832
Monoclonal antibody	Cancer	1057	1751
Growth hormone	Growth disorders	1614	1706
Monoclonal antibody	Various	789	1152
Plasminogen activator	Thrombotic disorders	638	642
Interleukin	Cancer, immunology	195	184
Growth factor	Wound healing	98	115
Therapeutic vaccine	Various	31	50
Other various proteins	Various	1834	2006
Total		**23 520**	**27 111**

to the market, and it is expected that in time they will outnumber the classical small molecules. The value of biopharmaceuticals is already in excess of $40 billion, compared to $12 billion just four years ago. Some of the most successful biopharmaceuticals are listed in Table 16.6.

16.5
Which are the Current and Future Markets?

The determination of the complete human genome sequence provoked high expectations. Genomics were expected to be an incentive for efficient discovery and development programs. One of the emerging ideas was that of a personalized medicine, which to some appears as the "holy grail" of healthcare. A very prominent and extremely successful example of the new drugs applied within such an approach is the breast cancer drug Herceptin®, which is addressed to the over-expressed *HER-2/neu* gene. Among breast cancer patients, the cancer can be attributed to over-expression of this gene and the encoded HER-2 receptor in only 40% of cases. When using an integrated diagnosis, these patients can be identified and treated successfully with Herceptin®. Another successful example is Gleevec®, which is used to treat chronic myeloid leukemia.

It is clear from Tables 16.4 and 16.5 that most of the newly approved medicines have been developed to address indications such as diabetes, hemophilia, myocardial infarction and various cancers, these being the major causes of death in the industrialized Western World. The spectrum of approvals and major indications has not changed very much over the past ten years. Hepatitis B and C remain the most targeted indications of new medicines: eight interferon-based products and eight pro-phylactic vaccines against hepatitis B were approved during the past three years. The amount of development activity which has been focused on this viral infection reflects, beyond any doubt, its global significance. Some 2 billion people are infected with hepatitis B, 350 million individuals suffer from lifelong chronic infections, and more than 1 million infected patients die each year from liver cirrhosis and/or liver cancer as sequelae of the infection (see Chapter 15).

New classes of biopharmaceuticals approved during the past three years include various bone morphogenetic proteins (BMP), engineered cytokine fusions, such as Enbrel® (Amgen), and engineered receptor antagonists such as Kineret® (Amgen).

Recombinant BMPs include BMP-7 and BMP-2, which have been developed as therapeutics to exploit their capability in bone and cartilage formation; for example, they can accelerate bone fusion in slow-healing fractures.

Enbrel® (developed by Immunex, which is now part of Amgen) represents one of the most interesting engineered biopharmaceuticals to date. Enbrel is a fusion protein comprising a recombinant p75 tumor necrosis factor (TNF) receptor joined to the Fc region of IgG. It is used to treat rheumatoid arthritis, and is thought to act by neutralizing the pro-inflammatory cytokine TNF with the Fc portion of the molecule, thereby facilitating aggregation and uptake by phagocytes.

Of special interest is Humira®. This product is the first approved human mono-clonal antibody that was generated by the phage display technique. This successful product development holds promise for further success in using this exciting technology.

Until 1994, Ortho Biotech's (Johnson & Johnson) Orthoclone OKT3, approved in 1986, was the only therapeutic monoclonal antibody in the market, with sales in 1993 totaling less than $60 million. Following the entry of abciximab (ReoPro®, developed by Centocor, now part of Johnson & Johnson) in 1994, sales of therapeutic antibodies doubled to $123 million. In 1998, with the spectacular launch of Genentech's rituximab (Rituxan®) and the continued growth of abciximab, sales of monoclonal antibodies tripled, from $324 million to $900 million. Monoclonal antibodies are, without doubt, successful product developments which have had a major impact on modern medicine. Nonetheless, further products are either in the "pipeline", or have already reached the status of an approved drug (see Table 16.4). Furthermore, existing monoclonal antibodies are currently tested for additional indications.

In the industrialized world, mortality due to infectious diseases has decreased steadily over the past 50 years, and is no longer considered a serious threat. However, just the opposite is true for the rest of the world, where for example more than 17 million people die from infectious diseases each year. Among the 42 million people on Earth infected by AIDS, 29 million live in Africa. Similarly, the incidence of malaria has quadrupled over the past five years worldwide, and with 40% of the world's population living in areas where malaria is endemic, up to 500 million new cases are reported each year (Table 16.7). The global market for anti-infectives is currently pegged at about $27 billion, and several biotech firms such as Acambis, Bavarian Nordic, or Human Genome Sciences are developing new antibodies to prevent or to treat the diseases.

Tab. 16.7 Infectious diseases, the big killer.

Disease	No. of cases	No. of deaths
The Plague, 6th century	142 million	100 million
Bubonic Plague	30 million	12 million
Spanish Flu, 20th century	1 billion	21 million
Malaria, 21st century	300–500 million	1.5–2 million/year
Tuberculosis, 21st century	8 million/year	2 million/year
AIDS, 21st century	6 million/year	3 million/year

16.6
The Clinical Development of Biopharmaceuticals

The clinical development of novel therapeutics, including recombinant proteins (biopharmaceuticals), is divided into three subsequent phases of increasing complexity and requirements of both time and resources. The initial testing of new therapeutics in humans occurs in Phase I, the objectives of which are to determine drug safety and to establish the pharmacokinetic and pharmacodynamic profiles of the product. This includes reporting any adverse effects, highest tolerated doses, rates of absorption, metabolism, and distribution of the product in the body after various doses. Testing may occur either in normal volunteers or in patients. When the drug has successfully passed Phase I, studies are conducted in Phase II to determine product efficacy in patients, and to provide additional safety information. Data from these pilot studies are used to design the pivotal Phase III studies, which provide precise measurements of safety and efficacy of the product in a targeted patient population. The exact indication is defined by the outcome of Phase III study data that are evaluated for approval by the FDA or alternative regulatory authorities.

In the case of biopharmaceuticals, the available data for clinical studies are still limited and exhibit a high variability in the design of the executed programs. The number of studies ranged between four and 23, the number of subjects between 390 and 22 270 in individual drug developments. However, these numbers are comparable to those reported for the clinical evaluation of traditional small molecules (Reichert and Paquette 2003).

The probabilities for clinical phase and regulatory review transitions leading to possible approval averaged for the cohort of all biopharmaceutical therapeutics in all countries were as follows: 89% for Phase I to II; 69% for Phase II to III; 72% for Phase III to any review; and 96% for any review to approval. Variations could be observed when the data were stratified according to the assigned primary category of therapeutics. The transition probabilities for five of eight categories followed the same general trend: the probability of Phase I to II transition was high (80–95%), decreased at the Phase II to III transition (50–70%), then increased slightly at the Phase III to review transition (65–90%), and was highest at the review to approval transition (95–100%) (Reichert and Paquette 2003). While it is too early to draw firm conclusions from these data, it seems that neither the success rates nor the failure rates differ very much from those recorded for the chemical entities of traditional drugs. Products developed for two of the therapeutic categories (endocrine and stimulation of blood cell production) were observed as notable exceptions to the general trend in transition probabilities. For both therapeutic categories, either the Phase II to III transitions was higher than expected, or the Phase III to review transition probability was lower than expected. The Phase II to III transition probability for biopharmaceutical therapeutics developed as endocrine treatments was exceptionally high (94%). The majority of the products in this category are biopharmaceutical versions of well-characterized and well-studied biological products, such as insulin or human growth hormone.

The Phase II to III transition probability is also high for products categorized as blood cell production stimulants, although this therapeutic category has a lower Phase III to review transition probability (67%) than one would expect from the general trend. The products that failed at Phase III were treatments for thrombocytopenia and/or neutropenia in cancer patients. Considering the trend in the phase transition, it is possible that in both therapeutic categories products advanced that should have failed in Phase II in the first place. They were then discontinued in Phase III for various reasons, such as the launch of alternative promising drugs or the failure to meet statistically significant end-points.

New biopharmaceutical drugs such as anti-neoplastic therapeutics appeared to have the lowest Phase II to III transition probability (41%), rendering such developments very risky. The majority of the attrition at this phase can be attributed to the discontinuation of interferon and interleukin products, as discussed earlier. Two interferons (IFNα-2a and IFNα-2b) and two interleukins (aldesleukin and denileukin) are approved for marketing in the United States. The decision to discontinue other anti-neoplastic interferon and interleukin products at Phase II might have been provoked to some extent by competitive products that were more advanced in clinical development.

The Phase III to review transition probability for biopharmaceuticals categorized as enzyme replacement therapies is exceptionally high, and reaches 100%. These products relate to biological counterparts of known function such as glucocerebrosidase, α-glucosidase, and α-galactosidase A. Since the biopharmaceutical versions simply replace the traditional biologicals, potential risks associated with the clinical development are much lower compared to those for novel biopharmaceuticals. However, difficulties in the development can still be encountered, most notably in targeting adequate quantities of the replacement product to the anticipated location in the body. Once these products have demonstrated preliminary safety and efficacy in Phase II, the transition from Phase III to approval has usually been relatively smooth. These products are therefore candidates of a relatively low risk profile.

The recently launched biopharmaceutical laronidase (developed by BioMarin Pharmaceuticals) is approved for the treatment of mucopolysaccharidosis I (MPS 1), a progressive, serious and life-threatening genetic disease characterized by lysosomal accumulation of complex carbohydrates. This accumulation is caused by a deficiency of functional α-L-iduronidase, a 628-amino acid enzyme. There are three categories of MPS I: mild (Schreie's syndrome); intermediate (Hurler–Schreie syndrome); and severe (Hurler's syndrome), all of which are provoked by different mutations in the α-L-iduronidase gene that impair (to different extents) the functionality of the encoded enzyme. Enzyme replacement therapy has been shown to reduce liver size and provide some clinical benefits to patients. Clearly, all syndromes can be treated with the same biopharmaceutical drug, but administered at different dose levels correlating with the severity of the deficiency.

On average, the time taken to perform clinical studies of biopharmaceutical drugs increased between the 1980s and the 1990s. Again, this trend probably relates to the fact that the early approved biopharmaceutical drugs were versions of established biological proteins. Such drugs are more likely to be successful and have relatively short clinical phases, as data for their biological counterparts are available. In the 1980s, the newly established biopharmaceutical companies focused on the development of such drugs which were expected to have a high probability of success. The extension of clinical phases in the 1990s might also be due to the significant technical advances that had been made in science and medicine. On one hand these advances enabled the generation of innovative novel recombinant therapeutics and to study them in humans, while on the other hand they provoked higher standards by the FDA and other regulatory authorities.

16.7
Drug Delivery and Modification of Proteins

Like all other drugs, biopharmaceuticals must comply with regulatory release criteria for identity, content, purity and level of degradation products, and content uniformity of the drug substance. Because of their large size, most biopharmaceuticals are poorly absorbed in the gastro-intestinal tract and are sensitive to degradation; therefore, they are usually administered parenterally. Consequently, the regulatory authorities require the products to be sterile, which in turn requires production in cleanroom facilities, which in turn leads to a substantial increase in production costs.

Biopharmaceuticals must also meet other special requirements. Product samples are not allowed to exceed defined maximal bacterial endotoxin levels. Products must have a certain shelf lifetime which is determined experimentally under specified storage conditions. A target shelf lifetime of at least 2 years or longer at 5 °C is generally desired for these products. Because of their large size and multiple functional groups, biopharmaceuticals are in general more sensitive to proteolytic degradation caused by a plethora of cellular proteases. As a result, the share and the structure of degradation products are very difficult to predict, and a number of different analytical techniques have to be employed to "fully" characterize a drug and the derived degraded polypeptide species. Other undesired effects include denaturation, precipitation, fragmentation, deamidation, disulfide scrambling, oligomerization, aggregation, cross-linking, hydrolysis, isomerization, racemization, imide formation, or oxidation. Other release criteria may include pH, osmolarity, residual moisture or solvent content, particle size distribution, and encapsulation efficiency. Degradation and undesired modifications are also all important during purification and downstream processes, and can have a major impact on the costs of biopharmaceuticals. Despite over 20 years of industrial experience, many further improvements are required in this area. Additionally, a careful selection of formulation approaches must be considered (Crommelin et al. 2003).

In most parenteral drugs based on small organic molecules, only a few excipients are used, primarily as tonicity and pH modifiers. However, a typical biopharmaceutical formulation contains more excipients. The presence of these is required to properly administer the drug and to ensure its stability and bioavailability. Some arbitrary functions of excipients include: inhibition of surface adsorption, substitution for a nascent complexing protein, conformational stabilization, cryoprotection, lyoprotection, inhibition of aggregation, thermoprotection, inhibition of isomerization, protection against oxidation, steric exclusion, tonification, and pH control (Crommelin et al. 2003). Clearly, no single excipient can meet all these requirements. Some excipients have multiple functions, and some may even counteract the effect of others. Therefore, a careful selection of excipients is crucial for success. Again, innovations in this area are very much needed.

Many biopharmaceutical show a very rapid clearance and a resultant short therapeutic half-life upon injection. This short half-life may not only be the result of rapid excretion – as is usually the case with the small-sized proteins – but also of metabolic degradation. Often, an extended persistence in blood or tissues is desirable. A number of candidate proteins where, in principle, special formulations that provide a controlled release would greatly improve the performance of a therapeutic, are listed in Table 16.8 (Ash and Ash 2002). A number of different approaches to such a controlled release have been taken in the past. Among the oldest strategies is the complexing of a compound with Zn^{2+}. In the case of insulin-Zn^{2+} complexes, the time span of insulin release could be increased to a few days. Later, polylactic acid-polyglycolic acid-based implants and microspheres were introduced which enabled protein release over a period of up to one month. Since then, hydrogel technologies have attracted attention as they are more "protein-friendly" than polylactic acid-polyglycolic acid, and provide for improved release control. Two approaches are presently being

Tab. 16.8 Some candidate biopharmaceuticals for controlled release.

Erythropoietin
Tumor necrosis factor receptor
Interleukin-2
Interleukin-1 receptor
Interleukin-4 receptor
Interferons
Granulocyte-macrophage colony-stimulating factor
Granulocyte colony-stimulating factor
Transforming growth factor-1
Acidic fibroblast growth factor and basic fibroblast growth factor
Nerve growth factor
Vascular endothelial growth factor
Bone morphogenic protein-2

investigated for release control in clinical tests, namely: the in-situ (at the site of injection) gelling of viscous solutions or dispersions of protein drugs; and the administration of hydrogel-based microspheres loaded with the biopharmaceutical.

16.8
Expression Systems for Commercial Drug Manufacture

Currently, for the manufacture of biopharmaceutical drugs, three different expression systems are well established that are also described in this book: bacteria; various yeasts; and mammalian cell lines, with *E. coli* and the yeasts *S. cerevisae* or *H. polymorpha* as the most commonly used microbes.

Details on how to engineer production strains based on these microbes have been provided in the previous chapters. Fermentation reactors can be large, with adjustable temperature, pH, oxygen and carbon dioxide controls for optimal growth of the cells. Large quantities of several hundreds of kilograms of therapeutic proteins can be made in this way. The third system that is widely employed is based on mammalian cells. Cell lines derived from different mammals including human, mouse, rabbit or rat are used for the production of therapeutic proteins. Again, the reader is referred to the respective chapter for a detailed description.

As biopharmaceutical drugs are manufactured using large vats or reactors, the manufacturing capacity is often measured in liters, and yields are often poor when compared to those obtained in other industrial processes. There is an ongoing debate as to whether the industry will face a shortage of production capacity, or not. This will depend very much on how many new biopharmaceuticals will enter the market, and how these new entries will be received. With costs which sometimes exceed $150 000 per patient per year, the burden on the healthcare systems might become too extensive and thus prevent the general use of such expensive drugs in chronic diseases. On the other hand, the costs per patient can most likely shrink with future improvements of the expression systems, with innovative design of purification procedures and novel routes of administration.

16.9
Will Demand Rise?

Despite the global economic crisis and stock market downturn, the biotech industry has enjoyed an overall healthy growth in recent years. Total biotech drug revenues increased from US $41.3 billion in 2002 to $35.9 billion the previous year 2003, a rise of 15%. This illustrates not only that the fundamental science underlying the industry is strong and promising, but also that the profits can be high.

Globally, there are now an estimated 4400 biotechnology companies, with the possibility of applying their R&D activities to the discovery of a whole range of new drugs. In 2002, a massive US $25 billion was invested in biotech R&D by government and private ventures. The drug discovery process has greatly benefited from

Tab. 16.9 Biopharmaceutical drugs losing patent protection by 2006.

Brand name	Generic name	Company	2001 global sales ($ m)	U.S. Patent expiration (year)
Epogen/Procrit	Erythropoetin α	Amgen, Johnson & Johnson	6803	2004
Novolin	Human Insulin	Novo Nordisk	1829	2005
Humulin	Human Insulin	Elli Lilly	1061	2003
Neupogen	Filgrastim	Amgen	1380	2006
Avonex	Interferon beta-1a	Biogen	972	2003
Cerezyme/Ceredase	Alglucerase	Genzyme	570	2001
Synagis	Palivizumab	MedImmune	668	2004
Humatrope	Somatropin	Elli Lilly	311	2003
Activase	Alteplase	Genentech, Boehringer Ingelheim	267	2005
Nutropin	Somatropin	Genentech	250	2003
Protropin	Somatrem	Genentech	250	2005

this investment. Over 350 biotech drugs are currently under clinical trials, with many more to follow. In case these drugs are successful in clinical trials and achieve FDA approval, the demand for manufacturing capacity will certainly rise.

More than 20 years have passed since the approval of the first biopharmaceutical products, and therefore a number of these first-generation biopharmaceuticals will shortly lose patent protection, as detailed in Table 16.9. This applies to more than ten drugs such as Novolin®, Humulin®, Avonex®, Activase® or Protropin® with global sales exceeding $8 billion. This loss or impending expiry of patent protection is expected not only to provide an opportunity for biogeneric equivalents, but also to spur further demand. Many new players – often spin-offs of traditional pharmaceutical companies – have set out to launch generic forms of these biopharmaceutical drugs.

16.10
Conclusions and Perspectives

Three years of an almost unprecedented downturn in capital markets has imposed a tremendous burden on the biotechnology industry by draining the sources for funding. The funding was not only needed for start-ups but the financial restrictions also pushed companies out of the market, sometimes just prior to a successful product launch. Although funding and financial support remains difficult, there are encouraging signs that biotechnology is regaining its former recognized position as a rewarding target for investment. The positive perspective is backed a value increase of biotech stocks. The NASDAQ biotechnology index for instance finalized last year with an impressive 45% from last January's starting price reflecting the underlying prosperity of this industrial branch.

Such good news re-established the confidence of investors. According to life science merchant bank Burill & Co, US biotech firms raised $4.6 billion in the finance markets in 4Q03 taking the year-end to $16.3 billion – 56% up on 2002 (see Table 16.10). Convertible debt continued to be the finance vehicle of choice for US biotech companies, which acquired almost $1.6 billion in debt financing in the fourth quarter of 2003, taking the total debt issued in 2003 to $7.2 billion – 36% above the 2002 figure.

However, only $453 million were generated within the anticipated IPO window. In addition, the seven NASDAQ-listed firms had only mixed and limited success. Elsewhere, the IPO markets opened slightly: in December, specialty drug developer Alchemia (Brisbane, Australia) floated in Australia, cancer gene therapy firm OncoTherapy Science (Tokyo) completed its raising of 17 billion Yen ($160 million) on the Tokyo stock market, and drug developer Sinclair Pharma (Goldalming, UK) raised $28 million on the London Stock Exchange.

A recent research report from the credit ratings agency Standard & Poor's (New York) predicts that US biotechnology industry revenues will grow by a further 23% this year, fuelled in particular by cancer medications. Not surprisingly though, a large share of the extra sales will be contributed by established giants such as Genentech, whose potential cancer blockbuster Avastin® recently received FDA approval.

Although the fourth quarter of 2003 saw some big merger and acquisition deals – notably the acquisitions of Amersham (London) and Esperion (Ann Arbor, Michigan, USA) by General Electric (Fairfield, CT, USA) and Pfizer (New York), respectively – analysts do not believe that a huge extent of general consolidation will be encountered.

The availability of a range of biogenerics will become reality in the near future, and the resulting reduction in the cost of goods should improve market opportunities for biopharmaceuticals in general. The yields for monoclonal antibodies and other recombinant proteins are likely to increase by new technologies and process components. This will help to relieve to some extent the pressure on fermentation capacity, but technological improvements in purification, or additional purification capacity will still be required. Further improvements in formulation and drug delivery will be necessary for a more economical use of the drug. This again will pave the way for further applications of generic biopharmaceutical drugs.

Novel biopharmaceutical drugs and techniques hold promise for the treatment of chronic and critical diseases. The trail-blazing biotechnology industry is regaining

Tab. 16.10 US biotechnology fundraising.

	4Q03	*3Q03*	*4Q02*
IPO	$453	$0	$0
Secondary public	$1369	$1374	$180
Private investment in public equity	$587	$676	$77
Debt/other	$1589	$2779	$567
Venture capital	$568	$670	$528
Partnering	$2918	$2556	$2354
Total	$7484	$8055	$3707

its key position to combat an ever-expanding range of diseases. This also includes some global targets, which are defined as Millennium Development Goals by the United Unions for the year 2015. One of the most important goals is to fight against HIV/AIDS, malaria, and other diseases most common in developing countries. These aims certainly must rely on biotechnological approaches if they are to be met on time, at bearable costs. The successful evolution of biotechnology during the recent past legitimates an optimistic view that this indeed will happen. One can be confident that the biopharmaceutical industry will not only be a driving economical force, but that it will make substantial contributions to improve human life.

References

ASH M, ASH I (2002) Textbook of pharmaceutical additives. Synapse Information Resources Inc, New York, USA

BINDRA JS, LEDNICER D (1983) Chronicles of drug discovery. John Wiley, New York, USA

BOS G, VERRIJK R, FRANSSEN O, BESEMER J, HENNINK WE, CROMMELIN DJA (2001) Hydrogels for controlled release of pharmaceutical proteins. Pharm Tech Europe 13: 64–74

BURRILL GS (2003) Biotech 2003, 17th annual report on the industry. Burrill & Company, San Francisco, USA

CROMMELIN DJA, BISSIG M, GOUVEIA W, TREDREE R (2003) Storage and handling of biopharmaceuticals: problems and solutions – A workshop discussion. Eur J Hosp Pharm 8: 89–93

GITSCHIER J, WOOD WI, GORALKA TM, WION KL, CHEN EY, EATON DH, VEHAR GA, CAPON DJ, LAWN RM (1984) Characterization of the human factor VIII gene. Nature 312: 326–330

IMS HEALTH (www.ims-health.com)

KÖHLER G, MILSTEIN C (1975) Continuous culture of fused cells secreting antibody of predefined specificity. Nature 256: 495–497

REICHERT JM, PAQUETTE C (2003) Clinical development of therapeutic recombinant proteins. BioTechniques 35: 176–185

THROMBER C (1986) The pharmaceutical industry. In: The chemical industry (Heaton A, Ed) Blackie Academic and Professional Publisher, London, UK

WATSON JD, CRICK FHC (1953) Molecular structure of nucleic acids. A structure for desoxyribosenucleic acid. Nature 171: 737–738

Subject Index

Production of Recombinant Proteins. Novel Microbial and Eucaryotic Expression Systems. Edited by Gerd Gellissen
Copyright © 2005 WILEY-VCH Verlag GmbH & Co. KGaA, Weinheim
ISBN: 3-527-31036-3

HPLC   301
Hsp60   73
Hsp60 chaperone   72
Hsp70   73
Hsp70 chaperone   72
Hsp70 proteins   166
human cytomegalovirus (CMV)   235
human designer cell lines   234
human epidermal growth factor (hEGF)   130
human factor XIIIa   181
human growth hormone   55, 364
human initiation factor 4A1 promoter   235
human pancreatic alpha-amylase   74
human respiratory synctial virus (RSV)   82
human serum albumin (HSA)   74 f., 126, 300
human ubiquitin C promoter   235
Hurler's syndrome   378
Hurler-Schreie syndrome   377
hydantoinase   13
hydrogel technologies   379
hydrogen peroxide   114
hydrolases   200
hydrolytic enzymes   191
hydrophobicity   30, 258
hygromycin   275
hygromycin B   99, 125, 170, 225
hygromycin B phosphotransferase   219
HylD   34
hyperallergenic structures   131
hyperbranching phenotype   206
hyperglycosylation   118, 154, 156, 166, 181
hyper-immune globulin   344
hyperosmolar medium   261
hyperthermophilic   59
hyphae   164, 305
hyphal compartments   225

i
IbpA   28
IbpB   28

ica   78
α-L-iduronidase   377
IFNα-2b   377
IGF-1   154
IgG responses   83, 344
IL-6   202 f.
IMAC   292
imide formation   378
immediate early promoters   235
immune globulin   341
immune response   329 f., 341
immunization program   342
immunoblot assay   333
immunogenic domain   346
immunogenicity   340 f., 347 f.
immunoglobulin light chain   293
immunoglobulins   265
immunotherapy   347
implants   379
incinerated vegetation   216
inclusion bodies   17, 27 f., 67, 291
inclusion body formation   27
incomplete vaccination   343
β-indoleacrylic acid   18
inducible promoter   173
induction times   291
infectious diseases   375
insect cells   331
insulin   5, 46, 72, 81, 153, 319, 363 f., 366
insulin-like growth factor   1   154
λ integrase   18
integration site   130
integration vectors   124
integrative vectors   176
intercellular adhesion (*ica*) gene   77
interferons   347, 373, 377
interleukin-2   347
interleukin-3   74
interleukin-6   200 ff.
interleukin-12   260
interleukins   29, 377
intracellular degradation   258
intradermal injection   343
introns   97, 168

secretion signals 72
secretion targeting signals 175
secretory pathway 67, 71, 80
sedimentation 247
seed fermentor 128, 336
selection markers 99
selenocysteine 26
sepsis 78
sequelae of the infection 375
2 μm sequence 120
serine 154
seroconversion 346
seroconversion rate 342
seroprotection 341
serotypes 326
serotyping 327
serum 330
serum conversion rates 343
serum of chronic carriers 331
sexual cycle 96
shake flasks 81
shake-flask cultures 150 f., 198, 298, 305
SHBs antigenemia 328
SHBsAg 346
Shine–Dalgarno (SD) sequence 21
shuttle vectors 148, 176
sialic acid 119, 154
sigma factors 14
signal peptidases 30
signal peptide 80
signal recognition particle 30
single chain antibodies 56
single chains Fv 265
single-cell proteins 164
single-chain antibodies 31, 61, 257
single-dose vaccines 345
λ site-specific recombination 14
size distribution 378
Skp/OmpH 31
small surface antigen 338
small surface proteins (SHBs) 322
sodium benzoate 54
sodium chloride 259, 305
sok 13

solid/liquid separation 338
Sordariales 216
sortase enzyme 82
N-source 301
southern analysis 205 f.
spheroplasts 96
spleen cDNA library 56
3'-splice site 97
5'-splice site 97
spores 303 ff.
sporulation 122, 206
Sp2/0 (myeloma) 234
SsrA RNA 22
ssrA tmRNA 25
ssrA-encoded peptide-tagging system 22
starch 91, 198
stationary cultures 305
stationary phase 297, 305
stem–loop structure 12, 26 f.
steric exclusion 379
sterilization in place (SIP) 245
Ste13 protein 153
stirred tank reactors (STR) 81, 245
stock market downturn 380
strain characterization 333
strains 33
streptococcal protein G 82
sublancin 168 74
substrate/translocase interactions 70
subtypes 319
subtypes of HBV 326
sucrose 164, 168, 172, 303
SUC2 172
sugar cane molasses 98
sulfuric acid 306
superoxide dismutase 291
SurA 31
surface antigen (HbsAg) 330
surface display 77, 82
surface protein of HBV 320
surfactants 305
SV40 early promoter 235
SV40 large T antigen 236
SV40 late transcript 235
SYN6 medium 296 f., 299